Polymer Science and Engineering

Polymer Science and Engineering

Editor: Bethany Ferguson

NY Research Press
New York

Published by NY Research Press
118-35 Queens Blvd., Suite 400,
Forest Hills, NY 11375, USA
www.nyresearchpress.com

Polymer Science and Engineering
Edited by Bethany Ferguson

International Standard Book Number: 978-1-63238-659-5 (Hardback)

Cataloging-in-Publication Data

Polymer science and engineering / edited by Bethany Ferguson.
p. cm.
Includes bibliographical references and index.
ISBN 978-1-63238-659-5
1. Polymers. 2. Polymer engineering. 3. Polymerization.
I. Ferguson, Bethany.
QD281.P6 P65 2019
547.28--dc23

Printed in China.

Contents

Preface

This book aims to highlight the current researches and provides a platform to further the scope of innovations in this area. This book is a product of the combined efforts of many researchers and scientists, after going through thorough studies and analysis from different parts of the world. The objective of this book is to provide the readers with the latest information of the field.

The science of polymers studies the properties, synthesis, design and characterization of synthetic polymers for the production of commercial products. The principal sub-disciplines of polymer physics, polymer chemistry and polymer characterization are concerned with the different aspects of the study of polymers. Synthetic polymers are divided into the two main groups- plastics and elastomers. Plastics refer to all malleable polymers and elastomers are polymers with viscoelasticity. Some examples of polymers are rubber, teflon, nylon, kevlar, etc. This book unfolds the innovative aspects of polymer science and engineering which will be crucial for the progress of this field in the future. Some of the diverse topics covered herein address the varied branches that fall under this category. As this field is emerging at a rapid pace, the contents of this book will help the readers understand the modern concepts and applications of the subject.

I would like to express my sincere thanks to the authors for their dedicated efforts in the completion of this book. I acknowledge the efforts of the publisher for providing constant support. Lastly, I would like to thank my family for their support in all academic endeavors.

Editor

Cellulose Fiber-Reinforced PLA versus PP

Nina Graupner and Jörg Müssig

The Biological Materials Group, Department of Biomimetics, HSB-City University of Applied Sciences, Neustadtswall 30, 28199 Bremen, Germany

Correspondence should be addressed to Nina Graupner; nina.graupner@hs-bremen.de

Academic Editor: Yulin Deng

The present study focuses on a comparison between different cellulose fiber-reinforced thermoplastics. Composites were produced with 30 mass-% lyocell fibers and a PLA or PP matrix with either an injection (IM) or compression molding (CM) process. Significant reinforcement effects were achieved for tensile strength, Young's modulus, and Shore D hardness by using lyocell as reinforcing fiber. These values are significantly higher for PLA and its composites compared to PP and PP-based composites. Investigations of the fiber/matrix adhesion show a better bonding for lyocell in PLA compared to PP, resulting in a more effective load transfer from the matrix to the fiber. However, PLA is brittle while PP shows a ductile stress-strain behavior. The impact strength of PLA was drastically improved by adding lyocell while the impact strength of PP decreased. CM and IM composites do not show significant differences in fiber orientation. Despite a better compaction of IM composites, higher tensile strength values were achieved for CM samples due to a higher fiber length.

1. Introduction

Petrochemical-based polyolefins like polypropylene (PP) are common matrices for natural fiber-reinforced composites [1–4]. An overview about natural fibers and their composites may be taken from [5–9]. Nowadays the main application area of cellulose fiber-reinforced thermoplastics can be found in the automotive industry for nonvisible parts like indoor panels, dashboards, wheel covers, and so on [10, 11]. In the last years the use of biobased matrices like polylactide (PLA) becomes more and more important [12, 13]. In the near future it is envisaged to establish natural fiber-reinforced biobased plastics apart from the automotive industry for higher loaded components.

The present study gives an overview about the characteristics of cellulose fiber-reinforced PP and PLA composites and a possibility of improving the brittle character of PLA with regenerated cellulose fibers. Two different production procedures—compression molding (CM) and injection molding (IM)—were used for composite production.

It is known that structure and properties of fibers and matrix are directly related to the composite characteristics. To elucidate the different mechanical properties of CM and IM cellulose fiber-reinforced PLA and PP composites, analyses of fiber orientation, compaction, and voids, fiber length distribution and fiber/matrix adhesion were carried out to clear up the following hypotheses:

(1) PLA has a higher hydrophilicity and polarity [14] than PP, which displays a more nonpolar and hydrophobic character [15]. Due to this fact the fiber/matrix interaction between lyocell and PLA is assumed to be better than between lyocell and PP.

(2) Because of the higher stiffness and strength of PLA as well as the assumed better fiber/matrix adhesion with cellulose compared to PP, tensile strength and Young's modulus of lyocell/PLA should result in higher values in comparison to lyocell/PP.

(3) CM should lead to higher composite tensile characteristics due to higher fiber length and more aligned (oriented) fibers in production direction.

(4) The brittleness of PLA leads to a lower composite impact strength compared to the more ductile PP.

2. Materials and Methods

Bright raw white Tencel® lyocell fibers with a fineness of 6.7 dtex (equivalent diameter 23.8 μm), a staple length of 60 mm, a density of 1.5 g/cm^3, and a tensile strength of 301 ± 53 MPa (median +/- mean arithmetic deviation) at a gauge length of 20 mm and a testing speed of 10 mm/min [16] were used for composite production. Fibers were supplied by Lenzing AG (Lenzing, Austria). PLA was used in fiber form (Ingeo fibers type SLN2660D) with a fineness of 6.7 dtex and a staple length of 64 mm produced by Eastern Textile Ltd. (Taipei, Taiwan) from a NatureWorks™ PLA 6202D (density: 1.24 g/cm^3, melting temperature: 160–170°C, and glass transition temperature: 60–65°C). Polypropylene Moplen EP 600V (PP) granules were supplied by LyondellBasell Polymers (LyondellBasell Industries Holdings, B.V., Rotterdam, NL) with a density of 0.9 g/cm^3, a Vicat softening temperature of 152°C and a melt flow rate ($MFR_{230;2,16}$) of 100 g/10 min.

Composites were produced from PLA and PP matrices with 30 mass-% lyocell as reinforcing fibers via compression (CM) and injection molding (IM). Previous studies have shown that a fiber mass fraction of 30% is best suited for reproducible production and comparison of IM and CM composites. With a fiber mass fraction of 20%, a nonuniform distribution of the reinforcing fibers may occur in CM samples, whereas injection molded composites with a fiber mass fraction of 40% often show a high proportion of voids (compare [16]).

Semifinished products were produced by mixing lyocell and PLA fibers during a carding process. The multilayer webs contain fibers with a preferred orientation in machine direction but not 100% unidirectional direction [16]. PP-based composites were produced from multilayer webs containing the reinforcing fibers with a preferred orientation and polymer foils, produced from PP granules, via a film stacking process [17, 18]. Prior to composite production the multilayer webs were predried at 103°C for 2 h. Composite boards were produced with a hot pressing process from these semifinished products. Plates for granule production for the IM process were manufactured in a similar manner with a reduced press-time (3 min instead of 5 min). These plates were shredded into granules for the following IM process [19]. More details about composite production are published in previous studies [16, 17]. Reference samples were produced from neat PLA and PP matrices via CM and IM. Prior to mechanical characterization of polymers and composites, test specimens were conditioned according to DIN EN ISO 291 [20] at 23°C and 50% relative humidity for at least 18 h.

Tensile tests (specimen type 1 A according to DIN EN ISO 527-2 [21]; gauge length: 100 mm; testing speed 2 mm/min; elongation measurement via VideoXtens, Zwick/Roell GmbH, Ulm, Germany) and unnotched Charpy impact tests (Charpy impact testing device type 5102, Zwick GmbH, Ulm, Germany) [22] of at least 5 specimens of matrices and composites as well as their fracture surfaces via SEM were investigated as described earlier in [16]. The apparent interlaminar shear strength (ILSS) of 10 CM lyocell/PP and lyocell/PLA composites was characterized with double-notched tensile tests as described in [23] (sample size: 60 × 15 × 2 mm^3; notch distance: 10 mm; testing speed: 0.5 mm/min; gauge length: 40 mm). Shore D hardness measurements on the composites were done with a Shore D durometer type Z7206.07/00 (Zwick GmbH & Co. KG, Ulm, Germany) at ten different positions.

Processed PLA and PP granules were investigated for their rheological properties with a Göttfert Rheograph 75 device (Göttfert Material Testing Machines GmbH, Buchen, Germany) at 180°C with a capillary diameter of 15 mm, a capillary length of 150 mm, and a nozzle diameter of 1 mm. The measurement was performed at eight different apparent shear rates (72, 180, 450, 1125, 2811, 7030, and 17579 1/s). The apparent viscosity with inlet pressure losses was determined.

To investigate the fiber length distribution, fibers were extracted from the composites using xylene and were analysed via image analysis (FibreShape software, IST AG, Vilters, Switzerland) as described in [24]. The investigation of fiber orientation and voids in the composite structure was carried out via synchrotron-radiation based μ-computer tomography (SR μ-CT) at DESY (Hamburg, Germany; see [25] for details). The SR μ-CT micrographs were evaluated as reported in detail in [25].

Test results were statistically analysed with the open source R software (http://www.r-project.org/). A Shapiro–Wilk test was used for the analysis to a normal distribution. To proof if there are significant differences between the results of variable samples, for normally distributed results with homogenous variances, the Tukey test, and for data which do not follow a normal distribution the Wilcoxon test was chosen. All tests were performed with a level of significance $\alpha = 0.05$. Significant differences are marked with different letters in the diagrams; an asterisk shows results which are not distributed normally.

3. Results and Discussion

3.1. Shore D Hardness. Shore D hardness values are shown for IM and CM neat matrices and their composites in Figure 1. The use of lyocell fibers does not significantly affect the hardness of PLA while the hardness of PP increased significantly by 6% because of the higher hardness of lyocell compared to PP. These findings are in agreement with results reported by Bourmaud and Baley [26]. The authors investigated the hardness of flax fibers and a PLA matrix via nanoindentation. Virgin flax fibers have shown a 1.8 times higher hardness than neat PLA. However, the hardness of the fibers decreases significantly after compression and injection molding to a similar value as reported for pure PLA. Considering PP and lyocell fibers, Lee et al. [27] have shown 5.6 times higher hardness for lyocell compared to PP analysed with nanoindentation carried out with an atomic force microscope (AFM).

3.2. Tensile Characteristics. Results have shown for both PP and PLA matrices higher values for IM matrices compared to CM matrices (see Table 1) which may be explained with the higher alignment of polymer chains in axial direction to the specimen due to higher pressure during IM [16]. The CM

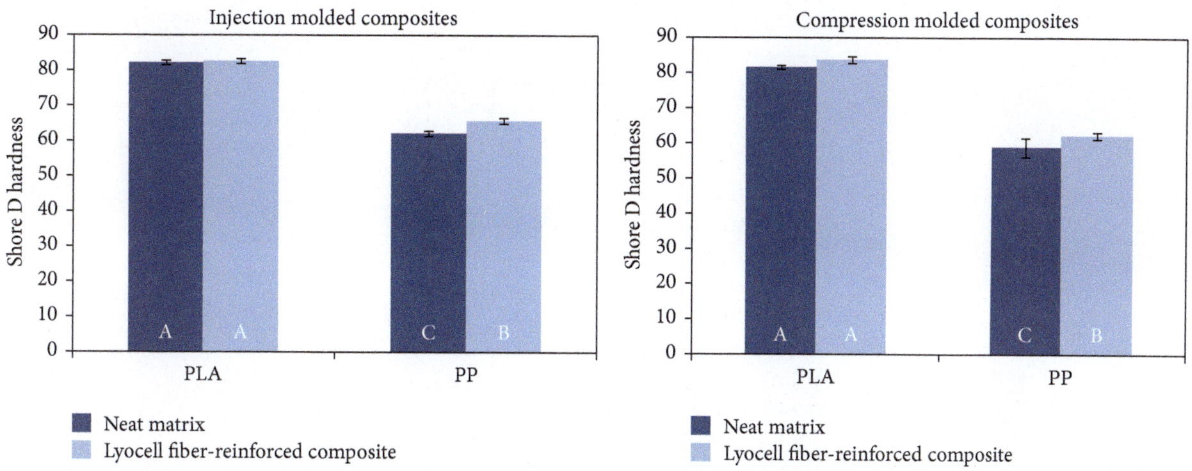

FIGURE 1: Shore D hardness of neat IM and CM matrices and lyocell fiber-reinforced composites (mean value ± standard deviation; different letters show significant differences).

TABLE 1: Tensile characteristics of PLA, PP, and their composites (mean value ± standard deviation).

Material	Tensile strength in MPa	Young's modulus in GPa	Elongation at break in %
Compression molded samples			
PP	13.5 (±1.0)	0.73 (±0.08)	>200
PLA	51.9 (±3.6)	3.44 (±0.23)	1.7 (±0.2)
Lyocell/PP	33.3 (±3.5)	2.50 (±0.29)	3.8 (±0.8)
Lyocell/PLA	87.6 (±8.1)	9.95 (±0.75)	3.6 (±0.4)
Injection molded samples			
PP	19.6 (±0.3)	1.40 (±0.22)	>200
PLA	59.9 (±0.2)	3.57 (±0.25)	2.0 (±0.2)
Lyocell/PP	31.4 (±1.0)	3.51 (±0.15)	2.7 (±0.4)
Lyocell/PLA	69.4 (±1.6)	7.92 (±0.69)	1.0 (±0.2)

process causes no specific flow direction and therefore no preferred orientation of the polymer chains. Differences by different degrees of crystallisation were excluded with microscopic investigations of thin sections in polarised light. In the whole consideration, IM composites display significantly lower strength values than CM composites.

In general, the use of lyocell leads to significant reinforcement effects of both CM and IM produced PLA and PP composites. Because of the lower tensile strength of neat PP compared to PLA the reinforcing effect by lyocell was higher in PP. Considering CM samples, the tensile strength of PLA increased by 70%, for PP even by 150%. For the Young modulus reinforcing effects of 196% for PLA and 240% for PP were observed (compare Table 1).

These reinforcing effects were lower for IM samples. Considering the tensile strength, the values increased by 20% for PLA and 60% for PP, respectively. For the Young modulus reinforcing effects of 120% and 150% were determined for PLA and PP composites compared to the pure polymers. The results have shown significantly higher values for PLA composites and confirm the hypothesis that lyocell leads to higher strength and stiffness values in PLA compared to PP.

As visible in Table 1 the elongation at break of neat PP is significantly higher than that of PLA while the PP-based composites show only a slightly higher elongation than PLA-based composites. The higher elongation of PP compared to PLA should influence the impact performance which is analysed and discussed in Section 3.5.

To find out the reasons which may affect the different behavior of IM and CM samples (e.g., the lower tensile strength of IM composites compared to CM composites although the strength of the neat IM matrices is higher) like fiber orientation, voids, viscosity of the polymer melt, fiber/matrix adhesion, and fiber length distribution, further experiments were carried out. The results of these investigations are described in Sections 3.3 and 3.4.

3.3. Fiber Orientation, Voids, and Viscosity. Since CM samples were produced from multilayer webs, it is assumed that the fiber orientation shows no significant differences whether PLA or PP is used as a matrix. In contrast to IM the flow processes of the matrix during our used CM technique are negligible. Interestingly, a previous study has showed no significant differences between the mean orientation angles of IM and CM 30% lyocell 6.7 dtex/PLA composites [16]. The hypothesis which stated a higher fiber orientation of CM composites reinforced with 30 mass-% lyocell must be discarded. In this study additional fiber orientation

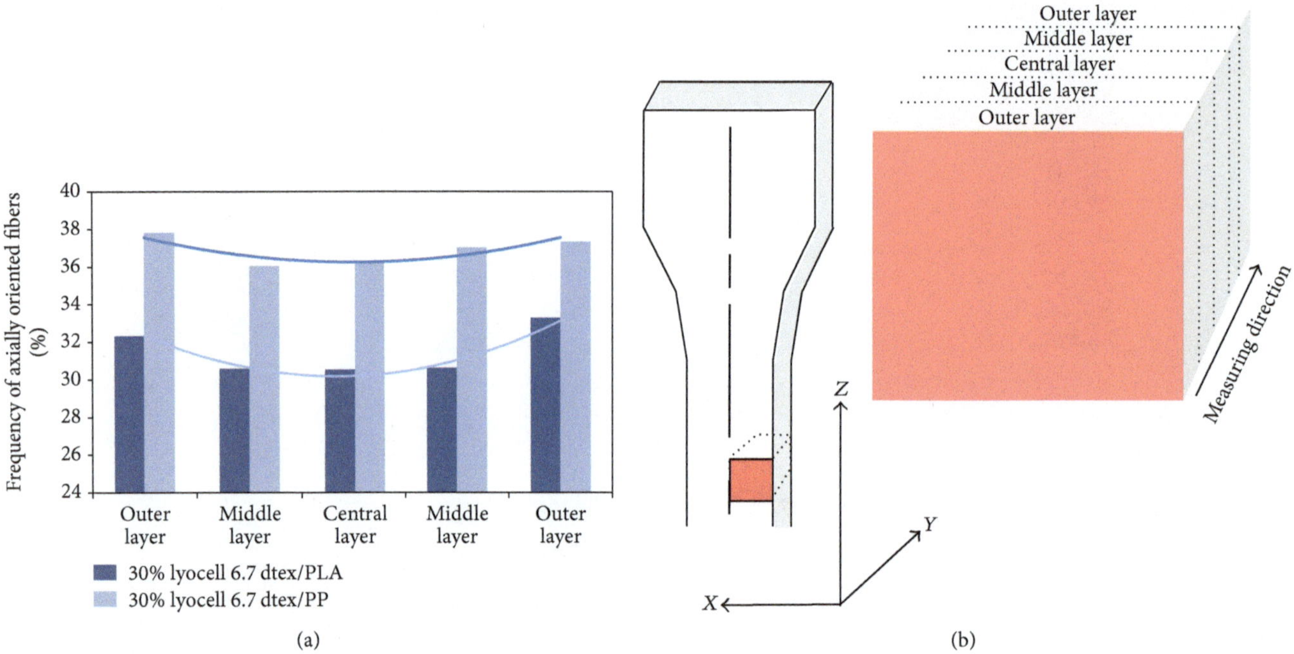

FIGURE 2: (a) Frequency of axially oriented fibers (0–10°) measured across a tensile test specimen of an IM PLA and IM PP composite (note interrupted y-axis). (b) Schematic of the measuring direction across sample height in the parallel section of a test specimen.

measurements were carried out with micrographs obtained from synchrotron-radiation based μ-computer-tomography for IM samples with either a PLA or a PP matrix [25]. Figure 2(a) shows the frequency of axially oriented fibers (0–10° fiber orientation angles) measured across the sample height in the parallel section of an IM lyocell/PLA and PP tensile test specimen (Figure 2(b)). The frequency of axially oriented fibers appears to be clearly higher in PP composites. Additionally, a higher number of axially oriented fibers were observed in the outer (shell) layers compared to the central (core) layer. The core layer usually shows a lower preferred fiber orientation and more fibers are oriented in perpendicular direction as reported in previous studies for IM samples [16, 28]. This trend is more evident for the PP composites (see Figure 2(a)) and may be explained with the viscosity of the matrices. The apparent viscosity was plotted logarithmically as a function of the apparent shear rate in Figure 3. PLA and PP display similar apparent viscosities at low shear rates but the higher the shear rate the higher the differences between PLA and PP. While the apparent shear rate decreases as a linear double logarithmic function for PP, for PLA a nonlinear decrease was observed with increasing shear rate. The viscosity of PP is clearly lower at higher shear rates resulting in a better flowability and alignment of fibers in flow direction during IM. The higher frequency of axially oriented fibers in PP should lead to a higher reinforcing effect as compared to PLA composites.

Since voids affect the mechanical characteristics of composites negatively, composites were additionally analysed for their void distribution on the basis of data obtained with the synchrotron-radiation based μ-CT [25]. No voids were found in the IM composites and only a few voids were detected in the edge regions of the CM composites (2.1% void volume). Therefore, the influence of voids is neglected in this study.

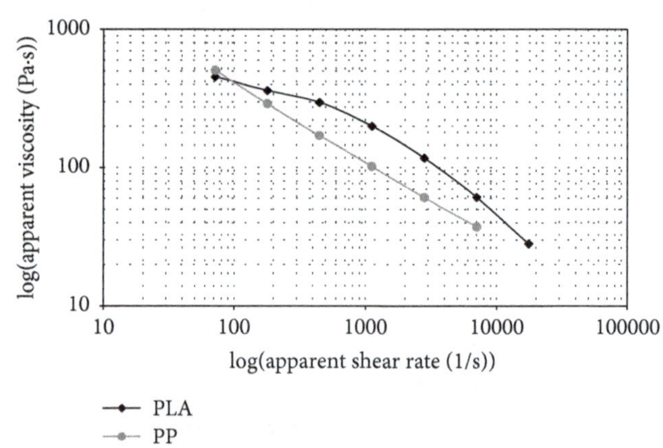

FIGURE 3: log-log plot of the apparent viscosity as a function of the apparent shear rate of neat PLA and PP.

3.4. Fiber/Matrix Adhesion and Fiber Length Distribution. In order to obtain a characteristic value for the fiber/matrix adhesion, double-notched tensile tests were performed for CM composites. Figure 4 shows a significant variation of the apparent interlaminar shear strength (ILSS) between lyocell/PP and lyocell/PLA. Despite the lower viscosity of PP and an assumed better wetting of the fibers with the matrix the higher value was measured for lyocell/PLA with 22.2 ± 1.9 MPa. Lyocell/PP resulted in a value of 11.6 ± 1.0. Hence, a lower critical fiber length (L_c) is required for lyocell

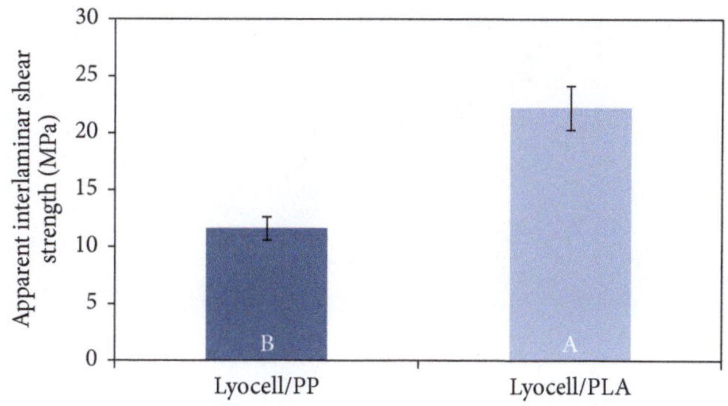

FIGURE 4: Apparent interlaminar shear strength (ILSS) of CM lyocell/PP and IM lyocell/PLA composites measured with double-notched tensile tests (mean value standard deviation, different letters show significant differences).

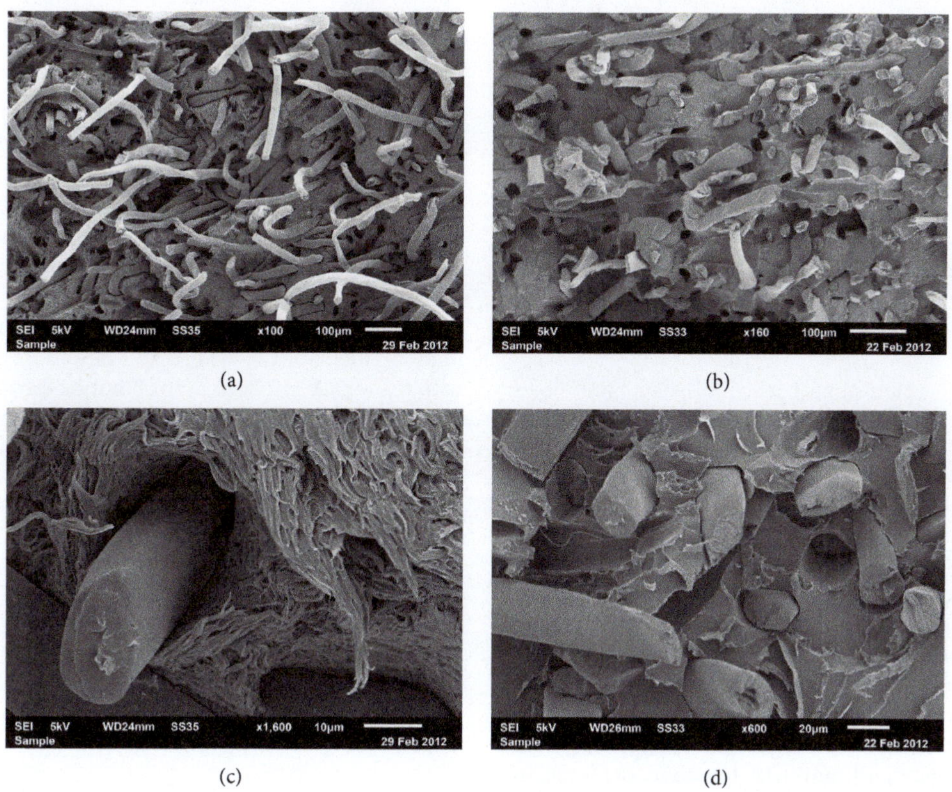

FIGURE 5: SEM micrographs of tensile fracture surfaces of IM lyocell/PP (a and c) and IM lyocell/PLA (b and d).

in PLA compared to lyocell in PP [29]. SEM investigations of tensile fracture surfaces shown in Figure 5 reveal clearly longer fiber pull-outs and larger gaps between lyocell fibers and PP indicating a worse bonding between fiber and matrix. Lower fiber pull-out lengths and smaller gaps are visible for lyocell in the PLA matrix revealing a better compatibility for lyocell/PLA.

The fiber length distribution was investigated for fibers extracted from the composites. Due to the high fiber length of CM composites it was not possible to measure the fiber length distribution with the image analysis procedure used in our study. The fibers were interlooped and not separable in as much quantity as necessary for a fiber length analysis. Figure 6 shows three-dimensional views of a CM and an IM composite obtained with the μ-CT analysis described in [25]. The fibers appear to be much longer in the CM composites. These μ-CT investigations as well as results obtained in a previous study [16] lead to the assumption that the fiber length is not affected by the used compression molding technique. The initial fiber length used for CM composites was 60 mm. In a previous study it was hypothesised that the fiber length (L) of CM samples is multiple higher than a fiber length of 3.0 mm which is necessary to reach the maximum reinforcement effect in a PLA composite [16]. In contrast to this the fiber length is drastically reduced in IM composites due to granule production and shear loading

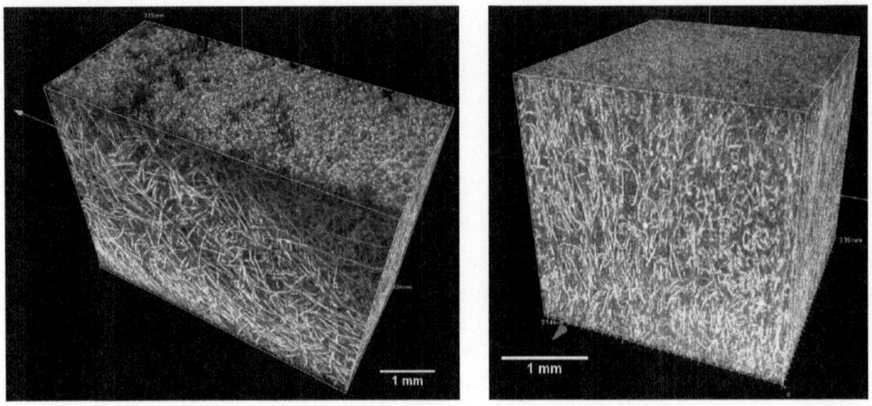

(a) (b)

FIGURE 6: Synchrotron-radiation based μ-CT micrographs of a compression molded (a) and an injection molded lyocell/PLA composite (b).

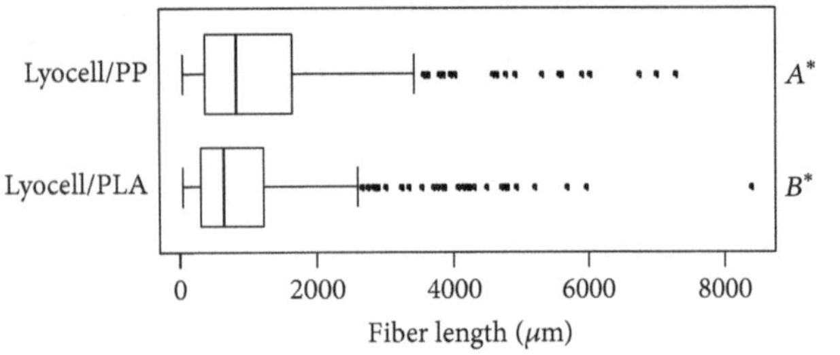

FIGURE 7: Box-Whisker plots of fiber length distribution of fibers extracted from IM lyocell/PP and IM lyocell/PLA composites (results do not follow a normal distribution; different letters show significant differences).

during the IM process. Granules were shredded to the same size from compression molded composite plates, assuming a similar initial fiber length in lyocell/PLA and lyocell/PP granules prior IM. The fiber length distribution in PP and PLA composites is shown in Figure 7 as Box-Whisker plots. According to the statistics, the median fiber length of lyocell extracted from PP is significantly higher than the median fiber length in the PLA composites. Lyocell fibers extracted from PP show a higher median value of 815 μm and an average fiber length of 1130 μm as compared to PLA (median fiber length: 661 μm, average fiber length: 983 μm). It is hypothesised that the fiber length is less reduced in a matrix of lower viscosity than in a matrix of higher viscosity.

The critical aspect ratio was evaluated from the critical fiber length L_c for lyocell/PP and lyocell/PLA with single fiber pull-out tests in a previous study [29]. The critical aspect ratio L_c/d (d = fiber diameter) was determined to be 21 for lyocell/PLA and 37 for lyocell/PP (the critical fiber aspect ratio describes the minimum aspect ratio which is required to achieve a reinforcement effect in a matrix). Due to the higher L_c a higher aspect ratio is necessary to achieve a reinforcement effect in a PP matrix. The mean aspect ratio(L/d) of the extracted fibers was calculated to be 28 for lyocell/PLA and 32 for lyocell/PP resulting in a 1.3 times higher aspect ratio for lyocell/PLA compared to the critical aspect ratio. The aspect ratio of lyocell in PP is by a factor of 0.9 lower. Thereby, lyocell/PLA contains 46% fibers with an aspect ratio higher than the critical aspect ratio while lyocell/PP only contains 33% fibers having an aspect ratio higher than the critical aspect ratio.

3.5. Calculation. The theoretical tensile strength was approximately calculated according to Kelly and Tyson [30] (see (1)) with the tensile strength of the composite σ_C, the fiber volume fraction V_F, the fiber tensile strength σ_F, the matrix volume fraction V_M, and the tensile strength of the matrix σ_M.

$$\sigma_C = \eta_L \cdot \eta_O \cdot V_F \cdot \sigma_F + V_M \cdot \sigma_M. \quad (1)$$

η_L is the fiber length efficiency factor (see (2) for $L \geq L_c$ and (3) for $L < L_c$). The fiber orientation efficiency factor η_O is calculated according to Krenchel [31] in (4) with a, the amount of fibers having an orientation angle ϕ. Aspects like voids and fiber/matrix adhesion are not considered in this model. Data used for this calculation are summarised in Table 2.

$$\eta_L = 1 - \frac{L_c}{2 \cdot L} \quad \text{for } L \geq L_c, \quad (2)$$

$$\eta_L = \frac{L}{2 \cdot L_c} \quad \text{for } L < L_c, \quad (3)$$

$$\eta_O = \sum a \cdot \left(\cos^4(\phi)\right). \quad (4)$$

As shown in Figure 8, the calculated tensile strength is higher than the measured values, especially for PP-based composites. The deviation between theoretical tensile strength and experimentally determined tensile strength was up to −10% for PLA composites while the deviation for PP composites was up to −61% despite the higher fiber orientation. It is assumed that this effect is mainly based on the lower fiber/matrix interaction between cellulose and PP compared to PLA as proven in Figures 4 and 5. Despite the previous discussed higher reinforcing effect for lyocell fibers

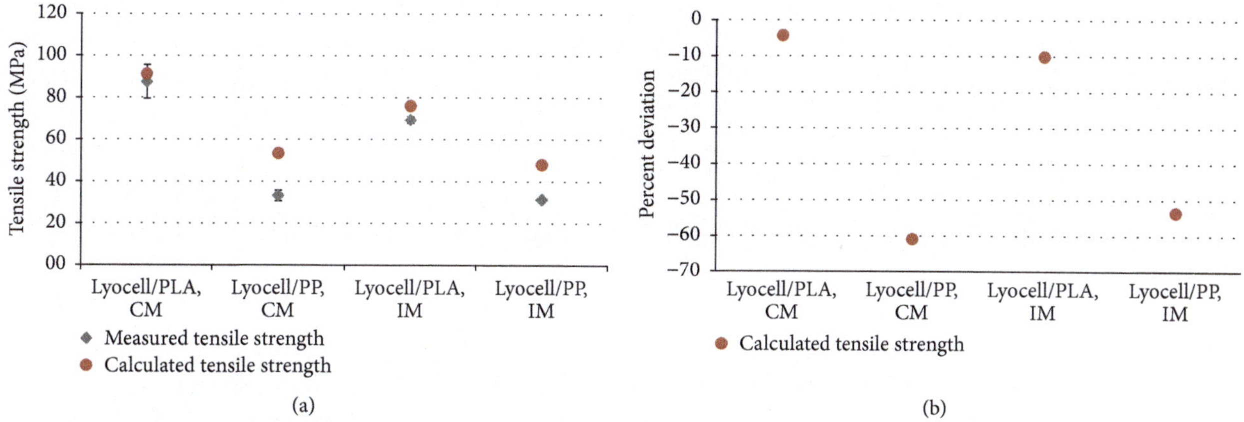

FIGURE 8: Measured tensile strength (mean value standard deviation) of CM and IM lyocell/PLA and lyocell/PP composites compared to calculated strength values according to Kelly and Tyson (a) and percent deviation between measured values and calculated values (b).

TABLE 2: Data used for tensile strength calculation with the Kelly-Tyson model according to (1).

Characteristic	PLA		PP	
	CM	IM	CM	IM
σ_F in MPa	301	301	301	301
σ_M in MPa	52	60	14	20
V_F in %	26	26	21	21
V_M in %	74	74	79	79
η_L	1.000	0.621	1.000	0.757
η_O	0.672	0.655	0.672	0.677

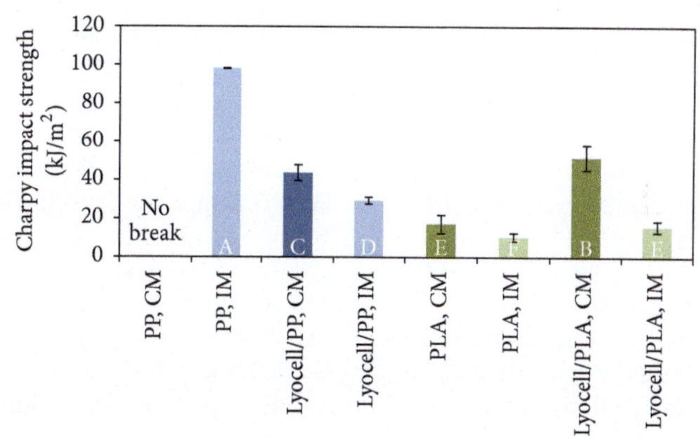

FIGURE 9: Unnotched Charpy impact strength of PP and PLA as well as their lyocell fiber-reinforced composites produced with CM and IM (mean value with standard deviation, different letters show significant differences; please note that either IM samples or CM samples should be compared with each other due to a different sample thickness of 4 mm for IM samples and 2 mm for CM samples).

in a PP matrix, the reinforcing potential of lyocell in a PLA matrix seems to be higher with respect to the tensile strength. The high deviations between measured and calculated tensile strength values make it necessary to develop a new model, especially for cellulose fiber-reinforced composites in future research activities (see [32]).

3.6. Unnotched Charpy Impact Strength. Figure 9 shows the unnotched Charpy impact strength of the different matrices and their composites manufactured with the CM and IM process. Due to the different thickness of CM and IM samples (2 mm for CM samples and 4 mm for IM samples), the results are not directly comparable. The impact strength is considerably affected by the sample thickness [33, 34]. Considering CM samples unlike the brittle PLA matrix no break occurred for neat PP due to the small sample thickness and the high flexibility of PP. The CM composites showed the highest impact strength for lyocell/PLA with 51.9 ± 6.5 kJ/m^2. This is an improvement by 300%. For lyocell/PP a value of 43.7 ± 4.1 kJ/m^2 was determined indicating an increasing brittleness compared to neat PP. The established hypothesis assuming increasing impact strength by the use of a more ductile matrix must be rejected for CM composites. This phenomenon may be justified with the fiber/matrix adhesion and the fiber length. In general, very good fiber/matrix adhesion leads to composites with good strength and stiffness, while weak fiber/matrix interaction may result in composites with a low strength. But the occurring fiber pull-outs lead to good energy absorption during fracture and composites with a good impact performance are the consequence [35, 36]. Due to the less hydrophobic character of PLA, the adhesion of lyocell fibers in a PLA matrix is considerably better than that in a PP matrix. Since the CM samples contain ductile fibers with $L \gg L_c$, more energy is dissipated during a fiber pull-out with a stronger bonding between fiber and matrix as observed for lyocell fibers in the PLA matrix (compare Figure 4).

With the IM samples, different trends were observed. While for PLA a reinforcing effect by 150% was determined, the impact strength of lyocell/PP was drastically reduced compared to pure PP. Nevertheless, it is significantly higher than the impact strength of lyocell/PLA. The different behavior can be explained with the stress-strain characteristics of lyocell, PLA and PP (compare Figure 10). While PLA has a low elongation and displays a brittle behavior, PP shows a

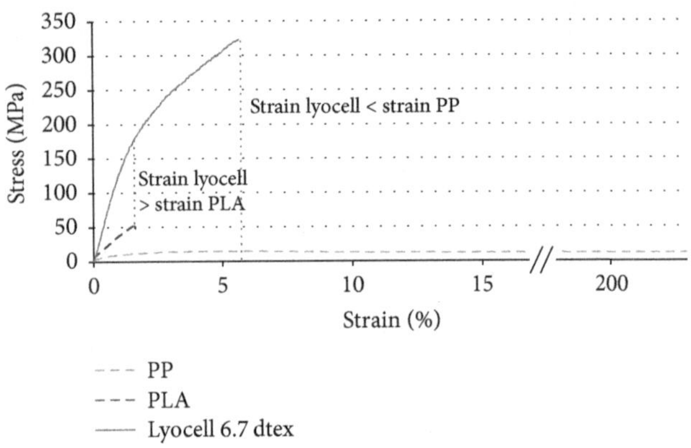

FIGURE 10: Typical stress-strain curves of neat PLA, neat PP, and lyocell fibers.

lower strength and stiffness but significantly higher ductility (Figure 10). Lyocell displays a significantly higher elongation than PLA and thus leads to an increase in impact strength for both IM and CM lyocell/PLA composites. However, compared to PP the elongation at break of lyocell is lower resulting in a reduction of the ductility of lyocell/PP. The higher impact strength of the IM PP composite compared to the IM PLA composite can be explained with the short fiber length and the different ductility of the neat matrices. In contrast to CM composites short fibers were used as reinforcement. The energy absorption by fiber pull-out is limited. A certain fiber length is necessary to absorb energy by fiber pull-out during crack propagation after impact. In this case the more ductile character of PP compared to PLA has a more important role as compared to long fiber-reinforced composites.

In the whole consideration the hypothesis of increasing impact strength by using more ductile lyocell fibers cannot be refuted for lyocell/PLA composites. Using a more ductile matrix like PP leads to an improvement of the impact strength of short fiber-reinforced composites compared to the use of a brittle PLA matrix because of the limited fiber pull-out during crack propagation after impact due to the short fiber length. In this case the ductility of the matrix is more important. In contrast to this in long fiber-reinforced composites the energy absorbing fiber pull-outs as a result of better fiber/matrix adhesion in lyocell/PLA lead to a significant improvement of impact strength compared to lyocell/PP with a weak fiber/matrix adhesion.

4. Summary and Conclusions

The results presented here demonstrate a high potential of PLA as a matrix for cellulose fiber-reinforced composites. Results have shown for both CM and IM composites significantly higher tensile strength and Young's modulus values for PLA-based composites compared to PP-based composites. However, the elongation at break of PLA is significantly lower, resulting in a brittle behavior, which in turn has a negative effect on the impact properties. By using lyocell as reinforcement the brittle nature of PLA decreases significantly while the impact strength of PP was reduced. Different impact behavior was observed for short fiber-reinforced IM composites compared to long fiber-reinforced CM composites. The high ductility of PP leads to higher impact strength for IM short fiber-reinforced lyocell/PP than for lyocell/PLA composites while a higher value was measured for CM long fiber-reinforced lyocell/PLA than for lyocell/PP composites because of higher energy absorption during fracture. Since no significant differences were measured with respect to the fiber orientation in IM and CM composites and the IM composites containing a lower volume fraction of voids the different behavior of CM and IM samples is predominantly based on the fiber length. Fibers in IM composites are too short to reach the maximum possible mechanical characteristics. Despite a better blending of lyocell fibers with PP due to the lower viscosity at higher shear rates, higher interfacial shear strength was measured for PLA-based composites. SEM micrographs show long fiber pull-outs with big gaps between lyocell and PP which appear to be smaller for lyocell in PLA leading to a better load transfer from the matrix to the fiber. This effect is based on the hydrophilic character of lyocell and the less hydrophobic character of PLA compared to PP. In the whole consideration the improved ductility of lyocell/PLA in combination with good tensile characteristics expands the possibilities for impact loaded construction materials for higher claimed products. In future research studies it should be checked whether it is possible to insert longer lyocell fibers in IM composites. Due to better wetting it is hypothesised that tensile and impact characteristics of IM composites may be enhanced above the characteristics of CM composites by using longer reinforcing fibers.

Disclosure

Parts of the results of this study were presented on the ICCE-22 (International Conference on Composites/Nano-Engineering, International Community for Composites/Nano-Engineering), 2014, and ICNF (International Conference on Natural Fibers), 2015.

Acknowledgments

The authors thank Professor Dr.-Ing. Gerhard Ziegmann (University of Technology, PuK, Clausthal-Zellerfeld, Germany) and Dipl.-Ing. (FH) Herbert Enzler (University of Technology, PuK, Clausthal-Zellerfeld, Germany) for their support during injection molding, Dr. Stefan Kirchberg (University of Technology, PuK, Clausthal-Zellerfeld, Germany) for the viscosity measurements, and M.S. Katharina Albrecht (University of Applied Sciences, Bremen, Germany) for fiber extraction. They also acknowledge the Deutsches Elektronen Synchrotron (DESY) and the Helmholtz Gesellschaft Geesthacht (HZG) for funding the measurements on the fiber orientation, in particular Dr. Fabian Wilde (HZG) for his support during our beam time.

References

[1] A. K. Bledzki and J. Gassan, "Composites reinforced with cellulose based fibres," *Progress in Polymer Science*, vol. 24, no. 2, pp. 221–274, 1999.

[2] J. K. Pandey, S. H. Ahn, C. S. Lee, A. K. Mohanty, and M. Misra, "Recent advances in the application of natural fiber based composites," *Macromolecular Materials and Engineering*, vol. 295, no. 11, pp. 975–989, 2010.

[3] J. Summerscales, N. P. J. Dissanayake, A. S. Virk, and W. Hall, "A review of bast fibres and their composites. Part 1. Fibres as reinforcements," *Composites Part A: Applied Science and Manufacturing*, vol. 41, no. 10, pp. 1329–1335, 2010.

[4] J. Summerscales, N. Dissanayake, A. Virk, and W. Hall, "A review of bast fibres and their composites. Part 2—composites," *Composites Part A: Applied Science and Manufacturing*, vol. 41, no. 10, pp. 1336–1344, 2010.

[5] J. Müssig, H. Fischer, N. Graupner, and A. Drieling, "Testing methods for measuring physical and mechanical fibre properties (plant and animal fibres," in *Industrial Applications of Natural Fibres - Structure, Properties and Technical Applications*, J. Müssig, Ed., pp. 269–310, John Wiley & Sons, Ltd, 2010.

[6] R. D. Anandjiwala and S. Blouw, "Composites from bast fibres - Prospects and potential in the changing market environment," *Journal of Natural Fibers*, vol. 4, no. 2, pp. 91–901, 2007.

[7] C. Baley, "A review of biocomposite development," *JEC Composites Magazine*, vol. 46, no. 46, pp. 32-33, 2009.

[8] T. Gurunathan, S. Mohanty, and S. K. Nayak, "A review of the recent developments in biocomposites based on natural fibres and their application perspectives," *Composites Part A: Applied Science and Manufacturing*, vol. 77, pp. 1–25, 2015.

[9] K. L. Pickering, M. G. A. Efendy, and T. M. Le, "A review of recent developments in natural fibre composites and their mechanical performance," *Composites Part A: Applied Science and Manufacturing*, vol. 83, pp. 98–112, 2016.

[10] G. Koronis, A. Silva, and M. Fontul, "Green composites: a review of adequate materials for automotive applications," *Composites Part B: Engineering*, vol. 44, no. 1, pp. 120–127, 2013.

[11] M. Barth and M. Carus, *Carbon footprint and sustainability of different natural fibre for biocomposites and insulation material. study providing data for the automotive and insulation industry*, 2015, http://www.bio-based.eu/ecology, http://bio-based.eu/download/?did=14089&file=0.

[12] P. K. Bajpai, I. Singh, and J. Madaan, "Comparative studies of mechanical and morphological properties of polylactic acid and polypropylene based natural fiber composites," *Journal of Reinforced Plastics and Composites*, vol. 31, no. 24, pp. 1712–1724, 2012.

[13] D. Garlotta, "A literature review of poly(lactic acid)," *Journal of Polymers and the Environment*, vol. 9, no. 2, pp. 63–84, 2001.

[14] H. Endres and A. Siebert-Raths, *Engineering Biopolymers: Markets, Manufacturing, Properties and Applications*, Carl Hanser Verlag, Munich, DE, 1st edition, 2011.

[15] G. Venkata Reddy, S. Venkata Naidu, and T. Shobha Rani, "A study on hardness and flexural properties of kapok/sisal composites," *Journal of Reinforced Plastics and Composites*, vol. 28, no. 16, pp. 2035–2044, 2009.

[16] N. Graupner, G. Ziegmann, F. Wilde, F. Beckmann, and J. Müssig, "Procedural influences on compression and injection moulded cellulose fibre-reinforced polylactide (PLA) composites: Influence of fibre loading, fibre length, fibre orientation and voids," *Composites Part A: Applied Science and Manufacturing*, vol. 81, pp. 158–171, 2016.

[17] N. Graupner and J. Müssig, "A comparison of the mechanical characteristics of kenaf and lyocell fibre reinforced poly(lactic acid) (PLA) and poly(3-hydroxybutyrate) (PHB) composites," *Composites Part A: Applied Science and Manufacturing*, vol. 42, no. 12, pp. 2010–2019, 2011.

[18] N. Graupner and J. Müssig, "The influence of lyocell and kenaf fibres on the mechanical characteristics of poly(lactic acid) (PLA) composites," *Journal of Biobased Materials and Bioenergy*, vol. 6, no. 4, pp. 500–507, 2012.

[19] F. Beckmann, "Verfahren zum Herstellen von Thermoplast gebundenem Naturfasermaterial in schütt- und rieselfähiger Form (2001), EP1307333B1".

[20] "DIN EN ISO 291:2005 – Kunststoffe - Normalklimate für Konditionierung und Prüfung, Deutsche Norm. - in German (2006)".

[21] "DIN EN ISO 527-2 – Kunststoffe: Bestimmung der Zugeigenschaften; Teil 2 Prüfbedingungen für Form- und Extrusionsmassen, Deutsche Norm. - in German (1996)".

[22] "DIN EN ISO 179:1996 – Kunststoffe - Bestimmung der Charpy-Schlagzähigkeit, Deutsche Norm. - in German (1997)".

[23] N. Graupner, H. Fischer, G. Ziegmann, and J. Müssig, "Improvement and analysis of fibre/matrix adhesion of regenerated cellulose fibre reinforced PP-, MAPP- and PLA-composites by the use of Eucalyptus globulus lignin," *Composites Part B: Engineering*, vol. 66, pp. 117–125, 2014.

[24] N. Graupner, K. Albrecht, G. Ziegmann, H. Enzler, and J. Müssig, "Influence of reprocessing on fibre length distribution, tensile strength and impact strength of injection moulded cellulose fibre-reinforced polylactide (PLA) composites," *Express Polymer Letters*, vol. 10, no. 8, pp. 647–663, 2016.

[25] N. Graupner, F. Beckmann, F. Wilde, and J. Müssig, "Using synchroton radiation-based micro-computer tomography (SR μ-CT) for the measurement of fibre orientations in cellulose fibre-reinforced polylactide (PLA) composites," *Journal of Materials Science*, vol. 49, no. 1, pp. 450–460, 2014.

[26] A. Bourmaud and C. Baley, "Effects of thermo mechanical processing on the mechanical properties of biocomposite flax fibers evaluated by nanoindentation," *Polymer Degradation and Stability*, vol. 95, no. 9, pp. 1488–1494, 2010.

[27] S.-H. Lee, S. Wang, G. M. Pharr, and H. Xu, "Evaluation of interphase properties in a cellulose fiber-reinforced polypropylene composite by nanoindentation and finite element analysis," *Composites Part A: Applied Science and Manufacturing*, vol. 38, no. 6, pp. 1517–1524, 2007.

[28] K. Albrecht, E. Baur, H. Endres, R. Gente et al., "Measuring fibre orientation in sisal fibre-reinforced, injection moulded polypropylene? Pros and cons of the experimental methods to validate injection moulding simulation," *Composites Part A: Applied Science and Manufacturing*, vol. 95, pp. 54–64, 2017.

[29] N. Graupner, J. Rößler, G. Ziegmann, and J. Müssig, "Fibre/matrix adhesion of cellulose fibres in PLA, PP and MAPP: A critical review of pull-out test, microbond test and single fibre fragmentation test results," *Composites Part A: Applied Science and Manufacturing*, vol. 63, pp. 133–148, 2014.

[30] A. Kelly and W. R. Tyson, "Fiber strengthened materials," in *High-strength materials*, V. F. Zackay, Ed., John Wiley & Sons, Berkeley, USA, 1964.

[31] H. Krenchel, *Fibre reinforcement: Theoretical and practical investigations of the elasticity and strength of fibre-reinforced materials [Ph.D. thesis]*, Technical University of Denmark, Laboratory of Structural Research, Copenhagen, DK, 1964.

[32] N. Graupner, G. Ziegmann, and J. Müssig, "Composite models for compression moulded long regenerated cellulose fibre-reinforced brittle polylactide (PLA)," *Composites Science and Technology*, 2017, In press.

[33] R. A. Malloy, "The design process and material selection," in *Plastic part design for injection moulding - an introduction*, R. A. Malloy, Ed., pp. 127–183, Carl Hanser Verlag, Munich, DE, 1994.

[34] L. J. Broutman and S. M. Krishnakumar, "Impact strength of polymers: 1. The effect of thermal treatment and residual stress," *Polymer Engineering & Science*, vol. 16, no. 2, pp. 74–81, 1976.

[35] J. L. Thomason and M. A. Vlug, "Influence of fibre length and concentration on the properties of glass fibre-reinforced polypropylene: 4. Impact properties," *Composites Part A: Applied Science and Manufacturing*, vol. 28, no. 3, pp. 277–288, 1997.

[36] N. Graupner, D. Labonte, H. Humburg, T. Buzkan et al., "Functional gradients in the pericarp of the green coconut inspire asymmetric fibre-composites with improved impact strength, and preserved flexural and tensile properties," *Bioinspiration & Biomimetics*, vol. 12, no. 2, Article ID 026009, 2017.

2

Joining of Incompatible Polymer Combinations by Form Fit using the Vibration Welding Process

Michael Wolf, Tobias Kleffel, Christoph Leisen, and Dietmar Drummer

Friedrich-Alexander-University (FAU), Institute of Polymer Technology, Am Weichselgarten 9, 91058 Erlangen, Germany

Correspondence should be addressed to Michael Wolf; wolf@lkt.uni-erlangen.de

Academic Editor: Eliane Espuche

The production of components consisting of various polymer types by welding is severely restricted and only possible for bonding compatible materials with melting points in a close range. Several modifications, such as the cross-linking of one joining partner, allow for circumventing the restrictions regarding the melting points but do not help in joining bonding incompatible materials. Investigations of dissimilar material combinations, especially from polymer-metal hybrid structures, show a high potential of connections based on form fits. Within the scope of this paper, the possibility of joining incompatible polymer combinations, such as polyamide 66 and high-density polyethylene, by micro form fit using the vibration welding process is analyzed. For this purpose, the generated bonding strength of the test specimen was determined by shear tests. Furthermore, the undercuts of the generated prestructures and the resulting bond of the test specimen were examined microscopically by computer-tomography. These investigations depict the high potential of joining incompatible polymer combinations by form fit in the vibration welding using prestructuring to generate undercuts.

1. Introduction

In order to increase the functionality of components, for example, in the automotive industry, it is often necessary to combine various polymer types. Otherwise, it is usually not possible to satisfy different requirements in media resistance, mechanical properties, and so on [1]. Such multimaterial combinations are particularly challenging for joining technologies with short cycle times, like welding [2]. For the creation of a bond between different polymers using welding technologies, a bonding compatibility of them is necessary [3–5]. This greatly restricts the material selection for these welding technologies. An overview of bonding compatible polymers which can form adhesion to each other is provided for joining technologies by Rotheiser [6] and, for the example of multicomponent injection molding, by Jaeger [7]. Furthermore, an approximately equal melting (semicrystalline thermoplastic) or softening temperature (amorphous thermoplastic) is required for the production of multimaterial joints by friction based welding [5]. The temperature in the joining zone usually does not increase by more than 10–20 K above the melting or softening temperature of the lower-melting component [5, 8, 9]. Menacher was able to show that a cross-linking of the material allows for a joining of polymers with different melting points, for example, high performance thermoplastics with standard thermoplastics, by vibration welding [10]. Nevertheless, the radiation cross-linking is an additional and time-intensive process step, which usually cannot be run in-house.

Investigations of dissimilar material combinations, especially from the field of polymer-metal hybrid structures, show a high potential of connections based on or provided by form fits. Such form fits are produced microscopically or macroscopically on the metal component, for example, with laser structuring [11–14], corundum blasting [15, 16], pickling [16], or by electrochemical treatments [17]. Thus, the aim was to transfer this strategy of joining polymer-metal hybrid structures by form fits to polymer-polymer joints by using existing joining technologies with short cycle times.

Here, the friction based vibration welding process shows a particularly high suitability. In addition to the short cycle times, vibration welding is also characterized by a high process flexibility and stability, energy efficiency, and high reproducibility. Compared to other joining methods, there is

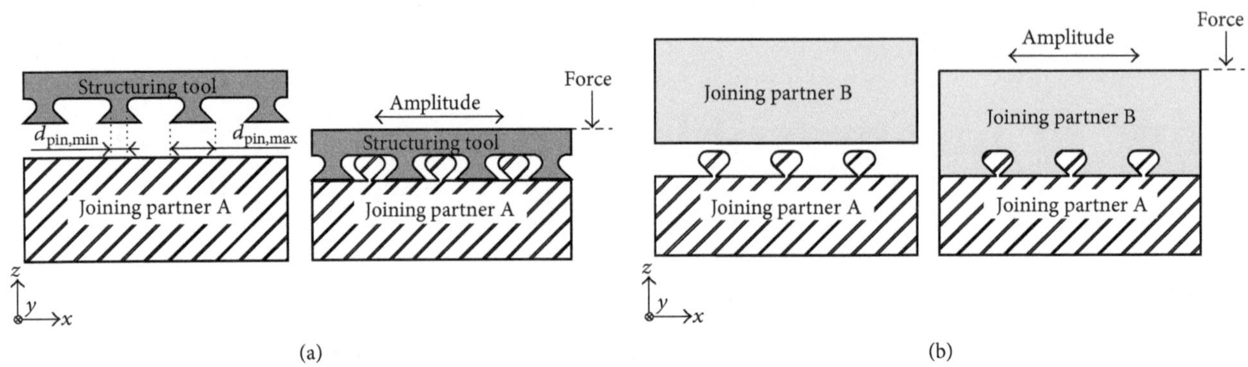

FIGURE 1: Process idea of joining incompatible polymer combinations by form fit using the vibration welding process. (a) Prestructuring process of joining partner A. (b) Welding process of joining partner A and joining partner B.

also no introduction of a foreign material [18]. The process principle of vibration welding is based on a friction-relative movement between the joining parts [3, 19]. During this movement, the parts are pressed together [18]. After a short time, the polymer melts at the contact surface and a melt film depending on the joining pressure is formed [18, 20]. After the end of the friction-relative movement, the joining parts are held together under a defined pressure until the joining area is cooled sufficiently and the material is frozen [5, 18]. The resulting weld strength is mainly depending on the joining pressure. A changing amplitude changes the resulting temperature profile in the melting layer but has only a small influence on the mechanical weld strength [20].

2. Materials and Methods

2.1. Schematically Process Idea. The idea of joining incompatible multimaterial combinations is based on adhesion by form fit. The bonding belongs to a prestructuring of joining partner A, which generates undercuts directly during the vibration welding process and a subsequent joining of the prestructured joining partner A with the joining partner B by using the same process, Figure 1. The joining partner A should have a significantly higher melting temperature than the joining partner B. This ensures a dimensional stability of the inserted prestructures of the joining partner A while the joining partner B is molten and can fill these structures. Other strategies for achieving a shape stability of the embedded prestructures, such as, for example, cross-linking, are also conceivable, if there is no sufficient difference in the melting points.

At the beginning of the process, the high-melting joining partner A (in our research polyamide 66) and the structuring tool with a defined structure are inserted into the vibration welding system. By means of a defined joining pressure and amplitude, the prestructure is formed into the joining partner A, Figure 1(a). In this case, the joining path should correspond approximately to the structural height of the structuring tool. When the desired welding depth is reached, the vibrational movement stops and the structuring tool is moved upwards out of the still molten polymer. As the oscillating head is moved by a magnet, this head always remains in the center position of the oscillating path at the end of the vibrational movement. This ensures that the undercuts are not destroyed when moving the structuring plate in z-direction, if the sum of the minimum pin width in x-direction ($d_{\text{pin,min}}$) and the amplitude a is equal to or greater than the maximum pin width in x-direction ($d_{\text{pin,max}}$; compare Figure 1):

$$d_{\text{pin,min}} + a \geq d_{\text{pin,max}}. \quad (1)$$

In a following step, the structuring tool is exchanged with the joining partner B (high-density polyethylene for our investigations). Afterwards, the conventional vibration welding takes place, Figure 1(b). The joining partners perform a frictional-relative movement to one another; the lower-melting joining partner B melts in the joining zone and fills the prestructures of the joining partner A under a defined joining pressure. Actual limitation of the necessity of similar melting temperatures of both materials for the welding process represents advantages for the joining based on prestructuring. Due to the fact that the temperature in the joining zone rises to values not more than 20 K above the melting temperature of the lower-melting joining partner B [5, 8, 9], the structures of the joining partner A remain dimensionally stable when a sufficient difference between the melting points of the two polymers exists. Therefore, the melting point of the joining partner A ($T_{M,\text{Joining Partner A}}$) should be clearly above the melting point of the joining partner B ($T_{M,\text{Joining Partner B}}$) plus 20 K:

$$T_{M,\text{Joining Partner A}} \gg T_{M,\text{Joining Partner B}} + 20\,\text{K}. \quad (2)$$

Depending on the choice of the process parameters, a more or less complete filling of the prestructures, especially of the undercuts, takes place. After reaching a selected welding depth, which should approximately correspond to the prestructure depth, the welding process is terminated and the cooling takes place under a defined holding pressure.

2.2. Materials. For our research, a commercially available polyamide 66 (PA66), type Durethan A30S (Lanxess Deutschland GmbH, Germany, Cologne) was used as joining partner A and structured with the structuring tool. As joining partner B, an also commercially available high-density

TABLE 1: Process parameter of injection molding process.

Material	Temperature profile Screw → nozzle [°C]	Mold temperature [°C]	Injection speed [mm/s]	Holding pressure [bar]	Holding pressure time [s]
PA66	250/260/270/280/280	80	50	500	30
HDPE	200/210/220/230/240	50	50	500	45

TABLE 2: Parameter settings of prestructuring process.

Welding partners	Amplitude [mm]	Joining pressure [N/mm^2]	Welding time [s]	Holding time [s]
Structuring tool/PA66	0.6	6.5	18	0

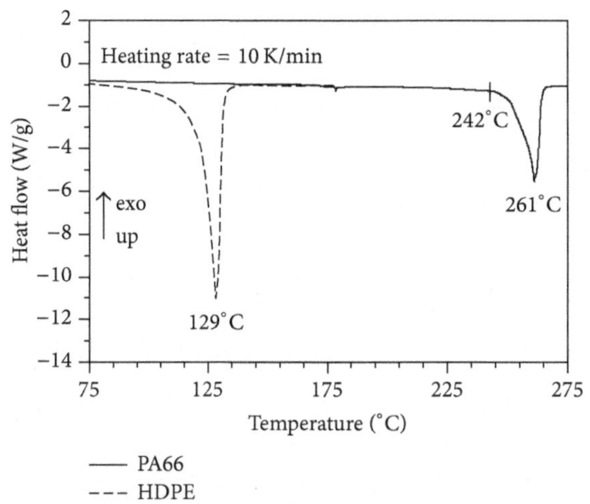

FIGURE 2: Melting behavior of the PA66 and the HDPE joining partner.

polyethylene (HDPE) type Lupolen 4261 (LyondellBasell Industries N.V., Netherlands, Rotterdam) was selected. Before the manufacturing process, the PA66 granular was dried at 80°C for 12 h.

The melting behavior of both materials was analyzed with DSC measurements (3 mg samples, 10 K/min heating rate; DSC Q 2000 (TA Instruments, US, New Castle)) as shown in Figure 2. The HDPE exhibits a melting peak temperature of about 129°C, whereas the melting peak of the PA66 is located at a temperature of 261°C. The difference of the melting peak temperatures between both materials is 132 K. A softening of the material can take place from the onset of the melting peak on and hence dimensional stability may no longer be given which is the case for temperatures above 242°C for the used PA66. The difference between the melting peak temperature of the HDPE and the onset temperature of the PA66 is measured to 113 K. In consequence, the onset temperature of the PA66 significantly exceeds the maximum temperature in the joining zone of approximately 20 K above the melting peak temperature of the low-melting polymer [8, 9]. Thus, a melting of the generated pin structures can be excluded and dimensional stability can be assumed.

The structuring tool for prestructuring the joining partner A was made of structural steel, type S235JR, and subsequently hardened in order to avoid the wear of the structures.

2.3. Experimental Procedure

2.3.1. Joining Partners. Plates with a geometrical dimension of 150 mm × 68 mm × 4 mm were manufactured via injection molding using an injection molding machine Allrounder 370V/800-315 (Arburg, Germany, Loßburg). The joining samples with the dimension of 70 mm × 50 mm × 4 mm were milled out of these plates. An extract of the used injection molding parameters for the PA66 and HDPE joining partners is given in Table 1.

2.3.2. Structuring Tool. The structuring tool, which was used for the prestructuring of the joining partner A, was milled out of a steel plate and had the dimension of 70 mm × 50 mm × 4 mm. On one side of the structuring tool, four rows with pin structures are located, consisting of 16 pins. Thus, the patterning surface features 64 pins which allows for the double number of undercut structures (two undercut sections per pin). The height of the pin structures is 1.75 mm which results from initial drillings with a diameter of 2.0 mm and a subsequent milling by a depth of 0.25 mm. The individual pins are spaced approximately 3 mm apart. The pin rows are separated by a 2.0 mm wide groove produced by milling. Figure 3 shows the pin based structuring tool which was used for the prestructuring and the inserting of the undercuts into the PA66.

2.3.3. Prestructuring of Joining Partner A. The prestructuring of joining partner A, in this study PA66, was performed on a Branson 2800 linear vibration welding machine (Branson Ultraschall, EMERSON Technologies GmbH & Co. OHG, Germany, Dietzenbach). The welding machine is equipped with an oscillation driver system, type M-522 H, with highly dynamic amplitude control. The system works with an amplitude up to 0.9 mm at a nominal resonance frequency of 240 Hz. For prestructuring of joining partner A, the structuring tool was vibrated into the PA66 plate with an amplitude of 0.6 mm and a joining pressure of 6.5 N/mm^2 for a time of 18 s, Table 2. A holding time to cool down the molten PA66 was not implemented; the structuring tool was moved out of the specimen directly after the end of vibration.

2.3.4. Vibration Welding. The welding tests were carried out after the prestructuring process. The same vibration welding machine was used for this purpose; only the structuring

Table 3: Parameter settings of vibration welding process.

Welding partners	Amplitude [mm]	Joining pressure [N/mm^2]	Welding time [s]	Holding time [s]
PA66 prestructured/HDPE	0.6	2.0	15	15
PA66 unstructured/HDPE	0.6	1.0/2.0/3.0	15	15

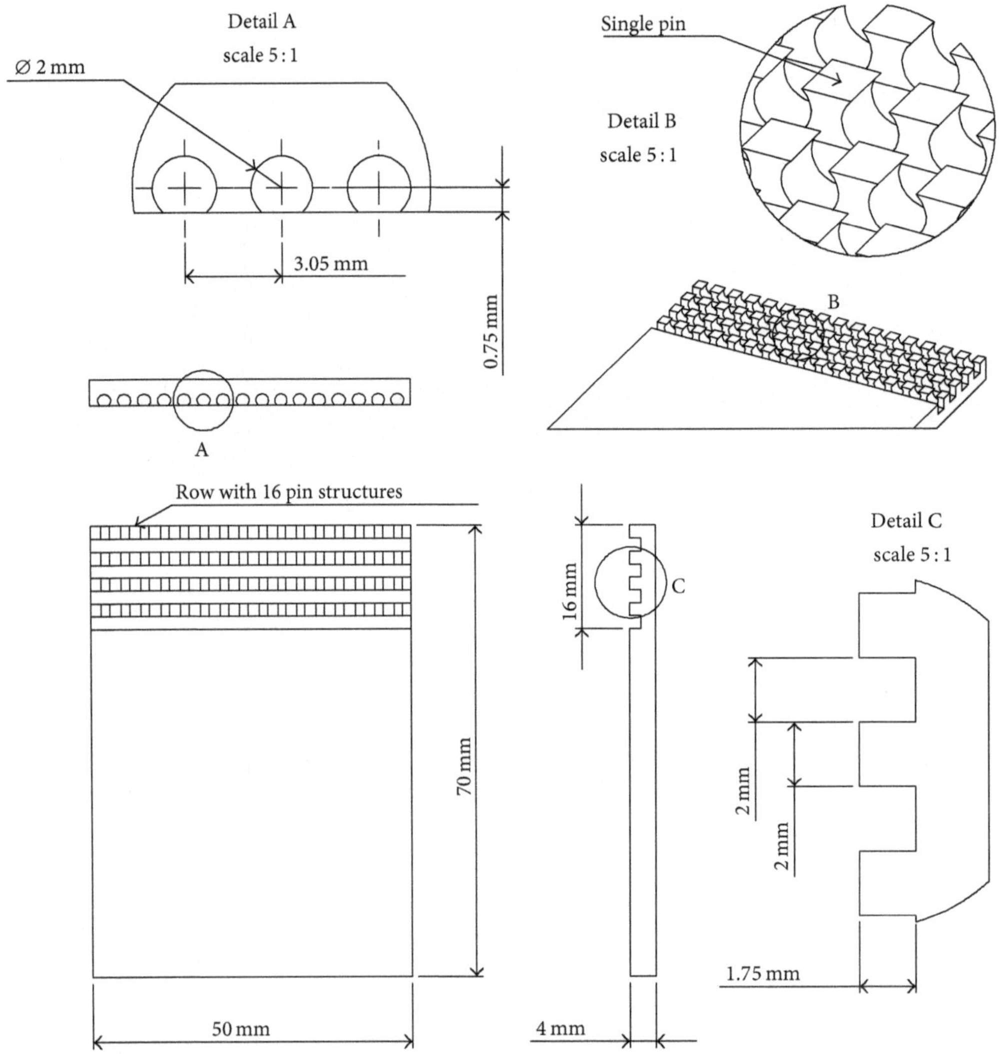

Figure 3: Geometrical dimensions of the used structuring tool with 64 pin structures involving undercut areas.

tool was exchanged with the joining partner B, for this investigation HDPE. The prestructured joining partner A remained in the machine. In the vibration welding tests, an amplitude of 0.6 mm with a joining pressure of 2.0 N/mm^2 and a welding time of 15 s were chosen, Table 3. The holding time was set to 15 s with a holding pressure equal to the joining pressure. The overlap length of the lap-joints was designed in such a way that the undercut structures of three pin rows ensure the bonding based on form fit between the two polymers, Figure 4. Therefore, a total number of 48 pins or rather 96 undercut sections are used for the interlocking connection. The use of three out of four pin rows is due to further structuring tools which only contain three pin rows and should be considered in additional investigations.

In addition to the welding tests with prestructured PA66 joining partners, unstructured PA66 and HDPE plates were joined by vibration welding. The parameters were chosen according to the previous experiments, Table 3. In addition, tests with a joining pressure of 1.0 and 3.0 N/mm^2 were made. The joining pressure was varied to exclude an influence of the pressure on the incompatibility of the used polymers in the welding process.

2.4. Analysis Methods

2.4.1. Computer-Tomography (CT). The CT analyses were done with a micro-radiography-computer-tomograph device, type subμ-CT (Fraunhofer Institute for Integrated Circuits (IIS) e.V., Germany, Erlangen). The measurements were

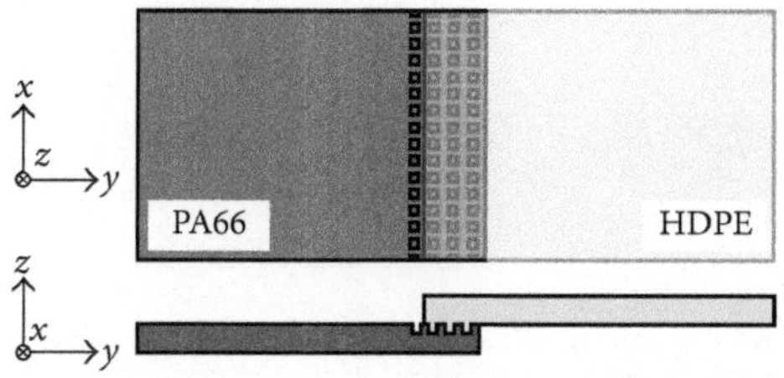

FIGURE 4: Overlap geometry of the vibration welding test specimen consisting of prestructured joining partner A (PA66) and joining partner B (HDPE).

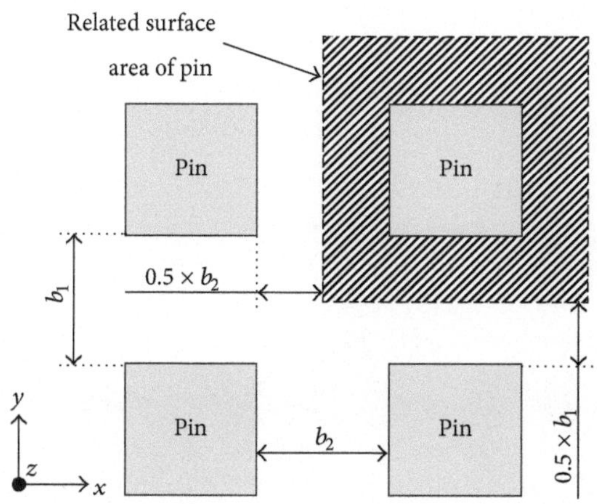

FIGURE 5: Definition of related surface area of the pins.

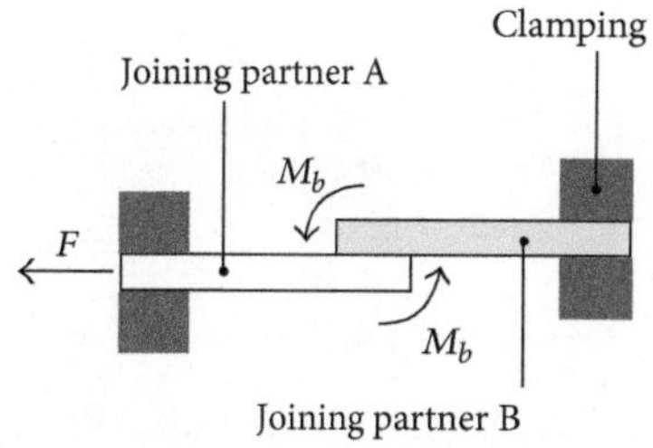

FIGURE 6: Loads on a single lap test specimen during the test.

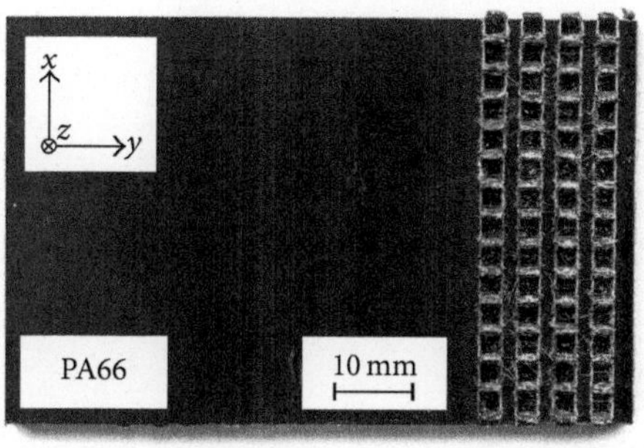

FIGURE 7: Photography of the prestructured joining partner A (PA66) with generated undercuts.

done on the whole test specimen and, in order to enable a higher resolution of the joining zone, also on cubes with a dimension of approximately 8 mm × 8 mm × 8 mm prepared out of the produced test specimens.

2.4.2. Mechanical Test: Shear Test. For the shear tests, a universal testing machine with computer-controlled test executions, type Zwick 1474 (Zwick GmbH & Co. KG, Germany, Ulm) was used. Each test was performed under standard conditions according to DIN EN ISO 219 ($T = 23^{\circ}$C, relative air humidity: 50%) with a pull-off speed of 1 mm/min. Before testing, the specimens were stored dry in a vacuum furnace at 23°C. The tensile force and strain as well as the force at failure for 10 samples were recorded. The tensile force per single pin and its related surface area, seen in Figure 5, is calculated by division of the measured force by the number of generated pins.

The performed shear tests serve as proof of concept and allow a first assessment of the potential of the joint based on form fit. As a result of the construction of the single lap test specimen, there is a bending moment M_b during the test, which leads to a load outside of the joining surface, Figure 6. Without a resulting undercut in z-direction, a failure of the test specimen would occur by a simple release of the pins without an application of a relevant bond strength.

3. Results and Discussion

3.1. Joining Process

3.1.1. Prestructuring Process. The prestructured joining partner A (PA66), Figure 7, shows a negative image of the pin rows of the structuring tool and thus undercut structures. The generated prestructure and the undercut areas can be seen in the computer-tomography tests, Figure 8. The undercuts are mainly located in the lower half of the pin structures which accords to the geometry of the structuring tool. The structure height of the pins extends over the entire area in the range between 1.7 and 1.9 mm, exemplarily shown in the illustrated section (Figure 8). The height of the prestructures corresponds relatively well to the pin geometry of the structuring tool, which has a height of 1.75 mm. The slight height deviations can be explained by the extraction of the structuring tool out of the still molten polymer. In this case, a slight deformation of the prestructures can take place.

3.1.2. Vibration Welding. The welding tests were carried out for the manufacturing of single lap test specimens. A lateral view in x-direction of the joining zone shows that the prestructures were filled during the vibration welding process by the molten HDPE material. The computer-tomography tests show a complete filling of the previously inserted structures across the entire joining area, Figure 9(a). In accordance with the plain prestructures, the polymer joint shows areas with undercuts. By filling these undercuts with vibration welding, areas are generated which cause a bonding by means of a form fit in z-direction, Figure 9(b). In addition, a form fit in x-direction occurs by a complete filling of the space

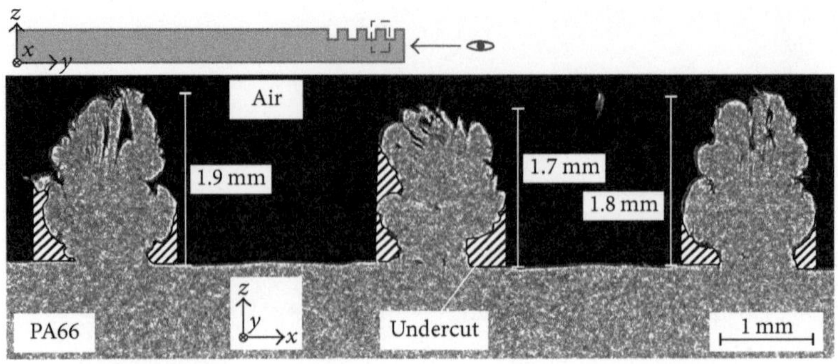

FIGURE 8: CT image of prestructured joining partner A (PA66) with generated undercuts.

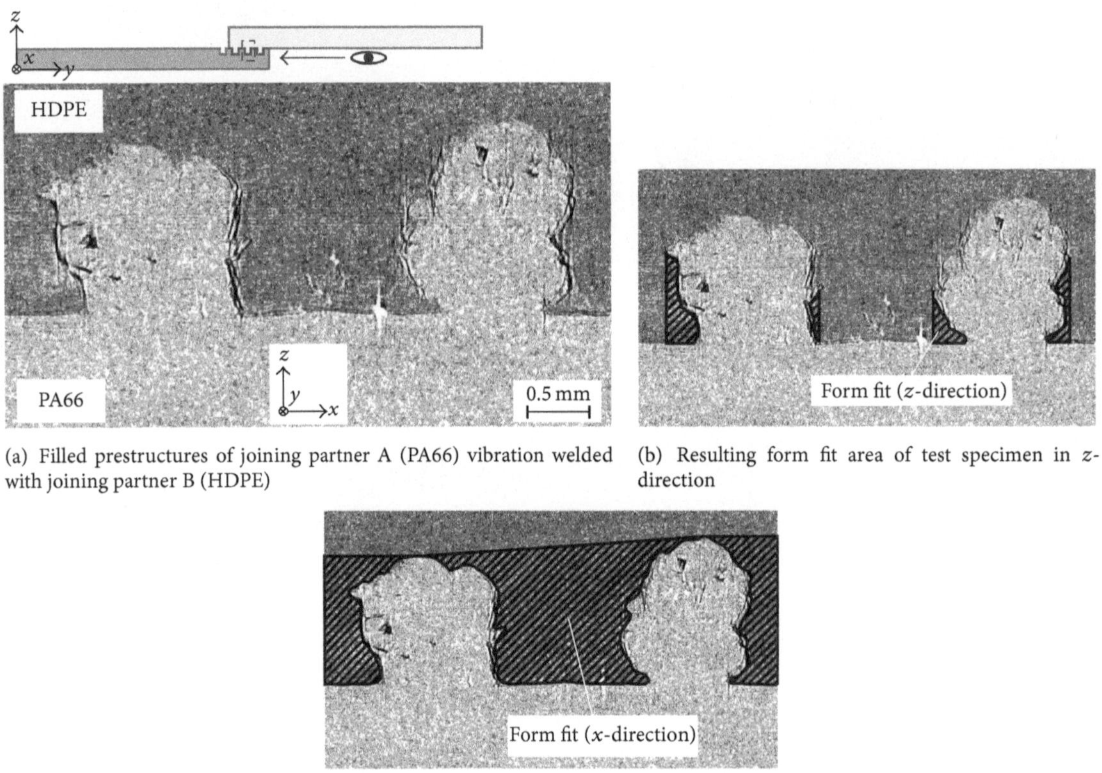

(a) Filled prestructures of joining partner A (PA66) vibration welded with joining partner B (HDPE)

(b) Resulting form fit area of test specimen in z-direction

(c) Resulting form fit area of test specimen in x-direction

FIGURE 9: CT image of PA66—HDPE test specimen, produced by prestructuring and subsequent vibration welding.

between the pins in z-direction, Figure 9(c). In comparison to the conventional welding trials which resulted in immediate failure of the joint during handling, the structure based joint of the incompatible polymers PA66 and HDPE is more resilient.

3.2. Failure Behavior: Tensile Force. The tensile force to displacement relationship is shown in Figure 10 for each sample individually. A nearly identical curve progression can be seen up to a tensile force of about 2000 N. The maximum tensile force attained by a single test specimen is 2350 N; the average is located at 2147 ± 123 N. For a single pin and its related surface area an average tensile force of 45 ± 3 N results. Due to the nonmeasurable bonding strength of unstructured specimens, it can be assumed that solely the pins and their existing form closure cause the bonding strength. The surrounding surface area probably has a negligibly small influence on the bonding strength.

Predominantly, adhesive failure behavior by the release of the form fit from the undercuts could be observed, Figure 11. Due to the nearly identical curve characteristic of the mechanical tests, it can be assumed that the joining by form fit based on a prestructuring process using the vibration welding process created a reproducible joining method for conventionally incompatible polymers. On the contrary to the state of the art [8, 9], condition is a dissimilar melting point of the joining partners or the ensuring of a shape

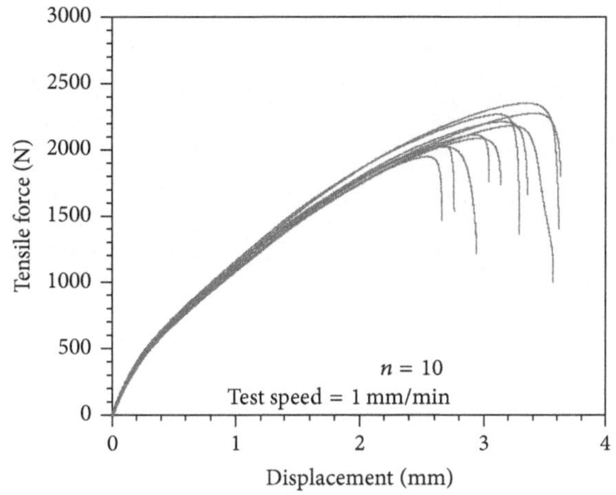

FIGURE 10: Tensile force of PA66—HDPE test specimen based on form fit, produced by prestructuring joining partner A (PA66) and subsequent welding with joining partner B (HDPE).

FIGURE 11: Adhesive fracture behavior of test specimen, captured by photography after mechanical test. (a) Section of joining partner A (PA66) after shear test. (b) Section of joining partner B (HDPE) after shear test.

stability of the inserted pin structures, for example, by cross-linking.

Tensile tests for unstructured welding tests could not be carried out. A failure occurred due to the bonding incompatibility of the polymers directly after the joining process without the application of a mechanical load.

4. Conclusion

The investigations have shown that the production of joints between bonding incompatible polymers by means of form fit using the vibration welding process is possible and results in components with reproducible shear behavior. Furthermore, we were able to demonstrate the high potential for the joining of conventionally incompatible polymer combinations by form fit with a prestructuring of one joining partner with undercut structures. This could allow for a significant expansion of the combinable polymer spectrum for friction based welding technologies, which is helpful for new applications or a further increase of the functionality of components. Furthermore, this new preconditioning method can potentially result in even higher bonding strengths values if there is additional bonding compatibility of the polymers, albeit the lower-melting component will not melt so that there is no direct molecular exchange in the welding layer.

In the usual implementation, the joining of PA66 and HDPE by welding is not possible due to the bonding incompatibility of both materials. Analysis on unstructured PA66–HDPE combinations showed a failure which occurred without the application of a mechanical load. Based on a prestructuring of the PA66 specimen and the generation of undercuts, a form fit between the PA66 and HDPE specimen could be created using vibration welding. This increased the bonding strength of the test specimen to a measurable level and, in a single case, allowed for an application of a mechanical load of up to 2350 N (2147 N average) to the polymer joint or, respectively, 45 N per single pin of the prestructure before a failure of the joint occurred.

Due to the shown high potential of the joining of with conventional methods incompatible polymer combinations by form fit based on prestructuring of one joining partner, further detailed fundamental investigations are planned. On the one hand, the interactions between the material, the process, and the prestructure as well as the resulting bond strength especially the behavior with pure tensile load of the generated undercuts perpendicular to the joining zone have to be analyzed. Furthermore, different structural geometries for the prestructuring and their introduction mechanism are to be examined numerically and experimentally. The investigations also have to be extended to encompass other polymer combinations and further friction based welding processes, such as ultrasonic welding.

References

[1] C. Hopmann and M. Weber, "New concepts for joining dissimilar thermoplastics using laser radiation," *Joining Plastics*, vol. 2, pp. 116–125, 2012.

[2] Deutscher Verband für Schweißen and verwandte Verfahren, "Herausforderungen an die Verbindungstechnik in Anwendungen aus Leichtbau und erneuerbaren Energien," *Joining Plastics*, vol. 6, pp. 99–107, 2012.

[3] G. W. Ehrenstein, *Handbuch Kunststoff-Verbindungstechnik*, Hanser, München, Germany, 2004.

[4] C. Bonten, *Beitrag zur Erklärung des Wirkmechanismus in Schweißverbindungen aus teilkristallinen Thermoplasten [Ph.D. thesis]*, Universität Gesamthochschule Essen, Essen, Germany, 1998.

[5] D. A. Grewell, A. Benatar, and J. Park, *Plastics and Composites Welding Handbook*, Hanser, München, Germany, 2003.

[6] J. Rotheiser, *Joining of Plastics—Handbook for Designers and Engineers*, Hanser, München, Germany, 1999.

[7] A. Jaeger, "Kombinationen potenzieren das Leistungsvermögen—Trends, Tendenzen und Perspektiven beim Mehrkomponenten-Spritzgießen," *Kunststoffe*, vol. 10, pp. 91–94, 1991.

[8] J. Vetter and G. W. Ehrenstein, "Biaxial vibration welding of polypropylene," in *Proceedings of the Annual Technical Conference - ANTEC, Conference Proceedings*, pp. 1344–1348, New York, NY, USA, 1999.

[9] P. J. Bates, J. MacDonald, V. Sidiropoulos, and M. Kontopoulou, "Comparison of experimental and analytical vibration welding meltdown-time profiles for nylon 66 and polypropylene," *Polymer Engineering and Science*, vol. 45, no. 6, pp. 789–797, 2005.

[10] M. Menacher, *Vibrationsschweißen von strahlenvernetztem Polyamid 66 [Ph.D. thesis]*, Friedrich-Alexander-Universität Erlangen-Nürnberg, Erlangen, Germany, 2015.

[11] P. Amend, S. Pfindel, and M. Schmidt, "Thermal joining of thermoplastic metal hybrids by means of mono- and polychromatic radiation," *Physics Procedia*, vol. 41, pp. 98–105, 2013.

[12] E. Rodríguez-Vidal, C. Sanz, C. Soriano, J. Leunda, and G. Verhaeghe, "Effect of metal micro-structuring on the mechanical behavior of polymer-metal laser T-joints," *Journal of Materials Processing Technology*, vol. 229, pp. 668–677, 2016.

[13] A. N. Fuchs, F. X. Wirth, P. Rinck, and M. F. Zaeh, "Laser-generated macroscopic and microscopic surface structures for the joining of aluminum and thermoplastics using friction press joining," *Physics Procedia*, vol. 56, pp. 801–810, 2014.

[14] E. E. Feistauer, R. P. M. Guimarães, T. Ebel, J. F. Dos Santos, and S. T. Amancio-Filho, "Ultrasonic joining: a novel direct-assembly technique for metal-composite hybrid structures," *Materials Letters*, vol. 170, pp. 1–4, 2016.

[15] G. Lucchetta, F. Marinello, and P. F. Bariani, "Aluminum sheet surface roughness correlation with adhesion in polymer metal hybrid overmolding," *CIRP Annals—Manufacturing Technology*, vol. 60, no. 1, pp. 559–562, 2011.

[16] P. Mitschang, R. Velthuis, S. Emrich, and M. Kopnarski, "Induction heated joining of aluminum and carbon fiber reinforced nylon 66," *Journal of Thermoplastic Composite Materials*, vol. 22, no. 6, pp. 767–801, 2009.

[17] T. Kleffel and D. Drummer, "Investigating the suitability of roughness parameters to assess the bond strength of polymer-metal hybrid structures with mechanical adhesion," *Composites Part B: Engineering*, vol. 117, pp. 20–25, 2017.

[18] V. K. Stokes, "Vibration welding of thermoplastics. Part I: phenomenology of the welding process," *Polymer Engineering & Science*, vol. 28, no. 11, pp. 718–727, 1988.

[19] H. Potente, P. Michel, and B. Ruthmann, "Eine Analyse des Vibrationsschweißens," *Kunststoffe*, vol. 77, pp. 711–716, 1987.

[20] A. K. H. Schlarb, *Zum Vibrationsschweißen von Polymerwerkstoffen [Ph.D., thesis]*, Universität-Gesamthochschule Kassel, Kassel, Germany, 1989.

3

Reversible Shaping of Microwells by Polarized Light Irradiation

Federica Pirani,[1,2] Angelo Angelini,[1] Francesca Frascella,[1] and Emiliano Descrovi[1]

[1]*Dipartimento di Scienza Applicata e Tecnologia, Politecnico di Torino, Corso Duca degli Abruzzi 24, 10129 Torino, Italy*
[2]*Centro per le Tecnologie Future Sostenibili, Istituto Italiano di Tecnologia, Corso Trento 21, 10129 Torino, Italy*

Correspondence should be addressed to Angelo Angelini; angelo.angelini@polito.it

Academic Editor: Nobuhiro Kawatsuki

In the last years, stimuli-responsive polymeric materials have attracted great interest, due to their low cost and ease of structuration over large areas combined with the possibility to actively manipulate their properties. In this work, we propose a polymeric pattern of soft-imprinted microwells containing azobenzene molecules. The shape of individual elements of the pattern can be controlled after fabrication by irradiation with properly polarized light. By taking advantage of the light responsivity of the azobenzene compound, we demonstrate the possibility to reversibly modulate a contraction-expansion of wells from an initial round shape to very narrow slits. We also show that the initial shape of the microconcavities can be restored by flipping the polarization by 90°. The possibility to reversibly control the final shape of individual elements of structured surfaces offers the opportunity to engineer surface properties dynamically, thus opening new perspectives for several applications.

1. Introduction

In recent years, the ability to pattern large areas of stimuli-responsive materials at the micro- and nanoscale has opened the opportunity to engineer surface structures and trigger peculiar properties such as complex optical functionalities [1] or surface properties [2] by light-matter interactions. Over the last two decades, the progress on micro- and nanoscale research has enabled the development of several patterning techniques, including soft lithography [3], nanoimprint lithography [4], interference lithography, and advanced photo lithography [5]. Micro- and nanofabrication technology finds applications in a plethora of fields, such as optics and photonics [6], microfluidics [7], and even biology [8].

However, despite such impressive achievements, most of the conventional patterning techniques result in static structures, whose features cannot be modified after fabrication. To overcome such limitation, Karageorgiev et al. proposed exploiting the photoresponsivity of azopolymers in a technique named Directional Photofluidization Lithography [9–11], although the term "photofluidization" is controversial as the fundamental mechanism underlying such effect is still under debate [12]. Based on the same approach, here we propose a reversible manipulation of prefabricated azopolymeric structures driven by polarized light.

Azopolymers are well known to transduce a visible light stimulus into a mechanical modification due to cyclic isomerization of the azobenzene molecule triggered by photon absorption [13]. The cyclic isomerization results in a mechanical modification of the polymer [14] and sometimes in an effective mass migration [15] that allows structuring the polymeric surface via direct illumination [16]. As mass-migration occurs parallel to the polarization of the incident radiation, prefabricated microstructures can be anisotropically deformed depending on the illumination conditions [17, 18]. In detail, we start with an array of microwells and show that the shape of the microwells can be finely controlled by exposing the film to properly polarized light. If linearly polarized radiation is used, the initial shape of microwells can be restored by turning the polarization by 90°. The technique presented here represents a valuable approach to tune the features of prestructured surfaces.

2. Materials and Methods

Sample Preparation. The azopolymer considered is an azo-polyelectrolyte (poly{1-[4-(3-carboxy-4-hydroxyphenylazo) benzenesulfonamido]-1,2-ethanediyl, sodiumsalt}) named as PAZO (molecular weight [Mw] ≈ 369,33 g/mol), supplied by Sigma-Aldrich. The PAZO is mixed with

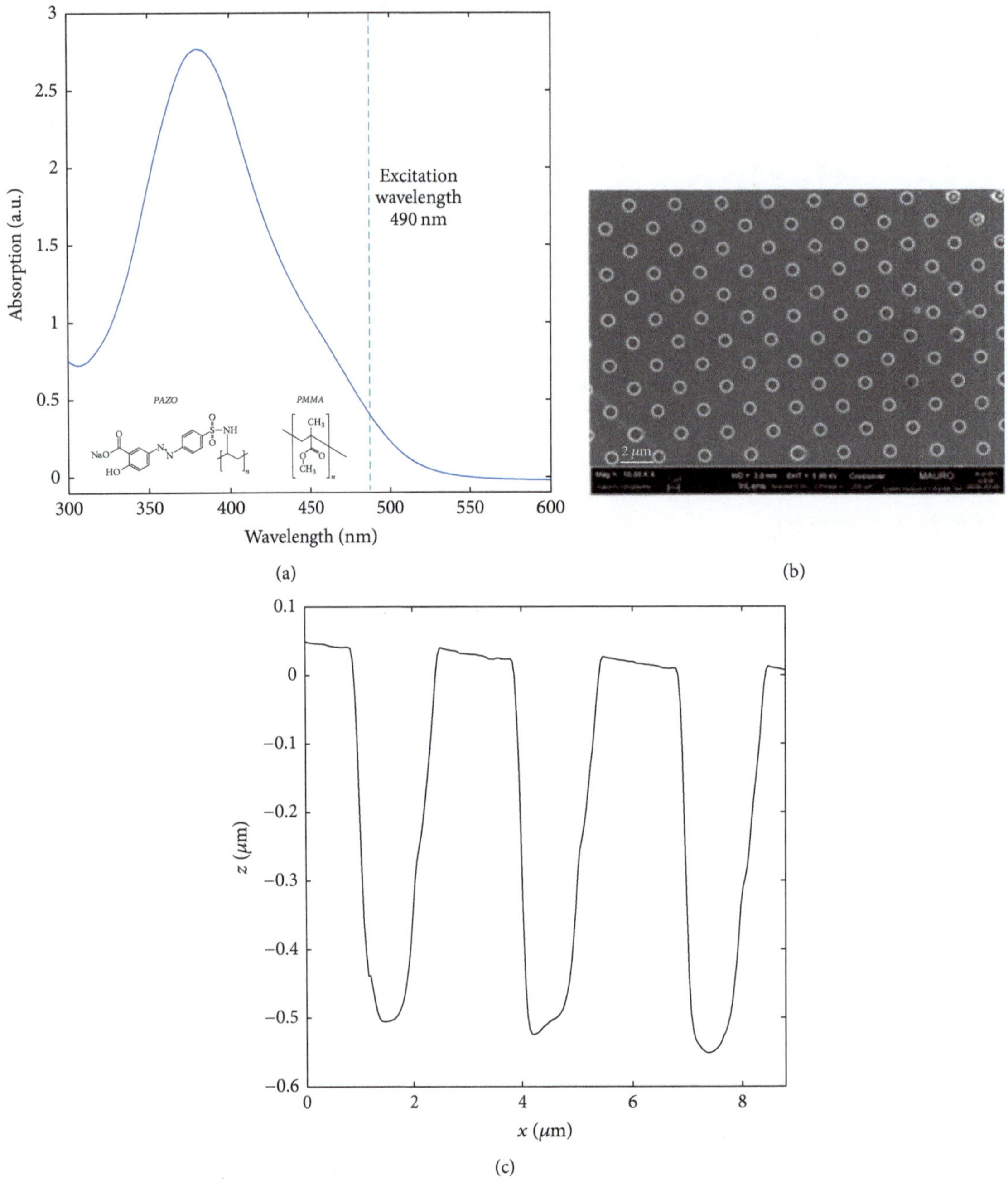

FIGURE 1: (a) Absorption spectrum of the PAZO-PMMA compound in the UV-Visible range. In the inset, sketch of the chemical structure of the two polymers. (b) SEM image of the array of microwells. (c) AFM profile of the microwells.

poly-(methyl methacrylate) (PMMA), resulting in a mixture prepared as follows: PMMA powder (molecular weight [Mw] ≈ 15000 g/mol by GPC®) is dissolved in N,N-dimethylformamide at a concentration of 4 wt%. Nine units of PAZO (Mw ≈ 369,33 g/mol), dissolved in methanol at concentration 25 mg/mL, are mixed to one unit of PMMA solution and then mechanically stirred and sonicated for few minutes until a homogeneous yellow blend is formed. Figure 1(a) shows the absorption spectrum of the PMMA-PAZO compound obtained and, in the inset, the molecular structure of PAZO and PMMA. We measured the glass transition temperature (Tg) of the polymeric compound by means of Differential Scanning Calorimetry (DSC, data not shown) and we found $Tg = 53.9°C$ as compared to $Tg = 60.7°C$ of the PAZO without PMMA. An amount of 15 μl of PAZO-PMMA is spin-casted onto a glass substrate, previously washed in acetone, rinsed with isopropanol, and dried with a nitrogen flow. A PDMS stamp (10 : 1 ratio of the prepolymer and the curing agent, provided by Sylgard® 184, Dow Corning, cured at 60°C for 2 hours) is soaked in ethanol for few minutes and then gently pressed on the PAZO-PMMA film on glass. The sample is placed in a

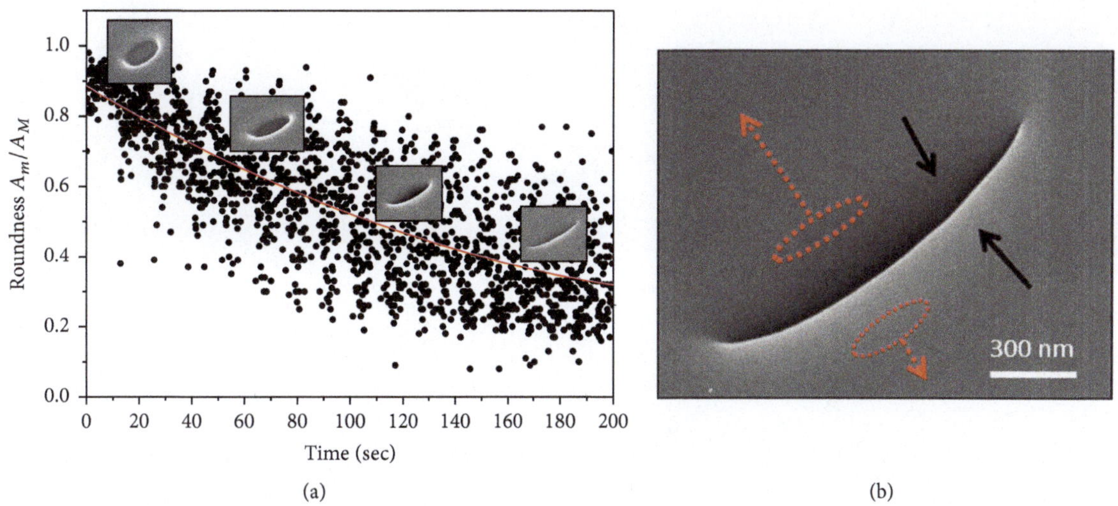

FIGURE 2: (a) Time-resolved roundness values for microwells during laser irradiation (black dots: roundness of individual wells, red line: mean roundness). In the insets, representative SEM images taken with sample tilted by 45° of four different time steps. (b) SEM image (sample tilted by 45°) of a single microwell after irradiation (black arrows: the direction of the pore reduction and red arrows: direction of the mass migration of the material).

vacuum desiccator for 2 hours to enhance patterning and let the polymer blend to fill in the PDMS template. As a next step, the sample is dried in oven, until the solvent is completely evaporated (3 hours at 60°C) before the final removal of the PDMS stamp. The SEM image in Figure 1(b) shows an array of azopolymeric microwells over a large area. The wells are circular with a diameter of 1 μm arranged as a squared lattice with periodicity $\Lambda = 3\,\mu$m. Atomic Force Microscopy analysis reveals circular wells with a diameter of 1 μm arranged as a squared lattice with periodicity $\Lambda = 3\,\mu$m. The depth, estimated by AFM analysis, is about 550 nm, as shown in Figure 1(c).

Optical Setup. The optical setup employed in this work is described in detail in [19]. Briefly, we employ a pulsed laser (laser temporal width ≈ 10 ps, wavelength 490 nm, and average intensity ≈ 50 mW) circularly polarized. The laser beam is slightly focused in such a way that the spot size on the sample surface is about 20 μm, thus resulting in a density of power of 1.6 kW/cm^2. We point out that the pulsed laser has been employed because of its tunable frequency that allows for a better matching with the absorption band of PAZO (see Figure 1(a)). Due to the slow thermal relaxation time of the isomerization transition (typically in the range of hundreds of ms [15]) as compared to the repetition rate of the pulsed laser (80 MHz), it seems reasonable to approximate the pulsed radiation to a CW light. The beam can be eventually filtered by a linear polarizer placed on a rotational stage to obtain linearly polarized light along arbitrary directions. A 100x objective (NA = 0.95) coupled with a tube lens allows imaging the sample surface on a CMOS camera (Thorlabs DCC3240M). In collection, the laser beam is filtered out by an edge filter (FEL550 from Thorlabs) along the imaging arm. In order to extract useful parameters, the images collected have been binarized by setting a proper threshold and subsequently analyzed by employing an ImageJ tool able to compute roundness and the overall area of the pores.

3. Results and Discussion

First, we want to investigate the effect of irradiation with linearly polarized light on the azopolymeric microstructures. In agreement with our previous experimental findings [19] and results from literature [20, 21], we observe a directional mass migration of the polymer along the polarization direction. For a quantitative analysis, the mean roundness parameter is evaluated for each frame as the ratio between the minor (A_m) and the major axis (A_M) averaged over 20 microwells within a single field-of view of the microscope. Initially, wells are almost circular, with a mean roundness of about 0.9 (Figure 2(a)). Upon linearly polarized irradiation, the wells deform into an elliptical shape (mean roundness down to 0.3). The SEM image in Figure 2(b), obtained keeping memory of the orientation of the sample during irradiation, reveals that the major axis of the ellipses is oriented perpendicular with respect to the polarization direction of light, confirming that the photoinduced mass migration of the material occurs along the electric field direction.

In order to show the reversibility of the deformation, we repeated the experiment with linearly polarized light. The microwells, initially showing circular shape (Figure 3(a)), turn into almost closed slits after about 220 seconds of irradiation (Figure 3(b)). Subsequently, we rotate the polarization of light by 90° and we observe a gradual restore of the circular shape (Figure 3(c)). The image analysis reveals that an average roundness of about 1 is achieved after further 200 seconds of irradiation. Such reversible modification is advantageously promoted by an improved stiffness of the polymeric mixture due to the presence of a stabilizing PMMA component, as

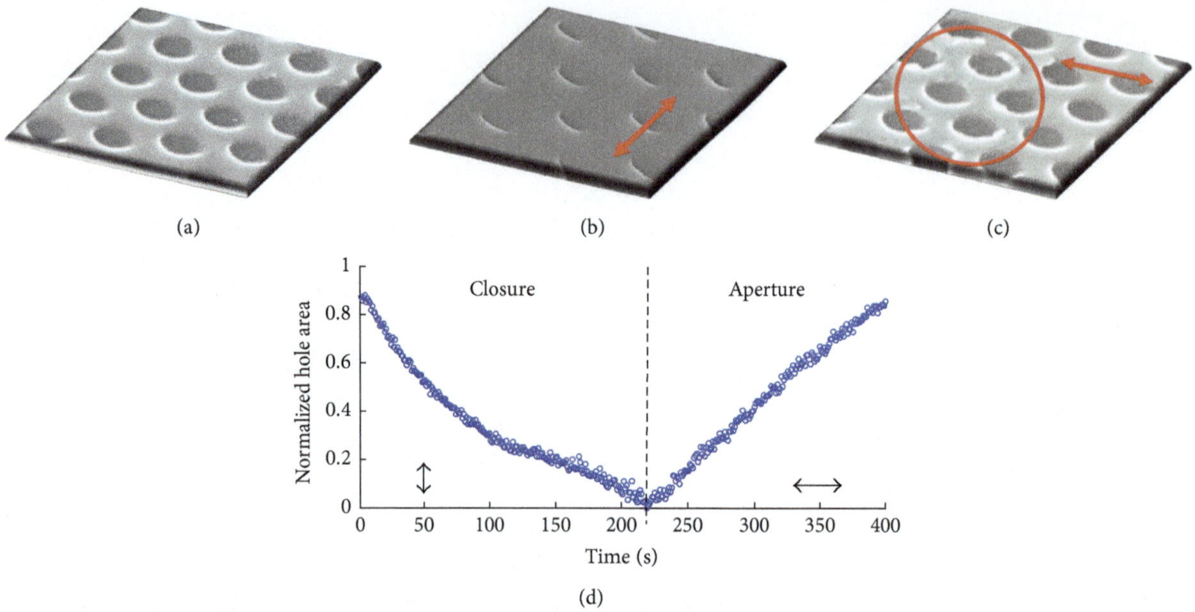

FIGURE 3: Photoreconfiguration of pristine microwells arrays by using linearly polarized light: (a) original concavity shape, (b) closed slits, and (c) restored shape. (d) Time-resolved hole area values for pores during laser irradiation with time-varying polarization states.

suggested elsewhere in other copolymer matrices containing azobenzenes [19–22]. We underline here that the role of PMMA seems to be crucial for reversibility, since with pure PAZO structures we were not able to restore the pristine shape.

In Figure 3(d) we plotted the average area of the holes versus time of irradiation. The graph shows that upon irradiation the holes area is actually reduced till a complete closing of the holes is obtained (black dashed line in the graph). Experimentally, we find that the polarization rotation by 90° triggers a recovery of the microwells area. Such effect is similar to photoinduced micropillar deformation reported elsewhere [17], but the fundamental mechanism underlying the observed phenomena is not straightforward and requires further investigation.

After probing the effect of linearly polarization on the holes shape, we aim to exploit circularly polarized light to reduce the pore size eventually until complete closing. Figure 4 shows the effect of circularly polarized light on an array of microwells. The graph indicates that irradiation with circular polarization does not affect the roundness of the microstructures that is kept constant around a mean value of 0.8. The inset in the graph shows optical images of a sequence of microwells individually addressed and irradiated with different doses (the position along the timeline corresponds to the irradiation time). The sequence shows that although the roundness is preserved, the wells tend to isotropically reduce their size until complete closure. In the case of circularly polarized light, we did not observe any reversibility related to the handedness of light polarization.

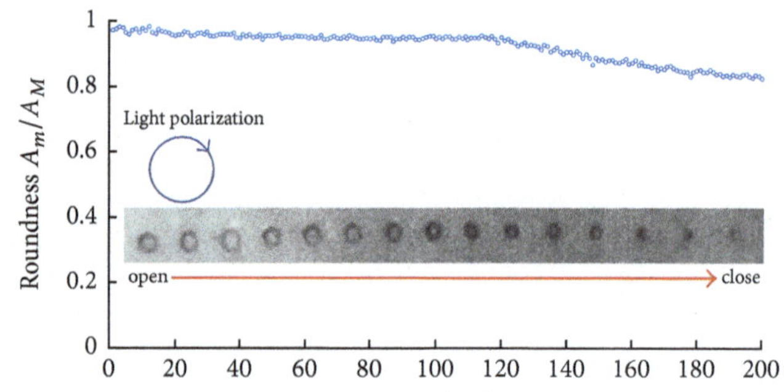

FIGURE 4: Time-resolved roundness evolution microwells during laser irradiation with circularly polarized light. In the inset, optical images of individual microwells are exposed to different light doses.

4. Conclusions

The presented work shows the possibility to exploit directional photofluidization to manipulate microstructures after fabrication. Starting from an array of circular microwells, we employ linearly polarized light to change them into elliptically shaped wells and eventually into narrow slits, depending on the light dose. Alternatively, the microwells can be reduced in size without changing the roundness till an almost complete closing is reached. Tuning the morphology of surface microstructures might provide a smart engineered platform where surface properties can be engineered on demand after fabrication [19, 20]. The light-induced contraction and expansion based reshaping of polymeric microstructures shows exciting potential for several applications including microfluidics and lithography and, in the last years, is showing its potential application in biology, where cells behavior can be regulated in response to material manipulation.

Acknowledgments

The authors gratefully thank Dr. Serena Ricciardi for technical support and fruitful discussions.

References

[1] A. Emoto, E. Uchida, and T. Fukuda, "Optical and physical applications of photocontrollable materials: Azobenzene-containing and liquid crystalline polymers," *Polymer*, vol. 4, no. 1, pp. 150–186, 2012.

[2] Z. L. Wu, R. Wei, A. Buguin et al., "Stimuli-responsive topological change of microstructured surfaces and the resultant variations of wetting properties," *ACS Applied Materials & Interfaces*, vol. 5, no. 15, pp. 7485–7491, 2013.

[3] Y. Xia and G. M. Whitesides, "Soft lithography," *Angewandte Chemie International Edition*, vol. 37, no. 5, pp. 550–575.

[4] M. Li, L. Chen, and S. Y. Chou, "Direct three-dimensional patterning using nanoimprint lithography," *Applied Physics Letters*, vol. 78, no. 21, pp. 3322–3324, 2001.

[5] D. S. Correra, L. De Boni, A. J. G. Otuka, V. Tribuzi, and C. R. Mendonça, "Two-photon polymerization fabrication of doped microstructures," in *Polymerization*, vol. 30, pp. 333–356, InTech, 2012.

[6] A. Arbabi, Y. Horie, M. Bagheri, and A. Faraon, "Dielectric metasurfaces for complete control of phase and polarization with subwavelength spatial resolution and high transmission," *Nature Nanotechnology*, vol. 10, no. 11, pp. 937–943, 2015.

[7] D. H. Kang, S. M. Kim, B. Lee, H. Yoon, and K.-Y. Suh, "Stimuli-responsive hydrogel patterns for smart microfluidics and microarrays," *Analyst*, vol. 138, no. 21, pp. 6230–6242, 2013.

[8] C. Rianna, A. Calabuig, M. Ventre et al., "Reversible holographic patterns on azopolymers for guiding cell adhesion and orientation," *ACS Applied Materials & Interfaces*, vol. 7, no. 31, pp. 16984–16991, 2015.

[9] P. Karageorgiev, D. Neher, B. Schulz et al., "From anisotropic photo-fluidity towards nanomanipulation in the optical near-field," *Nature Materials*, vol. 4, no. 9, pp. 699–703, 2005.

[10] S. Lee, H. S. Kang, and J.-K. Park, "Directional photofluidization lithography: Micro/nanostructural evolution by photofluidic motions of azobenzene materials," *Advanced Materials*, vol. 24, no. 16, pp. 2069–2103, 2012.

[11] S. Lee, H. S. Kang, and J.-K. Park, "High-resolution patterning of various large-area, highly ordered structural motifs by directional photofluidization lithography: Sub-30-nm line, ellipsoid, rectangle, and circle arrays," *Advanced Functional Materials*, vol. 21, no. 10, pp. 1770–1778, 2011.

[12] M. Saphiannikova and V. Toshchevikov, "Optical deformations of azobenzene polymers: Orientation approach vs. photofluidization concept," *Journal of the Society for Information Display*, vol. 23, no. 4, pp. 146–153, 2015.

[13] A. Natansohn and P. Rochon, "Photoinduced motions in azo-containing polymers," *Chemical Reviews*, vol. 102, no. 11, pp. 4139–4175, 2002.

[14] C. J. Barrett, J.-I. Mamiya, K. G. Yager, and T. Ikeda, "Photo-mechanical effects in azobenzene-containing soft materials," *Soft Matter*, vol. 3, no. 10, pp. 1249–1261, 2007.

[15] A. Priimagi and A. Shevchenko, "Azopolymer-based micro- and nanopatterning for photonic applications," *Journal of Polymer Science Part B: Polymer Physics*, vol. 52, no. 3, pp. 163–182, 2014.

[16] F. Frascella, A. Angelini, S. Ricciardi, F. Pirri, and E. Descrovi, "Surface-relief formation in azo-polyelectrolyte layers with a protective polymer coating," *Optical Materials Express* , vol. 6, no. 2, pp. 444–450, 2016.

[17] F. Pirani, A. Angelini, F. Frascella, R. Rizzo, S. Ricciardi, and E. Descrovi, "Light-driven reversible shaping of individual azo-polymeric micro-pillars," *Scientific Reports*, vol. 6, Article ID 31702, 2016.

[18] J. Vapaavuori, R. H. A. Ras, M. Kaivola, C. G. Bazuin, and A. Priimagi, "From partial to complete optical erasure of azobenzene-polymer gratings: Effect of molecular weight," *Journal of Materials Chemistry C*, vol. 3, no. 42, pp. 11011–11016, 2015.

[19] F. Pirani, A. Angelini, S. Ricciardi, F. Frascella, and E. Descrovi, "Laser-induced anisotropic wettability on azopolymeric microstructures," *Applied Physics Letters*, vol. 110, no. 10, Article ID 101603, 2017.

[20] S. L. Oscurato, F. Borbone, P. Maddalena, and A. Ambrosio, "Light-driven wettability tailoring of azopolymer surfaces with reconfigured three-dimensional posts," *ACS Applied Materials & Interfaces*, vol. 9, no. 35, pp. 30133–30142, 2017.

[21] A. Ambrosio, L. Marrucci, F. Borbone, A. Roviello, and P. Maddalena, "Light-induced spiral mass transport in azo-polymer films under vortex-beam illumination," *Nature Communications*, vol. 3, article 989, 2012.

[22] X. Zhou, Y. Du, and X. Wang, "Azo polymer janus particles and their photoinduced, symmetry-breaking deformation," *ACS Macro Letters*, vol. 5, no. 2, pp. 234–237, 2016.

4

Interaction of Polyaniline with Surface of Carbon Steel

Maria P. Sokolova,[1,2] Michael A. Smirnov,[3,4] Igor A. Kasatkin,[1] Ivan Yu. Dmitriev,[3] Natalia N. Saprykina,[3] Alexander M. Toikka,[1] Erkki Lahderanta,[2] and Galina K. Elyashevich[3]

[1] *Saint Petersburg State University, Universitetskaya Nab. 7-9, Saint Petersburg 199034, Russia*
[2] *Lappeenranta University of Technology, Skinnarilankatu 34, 53850 Lappeenranta, Finland*
[3] *Institute of Macromolecular Compounds, Russian Academy of Sciences, Bolshoy Pr. 31, Saint Petersburg 199004, Russia*
[4] *ITMO University, Saint Petersburg 197101, Russia*

Correspondence should be addressed to Maria P. Sokolova; pmarip@mail.ru

Academic Editor: Andrea Camposeo

The structure and barrier properties of the oxyhydroxide layers on a carbon steel surface covered with electroactive polyaniline were investigated. Two types of polymer structure differing in degree of macromolecular order were prepared by simultaneous (fast polymerization) or dropwise (slow polymerization) mixing of reagents. A larger amount of the most stable FeOOH modification was formed on steel covered with slowly polymerized sample during treatment in the corrosion-active medium. Amorphous rust products with weak barrier properties were observed in the sample prepared by fast polymerization. Additionally, barrier activity of dedoped polyaniline was studied with SEM, WAXD, and electrochemical methods.

1. Introduction

Due to their unique physicochemical properties and easy synthesis, the electroconducting polymers attract considerable attention as potential components of electronic devices [1], power sources [2, 3], chemical sensors [4], absorbents of electromagnetic radiation [5], and also corrosion protective coatings [6, 7]. Their advantages include high electrical conductivity, low cost, and possibility of reducing environmental pollution [8, 9]. Redox and catalytic activity of these materials promote formation of an oxyhydroxide layer on the polymer-metal interface [10–12], which passivates the metal surface and prevents it from corrosion; on the other hand, it causes instability of the electrical characteristics of the contact between the polymeric and metal phases [13]. Currently, the research in the field was mainly focused on the electrochemical investigations of the corrosion processes [14–18], but the oxide layer composition was not determined because of difficulties in applying ex situ experimental techniques [19, 20]. For example, X-ray photoelectron spectroscopy (XPS) measurements [9, 11, 21] are performed in vacuum where dehydration of the passive films takes place. Moreover, structure of the passive films is strongly affected by water content which changes dramatically under the conditions of XPS experiment [22, 23]. At the same time, X-ray diffraction (XRD) phase analysis can easily be performed at ambient conditions, which prevent dehydration of the passive film, and provides reliable information on its composition and crystalline structure [20]. Recently, this approach was used for investigation of polyaniline (PANI) ability to protect magnesium surface [24]. FTIR, Raman, and Mossbauer spectroscopy must be mentioned as an alternative for XRD method for studying structure of corrosion products at ambient conditions [25, 26]. But in certain cases it is difficult to distinguish bands connected with inorganic phase on the background of spectrum of conducting polymer. The same problems can arise when using Raman spectroscopy. Additionally, irradiation of sample with laser beam can cause heating and changing of the structure of corrosion products (e.g., dehydration of iron oxyhydroxides). In comparison with Mossbauer spectroscopy, which can give quantitative information about phases of corrosion products of iron, XRD is more simple method. Additionally, the XRD patterns are well known for wide number of iron oxides and oxyhydroxides (including crystalline modifications, i.e., goethite, lepidocrocite, and akaganeite) and quantitive phase analysis of corrosion products is quite simple.

It needs to be pointed out that in some cases it is possible to use pure conducting polymer as a protective coating. It was proposed [27] that in this case conducting polymer acts as a catalyst for reduction of oxygen with iron, which results in formation of passive oxide layer on the surface of metal. Because of porosity and low mechanical stability of coating prepared with this simplest method, other approaches were elaborated: using of conducting polymer as a primer with conventional top-coat, blending with a conventional polymer coating, and addition of low concentration of conducting polymer to the conventional organic coatings [28]. It was shown that conducting polymer primer in combination with conventional top-coat demonstrates better performance than top-coat alone [29]. At the same time, blending of conducting polymer with conventional resins allows avoiding multilayer coating. It was reported that blended coatings formed by addition of equal amount of PANI or poly-*o*-ethoxyaniline to the water soluble epoxy or polymethylmethacrylate showed better corrosion protection of iron than single component coatings [30]. Additionally, it was shown that in some cases introduction of very small amount of conducting polymer can lead to the increasing of anticorrosion properties of conventional organic coatings. In work [31], it was shown that addition of less than 0.6 weight% of polyaniline to the alkyd resin leads to the improvement of corrosion resistance of the coating. Hybrid composite coating approach based on the simultaneous incorporation of inorganic (10 wt% of ZnO) and PANI nanoparticles (2–5 wt%) into poly(vinyl acetate) matrix was demonstrated in work [32] as a method for preparation of enhanced corrosion protective coatings.

Although various conducting polymers such as polypyrrole [33], polythiophene [34], and their derivatives [33, 35] are proposed as components of corrosion protection coatings, PANI and its derivatives are the most studied electroconducting polymers for corrosion protection [36] due to simplicity of their synthesis and high redox activity [37]. Polyaniline is a semicrystalline polymer with a typical degree of crystallinity (χ_{cr}) of about 5–50%; it can form at least two polymorphs [38–40]. Nanostructure and chain ordering in this polymer are strongly dependent on many synthesis parameters such as temperature, reagent concentrations, acidity of the reaction medium, and method of reagent mixing. PANI may be prepared in the form of nanoparticles with different shapes: nanospheres or granules [41–45], nanosheets [46, 47], nanofibers [48, 49], and nanotubes [50–52]. The mechanisms of PANI macromolecules ordering and nanostructure formation under different conditions are discussed in literature [53, 54]. In a number of publications [14, 17, 18, 55], the effects of dopants and of the composition of PANI derivatives [11, 15, 56–58] on the electrochemical properties of barrier layers formed under the deposition of polymer onto the metal surface were reported. At the same time, influence of polymer crystallinity and of its nanostructure on barrier properties was not studied in detail.

The aim of this work is to investigate chemical composition, crystalline structure, and barrier properties of inorganic films on the carbon steel/polyaniline interface formed with PANI prepared by two different procedures: dropwise mixing of reagents (slow polymerization) and simultaneous addition of an oxidant to the monomer solution (fast polymerization). The effect of dedoping of PANI on the products of its interaction with the steel surface is also discussed.

2. Experimental

2.1. Preparation of the Samples. Ammonium peroxydisulfate (APS) and aniline were purchased from Sigma-Aldrich and used as received. Sodium chloride and sulfuric acid were obtained from Vekton (Russia). Carbon steel (steel 08 GOST 1050-88) plates were polished with abrasive paper (P500 grit) and then washed with water and alcohol. All solutions were prepared in distilled water.

PANI dispersions were synthesized by chemical oxidative polymerization of aniline monomer ($0.15\,mol\,L^{-1}$) in the H_2SO_4 water solution ($1\,mol\,L^{-1}$) with APS as an oxidant. The mole ratio of monomer to oxidant was equal to 1:1.125. The monomer and oxidant solutions were thermostated separately at 35°C and then mixed together either simultaneously (sample P1) or dropwise (sample P2). After mixing, the solutions were left for 2 hours on the magnetic stirrer. Then the polymer was filtered and washed with the sulfuric acid solution in order to remove the residual ammonium sulfate. The polymer samples were dried in vacuum at room temperature during 48 hours. For dedoping of PANI, the powder of P2 was dispersed in NH_4OH water solution with concentration of $1\,mol\,L^{-1}$ and stirred for 1 hour. Then the solid polymer phase was separated by filtration, washed, and dried. The dedoped sample is denoted in text as P2d.

The polymer powders were dispersed in water under the ultrasonic treatment and casted onto metal surfaces (4-5 $mg\,cm^{-2}$) with subsequent drying on air. Then the samples were placed in the corrosion-active 3.5 wt% NaCl solution in water. Those samples are denoted in text as "Fe-xxx," where "xxx" is P1, P2, or P2d.

2.2. Characterization Methods

2.2.1. Scanning Electron Microscopy. PANI powders were dispersed in water in the ultrasonic bath, casted on the metallic table in the same way as for the corrosion experiment, and studied with scanning electron microscope ZEISS MERLIN (Germany) at the voltage of 10 kV. Due to electroconductivity of PANI, the high quality images of the polymeric layers could be obtained without metal covering.

2.2.2. Wide-Angle X-Ray Diffraction. Initial PANI powders and oxide layers on the steel surface were investigated with wide-angle X-ray diffraction (WAXD) using a D2 PHASER diffractometer (Bruker, Germany) equipped with a CoK_α radiation source ($\lambda = 1.79$ Å) at scattering angles $2\theta = 5°–60°$, step of 0.05°, and 2 s time per step. The XRD patterns were obtained immediately after deposition of PANI (prior to corrosion experiments) and after 8 days of treatment

FIGURE 1: SEM images of PANI layers on metal surface. PANI prepared with simultaneous P1 (a, b) and dropwise P2 (c, d) addition of oxidant. Images (e) and (f) correspond to the dedoped polymer P2d.

in the corrosion-active medium. The Bruker TOPAS 4.2 software was used for quantitative phase analysis.

The degree of crystallinity of the samples (χ_{cr}, %) was calculated according to the equation:

$$\chi_{cr} = \frac{\int_0^\infty I_{cr}(q)\,dq}{\int_0^\infty [I_{cr}(q) + I_{am}(q)]\,dq}, \quad (1)$$

where I_{am} and I_{cr} are diffraction intensities in the amorphous and crystalline regions, respectively, and q is the length of the scattering vector.

2.2.3. Electrochemical Measurements. Barrier properties of the films were characterized by potentiodynamic experiment on the steel electrode with PANI coating using the P30-J potentiostat ("Elins," Russia). The experiments were carried out in a three-electrode cell with silver-chloride reference and Pt counter electrodes at 1st, 4th, and 8th days after immersion of the samples into the corrosion-active medium. The values of polarization resistance (R_p) were calculated from the slope of the voltage-current dependence in the range of ±20 mV relative to the open-circuit electrochemical potential of the working electrode. All experiments were performed at room temperature.

3. Results and Discussion

3.1. Morphology of PANI Layers. Morphology of the polymer particles formed in the samples P1 and P2 (simultaneous and dropwise addition of the oxidant to the monomer solution, resp.) studied with scanning electron microscopy (SEM) is shown in Figure 1.

The sample P1 contains a large amount of smooth rod-like particles with the length in the range of 150–250 nm and diameter of about 50–100 nm (Figure 1(b)) which are uniformly distributed (Figure 1(a)) on the surface.

Dropwise addition of the oxidant to the monomer solution leads to the considerable changes in the polymer morphology. The SEM images of the sample P2 taken with a different magnification are shown in Figures 1(c) and 1(d). As compared to the sample P1, the length of the rod-like particles increases to 200–1000 nm, but the diameter remains approximately the same (70–110 nm). At the same time, the surface of the rods in P2 becomes rough that may be explained by the stepped character of the reaction during dropwise addition of the oxidant. It is seen in Figure 1(d) (insert) that the flake-like morphological elements are also present in the sample. Figure 1(c) shows that the P2 polymer layer contains larger amount of hollow space than the P1 layer (Figure 1(a)). The increased length of the particles may be related to the more effective polymer chain ordering in the slowly polymerized material.

Dedoping of PANI leads to the appearance of the particles with a less pronounced rod-like morphology as seen in Figure 1(e). The particles have a lower aspect ratio, and their sizes are in the range of 50–200 nm. The flake-like morphological elements can be observed also in the P2d sample (marked with arrow in Figure 1(f)).

3.2. X-Ray Study of PANI Powders. The crystalline structure of the PANI samples was studied by WAXD. The results are presented in Figure 2. Wide diffraction peaks are observed for all samples and reveal their semicrystalline nature. WAXD pattern of the PANI powder prepared by simultaneous mixing of reagents (Figure 2(a)) shows two reflections at 2θ = 19.4° and 25°, which correspond to the interplanar distances of d = 4.6 and 3.5 Å, respectively. Similar results were obtained for the samples doped with H_2SO_4 by the authors

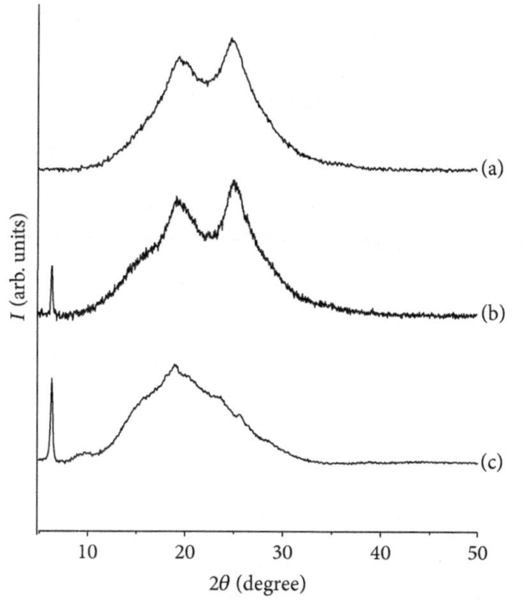

FIGURE 2: X-ray diffraction patterns of PANI powders prepared by simultaneous mixing of reagents, P1 (a), or dropwise addition of the oxidant, P2 (b), and by dedoping of P2 sample, P2d (c).

of [44, 49] who attributed the first peak to the characteristic distance between two adjacent chains of PANI and the second to the distance between the (110) planes.

When the method of mixing of reagents is changed to the dropwise addition of the oxidant (P2), the pattern of the polymer demonstrates an increased intensity of the peaks at $2\theta = 19.4^\circ$ and 25° (Figure 2(b)), additional sharp peak at $2\theta = 6.4^\circ$ ($d = 13.9$ Å), and a shoulder with the maximum at about $2\theta = 15.5^\circ$ ($d = 5.7$ Å). This suggests an increased ordering of the polymer molecules that agrees with a higher degree of crystallinity of the P2 sample in comparison with P1 (31 and 23%, resp.). In work [59], the sharp peak at $2\theta = 6.4^\circ$ was attributed to the 2D stacking arrangement of the polymer chains with the dopant ions intercalated between the stacks. The authors of [52] suppose that the interplanar distance of 13.9 Å could be attributed to the oligoanilines packed into the nanosheets with a lamellar arrangement of the molecules. Similar packing is observed in the case of large dopant anions which are able to form lamellar structures [47, 51, 60]. Presence of 2D-nanostructures is confirmed with SEM images shown in Figure 1(d).

The WAXD pattern of the dedoped sample P2d is shown in Figure 2(c). It is seen that the intensity of the reflection at $2\theta = 25^\circ$, which is attributed to the periodicity in the arrangement of dopant ions [44], is significantly reduced in comparison with the P2 sample. The degree of crystallinity also decreases after dedoping down to 9% that means amorphization of the polymer. At the same time, the peaks centered at $2\theta = 6.4^\circ$ and 19.4° and the shoulder at 15.5° remain at the same positions indicating that the molecular packing of the parent doped structure is preserved in some parts of the sample. It can be noted that the decrease in the intensity of the reflex at $2\theta = 25^\circ$ in the dedoped sample is associated with disappearance of the rod-like particles with high aspect ratio (Figure 1). Thus, the WAXD and SEM results evidence that the formation of particles with high aspect ratio is definitely related to the interaction between the PANI chains and the dopant ions.

3.3. Chemical Composition of Barrier Inorganic Layers Prior to and after Corrosion Experiments. Figure 3(a) shows the WAXD patterns of the PANI-covered samples taken prior to the corrosion experiment. The patterns obtained after keeping the samples in the 3.5% NaCl solution during 8 days are shown in Figure 3(b). The peaks corresponding to the different phases are marked in the patterns. Crystallite sizes in the inorganic component calculated from the integral breadths of the diffraction peaks are listed in Table 1.

It was found that both of the doped PANI samples P1 and P2 (Figure 3(a), patterns (1) and (2)) start to react with the support immediately after the steel plate was covered by the polymer dispersion. The product of the reaction is the hydrous iron sulfate in two crystalline modifications: $FeSO_4 \times 7H_2O$ (melanterite) and $FeSO_4 \times 4H_2O$ (rozenite). This fact may be explained by the presence of an excess of dopant acid in the polymer layer that was determined by elemental analysis. The degree of PANI doping (the ratio between the number of acid molecules and nitrogen atoms in the polymer chain) was 0.90 and 0.74 for the P1 and P2 samples, respectively. The steel surface coated by the dedoped PANI layer remained lustrous after drying of polymer dispersion, and no inorganic phases were detected with XRD (Figure 3(a), pattern (3)). Thus, in contrast with the samples prepared with the doped polymers, the dedoped PANI does not immediately react with Fe during the deposition of polymer dispersion, due to the absence of dopant acid which is needed for providing counterions for the ferrous cations.

The phase composition of the inorganic layer after corrosion experiment appears to be different for all samples (Table 1). Not more than 5% of NaCl was detected in each sample (not shown in Table 1), which may be considered as an impurity coming from the corrosive solution used in our experiments. The broad peaks in the Fe-P1 WAXD pattern (Figure 3(b), pattern (1)) belong to α-FeOOH (goethite) and $FeSO_4 \times 4H_2O$ (rozenite). Goethite can appear as a product of iron oxidation reaction which is catalyzed by PANI according to the scheme suggested by Wessling [61]. A quantitative analysis of the Fe-P1 sample was not possible because of the low degree of crystallinity.

In the samples Fe-P2 and Fe-P2d, the sharp diffraction peaks are observed (Figure 3(b), patterns (2) and (3)). This means that relatively well-crystallized inorganic phases were formed under the coatings of PANI synthesized by dropwise addition of the oxidant. As it was mentioned above, the slow polymerization reaction allows formation of highly ordered rod-like and flake-like structures of the conducting polymer (Figure 2(d)). This could promote formation of more dense modification of iron oxyhydroxide and lead to an increase in the sizes of crystallites in the inorganic layers. As seen in Table 1, goethite is the dominating phase (81%) in the sample Fe-P2, while in the Fe-P2d the barrier layer is mostly formed by lepidocrocite (73%). Goethite is the denser oxyhydroxide

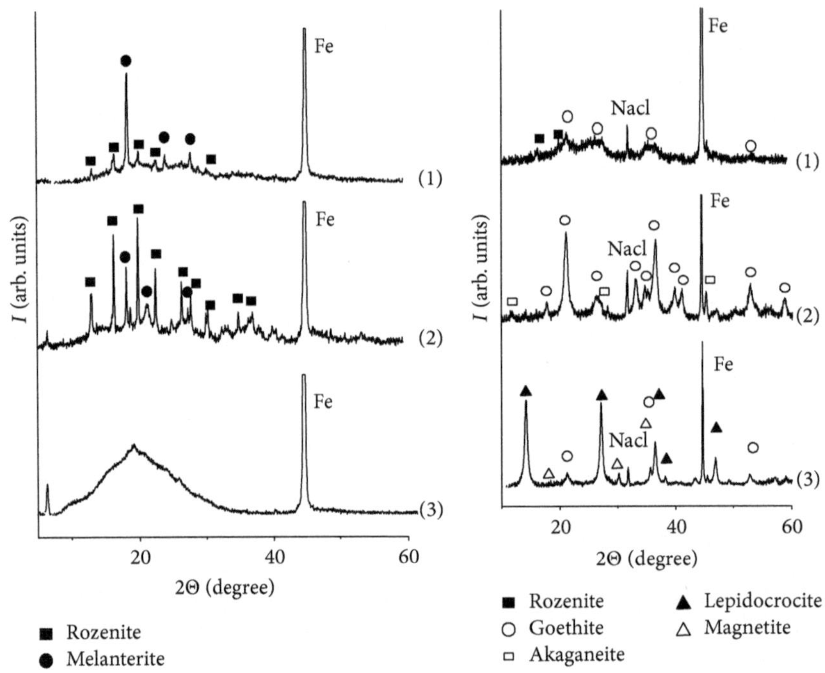

FIGURE 3: X-ray diffraction patterns of the samples Fe-P1 (1), Fe-P2 (2), and Fe-P2d (3) taken prior to (a) and after (b) corrosion experiments.

TABLE 1: Composition (in volume%) and structural parameters ofinorganic layers at the PANI-Fe interface before and after corrosion experiment.

Sample	Phase	Volume, %	Crystallite size, nm
Fe-P1/before	Melanterite ($FeSO_4 \times 7H_2O$)	85	34
	Rozenite ($FeSO_4 \times 4H_2O$)	15	75
Fe-P1/after	Goethite (α-FeOOH)	—	—
	Rozenite ($FeSO_4 \times 4H_2O$)	—	—
Fe-P2/before	Rozenite ($FeSO_4 \times 4H_2O$)	81	61
	Melanterite ($FeSO_4 \times 7H_2O$)	9	54
Fe-P2/after	Goethite (α-FeOOH)	81	13
	Akaganeite (β-FeOOH)	14	12
Fe-P2-b/before		No inorganic phases detected	
Fe-P2-b/after	Lepidocrocite (γ-FeOOH)	73	19
	Goethite (α-FeOOH)	14	11
	Magnetite (Fe_3O_4)	9	25

phase in comparison with lepidocrocite. Taking into account the highest degree of crystallinity in the doped P2 sample, it may be supposed that the appearance of high amount of goethite is related to the high degree of ordering in the polymer chain packing.

3.4. Barrier Properties of Oxide Layers. Polarization curves of the steel electrodes with the PANI coatings were measured at the 1st, 4th, and 8th day of their keeping in the corrosion-active medium. The values of R_p are presented in Table 2. The polarization resistance of pure steel decreased from 37.1 to 22.5 kOhm during the corrosion experiment. This may be explained by an increased surface roughness as a result of the electrode dissolution.

TABLE 2: Polarization resistance R_p (in kOhm) at different durations of corrosion experiment.

Sample	R_p at 1st day	R_p at 4th day	R_p at 8th day
Fe	37.1	24.4	22.5
Fe-P1	7.9	18.1	29.9
Fe-P2	12.4	16.4	83.3
Fe-P2d	1.4	41.4	67.8

The R_p values of all coated samples in the 1st day were lower than of pure steel. The R_p value is inversely proportional to the sum of all currents of all electrochemical processes occurring on the electrode surface. In the initial period, the electrical contact between Fe support and PANI layer is good that leads to a fast oxidation of steel reflecting the catalytic activity of the conducting polymer in this process. It can be seen that the dedoped polymer is more active in iron oxidation because it demonstrates the lowest initial R_p value in comparison with the Fe-P1 and Fe-P2 samples. This can be explained by two reasons. (1) As it was shown by WAXD, in the case of doped polymers, the layer of ferrous sulfate, which was formed even before the sample immersion into the corrosion-active medium, can increase the electrical resistance between the conducting polymer and the metal surface. (2) The doped polymers are characterized by a denser chain packing that leads to the slower diffusion of oxygen molecules and ions to the metal surface.

For the steel samples covered with PANI, the value of R_p increases with time due to the formation of passive layers. On 8th day, the R_p value for the Fe-P1 sample is only slightly higher than that for the uncovered steel. This may be explained by the formation of amorphous inorganic oxide layers which provide only a weak corrosion protection. The samples Fe-P2 and Fe-P2d demonstrate the high values of R_p (see Table 2) due to the crystalline oxide layers. The protection activity of Fe-P2 is higher than that of Fe-P2d that can probably be related to their structural features. The amount of goethite in the Fe-P2 sample is 81% while in Fe-P2d its concentration is only 14%. At the same time, α-FeOOH has a maximum density (Table 1) and is the most thermodynamically stable polymorph [62, 63]. Goethite is considered to be the most effective product for iron protection. Authors of [64, 65] have introduced the protective ability index of inorganic layers (PAI) which is calculated as the relation between the amounts of α-FeOOH and other components of a barrier layer. In our case, the indexes are 5.8 and 0.17 for the samples Fe-P2 and Fe-P2d, respectively, which is in agreement with electrochemical measurements. At the same time, the stronger and weaker protection of steel covered with P2 and P2d, respectively, are associated with the size of inorganic crystallites: 13 and 19 nm for the main part of the inorganic phase in the Fe-P2 and Fe-P2d samples, respectively (see Table 1). It is in accordance with [62] where it was found that a decrease in crystallite size in oxyhydroxides leads to the improvement of their barrier properties.

The polarization curves measured after 8 days of exposure in the corrosion-active medium are shown in Figure 4. It can be seen that the curves for Fe-P2 and Fe-P2d are located below the ones for Fe and Fe-P1 that agrees with the values of R_p listed in Table 2. The slope of cathodic branches for the Fe-P2 and Fe-P2d samples is higher compared to Fe and Fe-P1. This may be related to a possible reduction of the dense oxyhydroxide layer during polarization measurements.

4. Conclusions

Polyaniline dispersions were synthesized by two procedures: simultaneous (fast polymerization) and dropwise (slow polymerization) mixing of reagents. Lower polymerization rate permits to achieve a higher ordering in the arrangement of polymer chains, which leads to the increase in aspect ratio of the polymer particles. The coatings prepared from a polyaniline doped with sulfuric acid promote the formation of ferrous sulfate on the carbon steel surface. Different crystalline modifications of iron oxyhydroxides are formed during treatment of the samples in the corrosion-active solution (3.5% NaCl in water). Deposition of a more ordered polymer onto the steel surface leads to the formation of a large amount (81%) of the crystalline goethite phase which is the most thermodynamically stable rust product. Covering the steel with PANI prepared by common procedure (fast polymerization) leads to the formation of amorphous rust layer. For the inorganic layer formed under the polymer prepared by dropwise oxidant addition (slow polymerization), the electrochemical measurements show highest barrier properties which is confirmed by the data of structural investigations.

In a highly ordered dedoped PANI, the degree of the polymer crystallinity decreases from 31 to 9%, and the rod-like structures disappear. Compared with the doped polyaniline, the dedoped sample displayed lower protective properties, which were still considerably higher than those of the doped PANI prepared by simultaneous mixing of reagents.

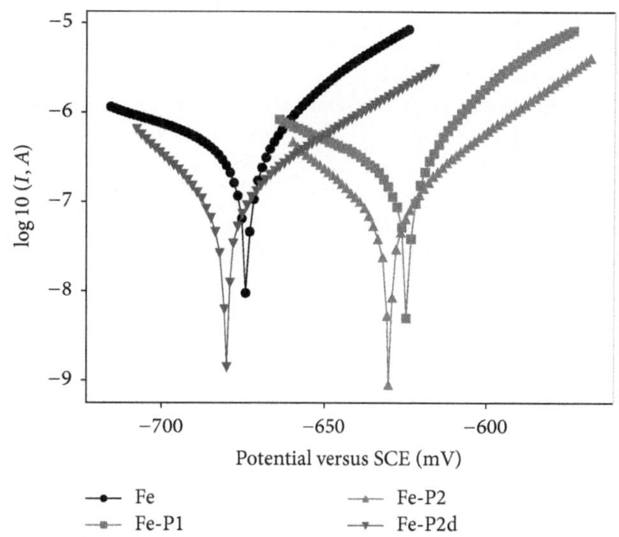

FIGURE 4: Polarization curves for pure Fe covered with PANI dispersions at 8th day of keeping in 3.5% NaCl.

Acknowledgments

This work was supported by Russian Foundation for Basic Research (Grant 14-03-31411 mol_a). M. Sokolova acknowledges Saint Petersburg State University for the Research Grant (no. 12.50.1195.2014). The experimental work was facilitated by the equipment of the Center for X-Ray Diffraction Studies of Nanotechnology Interdisciplinary Center and Centre for Innovative Technologies of Composite Nanomaterials of Saint Petersburg State University.

References

[1] B. Yaman, I. Terkesli, K. M. Turksoy, A. Sanyal, and S. Mutlu, "Fabrication of a planar water gated organic field effect transistor using a hydrophilic polythiophene for improved digital inverter performance," *Organic Electronics: Physics, Materials, Applications*, vol. 15, no. 3, pp. 646–653, 2014.

[2] M. A. Smirnov, M. P. Sokolova, N. V. Bobrova, I. A. Kasatkin, E. Lahderanta, and G. K. Elyashevich, "Capacitance properties and structure of electroconducting hydrogels based on copoly(aniline - P-phenylenediamine) and polyacrylamide," *Journal of Power Sources*, vol. 304, pp. 102–110, 2016.

[3] M. A. Smirnov, M. P. Sokolova, P. Geydt et al., "Dual doped electroactive hydrogelic fibrous mat with high areal capacitance," *Materials Letters*, vol. 199, pp. 192–195, 2017.

[4] H. J. N. P. D. Mello, T. Heimfarth, and M. Mulato, "Influence of the physical-chemical properties of polyaniline thin films on the final sensitivity of varied field effect sensors," *Materials Chemistry and Physics*, vol. 160, pp. 257–263, 2015.

[5] O. A. Andreeva, L. A. Burkova, M. A. Smirnov, and G. K. El'yashevich, "Correlation between IR spectra and electric conductivity of polyethylene-polypyrrole composites," *Polymer Science—Series B*, vol. 48, no. 11-12, pp. 331–334, 2006.

[6] B. N. Grgur, A. R. Elkais, M. M. Gvozdenović, S. Ž. Drmanić, T. L. Trišović, and B. Z. Jugović, "Corrosion of mild steel with composite polyaniline coatings using different formulations," *Progress in Organic Coatings*, vol. 79, pp. 17–24, 2015.

[7] Y. Li, H. Zhang, X. Wang, J. Li, and F. Wang, "Growth kinetics of oxide films at the polyaniline/mild steel interface," *Corrosion Science*, vol. 53, no. 12, pp. 4044–4049, 2011.

[8] G. K. Elyashevich, M. A. Smirnov, I. S. Kuryndin, and V. Bukošek, "Electroactive composite systems containing high conductive polymer layers on poly(ethylene) porous films," *Polymers for Advanced Technologies*, vol. 17, no. 9-10, pp. 700–704, 2006.

[9] P. J. Kinlen, D. C. Silverman, and C. R. Jeffreys, "Corrosion protection using polyaniline coating formulations," *Synthetic Metals*, vol. 85, no. 1-3, pp. 1327–1332, 1997.

[10] P. Li, T. C. Tan, and J. Y. Lee, "Corrosion protection of mild steel by electroactive polyaniline coatings," *Synthetic Metals*, vol. 88, no. 3, pp. 237–242, 1997.

[11] A. A. Hermas, "XPS analysis of the passive film formed on austenitic stainless steel coated with conductive polymer," *Corrosion Science*, vol. 50, no. 9, pp. 2498–2505, 2008.

[12] Y. Li, H. Zhang, X. Wang, J. Li, and F. Wang, "Role of dissolved oxygen diffusion in coating defect protection by emeraldine base," *Synthetic Metals*, vol. 161, no. 21-22, pp. 2312–2317, 2011.

[13] J. Fang, K. Xu, L. Zhu, Z. Zhou, and H. Tang, "A study on mechanism of corrosion protection of polyaniline coating and its failure," *Corrosion Science*, vol. 49, no. 11, pp. 4232–4242, 2007.

[14] J. E. P. da Silva, S. I. C. de Torresi, and R. M. Torresi, "Polyaniline acrylic coatings for corrosion inhibition: the role played by counter-ions," *Corrosion Science*, vol. 47, no. 3, pp. 811–822, 2005.

[15] S. K. Shukla, M. A. Quraishi, and R. Prakash, "A self-doped conducting polymer 'polyanthranilic acid': an efficient corrosion inhibitor for mild steel in acidic solution," *Corrosion Science*, vol. 50, no. 10, pp. 2867–2872, 2008.

[16] J. R. Santos Jr., L. H. C. Mattoso, and A. J. Motheo, "Investigation of corrosion protection of steel by polyaniline films," *Electrochimica Acta*, vol. 43, no. 3-4, pp. 309–313, 1998.

[17] A. Cook, A. Gabriel, D. Siew, and N. Laycock, "Corrosion protection of low carbon steel with polyaniline: passivation or inhibition?" *Current Applied Physics*, vol. 4, pp. 133–136, 2004.

[18] A. R. Elkais, M. M. Gvozdenović, B. Z. Jugović, and B. N. Grgur, "The influence of thin benzoate-doped polyaniline coatings on corrosion protection of mild steel in different environments," *Progress in Organic Coatings*, vol. 76, no. 4, pp. 670–676, 2013.

[19] J. W. Schultze and M. M. Lohrengel, "Stability, reactivity and breakdown of passive films. Problems of recent and future research," *Electrochimica Acta*, vol. 45, no. 15-16, pp. 2499–2513, 2000.

[20] M. F. Toney, A. J. Davenport, L. J. Oblonsky, M. P. Ryan, and C. M. Vitus, "Atomic structure of the passive oxide film formed on iron," *Physical Review Letters*, vol. 79, no. 21, pp. 4282–4285, 1997.

[21] M. Fahlman, S. Jasty, and A. J. Epstein, "Corrosion protection of iron/steel by emeraldine base polyaniline: an X-ray photoelectron spectroscopy study," *Synthetic Metals*, vol. 85, no. 1–3, pp. 1323–1326, 1997.

[22] J. O. M. Bockris, "Spectroscopic observations on the nature of passivity," *Corrosion Science*, vol. 29, no. 2-3, pp. 291–312, 1989.

[23] J. Kruger, "The nature of the passive film on iron and ferrous alloys," *Corrosion Science*, vol. 29, no. 2-3, pp. 149–162, 1989.

[24] Y. Luo, Y. Sun, J. Lv, X. Wang, J. Li, and F. Wang, "Transition of interface oxide layer from porous $Mg(OH)_2$ to dense MgO induced by polyaniline and corrosion resistance of Mg alloy therefrom," *Applied Surface Science*, vol. 328, pp. 247–254, 2015.

[25] J. A. Jaén, J. Iglesias, and C. Hernández, "Analysis of short-term steel corrosion products formed in tropical marine environments of Panama," *International Journal of Corrosion*, vol. 2012, Article ID 162729, 11 pages, 2012.

[26] A. V. R. Kumar and R. Balasubramaniam, "Corrosion product analysis of corrosion resistant ancient Indian iron," *Corrosion Science*, vol. 40, no. 7, pp. 1169–1178, 1998.

[27] B. Wessling and J. Posdorfer, "Corrosion prevention with an organic metal (polyaniline): corrosion test results," *Electrochimica Acta*, vol. 44, no. 12, pp. 2139–2147, 1999.

[28] P. P. Deshpande, N. G. Jadhav, V. J. Gelling, and D. Sazou, "Conducting polymers for corrosion protection: a review," *Journal of Coatings Technology Research*, vol. 11, no. 4, pp. 473–494, 2014.

[29] A. J. Dominis, G. M. Spinks, and G. G. Wallace, "Comparison of polyaniline primers prepared with different dopants for corrosion protection of steel," *Progress in Organic Coatings*, vol. 48, no. 1, pp. 43–49, 2003.

[30] J. H. Huh, E. J. Oh, and J. H. Cho, "Investigation of corrosion protection of iron by polyaniline blend coatings," *Synthetic Metals*, vol. 137, no. 1–3, pp. 965-966, 2003.

[31] J. I. I. Laco, F. C. Villota, and F. L. Mestres, "Corrosion protection of carbon steel with thermoplastic coatings and alkyd resins containing polyaniline as conductive polymer," *Progress in Organic Coatings*, vol. 52, no. 2, pp. 151–160, 2005.

[32] R. C. Patil and S. Radhakrishnan, "Conducting polymer based hybrid nano-composites for enhanced corrosion protective coatings," *Progress in Organic Coatings*, vol. 57, no. 4, pp. 332–336, 2006.

[33] D. E. Tallman, G. Spinks, A. Dominis, and G. G. Wallace, "Electroactive conducting polymers for corrosion control: part 1. General introduction and a review of non-ferrous metals," *Journal of Solid State Electrochemistry*, vol. 6, no. 2, pp. 73–84, 2002.

[34] U. Riaz, C. Nwaoha, and S. M. Ashraf, "Recent advances in corrosion protective composite coatings based on conducting polymers and natural resource derived polymers," *Progress in Organic Coatings*, vol. 77, no. 4, pp. 743–756, 2014.

[35] E. Armelin, R. Oliver, F. Liesa, J. I. Iribarren, F. Estrany, and C. Alemán, "Marine paint fomulations: conducting polymers as anticorrosive additives," *Progress in Organic Coatings*, vol. 59, no. 1, pp. 46–52, 2007.

[36] P. Zarras and J. D. Stenger-Smith, "Electro-active polymer (EAP) coatings for corrosion protection of metals," in *Handbook of Smart Coatings for Materials Protection*, A. S. H. Makhlouf, Ed., pp. 328–369, Woodhead Publishing, 2014.

[37] G. G. Wallace, G. M. Spinks, L. A. P. Kane-Maguire, and P. R. Teasdale, *Conductive Electroactive Polymers: Intelligent Polymer Systems*, CRC Press, London, UK; Taylor & Francis Group, New York, NY, USA, 2009.

[38] J. P. Pouget, M. E. Jozefowicz, A. J. Epstein, X. Tang, and A. G. MacDiarmid, "X-ray structure of polyaniline," *Macromolecules*, vol. 24, no. 3, pp. 779–789, 1991.

[39] M. E. Jozefowicz, R. Laversanne, H. H. S. Javadi et al., "Multiple lattice phases and polaron-lattice spinless-defect competition in polyaniline," *Physical Review B*, vol. 39, no. 17, pp. 12958–12961, 1989.

[40] M. E. Józefowicz, A. J. Epstein, J.-P. Pouget et al., "X-ray structure of polyanilines," *Synthetic Metals*, vol. 41, no. 1-2, pp. 723–726, 1991.

[41] H.-W. Park, T. Kim, J. Huh, M. Kang, J. E. Lee, and H. Yoon, "Anisotropic growth control of polyaniline nanostructures and their morphology-dependent electrochemical characteristics," *ACS Nano*, vol. 6, no. 9, pp. 7624–7633, 2012.

[42] E. A. Sanches, J. C. Soares, A. C. Mafud, G. Trovati, E. G. Fernandes, and Y. P. Mascarenhas, "Structural and morphological characterization of chloride salt of conducting poly(o-methoxyaniline) obtained at different time synthesis," *Journal of Molecular Structure*, vol. 1039, pp. 167–173, 2013.

[43] J. Stejskal, M. Spirkova, A. Riede, M. Helmstedt, P. Mokreva, and J. Prokes, "Polyaniline dispersions 8. The control of particle morphology," *Polymer*, vol. 40, no. 10, pp. 2487–2492, 1999.

[44] D. Zhu, J. Zhang, C. Xu, and M. Matsuo, "The frequency-dependence conduction of polyaniline based on their para-crystalline structures," *Synthetic Metals*, vol. 161, no. 17-18, pp. 1820–1827, 2011.

[45] A. A. Ganash, F. M. Al-Nowaiser, S. A. Al-Thabaiti, and A. A. Hermas, "Protection of stainless steel by the electrodeposition of polyaniline/poly(o-phenylenediamine) composite layers," *Journal of Solid State Electrochemistry*, vol. 17, no. 3, pp. 849–860, 2013.

[46] Z. D. Zujovic, C. Laslau, G. A. Bowmaker et al., "Role of aniline oligomeric nanosheets in the formation of polyaniline nanotubes," *Macromolecules*, vol. 43, no. 2, pp. 662–670, 2010.

[47] T. Jana, J. Chatterjee, and A. K. Nandi, "Sulfonic acid doped thermoreversible polyaniline gels. 3—structural investigations," *Langmuir*, vol. 18, no. 15, pp. 5720–5727, 2002.

[48] E. A. Sanches, J. C. Soares, A. C. Mafud, E. G. R. Fernandes, F. L. Leite, and Y. P. Mascarenhas, "Structural characterization of Chloride Salt of conducting polyaniline obtained by XRD, SAXD, SAXS and SEM," *Journal of Molecular Structure*, vol. 1036, pp. 121–126, 2013.

[49] H. Zhang, X. Wang, J. Li, and F. Wang, "Facile synthesis of polyaniline nanofibers using pseudo-high dilution technique," *Synthetic Metals*, vol. 159, no. 14, pp. 1508–1511, 2009.

[50] C. Laslau, B. Ingham, Z. D. Zujovic et al., "Synchrotron X-ray scattering reveals early-stage crystallinity during the self-assembly of polyaniline nanotubes with rectangular cross-sections," *Synthetic Metals*, vol. 161, no. 23-24, pp. 2739–2742, 2012.

[51] L. J. Zhang, Y. Z. Long, Z. J. Chen, and M. X. Wan, "The effect of hydrogen bonding on self-assembled polyaniline nanostructures," *Advanced Functional Materials*, vol. 14, no. 7, pp. 693–698, 2004.

[52] J. K. Park, S. S. Jeon, and S. S. Im, "Effect of 4-sulfobenzoic acid monopotassium salt on oligoanilines for inducing polyaniline nanostructures," *Polymer*, vol. 51, no. 14, pp. 3023–3030, 2010.

[53] J. Stejskal, I. Sapurina, M. Trchová, E. N. Konyushenko, and P. Holler, "The genesis of polyaniline nanotubes," *Polymer*, vol. 47, no. 25, pp. 8253–8262, 2006.

[54] J. Stejskal, I. Sapurina, and M. Trchová, "Polyaniline nanostructures and the role of aniline oligomers in their formation," *Progress in Polymer Science*, vol. 35, no. 12, pp. 1420–1481, 2010.

[55] M. A. Smirnov, M. P. Sokolova, I. Y. Dmitriev, E. N. Vlasova, E. Y. Rozov, and G. K. El'yashevich, "Barrier properties and structure of inorganic layers at polyaniline-steel interface," *Russian Journal of Applied Chemistry*, vol. 88, no. 7, pp. 1168–1173, 2015.

[56] H. Bhandari, V. Choudhary, and S. K. Dhawan, "Influence of self-doped poly(aniline-co-4-amino-3-hydroxy-naphthalene-1- sulfonic acid) on corrosion inhibition behaviour of iron in acidic medium," *Synthetic Metals*, vol. 161, no. 9-10, pp. 753–762, 2011.

[57] T. Mousavinejad, M. R. Bagherzadeh, E. Akbarinezhad, M. Ahmadi, and M. J.-F. Guinelac, "A novel water-based epoxy coating using self-doped polyaniline-clay synthesized under supercritical CO_2 condition for the protection of carbon steel against corrosion," *Progress in Organic Coatings*, vol. 79, pp. 90–97, 2015.

[58] M. P. Sokolova, N. V. Bobrova, I. Y. Dmitriev et al., "Anticorrosion activity of aniline–aniline-2-sulfonic acid copolymers on the steel surface," *Russian Journal of Applied Chemistry*, vol. 89, no. 3, pp. 432–438, 2016.

[59] R. Murugesan and E. Subramanian, "Effect of organic dopants on electrodeposition and characteristics of polyaniline under the varying influence of H_2SO_4 and $HClO_4$ electrolyte media," *Materials Chemistry and Physics*, vol. 80, no. 3, pp. 731–739, 2003.

[60] L. Shi, X. Wu, L. Lu, X. Yang, and X. Wang, "Intercalated polyaniline nanosheets prepared from lyotropic liquid crystalline solutions and their capacitive performance," *Synthetic Metals*, vol. 160, no. 9-10, pp. 989–995, 2010.

[61] B. Wessling, "Passivation of metals by coating with polyaniline: corrosion potential shift and morphological changes," *Advanced Materials*, vol. 6, no. 3, pp. 226–228, 1994.

[62] J. Wei, X. X. Fu, J. H. Dong, and W. Ke, "Corrosion evolution of reinforcing steel in concrete under dry/wet cyclic conditions contaminated with chloride," *Journal of Materials Science and Technology*, vol. 28, no. 10, pp. 905–912, 2012.

[63] Y.-L. Zhou, J. Chen, and Z.-Y. Liu, "Corrosion behavior of rusted 550 MPa grade offshore platform steel," *Journal of Iron and Steel Research International*, vol. 20, no. 3, pp. 66–73, 2013.

[64] T. Kamimura, S. Hara, H. Miyuki, M. Yamashita, and H. Uchida, "Composition and protective ability of rust layer formed on weathering steel exposed to various environments," *Corrosion Science*, vol. 48, no. 9, pp. 2799–2812, 2006.

5

Synthesis of Poly-(R-hydroxyalkanoates) by *Cupriavidus necator* ATCC 17699 Using Mexican Avocado *(Persea americana)* Oil as a Carbon Source

Araceli Flores-Sánchez,[1] Ma. del Rocío López-Cuellar,[2] Fermín Pérez-Guevara,[3] Ulises Figueroa López,[1] José Mauricio Martín-Bufájer,[1] and Berenice Vergara-Porras[1]

[1]*Escuela de Ingeniería y Ciencias, Tecnológico de Monterrey, Campus Estado de México, Carretera Lago de Guadalupe Km 3.5, Margarita Maza de Juárez, Atizapán de Zaragoza, MEX, Mexico*
[2]*Cuerpo Academico de Biotecnología Agroalimentaria, Instituto de Ciencias Agropecuarias, Universidad Autónoma del Estado de Hidalgo, Pachuca, HGO, Mexico*
[3]*Departamento de Biotecnología y Bioingeniería, Centro de Investigación y Estudios Avanzados (CINVESTAV), Avenida IPN 2508, Zacatenco, Gustavo A. Madero, Ciudad de México, Mexico*

Correspondence should be addressed to Berenice Vergara-Porras; vergarabp@itesm.mx

Academic Editor: Raffaele Cucciniello

Poly-R-hydroxyalkanoates (PHAs) are polymers produced by a vast number of bacterial species under stress conditions. PHAs exhibit different thermal and mechanical properties that depend on their molecular structure. In this work, PHAs were produced using avocado oil as the carbon source. *Cupriavidus necator* H16 was cultured in three-stage fermentation using fructose during the cell growth stages and avocado oil during the PHA synthesis stage. Different concentrations of avocado oil were used during the third stage to test the incorporation of various monomeric units into the PHAs. Biomass and PHA production were measured during the fermentation. DSC, FTIR, and gas chromatography analysis aided the PHA characterization. Different proportions of 3-hydroxyvalerate were present in the 3-hydroxybutyrate main chain depending on the concentration of avocado oil. The results suggest that avocado oil is a viable new substrate for PHA production.

1. Introduction

Poly-R-hydroxyalkanoates (PHAs) are polymers synthesized by a large number of bacterial species as a response to unbalanced nutritional conditions [1]. PHAs are thermoplastic polyesters of R-hydroxy alkanoic acids and accumulate intracellularly as granules that exhibit different properties depending on their chemical composition [2, 3]. A single monomer that forms the chain of PHAs typically contains from 3 to 15 carbon atoms [4], but the final chemical composition of PHAs is related to the synthesizer microorganism, the carbon source, the culture conditions, and the specificity of the PHA-synthase enzyme [5–7]. Homopolymers, copolymers, or terpolymers of PHAs can be obtained; for example, PHA copolymers can be synthesized from a combination of different substrates [8]. The thermal properties of PHAs, such as melting temperature and degree of crystallinity, depend on the length of the PHA monomeric units. Monomers containing more than five carbon atoms significantly decrease the polymer melting temperature, as well as the degree of crystallinity [9].

Many PHAs have main chains formed from monomers with different numbers of carbon atoms. Short-length-chain PHAs (PHA_{slc}) consist of monomers ranging from 3 to 5 carbons, whereas medium-length-chain PHAs (PHA_{mlc}) are formed from monomers containing 6 to 14 carbon atoms [4, 8]. One PHA_{slc}, poly(3-hydroxybutyrate) (PHB), is the most common PHA and was first identified by Maurice Lemoigne in 1926 [1, 4]. PHB biodegradability and biocompatibility make it an attractive material; however, its

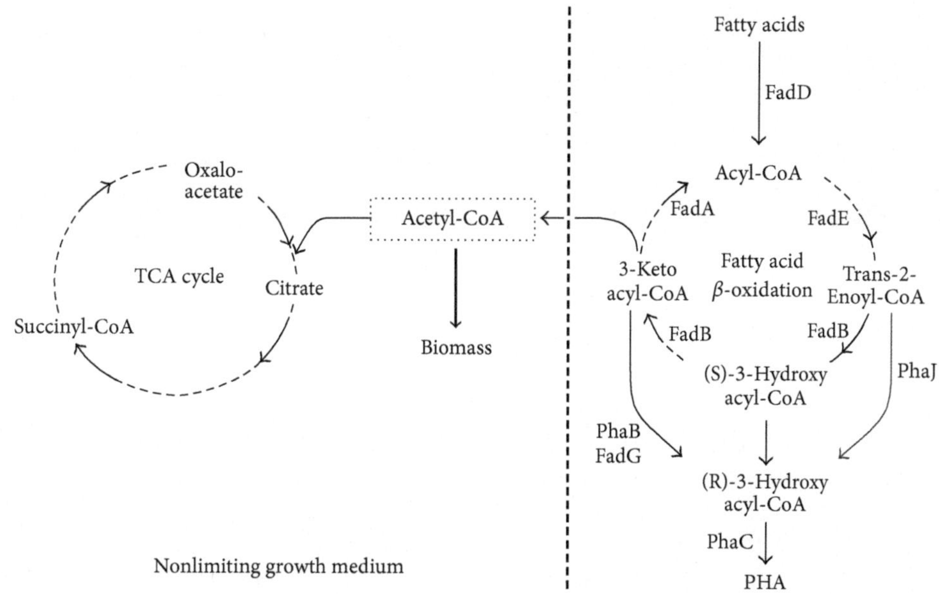

FIGURE 1: A simplified representation of fatty acid metabolism in *Cupriavidus necator.* In a nonlimiting growth media (left side), the tricarboxylic acid (TCA) cycle is active and the main product is biomass itself. Under unbalanced conditions (right side), caused by a noncarbon nutrient deficiency, C-source via fatty acid β-oxidation flows towards the poly-R-hydroxyalkanoate (PHA) biosynthesis pathway.

brittleness and limited degree of crystallinity have restricted its possible applications. The melting point of PHB (≈175°C) is also very close to its decomposition temperature (≈180°C), which creates challenges for thermal processing due to a narrow processing window [10]. Copolymers such as poly-(hydroxybutyrate-co-hydroxyvalerate) (PHBV) that contain (3)-hydroxybutyrate (3HB) and (3)-hydroxyvalerate (3HV) units in the chain show lower molecular weights and lower melting temperatures when compared to a PHB homopolymer. Babel and Steinbüchel [11] reported melting temperatures of 170, 162, 150, 145, and 137 when 3HV was in a 3, 9, 14, 20, and 25 mol% content, respectively.

Renewable carbon sources, such as sucrose, cellulose, and triacylglycerol, have served as substrates for PHA synthesis. Extensive studies have been conducted on the use of inexpensive substrates, including starch, glycerol, soybean oil, sugar cane bagasse, molasses, and activated sludge, to reduce the production cost of PHB [5, 12–14]. Similarly, byproducts from the food and the agroalimentary industry, methane, mineral oil, and lignite have been used to synthesize PHBV copolymers [1, 4, 15, 16]. Therefore, the carbon source and the microorganism consumption affinity are of great importance for the production of specific PHAs.

One organism that has been extensively used in the synthesis of PHAs is *Cupriavidus necator* (formerly *Ralstonia eutropha*), due to its versatility to accumulate polymer in amounts as high as 90% of its dry cell weight (DCW) [6]. The ability of *C. necator* to synthesize PHB and PHBV, as well as other PHAs, has been previously reported [6, 20, 21]. The limiting productions costs had led to proposals for the use of cheaper carbon sources, such as organic debris, wastewater, or even vegetable oils [22]. The use of complex carbon sources could also extend the incorporation of 3HV units to the main chain or even the synthesis of PHA_{mlc} [8].

The use of fatty acids, such as those present in vegetable oils, as a carbon source drives the β- fatty acid oxidation metabolic pathway in *C. necator*. The PHA synthesis in *C. necator* is highly associated with growth conditions and is mediated by the acetyl-CoA precursor [23, 24]. In balanced nutritional environments, fatty acids provide precursors, free energy, and cofactors for cell growth and maintenance and for macromolecular synthesis. By contrast, noncarbon nutrient limitation leads to inhibition of the central enzymes of the tricarboxylic cycle (TCA cycle). Consequently, the acetyl-CoA is channelled towards the PHA synthesis [23–25], as depicted in Figure 1.

Mexican avocado *(Persea americana)* is a lipid-rich fruit that occupies a prominent place in the market [26]. In 2015, 51.4% of the globally commercialized avocado was produced in Mexico [26]. During its cultivation, a high amount of waste material is produced. For example, in Mexico, nearly 54% of the annual avocado production is considered as waste. Moreover, the farming, packing, transportation, and commercialization stages are also important sources of avocado wastes [27]. Furthermore, fruit peel and seed, representing 12 to 15% and 20 to 27% of fruit weight, respectively, are currently discarded. Only the fruit pulp is destined for human consumption [28].

The fatty acids composition in Mexican avocado mainly includes palmitic, stearic, oleic, linoleic, heptanoic, nonanoic, and heptadecanoic acids [26]. The high content of fatty acid and the amount of waste generated from its cultivation identify avocado as a possible and sustainable carbon source for the biosynthesis of PHAs. As a first attempt, and to standardize the chemical composition of the substrate, the viability of PHA biosynthesis from avocado oil by *C. necator* was tested. The biopolymers synthesized from different oil contents were thermally and chemically characterized to demonstrate the feasibility of using this oil as an alternative substrate for PHA production.

2. Materials and Methods

2.1. Strain, Media, and Materials. C. necator H16 (ATCC 17699) was grown in mineral medium supplemented with fructose for 24 h, at 30°C and 200 rpm, for seed culture preparation. The medium contained, per litre of water, 10 g fructose, 3.70 g $(NH_4)_2SO_4$, 0.40 g $MgSO_4$, 6.36 g $Na_2HPO_4{\cdot}7H_2O$, 2.70 g KH_2PO_4, and 1.0 g nutrient broth.

The growth medium contained, per litre of water, 10 g fructose, 1.57 g NH_4SO_4, 5.66 g $NaH_2PO_4{\cdot}12H_2O$, 1.5 g KH_2PO_4, 0.2 g $MgSO_4{\cdot}7H_2O$, 10 mg $CaCl_2{\cdot}2H_2O$, 20 mg $FeSO_4{\cdot}7H_2O$, and 1 mL of trace element solution (0.3 g H_3BO_3, 0.2 g $CoCl_2{\cdot}6H_2O$, 0.1 g $ZnSO_4{\cdot}7H_2O$, 30 mg $MnCl_2{\cdot}4H_2O$, 30 mg $Na_2MoO4{\cdot}2H_2O$, 20 mg $NiCl_2{\cdot}6H_2O$, and 10 mg $CuSO_4{\cdot}5H_2O$, in HCl 0.1 N solution); the pH was adjusted to 7.

Commercial Mexican avocado oil for PHA synthesis was obtained from a single batch (the same production number) (Ahuacatlán, Mexico) to ensure a homogeneous chemical composition of the substrate.

2.2. Fermentation Studies

2.2.1. Inoculum Preparation. Experiments were conducted in duplicate using 200 mL of growth media in 500 mL flasks at 30°C, pH 7.0, in an incubator with rotational agitation at 200 rpm (New Brunswick Innova 4300, USA). A 10% v/v of the seed culture was used to inoculate the growth medium to obtain 0.13 g L^{-1} (±0.1) of initial biomass (X) in the growth medium.

2.2.2. Fermentation. A fermentation procedure consisting of three different stages was carried out as follows.

Stage 1. Batch cultivation at an initial carbon/nitrogen (C/N) ratio of 14 using the growth medium: carbon depletion in the medium (3 g L^{-1} of fructose) determined the length of the stage.

Stage 2. A fed-batch stage to increase biomass density at C/N ratio of 6.5: two additions of fructose and ammonium were made. The time of addition was determined as the point when the fructose remaining in the media reached approximately 3 g L^{-1}.

Stage 3. PHA production under nitrogen limitation: avocado oil was added to the culture at the beginning of the stage, at 30 h. Different concentrations were tested: 5, 10, 15, 20, and 25% (v/v).

Control experiments consisted of additional flasks prepared using fructose as the carbon source for the three-stage fermentation.

2.2.3. Analytical Procedures. Fermentation samples were taken every two hours and immediately centrifuged at 10000 rpm for 10 min at 4°C. Fructose and ammonium were analysed in the supernatant, and the bottom pellet (biomass) was washed thoroughly with distilled water before lyophilizing for gravimetric estimation of the dry cell weights (DCW).

Fructose consumption in the fermentation media was quantified using a 3,5-dinitrosalicylic acid (DNS) method [29]. Ammonium consumption was analysed according to the protocol of Weatherburn [30].

The intracellular polymer was extracted from the lyophilized biomass using chloroform (1 g of biomass per 50 mL of solvent) at 60°C for 30 min with constant stirring. After incubation, PHA dissolved in the chloroform phase was filtered to eliminate cellular debris and then precipitated with hexane. The residual solvent in the polymer was removed by evaporation [31, 32].

Dimensionless biomass yield ($Y_{x/s}$) was estimated as the ratio of the amount of biomass produced to the amount of total substrate consumed. This was calculated at the end of Stages 1 and 2. Productivity was estimated at the end of the fermentation as the final PHA concentration achieved divided by the total cultivation time required to attain that concentration. Residual biomass was also calculated at the end of fermentation as final produced biomass (CDW) minus PHA concentration.

2.3. PHA Characterization

2.3.1. Gas Chromatography. Fatty acid methyl esters were derived from acid methanolysis of PHA at 100°C for 4 h by incubating 100 mg of PHA, 2 mL of chloroform, 2 mL of methanol (20% of HCl), and benzoic acid (as an internal standard) in borosilicate glass tubes with screw caps at 100°C for 4 h. After cooling, distilled water was added (1 mL), the tubes were vortexed for 60 s, and the lower phase containing the resulting methyl esters was recovered for analysis [33]. Fatty acid methyl esters (1 μL) were analysed on a gas chromatograph (SRI Instruments, Model 310, USA) equipped with a flame ionization detector (FID) and a 6 ft. × 1/8 in. silica gel column. Nitrogen at 30 mL min^{-1} was used as carrier gas and the injector and detector were set at 220 and 170°C, respectively. Reference standards were poly(hydroxybutyrate) and the copolymer poly(hydroxybutyrate-co-hydroxyvalerate) [12 mol% hydroxyvalerate] (Goodfellow, UK).

2.3.2. Fourier Transform Infrared (FTIR) Spectroscopy. FTIR was performed within wavenumber ranges from 600 to 4000 cm^{-1} (BUCK Scientific, model 530, USA). PHA was dissolved in chloroform before pouring the solution onto KBr plates to form the polymer films.

2.3.3. Differential Scanning Calorimetry (DSC). DSC curves were obtained using a differential scanning calorimeter

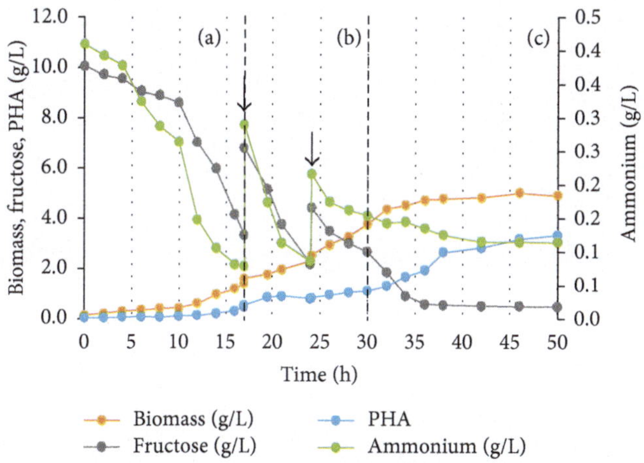

FIGURE 2: Profiles of fructose (grey circles), ammonium (green circles), biomass (orange circles), and poly-R-hydroxyalkanoate (PHA) (blue circles) production during the cultivation of *Cupriavidus necator* on fructose and avocado oil (20% v/v) in three-stage fermentation. Stage 1: batch cultivation (a), Stage 2: fed-batch stage (additions indicated by black arrows) (b), and Stage 3: PHA accumulation (c). Oil addition occurs at 30 h.

(Mettler Toledo DSC 823) according to López-Cuellar et al. [6]. Approximately 4 mg of the specimens was crimped in aluminium pans. The samples were evaluated under the following conditions: dynamic nitrogen atmosphere of 50 mL min^{-1}, a heating rate of 10°C min^{-1}, and an extended temperature range (−40 to 200°C). Two runs under same conditions were carried out; the first run erased the thermal history of sample. Thermograms obtained during the second run were analysed to determine the melting temperature (T_m) and melting enthalpy (ΔH_m) of the PHAs. Pure PHB (Goodfellow, UK) served as reference standard.

3. Results

3.1. Fermentation Studies. The representative profiles of the PHAs synthesized by *C. necator* H16 using fructose and avocado oil as carbon sources are depicted in Figure 2.

Stage 1 (Figure 2(a)), conducted as a batch cultivation, lasted 17 h and was initiated by adding 10 g L^{-1} of fructose and 0.42 g L^{-1} of ammonium to the medium. A lag phase of 10 h occurred. The concentration of substrate (i.e., fructose during Stage 1) decreased from 10 to 3.3 g L^{-1} (±0.10), whereas the biomass density, measured gravimetrically as DCW, increased from 0.13 to 1.44 g L^{-1} (±0.06) to achieve a growth yield ($Y_{x/s}$) of 0.19.

Stage 2 (Figure 2(b)), conducted as a fed-batch cultivation to increase cellular density, lasted about 12.5 h. Fructose as substrate was added on two different occasions: at 17.5 h and 24 h (Figure 2, black arrows). The first addition consisted of 3 g L^{-1} of fructose and 0.21 g L^{-1} of ammonium. For the second addition, the medium was supplemented with 2 g L^{-1} of fructose and 0.12 g L^{-1} of ammonium. During Stage 2, the biomass density increased from 1.44 (±0.06) to 3.79 g L^{-1} (±0.09), consuming about 5.63 g L^{-1} (±0.03) of fructose. At the end of Stage 2, an average $Y_{x/s}$ of 0.42 was achieved, matching the theoretical yield when simple sugars are used as carbon sources ($Y_{x/s}$ of 0.30–0.40) [8]. In addition, a slight accumulation of PHA was observed, but this only represented less than 30% of the DCW, in agreement with the balanced nutrient conditions.

Synthesis of PHAs (Stage 3) was first observed at 30 h (Figure 2(c)). The fructose remaining in the medium was about 1.9 g L^{-1} at the beginning of this stage. Avocado oil was added to the flasks in a single addition at the beginning of the stage to induce polymer synthesis. The avocado oil concentrations tested were 0 (used as a control), 5, 10, 15, 20, and 25% (v/v). Stage 3 lasted 20 h. During this time, nitrogen levels remained around 0.1 g L^{-1} to generate cellular stress and to promote polymer accumulation [6]. A rapid consumption of fructose was observed at the beginning of the stage and, by the end of the stage, fructose was barely detectable in the medium (<0.5 g L^{-1}).

The results of the 50 h, three-stage fermentation are summarized in Table 1. A significant amount of PHA accumulation was observed, ranging from 59 to 70% of the DCW. From the results, a positive trend for PHA accumulation was observed in flasks with avocado oil concentrations of 5, 10, 15, and 20% v/v. The highest PHA concentration was reached when the oil in the media was 20% (v/v), with PHA values of 3.48 g L^{-1} (±0.04) achieved, which represented 70.8% of the accumulated PHA in terms of DCW. However, flasks with and oil concentration of 25% (v/v) showed a decrease in PHA accumulation efficiency, reaching about 3.07 g L^{-1} (±0.02). The overall productivity of the experiments fluctuated between 0.053 and 0.070 g L^{-1} h^{-1}.

Conversely, control experiments (0% fed oil) produced 77% of the polymer, almost reaching the typical 80–85% PHA accumulation reported for the strain [34].

3.2. PHAs Composition Analysis through Gas Chromatography. From gas chromatography, the chemical composition of the PHAs was determined using benzoic acid (internal standard) as evaluation base. Representative chromatograms of the evaluated methyl esters are presented in Figure 3. The retention times for the reference standards and the samples were consistent (5.516 min for 3-hydroxybutyrate [3HB], 7.150 min for 3-hydroxyvalerate [3HV]).

The most abundant monomer detected in all samples was 3HB monomer ranging from 92.8 to 98.94% for samples fed with avocado oil, as summarized in Table 2. In all cases, 3HV monomeric units was also recognized, ranging from 1 to 7 mol%. The highest hydroxyvalerate amounts were found in the PHA synthesized from 20% (v/v) of oil, matching the biomass profiles. An unusual pattern was observed, wherein the accumulation of 3HV units was highly dependent on the avocado oil concentration, reaching a maximum value of 7% (Figure 2(c)). Other 3-hydroxy-acids containing more than five carbons were also identified in some PHAs, but in minimal quantities (i.e., less than 0.16 mol%). Thus, avocado oil could promote the synthesis of PHAs containing 3HB and significant fractions of 3HV monomers.

3.3. Functional Group Identification by Infrared Spectroscopy (FTIR). The spectra recorded from the PHAs, depicted in

TABLE 1: Final yields of poly-R-hydroxyalkanoates obtained from a three-stage fermentation of *Cupriavidus necator*.

Substrate	[a]Av. oil (% v/v)	[b]Biomass ($g\,L^{-1}$)	PHA ($g\,L^{-1}$)	[c]Residual biomass ($g\,L^{-1}$)	PHA (%)	Productivity ($g\,L^{-1}\,h^{-1}$)
Fructose	—	5.25 (±0.01)	4.04 (±0.02)	1.21 (±0.009)	76.87 (±1.03)	0.081
Fructose, avocado oil	5	4.45 (±0.02)	2.64 (±0.03)	1.82 (±0.012)	59.21 (±1.12)	0.053
	10	4.63 (±0.05)	3.14 (±0.04)	1.49 (±0.017)	67.69 (±1.38)	0.063
	15	4.74 (±0.04)	3.27 (±0.03)	1.47 (±0.013)	69.04 (±1.43)	0.065
	20	4.91 (±0.04)	3.48 (±0.04)	1.43 (±0.015)	70.83 (±1.31)	0.070
	25	4.61 (±0.03)	3.07 (±0.06)	1.54 (±0.011)	66.63 (±1.29)	0.061

[a]Av. oil: avocado oil concentrations (% v/v). [b]Biomass: measured gravimetrically. [c]Residual biomass: biomass concentration after PHA extraction.

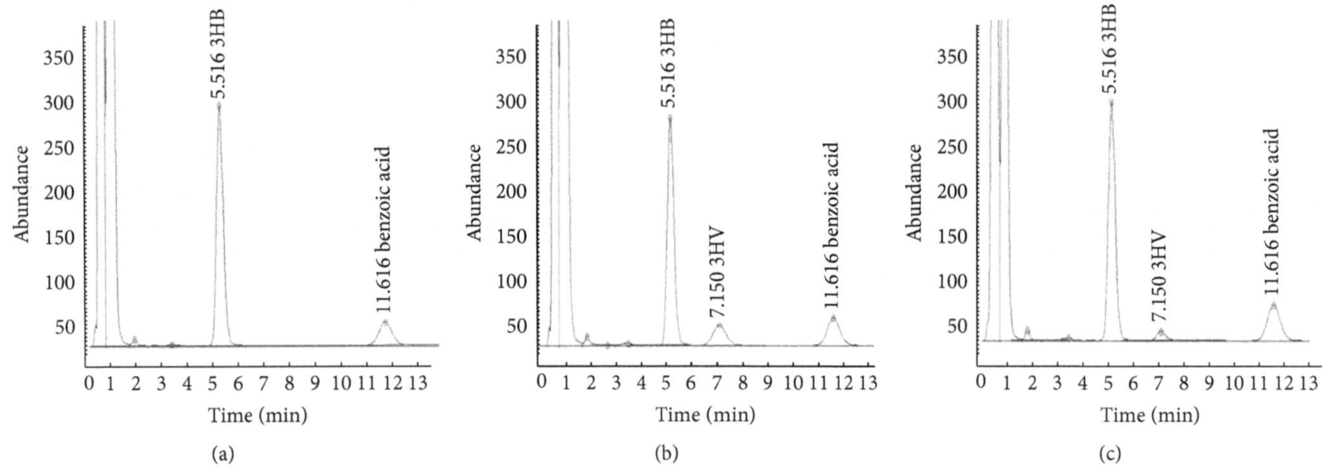

FIGURE 3: Chromatogram obtained from gas chromatography (GC) of (a) poly(3-hydroxybutyrate) (PHB) standard, (b) poly(hydroxybutyrate-co-hydroxyvalerate) (PHBV) standard with 12 mol% HV content, and (c) polymer produced from 20% v/v of avocado oil (7 mol% of HV). Benzoic acid was used as internal standard.

Figure 4, were similar to the PHBV spectra reported in previous studies [35]. The PHAs exhibited the particular chemical bonds of PHAs and replicated the absorption spectra among the samples.

The most prominent peak, located around 1720 cm^{-1}, was related to the ester carbonyl group (C=O). The bands located in the region of 2800 to 3000 cm^{-1} corresponded to the methyl-methylene groups. The presence of these peaks was due to the symmetric and asymmetric stretching of the CH_3 and CH_2 groups and these peaks were related to the monomeric units in the lateral chain. Besides the C=O group, an asymmetrical C-H bending vibration in CH_3 group shows an absorption band at 1453 cm^{-1}, whereas C-O-H bond shows a peak at 1378 [36]. Other peaks were recorded in the 1100 to 1290 cm^{-1} region; bands found around 1176, 1221, and 1270 cm^{-1} were attributed to polymer crystalline structures, and their presence was related to the C-O-C functional group [24].

3.4. Thermal Properties by DSC. The recorded thermograms obtained during the second run of the DSC analysis of the PHAs are shown in Figure 5. The melting points of the synthesized PHAs ranged from 159 to 173°C, whereas the measured ΔH_m was between 51 and 57 $J\,g^{-1}$ (Table 2). A decreasing trend was observed for the T_m value with increasing oil concentration. The lowest T_m and ΔH_m values were reached at 20% v/v, at 159°C and 51.81 $J\,g^{-1}$, respectively. An oil concentration above 20% v/v caused an increase in T_m of the synthesized PHA. Traces of monomers with a higher number of carbons were detected in some fed oil samples; however, these monomers were not identified, as mentioned in Table 2.

The thermal properties of the control experiment (0% fed oil) were also estimated and compared against a PHB reference standard. T_m of the sample fed with 0% oil was 175.16°C and ΔH_m value was 67.05 $J\,g^{-1}$. T_m of the PHB reference standard was 176.3°C and ΔH_m value was 75.48 $J\,g^{-1}$, in agreement with previous reports [37, 38] and confirming the production of pure PHB in the controls.

4. Discussion

Different PHAs were synthesized by *C. necator* H16 in a flask system using fructose and avocado oil as substrates. Fermentation was conducted using a three-stage process in order to increase biomass densities and polymer accumulation, without triggering the substrate inhibition reported to occur at levels of fructose higher than 10 $g\,L^{-1}$ [39]. Different oil

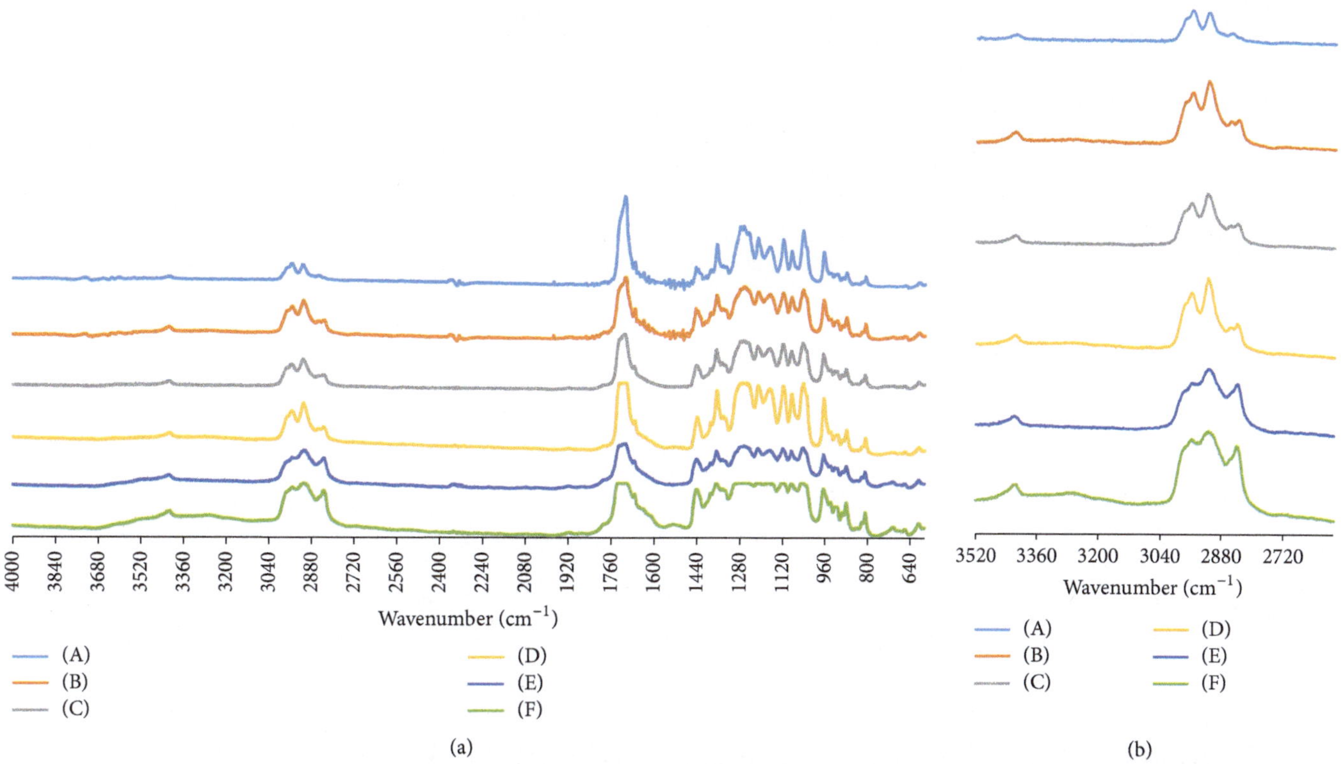

FIGURE 4: Fourier transform infrared spectroscopy (FTIR) spectra of the poly-R-hydroxyalkanoates (PHAs) produced by the addition of avocado oil: (A) 0 (control); (B) 5, (C) 10, (D) 15, (E) 20, and (F) 25 percent [v/v]. (a) 600 to 4000 cm^{-1}, (b) 2500 to 4000 cm^{-1}.

TABLE 2: Thermal properties and chemical composition of the poly-R-hydroxyalkanoates (PHAs) produced in *Cupriavidus necator.*

Substrate	[a]Av. oil (% v/v)	T_m (°C)	ΔH_m (J/g)	Monomeric composition of PHAs[b] (mol%)		
				3HB	3HV	3HA
Fructose	—	175.15	67.05	100.00	—	—
Fructose, avocado oil	5	173.22	56.68	98.94	1.06	
	10	169.71	55.32	97.15	2.77	0.08
	15	168.22	54.29	96.64	3.27	0.09
	20	159.57	51.81	92.83	7.01	0.16
	25	164.31	53.26	95.14	4.75	0.11

[a]Av. oil, avocado oil tested concentrations; [b]3HB, 3-hydroxybutyrate; 3HV, 3-hydroxyvalerate; 3HA, unidentified hydroxyl acid monomeric unit with more than 5 carbons.

concentrations (5, 10, 15, 20, and 25% v/v) were evaluated for PHA synthesis during Stage 3. The control experiments, using fructose as the carbon source during the overall process, complemented the studies.

Previous flask studies have typically been conducted in batch mode [12, 17–20], with overall productivities ranging from 0.025 to 0.08 $g\,L^{-1}\,h^{-1}$ (Table 3). In the current study, remarkable productivity was obtained, ranging from 0.05 to 0.07 $g\,L^{-1}\,h^{-1}$, for the evaluated oil concentrations (% v/v). Similar studies have achieved productivities between 0.025 and 0.05 $g\,L^{-1}\,h^{-1}$ [17, 19]. The highest productivity previously reported was reached using simple sugar as a carbon source [12] and resembled the productivity reported here for control samples (0.081 $g\,L^{-1}\,h^{-1}$). Hence, the feasibility of the three-stage fermentation was confirmed and suggested an interconnection with the operational mode of the system and the strain's affinity for the substrates [34].

A maximum biomass yield was reached with 20% v/v oil in the medium. Flasks with 25% v/v oil showed a decrease in cellular density and polymer accumulation. This yield decrease could be related to oxygen transfer limitations or a substrate inhibition [40]. Possibly, large oil amounts reduce

Table 3: Comparison of studies reporting PHAs production in *Cupriavidus necator*.

Strain	Substrate	Scale	Control strategy	PHA[1] produced	Biomass ($g L^{-1}$)	PHA ($g L^{-1}$)	PHA (%)	Productivity ($g L^{-1} h^{-1}$)	Reference
C. necator	Plant oils[2]	Flask	Batch	P(3HB)	3.6–4.3	2.9–3.4	79–81	0.04–0.05	Fukui and Doi [17]
C. necator (recombinant, *Aeromonas caviae*)	Plant oils[2]			P(3HB-co-3HHX)	3.5–3.6	2.7–2.9	76–81	0.04	
C. necator DSM 541	Fructose	Flask	Batch	P(3HB)	3.4	n.a	55	n.a	Dennis et al. [18]
	Palmitate			P(3HB-co-3HV-co-3HHX)	0.51	n.a	58	n.a	
	Oleate			P(3HB-co-3HHX)	1.44	n.a	57	n.a	
C. necator	Centrifuged fermented organic waste	Flask	Batch	P(3HB-co-3HV)	2.77	1.13	40.0	0.025	Ganzeveld et al., [19]
C. necator	Bagasse hydrolysate	Flask	Batch	n.a	6	3.9	65	0.08	Yu and Stahl [12]
C. necator	Palm oil	Flask	Batch	P(3HB-co-3HV)	3.6	2.66	74	n.a	Liu et al. [20]
C. necator	Fructose[3]	Flask	Fed batch	P(3PHB)	5,25	4.04	76.87	0.081	This study
	Fructose, avocado oil			P(3HB-co-3HV)	4.45–4.91	2.64–3.48	59–70	0.053–0.07	

[1]3PHB, 3 hydroxybutyrate; 3HV, 3 hydroxyvalerate; 3HHX, 3-hydroxyhexanoate. [2]Olive oil, corn oil and palm oil tested once at a time. [3]Control experiments. n.a: Not available.

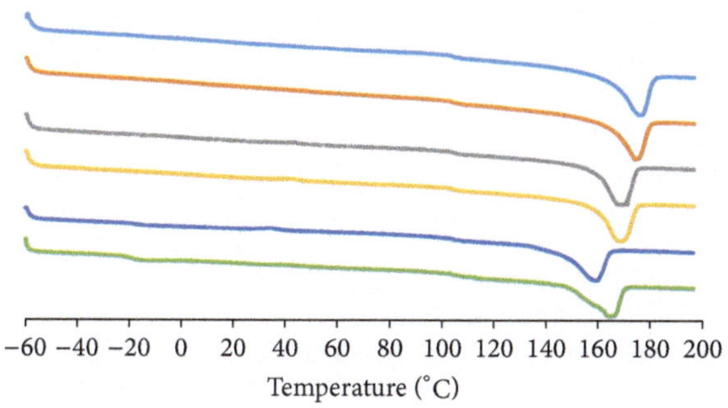

Figure 5: Thermograms obtained during the second run of differential scanning calorimetry (DSC) of the poly-R-hydroxyalkanoates (PHAs) produced by the addition of avocado oil at (a) 0 (control), (b) 5, (c) 10, (d) 15, (e) 20, and (f) 25 percent [v/v].

oxygen transfer to the cells, thereby delaying synthesis and accumulation of polymers. However, further studies on larger scales are required to analyse the oxygen transfer dynamics.

C. *necator* H16 is well known to exhibit a preference for synthesizing PHAs containing 3HB units as the most abundant monomer regardless of the carbon source, even from vegetable oils. Some studies have confirmed the synthesis of pure PHB when C. *necator* is grown in vegetable oils [41]. Conversely, Du et al. [42] achieved the production of PHBV using fatty acids from food scraps, whereas Dennis et al. [18] demonstrated that the C. *necator* synthase could accept C6 substrates (Table 3). A few studies have also identified larger monomeric units when using vegetable oils as carbon sources [6, 21].

In the present study, PHAs were composed mainly of 3HB monomers, followed by 3HV ranging from 1 to 7 mol%, and small quantities of 3HA (more than 5 carbons). Interestingly, all samples fed with avocado oil contained identifiable 3HV monomers and the presence of these monomers was highly correlated with the oil concentration in the medium. In some manner, the particular fatty acid composition of the avocado oil seems to promote the formation of 3HV precursors. Although Ganzeveld et al. [19] used centrifuged fermented organic waste and Liu et al. [20] employed palm oil to produce the same copolymer reported in this work (PHBV), the incorporation of 3HV monomers has not been sufficiently investigated when vegetable oils are used as the substrate (Table 3).

The thermal properties of the PHAs were enhanced by the presence of 3HV monomers in the polymer. As reported by Babel and Steinbüchel [11], T_m depends on the percentage of 3HV incorporated into the polymer. The melting temperature of the synthesized PHAs ranged from 159 to 173°C, in accordance with the 3HV mol% content (1 to 7 mol%). The presence of 3HV units in the polymer and even the small quantities of other 3HA improved the general thermal properties of PHA. Consequently, avocado oil as the substrate for PHA synthesis promoted the production of a more versatile material than what was obtained with pure PHB [43, 44], as demonstrated by DSC, FTIR, and GC analysis.

5. Conclusions

C. necator H16 is capable of incorporating small quantities of 3HV units into a PHA copolymer when avocado oil is used as substrate for PHA synthesis. The PHAs exhibited different monomeric compositions and properties, depending on the concentration of added oil. However, the highest yield, with a greater incorporation of larger monomer units (HV and more), was obtained when 20% (v/v) oil was added. Incorporation ranging from 1 to 7 mol% of 3HV monomeric units into the polymeric main chain was demonstrated. The results confirmed the feasibility of using avocado oil as a renewable carbon source for PHA production processes.

Disclosure

Partial results of this manuscript were presented as an abstract at the 9th Congress of FEBiotec (Annual Congress of Biotechnology).

Acknowledgments

Araceli Flores-Sánchez received grant-aided support from CONACyT (no. 417745). This research was partially funded by CONACyT (CONACyT-INFR-2015-254437 and CONACyT-CB-2014-239553). The authors acknowledge Alba-Flores Joel's (CINVESTAV) technical assistance during experimental development. Support of Piliado-Hernández D.M. (ITESM) and González-Bret K. during the writing of this manuscript was also appreciated.

References

[1] C. S. K. Reddy, R. Ghai, Rashmi, and V. C. Kalia, "Polyhydroxyalkanoates: an overview," *Bioresource Technology*, vol. 87, no. 2, pp. 137–146, 2003.

[2] M. V. Arcos-Hernández, B. Laycock, B. C. Donose et al., "Physicochemical and mechanical properties of mixed culture polyhydroxyalkanoate (PHBV)," *European Polymer Journal*, vol. 49, no. 4, pp. 904–913, 2013.

[3] J. Yu and Y. Si, "Metabolic carbon fluxes and biosynthesis of polyhydroxyalkanoates in Ralstonia eutropha on short chain fatty acids," *Biotechnology Progress*, vol. 20, no. 4, pp. 1015–1024, 2004.

[4] N.-S. Lau, D. H.-E. Ch'ng, K.-H. Chia, Y.-M. Wong, and K. Sudesh, "Advances in polyhydroxyalkanoate (PHA): Unraveling the development and new perspectives," *Journal of Biobased Materials and Bioenergy*, vol. 8, no. 2, pp. 118–129, 2014.

[5] E. Gasser, P. Ballmann, S. Dröge, J. Bohn, and H. König, "Microbial production of biopolymers from the renewable resource wheat straw," *Journal of Applied Microbiology*, vol. 117, no. 4, pp. 1035–1044, 2014.

[6] M. R. López-Cuellar, J. Alba-Flores, J. N. G. Rodríguez, and F. Pérez-Guevara, "Production of polyhydroxyalkanoates (PHAs) with canola oil as carbon source," *International Journal of Biological Macromolecules*, vol. 48, no. 1, pp. 74–80, 2011.

[7] R. Moita Fidalgo, J. Ortigueira, A. Freches et al., "Bio-oil upgrading strategies to improve PHA production from selected aerobic mixed cultures," *New Biotechnology*, vol. 31, no. 4, pp. 297–307, 2014.

[8] K. Sudesh, H. Abe, and Y. Doi, "Synthesis, structure and properties of polyhydroxyalkanoates: biological polyesters," *Progress in Polymer Science*, vol. 25, no. 10, pp. 1503–1555, 2000.

[9] B. Laycock, P. Halley, S. Pratt, A. Werker, and P. Lant, "The chemomechanical properties of microbial polyhydroxyalkanoates," *Progress in Polymer Science*, vol. 39, no. 2, pp. 397–442, 2014.

[10] L. M. W. K. Gunaratne, R. A. Shanks, and G. Amarasinghe, "Thermal history effects on crystallisation and melting of poly(3-hydroxybutyrate)," *Thermochimica Acta*, vol. 423, no. 1-2, pp. 127–135, 2004.

[11] W. Babel and A. Steinbüchel, *Biopolyesters. Special issue of advances in Biochem. Eng. Biotechnology*, Springer-Verlag, Berlin, Germany, 2001.

[12] J. Yu and H. Stahl, "Microbial utilization and biopolyester synthesis of bagasse hydrolysates," *Bioresource Technology*, vol. 99, no. 17, pp. 8042–8048, 2008.

[13] F. Fang, H. Jiang, J. Wang, and H.-Q. Yu, "Identifying the influential priority of the factors governing PHB production by activated sludge with integration of uniform design and grey relational analysis," *Separation and Purification Technology*, vol. 136, pp. 111–114, 2014.

[14] S. Anterrieu, L. Quadri, B. Geurkink et al., "Integration of biopolymer production with process water treatment at a sugar factory," *New Biotechnology*, vol. 31, no. 4, pp. 308–323, 2014.

[15] M. Venkateswar Reddy, Y. Yajima, Y. Mawatari, T. Hoshino, and Y.-C. Chang, "Degradation and conversion of toxic compounds into useful bioplastics by Cupriavidus sp. CY-1: relative expression of the PhaC gene under phenol and nitrogen stress," *Green Chemistry*, vol. 17, no. 9, pp. 4560–4569, 2015.

[16] A. F. Mohidin Batcha, D. M. R. Prasad, M. R. Khan, and H. Abdullah, "Biosynthesis of poly(3-hydroxybutyrate) (PHB) by Cupriavidus necator H16 from jatropha oil as carbon source," *Bioprocess and Biosystems Engineering*, vol. 37, no. 5, pp. 943–951, 2014.

[17] T. Fukui and Y. Doi, "Efficient production of polyhydroxyalkanoates from plant oils by Alcaligenes eutrophus and its recombinant strain," *Applied Microbiology and Biotechnology*, vol. 49, no. 3, pp. 333–336, 1998.

[18] D. Dennis, M. McCoy, A. Stangl, H. E. Valentin, and Z. Wu, "Formation of poly(3-hydroxybutyrate-co-3-hydroxyhexanoate) by PHA synthase from Ralstonia eutropha," *Journal of Biotechnology*, vol. 64, no. 2-3, pp. 177–186, 1998.

[19] K. J. Ganzeveld, A. Van Hagen, M. H. Van Agteren, W. De Koning, and A. J. M. S. Uiterkamp, "Upgrading of organic waste: production of the copolymer poly-3-hydroxybutyrate-co-valerate by Ralstonia eutrophus withorganic waste as sole carbon source," *Journal of Cleaner Production*, vol. 7, no. 6, pp. 413–419, 1999.

[20] Q. Liu, G. Luo, X. R. Zhou, and G.-Q. Chen, "Biosynthesis of poly(3-hydroxydecanoate) and 3-hydroxydodecanoate dominating polyhydroxyalkanoates by β-oxidation pathway inhibited Pseudomonas putida," *Metabolic Engineering*, vol. 13, no. 1, pp. 11–17, 2011.

[21] A. Rathinasabapathy, B. A. Ramsay, J. A. Ramsay, and F. Pérez-Guevara, "A feeding strategy for incorporation of canola derived medium-chain-length monomers into the PHA produced by wild-type Cupriavidus necator," *World Journal of Microbiology and Biotechnology*, vol. 30, no. 4, pp. 1409–1416, 2014.

[22] M. A. Hassan, L.-N. Yee, P. L. Yee et al., "Sustainable production of polyhydroxyalkanoates from renewable oil-palm biomass," *Biomass and Bioenergy*, vol. 50, pp. 1–9, 2013.

[23] S. Magdouli, S. K. Brar, J. F. Blais, and R. D. Tyagi, "How to direct the fatty acid biosynthesis towards polyhydroxyalkanoates production?" *Biomass and Bioenergy*, vol. 74, pp. 268–279, 2015.

[24] L. Shi and B. P. Tu, "Acetyl-CoA and the regulation of metabolism: mechanisms and consequences," *Current Opinion in Cell Biology*, vol. 33, pp. 125–131, 2015.

[25] R. A. J. Verlinden, D. J. Hill, M. A. Kenward, C. D. Williams, and I. Radecka, "Bacterial synthesis of biodegradable polyhydroxyalkanoates," *Journal of Applied Microbiology*, vol. 102, no. 6, pp. 1437–1449, 2007.

[26] M. d. Galvão, N. Narain, and N. Nigam, "Influence of different cultivars on oil quality and chemical characteristics of avocado fruit," *Food Science and Technology (Campinas)*, vol. 34, no. 3, pp. 539–546, 2014.

[27] L. González, *México es productor global montruo de aguacate:Grayeb*, The Economist, 2015.

[28] M. L. Dreher and A. J. Davenport, "Hass avocado composition and potential health effects," *Critical Reviews in Food Science and Nutrition*, vol. 53, no. 7, pp. 738–750, 2013.

[29] G. L. Miller, "Use of dinitrosalicylic acid reagent for determination of reducing sugar," *Analytical Chemistry*, vol. 31, no. 3, pp. 426–428, 1959.

[30] M. W. Weatherburn, "Phenol-hypochlorite reaction for determination of ammonia," *Analytical Chemistry*, vol. 39, no. 8, pp. 971–974, 1967.

[31] I. Chodak, "Polyhydroxyalkanoates: properties and modification for high volume applications," in *Degradable Polymers*, pp. 295–319, Springer, Dordrecht, 2002.

[32] M. Koller, A. Salerno, M. Dias, A. Reiterer, and G. Braunegg, "Modern biotechnological polymer synthesis: a review," *Food Technology and Biotechnology*, vol. 48, no. 3, pp. 255–269, 2010.

[33] K. Ichihara and Y. Fukubayashi, "Preparation of fatty acid methyl esters for gas-liquid chromatography," *Journal of Lipid Research*, vol. 51, no. 3, pp. 635–640, 2010.

[34] W. Babel and A. Steinbüchel, *Biopolyesters*, vol. 71, Springer, 2003.

[35] T. Mumtaz, S. Abd-Aziz, P. L. Yee, W. M. Z. W. Yunus, Y. Shirai, and M. A. Hassan, "Synthesis, characterization, and structural properties of intracellular copolyester poly(3-hydroxybutyrate-co-3-hydroxyvalerate) produced by comamonas sp. EB 172 from renewable resource," *International Journal of Polymer Analysis and Characterization*, vol. 15, no. 6, pp. 329–340, 2010.

[36] K. Hong, S. Sun, W. Tian, G. Q. Chen, and W. Huang, "A rapid method for detecting bacterial polyhydroxyalkanoates in intact cells by Fourier transform infrared spectroscopy," *Applied Microbiology and Biotechnology*, vol. 51, no. 4, pp. 523–526, 1999.

[37] T. Furukawa, H. Sato, R. Murakami et al., "Structure, dispersibility, and crystallinity of poly(hydroxybutyrate)/ poly(L-lactic acid) blends studied by FT-IR microspectroscopy and differential scanning calorimetry," *Macromolecules*, vol. 38, no. 15, pp. 6445–6454, 2005.

[38] C.-Y. Loo, W.-H. Lee, T. Tsuge, Y. Doi, and K. Sudesh, "Biosynthesis and characterization of poly(3-hydroxybutyrate-co-3- hydroxyhexanoate) from palm oil products in a Wautersia

eutropha mutant," *Biotechnology Letters*, vol. 27, no. 18, pp. 1405–1410, 2005.

[39] B. S. Kim, S. C. Lee, S. Y. Lee, H. N. Chang, Y. K. Chang, and S. I. Woo, "Production of poly(3-hydroxybutyric-co-3-hydroxyvaleric acid) by fed-batch culture of Alcaligenes eutrophus with substrate control using on-line glucose analyzer," *Enzyme and Microbial Technology*, vol. 16, no. 7, pp. 556–561, 1994.

[40] F. Garcia-Ochoa and E. Gomez, "Bioreactor scale-up and oxygen transfer rate in microbial processes: an overview," *Biotechnology Advances*, vol. 27, no. 2, pp. 153–176, 2009.

[41] F. C. Oliveira, D. M. G. Freire, and L. R. Castilho, "Production of poly(3-hydroxybutyrate) by solid-state fermentation with Ralstonia eutropha," *Biotechnology Letters*, vol. 26, no. 24, pp. 1851–1855, 2004.

[42] G. Du, L. X. L. Chen, and J. Yu, "High-efficiency production of bioplastics from biodegradable organic solids," *Journal of Polymers and the Environment*, vol. 12, no. 2, pp. 89–94, 2004.

[43] A. J. Anderson and E. A. Dawes, "Occurrence, metabolism, metabolic role, and industrial uses of bacterial polyhydroxyalkanoates," *Microbiological Reviews*, vol. 54, no. 4, pp. 450–472, 1990.

[44] S. Bengtsson, A. R. Pisco, M. A. M. Reis, and P. C. Lemos, "Production of polyhydroxyalkanoates from fermented sugar cane molasses by a mixed culture enriched in glycogen accumulating organisms," *Journal of Biotechnology*, vol. 145, no. 3, pp. 253–263, 2010.

6

Enhancement of the Wettability and Lubrication of Shale Rock via Nanoemulsions

Quan Xu,[1] Qingqiao Li,[1] Xu Wu,[1] Bofan Li,[1] Erdong Yao,[1] Yuan Li,[1] Fujian Zhou,[1] Yan Luo,[2] and Wei Cai[1]

[1] *State Key Laboratory of Heavy Oil Processing, China University of Petroleum (Beijing), Beijing 102249, China*
[2] *Department of Chemical Engineering, West Virginia University, Morgantown, WV 26506, USA*

Correspondence should be addressed to Quan Xu; xuquan@cup.edu.cn

Academic Editor: Chaoqun Zhang

Nanoemulsions have been widely used as additives for drilling fluids in recent years. With the development of nanotechnology, multifunctional nanomaterials have been added to nanoemulsions. The improvement of wettability of the surfaces, alteration of oil-wet on shale rock surfaces, and environmentally friendly conditions are considered as the future development directions of nanoemulsions. In this work, a novel nanoemulsion was prepared by using hydrocarbon-based polyoxyethylene ether, oil (hydrocarbon), distilled water, and formation crude oil as the main raw materials. The shale rocks before and after immersion with as-prepared nanoemulsion were characterized by contact angle measurement, atomic force microscope (AFM), and Fourier transform infrared spectroscopy (FTIR). It is clearly observed that the nanoemulsion greatly improved the wettability of the sandstone and rock surface by forming a layer of active agent film on the surface of the rock. The as-prepared nanoemulsion had good ability to curb the anticollapse and lubricate and protect the oil and gas layer.

1. Introduction

Nanoemulsion is a transparent/translucent (50–200 nm) or milky white emulsion (about 500 nm). It also has developed as small emulsion, ultrafine emulsion, latex liquid, unstable emulsion, and submicron emulsion. The introduction of nanomaterials in the drilling completion fluid can solve urgent issues, such as stability performance and environmentally friendly composites [1]. When the nanoemulsion was applied to drilling, solid particles and the filtrate from the drilling and completion were prevented from the deep formation to avoid permanent blockage in the oil and gas flow channel [2]. This played very important role in the protection of oil and gas layer. The current pilot exploration research showed that in the field of oil and drilling fluid nanoemulsion had a very strong attraction.

In this work, rock with a porosity of 11.6% and a permeability of 0.0001–1 mD was derived from Jimusaer dense oil reservoir [3–7]. The average density and viscosity of the crude oil were, respectively, 0.888 g/cm^3 and 73.45 mPa·s. The freezing point was 24.84°C, which belonged to high temperature and high freezing point. The basic information about Jimusaer dense oil reservoir was listed in Table 1.

In this work, the surface of the rock in Jimusaer oil field was modified by nanoemulsion due to the improved wettability of the rock surface and formed active agent film. This provided a new idea for the optimization of sandstone drilling fluid and the improvement of the stability of sandstone gas shaft [8–10].

2. Experiment

2.1. Materials. Alkyl polyoxyethylene ether, oil *(hydrocarbons)*, distilled water, stratified crude oil, and rock samples were obtained from Changji Oilfield.

2.2. Experiment. The experiments were processed as follows:

(1) Diluted crude oil: a diluted crude oil with a viscosity of 10.83 mPa·s was obtained by mixing crude oil with kerosene at a ratio of 1 : 1.25.

TABLE 1: The basic information of the Jimusaer dense oil reservoir.

Block		Changji oilfield	
Oil layer	Lauza ditch group on the desert		Luca ditch group under the desert
Thickness of the reservoir (m)	16 (thin interbed)		20 (thin interbed)
Lithology	Debris cloud, muddy siltstone, and silt sandstone		Cloudy siltstone
Natural cracks	Not developed		Not developed
Porosity (%)	11.0		11.6
Permeability (mD)	0.012		0.010
Oil saturation (%)	78.7		80.2
Formation pressure coefficient		1.273~1.32	
Reservoir temperature (°C)	86.6		93
Crude oil viscosity (mPa·s, 50°C)	64.3		204.6
Stratified crude oil viscosity (mPa·s)	10.58		21.5 (calculation)
Ground crude oil density (g/cm^3)	0.888		0.908

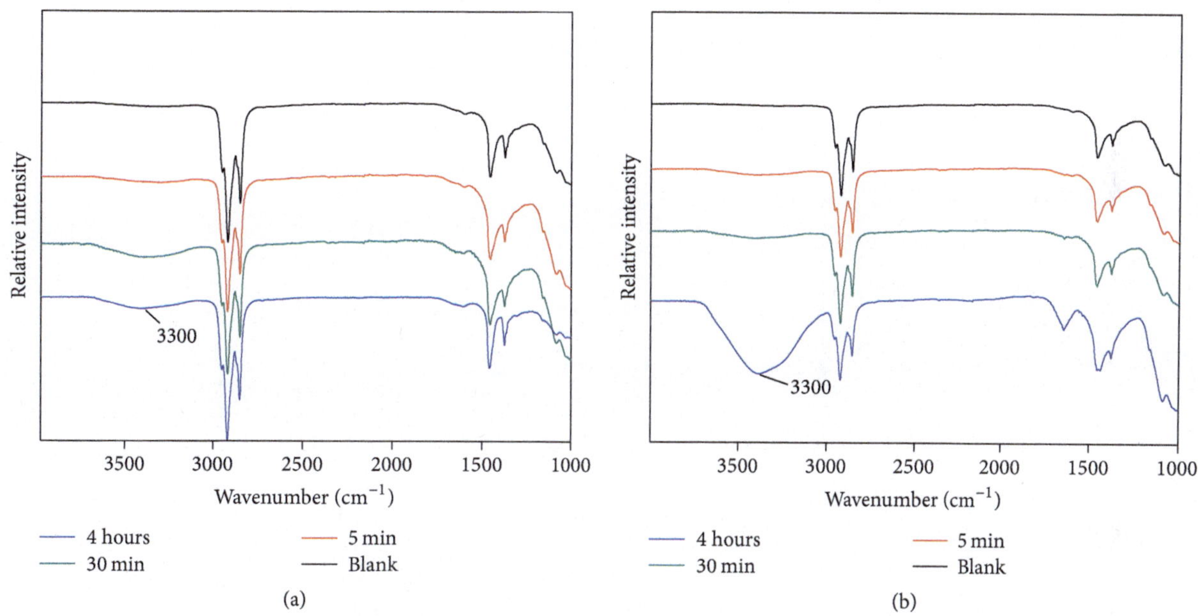

FIGURE 1: FTIR spectra of soaked sandstone samples: (a) sandstone soaked under NaCl solution for different immersion times and (b) sandstone soaked with nanoemulsions at different immersion time.

(2) Nanoemulsion mother liquor configuration: a certain amount of the surfactant (hydrocarbyl polyoxyethylene ether), alcohol, and oil (hydrocarbons) were mixed. Subsequently, distilled water was added to generate microemulsion mother liquor.

(3) Nanoemulsion configuration: nanoemulsion was obtained by mixing microemulsion mother liquor and 2% KCl solution at the ratio of 0.1 : 100.

(4) Rock soaking: the core was cut into 1 cm thick and 2.5 cm diameter with 1000 mesh sandpaper polished smooth; then, it was soaked with diluted oil for 10 days.

(5) Remove the rock and immerse it into nanoemulsion.

(6) The soaked rock was analyzed by FTIR, atomic force microscopy (AFM) and contact angle.

3. Results and Discussion

3.1. FTIR. Rocks soaked in NaCl solution and nanoemulsions for different immersion time (blank sample, 5 min, 30 min, and 4 h) were characterized by FTIR (Figures 1(a) and 1(b)). It is clearly seen that a strong peak at 3300 cm^{-1} occurred after nanoemulsion soaking [11–15]. The longer the soaking time, the stronger the relative intensity of the peak. This peak corresponded to a group with strong polarity, indicating that the nanoemulsion can greatly improve the wetting of the sandstone surface [15–18].

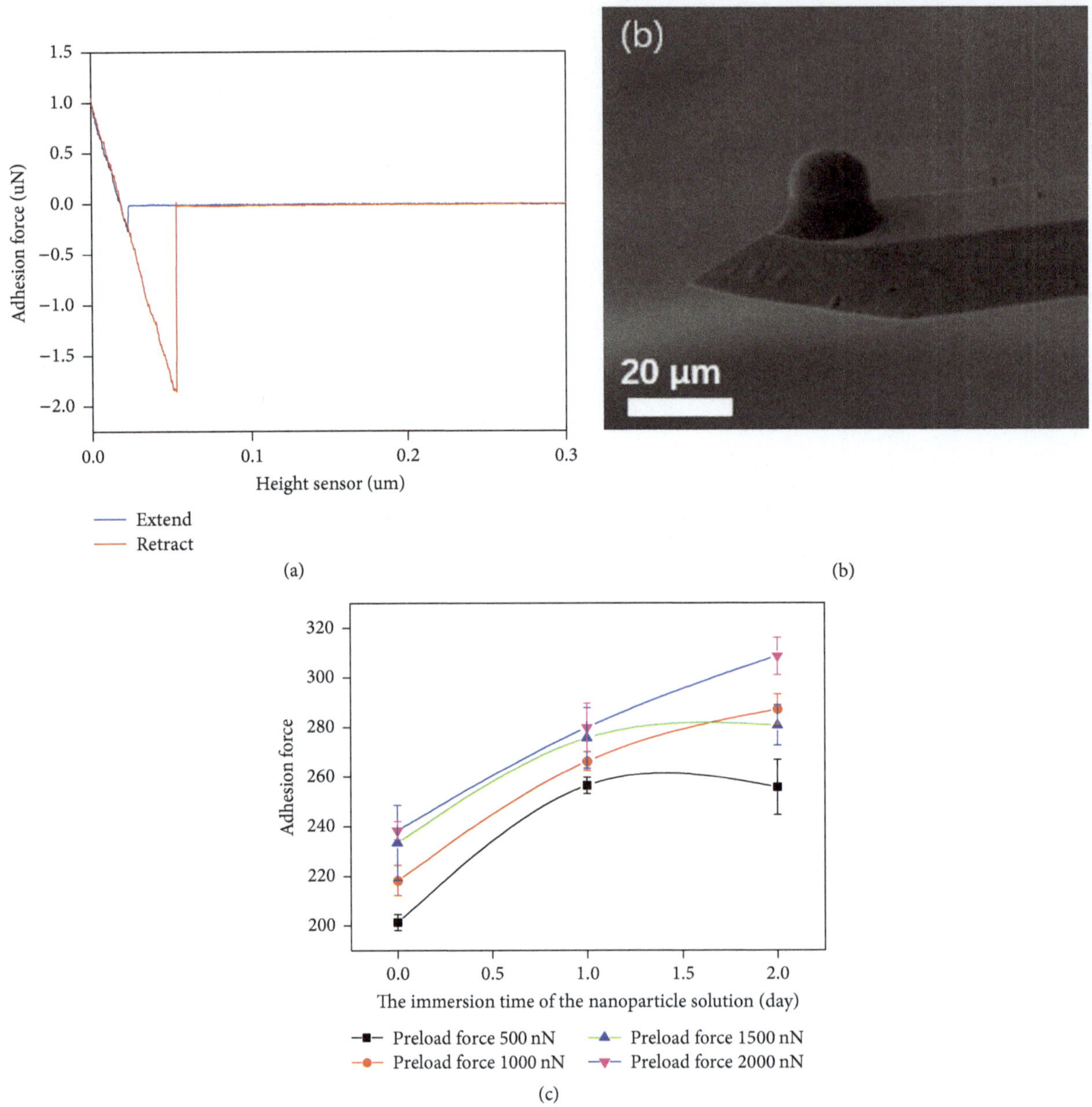

FIGURE 2: (a) A typical force curve data, (b) a typical AFM tip probe, and (c) adhesion force between rocks and probes under different soaking time with nanoemulsions.

3.2. AFM Adhesion Test. Figure 2 gives the adhesion forces between probes and rock surfaces under different nanoemulsion soaking time. All the tests were performed at a temperature of 23°C with humidity 23%. The same positions on the rocks were tested (experimental conditions: ramp size 500 nm, reserved velocity 1000 nm/s, preload time 0 s, and temperature 19°C) [19]. Figures 2(a) and 2(b) are a typical adhesive force curve and tip probe detected under AFM. It can be seen from Figure 2(a) that the adhesion force between particle and substrate increased with the increase of soaking time. The adhesive force shown in Figure 2(c) also increased as the loading force increased from 500 to 2000 nN, indicating that the surfactant in the nanoemulsion was effectively spread on the shale rock surface and formed a layer of active agent film with high stability performance.

3.3. Contact Angle. Wettability refers to the tendency of another fluid to expand or adhere to the solid surface when there is a mixed fluid [20]. It is the interaction between oil and water and reservoir rocks under reservoir conditions. According to the affinity between the reservoir rock surface and water and oil, the reservoir wetting property is divided into water wetting, oil wetting, neutral wetting, and subwetting. Reservoir wettability is the main factor to control the flow and distribution of fluid in reservoirs [21]. Therefore, the study of reservoir wettability is of great significance to the

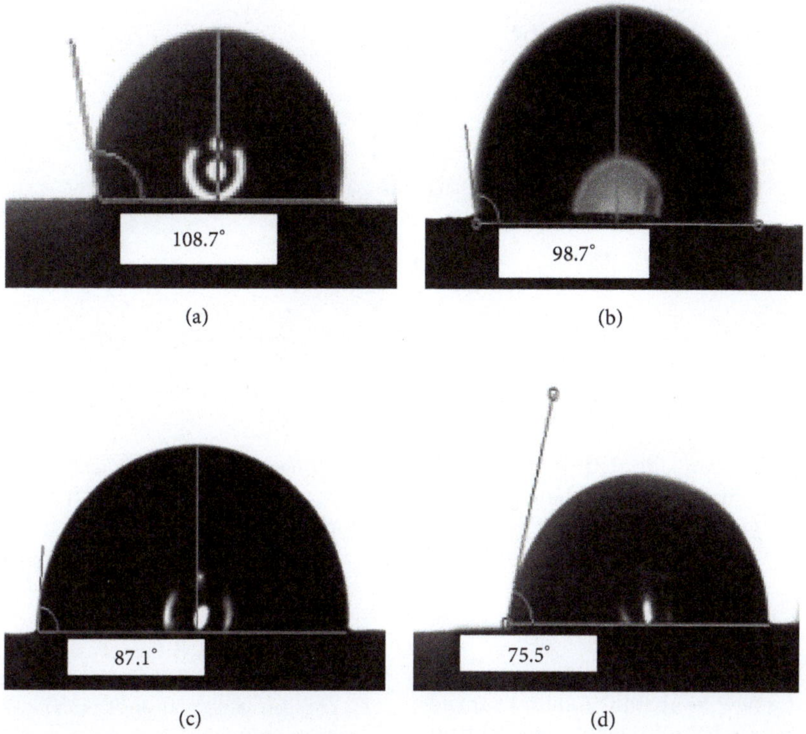

FIGURE 3: Contact angle of water droplets in air; rocks soaked in the nanoemulsion for various times, (a) 1 day, (b) 3 days, (c) 7 days, and (d) 9 days.

development of oil field and the final recovery rate of crude oil [22].

The contact angle method is mainly used for the determination of wettability of pure fluid and artificial core systems. The wettability of the system is usually defined according to the determination of the angle θ of the water on the solid surface. Generally, $\theta < 75°$, $105° > \theta > 75°$, and $\theta > 105°$ are defined as water wetting, neutral wetting, and oil wetting, respectively.

Figures 3(a)–3(d) are the contact angle of the rock sample immersed in the nanoemulsions for different soaking times. It can be clearly seen that the oil was wetted when the rock was soaked in the nanoemulsion for 1 day (Figure 3(a)) but weakly wetted when immersed for 3, 7, and 9 days (Figures 3(b)–3(d)). The graph inferred that the wetting of the rock surface changed with the immersion time.

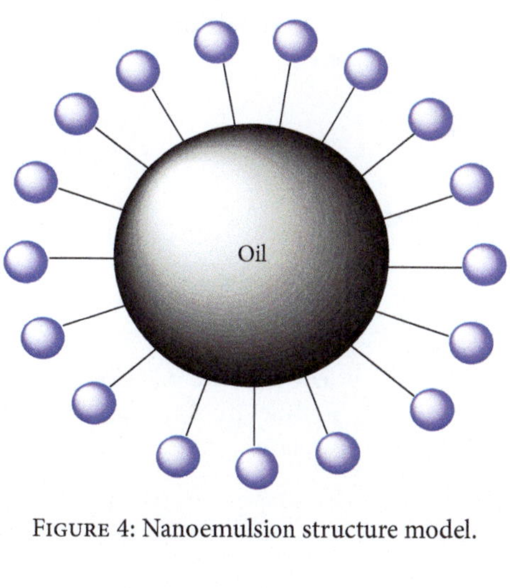

FIGURE 4: Nanoemulsion structure model.

3.4. The Schematic of the Reaction. Nanoemulsions are usually thermodynamically stable isotropic dispersions of oil, water, surfactants, and cosurfactants spontaneously formed. Figure 4 is the model of nanoemulsion structure [23]. From the microscopic principle analysis, the nanoemulsion is a heterogeneous dispersion, which consists of a continuous aqueous phase and a polymer dispersed in water in a microsphere form. During the emulsion formation process, the water is gradually volatilized or absorbed by the porousness of the surface. Then, the microspheres of the polymer slowly approach and finally contact each other until the polymer particles fuse with each other to form a continuous uniform coating [24].

Figure 5 is the schematic diagram of shale rock soaked in nanoemulsions. It can be seen that when the nanoemulsions are dispensed onto water surface under gravity, it can mix with rock in two ways; one is spreading on the surface and the other is infiltration into rocks. Since the infiltration speed of water is much higher than that of nanoemulsions,

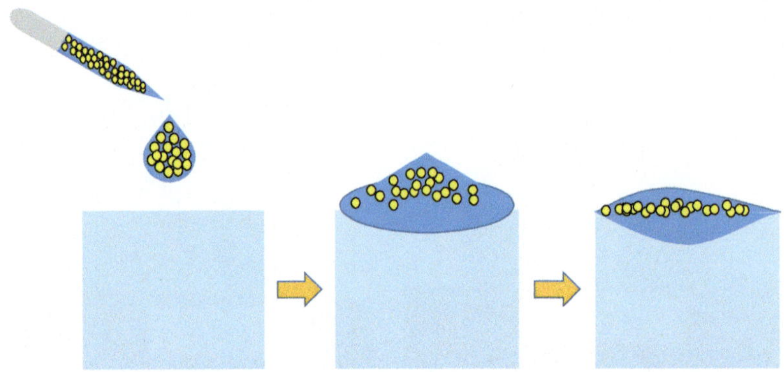

FIGURE 5: Schematic diagram of sandstone soaked in nanoemulsion.

thus nanoemulsions start to get concentrated under surface tension and form a solid thin film on top of the shale rock surfaces. That is the reason why the wettability of the shale rocks was changed [25–28].

4. Conclusions

In this work, we designed a new nanoemulsion and proved that the nanoemulsions can greatly improve the wettability of the sandstone surface and form a layer of active agent film on the surface of the rock. The drilling fluid used in the oil field had good ability to curb the anticollapse and lubricate and protect the oil and gas. AFM and FTIR results represent a thin layer of film formed on the surface of the shale rock. Those findings may guide the scientists and engineers regarding designing nanoemulsions with good wettability and stability properties.

Acknowledgments

The authors thank the National Science Foundation (no. 51505501), Beijing Nova Program (no. Z171100001117058), Tribology Science Fund of State Key Laboratory of Tribology (no. SKLTKF16A06), and Science Foundation of China University of Petroleum (Beijing) (nos. C201603, 2462014YJRC011) for the support.

References

[1] T. Tadros, "Application of rheology for assessment and prediction of the long-term physical stability of emulsions," *Advances in Colloid and Interface Science*, vol. 108-109, pp. 227-258, 2004.

[2] K. Meleson, S. Graves, and T. G. Mason, "Formation of concentrated nanoemulsions by extreme shear," *Soft Materials*, vol. 2, no. 2-3, pp. 109-123, 2004.

[3] N. Sadurní, C. Solans, N. Azemar, and M. J. García-Celma, "Studies on the formation of O/W nano-emulsions, by low-energy emulsification methods, suitable for pharmaceutical applications," *European Journal of Pharmaceutical Sciences*, vol. 26, no. 5, pp. 438–445, 2005.

[4] P. Fernandez, V. André, J. Rieger, and A. Kühnle, "Nano-emulsion formation by emulsion phase inversion," *Colloids and Surfaces A: Physicochemical and Engineering Aspects*, vol. 251, no. 1–3, pp. 53–58, 2004.

[5] P. Heunemann, S. Prévost, I. Grillo, C. M. Marino, J. Meyer, and M. Gradzielski, "Formation and structure of slightly anionically charged nanoemulsions obtained by the phase inversion concentration (PIC) method," *Soft Matter*, vol. 7, no. 12, pp. 5697–5710, 2011.

[6] D. Morales, J. M. Gutiérrez, M. J. García-Celma, and Y. C. Solans, "A study of the relation between bicontinuous microemulsions and oil/water nano-emulsion formation," *Langmuir*, vol. 19, no. 18, pp. 7196–7200, 2003.

[7] A. Forgiarini, J. Esquena, C. González, and C. Solans, "Formation of nano-emulsions by low-energy emulsification methods at constant temperature," *Langmuir*, vol. 17, no. 7, pp. 2076–2083, 2001.

[8] O. Sonneville-Aubrun, J.-T. Simonnet, and F. L'Alloret, "Nanoemulsions: a new vehicle for skincare products," *Advances in Colloid and Interface Science*, vol. 108-109, pp. 145–149, 2004.

[9] S. M. Jafari, Y. He, and B. Bhandari, "Optimization of nano-emulsions production by microfluidization," *European Food Research and Technology*, vol. 225, no. 5-6, pp. 733–741, 2007.

[10] D. Guzey and D. J. McClements, "Formation, stability and properties of multilayer emulsions for application in the food industry," *Advances in Colloid and Interface Science*, vol. 128–130, pp. 227–248, 2006.

[11] C. Solans, P. Izquierdo, and J. Nolla, "Nano-emulsions," *Current Opinion in Colloid & Interface Science*, vol. 10, no. 3, pp. 102–110, 2005.

[12] V. G. Babak and M.-J. Stebe, "Highly concentrated emulsions: physicochemical principles of formulation," *Journal of Dispersion Science and Technology*, vol. 23, no. 1–3, pp. 1–22, 2002.

[13] H. Karbstein and H. Schubert, "Developments in the continuous mechanical production of oil-in-water macro-emulsions," *Chemical Engineering and Processing: Process Intensification*, vol. 34, no. 3, pp. 205–211, 1995.

[14] Q. Xu, M. Li, L. Zhang, J. Niu, and Z. Xia, "Dynamic adhesion forces between microparticles and substrates in water," *Langmuir*, vol. 30, no. 37, pp. 11103–11109, 2014.

[15] Q. Xu, M. Li, J. Niu, and Z. Xia, "Dynamic enhancement in adhesion forces of microparticles on substrates," *Langmuir*, vol. 29, no. 45, pp. 13743–13749, 2013.

[16] M. Kahlweit, G. Busse, B. Faulhaber, and J. Jen, "Shape changes of globules in nonionic microemulsions," *The Journal of Physical Chemistry C*, vol. 100, no. 36, pp. 14991–14994, 1996.

[17] S. Engelskirchen, N. Elsner, T. Sottmann, and R. Strey, "Triacylglycerol microemulsions stabilized by alkyl ethoxylate surfactants—a basic study. Phase behavior, interfacial tension and microstructure," *Journal of Colloid and Interface Science*, vol. 312, no. 1, pp. 114–121, 2007.

[18] K. Aramaki, K. Ozawa, and H. Kunieda, "Effect of temperature on the phase behavior of ionic-nonionic microemulsions," *Journal of Colloid and Interface Science*, vol. 196, no. 1, pp. 74–78, 1997.

[19] M. A. Pes, K. Aramaki, N. Nakamura, and H. Kunieda, "Temperature-insensitive microemulsions in a sucrose monoalkanoate system," *Journal of Colloid and Interface Science*, vol. 178, no. 2, pp. 666–672, 1996.

[20] A. C. Lam and R. S. Schechter, "A study of diffusion and electrical conduction in microemulsions," *Journal of Colloid and Interface Science*, vol. 120, no. 1, pp. 42–55, 1987.

[21] H. Schott, A. E. Royce, and S. K. Han, "Effect of inorganic additives on solutions of nonionic surfactants (VII): cloud point shift values of individual ions," *Journal of Colloid and Interface Science*, vol. 98, no. 1, pp. 196–201, 1984.

[22] A. Maestro, I. Sole, C. Gonzalez, C. Solans, and J. M. Gutierre, "Influence of the phase behavior on the properties of ionic nanoemulsions prepared by the phase inversion composition method," *Colloid Interface Science*, vol. 327, no. 2, pp. 433–439, 2008.

[23] Q. Li, Y. Luo, B. Hou, and Z. Han, "Synthesis and mechanism of flocculating-decolorizing agent PAD used for polymer-sulphonated drilling wastewater," *Journal of Residuals Science and Technology*, vol. 13, no. 2, pp. 135–143, 2016.

[24] B. Hou, Q. Li, and Y. Luo, "Synthesis and mechanism of a novel organosilicone used as a selective plugging agent," *Journal of Residuals Science and Technology*, vol. 12, no. 3, pp. 149–156, 2015.

[25] K. S. Sharma, S. R. Patil, and A. K. Rakshit, "Study of the cloud point of C12En nonionic surfactants: effect of additives," *Colloids and Surfaces A: Physicochemical and Engineering Aspects*, vol. 219, no. 1–3, pp. 67–74, 2003.

[26] T. Gu, S. Qin, and C. Ma, "The effect of electrolytes on the cloud point of mixed solutions of ionic and nonionic surfactants," *Journal of Colloid and Interface Science*, vol. 127, no. 2, pp. 586–588, 1989.

[27] C. Zhang, H. Wu, and M. R. K. Kessler, "High bio-content polyurethane composites with urethane modified lignin as filler," *Polymer*, vol. 69, no. 1, pp. 52–57, 2015.

[28] C. Zhang, Y. Xia, R. Chen, S. Huh, P. A. Johnston, and M. R. Kessler, "Soy-castor oil based polyols prepared using a solvent-free and catalyst-free method and polyurethanes therefrom," *Green Chemistry*, vol. 15, no. 6, pp. 1477–1484, 2013.

7

CO_2 Induced Foaming Behavior of Polystyrene near the Glass Transition

Salah Al-Enezi

Petroleum Research Centre, Kuwait Institute for Scientific Research, Safat, Kuwait

Correspondence should be addressed to Salah Al-Enezi; senezi@kisr.edu.kw

Academic Editor: Miriam H. Rafailovich

This paper examines the effect of high-pressure carbon dioxide on the foaming process in polystyrene near the glass transition temperature and the foaming was studied using cylindrical high-pressure view cell with two optical windows. This technique has potential applications in the shape foaming of polymers at lower temperatures, dye impregnation, and the foaming of polystyrene. Three sets of experiments were carried out at operating temperatures of 50, 70, and 100°C, each over a range of pressures from 24 to 120 bar. Foaming was not observed when the polymer was initially at conditions below T_g but was observed above T_g. The nucleation appeared to occur randomly leading to subsequent bubble growth from these sites, with maximum radius of 0.02–0.83 mm. Three models were applied on the foaming experimental data. Variable diffusivity and viscosity model (Model C) was applied to assess the experimental data with the WLF equation. The model shows very good agreement by using realistic parameter values. The expansion occurs by diffusion of a dissolved gas from the supersaturated polymer envelope into the bubble.

1. Introduction

The use of high-pressure carbon dioxide as an efficient medium for polymer processing due to its chemical and economic advantages [1–3]. The unique properties associated with CO_2 have been explored for a long time in fields of polymer processing as they lower the glass transition temperature. In foaming process, the polymer is saturated with CO_2 in the supercritical state followed by equilibrium swelling state. After saturation, the sample transformed into a rubbery state and depressurized to atmospheric pressure to take the advantage of the swelling and plasticization of the polymer, which reduces the glass transition temperature (T_g) or melting temperature (T_m) due to its high solubility in polymers [4]. The chemical structure of polymers greatly impacts CO_2 solubility and diffusivity. The CO_2-philic groups such as carbonyl or ether groups in the backbone or on side chains of a polymer interact with CO_2 and help with the dissolution of CO_2 within the polymer [5, 6]. Gas foaming is one of the most generally used techniques for the fabrication of porous polymers and the use of supercritical CO_2 for generating porous polymeric foams has generated significant interest over the years [7–10]. The bubble growth data are important for understanding the foam growth. Recently many researchers investigated the effect of pressure drop rate (PDR) on the foam morphology, density, and cell size distribution in nanocellular foams [11–13]. It is important to note that an increase of the PDR increases the rate of stable nuclei formation, at the same time reducing the chances for the foaming agent to fill the freshly formed bubbles [13]. Bubble formation and foaming are caused by CO_2, earlier dissolved in the polymer under pressure, to be released upon depressurization. This has been well reported in the literature and a number of models for bubble growth have been presented by many authors [14–18]. The bubble growth data are important for understanding the foam growth. Foam modelling can by divided into three stages: bubble nucleation, bubble growth, and possibly coarsening. Bubble growth is the most thoroughly studied of the three. Many models have been proposed for radial growth of a single bubble in an infinite expanse of melt, and most of the models are similar. Viscosity, diffusivity, and surface tension are potentially significant to foaming. All three factors directly affect the growth of gas bubbles and therefore will have an impact on the bubble-size and ultimate properties of the foamed product [16].

In this paper, the bubble radius of polystyrene/CO_2 is measured in situ using a specially constructed high-pressure view cell with 2 optical windows, connected to a digital

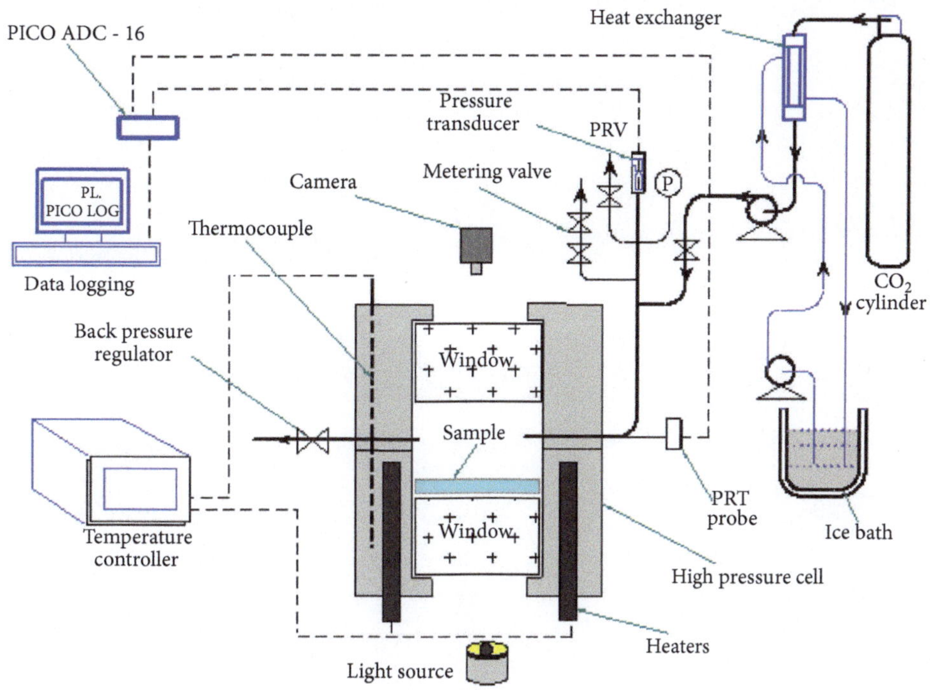

FIGURE 1: Experimental set-up showing connections to the viewing cell.

still camera with a high magnification lens to monitor and record the bubbles growing, to compare the results against bubble growth models. The experimental data are analysed by measuring bubble radii at different times using image analysis software. In the present work, three foam models were applied along with foaming experimental data. Model C was mainly applied to assess the variation of viscosity and diffusivity model with the WLF equation.

2. Methods and Materials

The polystyrene (Mw 184,000 g/mol, T_g 102.6°C) sheets with a thickness of 1 mm were obtained from Polybron Ltd., UK. The disc shaped test samples with a diameter of 20 mm are cut from the sheet using specimen cutting machine Charly4U. The pressure view cell is specially designed to investigate polystyrene, PS foaming, and bubble growth. The cylindrical cell was constructed from 316 stainless steel (with dimensions of 95 mm height and 65 mm diameter) with two window ports incorporated so that the sample can be continuously viewed through one window whilst using the second window for illumination (Figure 1).

The cell has high-pressure connections to work as inlet and outlet ports for CO_2 transfer. Each window has two O-rings (silicon ring can work up 200°C), one in the face and one for radial sealing. The cell was made in two sections and can be opened from the center. Around the sample there are three heating cartridges and a thermocouple fitted in the wall of the cell. With this cell it is also possible to control and record all pressure and temperature data during an experiment and at the same time obtain digital pictures of the experiments progress. The heaters are heated up to a temperature of 600°C.

The experimental procedure was performed according to the following method:

(a) The polymer sample was placed in the cell; then the cell was sealed.

(b) Carbon dioxide was introduced into the cell from a supply cylinder, using a pump where necessary to reach the desired pressure.

(c) Isothermal tests were performed at 70 and 100°C with different pressure range (24, 44, 63, 83, 100, and 120 bar) for 2 hours; each test was done in a separate experiment.

(d) Heating was then stopped and the cell depressurized over the course of 3 minutes.

(e) The bubble nucleation and growth were recorded by the optical microscope and digital camera (1 frame/ 3 sec).

(f) After the cell had completely cooled, the samples were removed.

(g) The bubble radius was then analysed by image analysis software (Image-J).

3. Results and Analyses

3.1. Effect of Conditioning Time. Experiments were performed on disk samples (1 mm) and diameter (20 mm) with

FIGURE 2: Photographs of samples of PS depressurized after holding for different times (30 to 180 min) at 100°C and 54 bar (1 mm thickness × 20 mm diameter).

different conditioning times of 30, 60, 120, 160, and 180 minutes with the same pressure (54 bar) and temperature (100°C), and the corresponding photographs of the depressurized samples were showed in Figure 2. This was in order to assess the time required for CO_2 to satisfactorily equilibrate in the sample prior to depressurization. As the conditioning time is increased up to 120 minutes, it can be seen that the radius of the foamed sample increases and there is less interstitial space between the bubbles. Both indicate a greater bubble volume as the conditioning time is increased. This must be caused by more CO_2 diffusing into the sample at the longer conditioning times indicating that the 30 minutes and 60 minutes times are insufficient to allow equilibration of CO_2 in the sample to occur. Sample conditioned at 120 min achieved maximum bubble foaming. It can also be seen that the 160- and 180-minute sample grew almost to the same size with notable color change due to degradation compared to the 120-minute sample. Therefore, it was decided that the 120-minute conditioning time was sufficient for our experiments.

3.2. Effect of Impingement. During the analysis of the bubble growth data, it was found that bubbles in the samples grow at different rates according to the proximity of other bubbles. One sample has been selected as an example, 24 bar and 100°C, to illustrate this effect. The following photo has been taken after 600 seconds at 0 bar (Figure 3) and shows four bubbles, one of which is "free" and the others are touching each other.

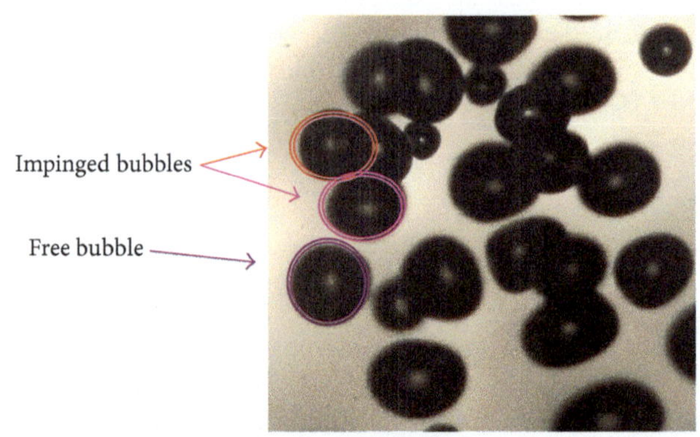

FIGURE 3: Photograph of a PS (1 mm thick × 20 mm diameter) sample 600 sec after the onset of depressurization after holding at 24 bar (CO_2 pressure) and 100°C for 2 hours, the photograph shows the location of three impinged bubbles and one free bubble that are considered here (scale of images is 6 × 6 mm).

3.3. Conditions Effect of Pressure and Temperature. Three main operating temperatures were selected (50, 70, and 100°C) each over a range of pressures from 24 to 120 bar. For the 100°C experiments, the polymer will be in the

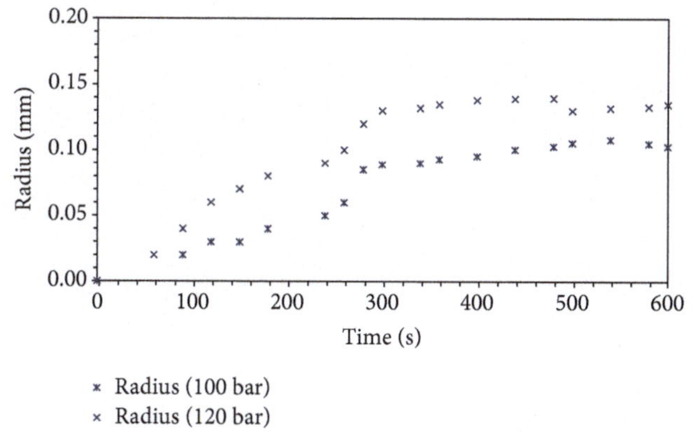

FIGURE 4: Bubble radius of PS/CO_2 during depressurization after holding at 100 and 120 bar at 50°C for 2 hours.

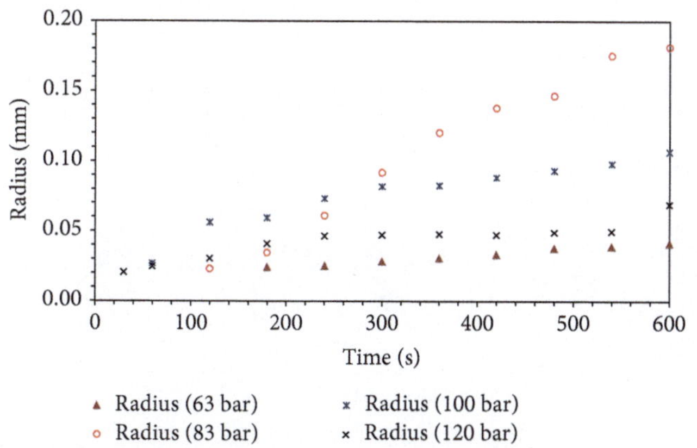

FIGURE 5: Bubble radius of PS/CO_2 during depressurization after holding at 63 to 120 bar at 70°C for 2 hours.

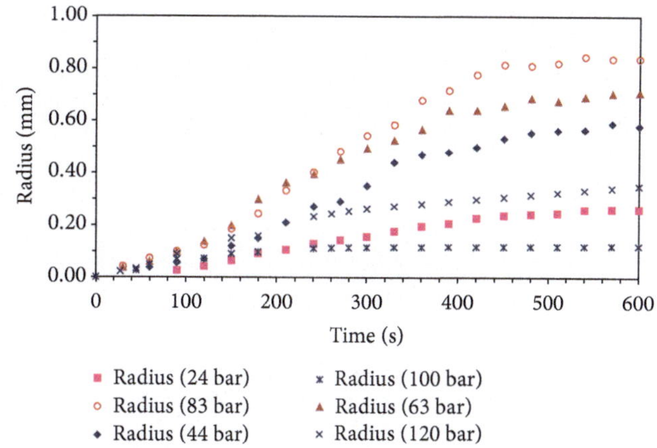

FIGURE 6: Bubble radius of PS/CO_2 during depressurization after holding at 24 to 120 bar at 100°C for 2 hours.

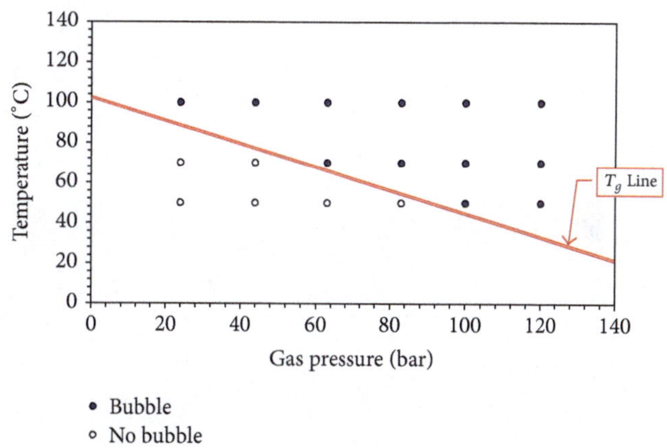

FIGURE 7: Incident of bubble formation and T_g line.

rubbery state at the onset of depressurization and much of the subsequent time period. However, most of the 50°C experiments (below 100 bar) and 70°C experiments (below 63 bar) will be with the polymer entirely within the glassy state.

Depressurization experiments at 50°C and 70°C up at low pressure show no sign of any bubble formation. These correspond to conditions below T_g of PS. However, at higher pressures for 50°C and 70°C bubbles start to appear after 30 sec, with maximum radius of 0.02–0.18 as the radius calculation shown in Figures 4 and 5. The third set of experiments was performed at a higher temperature (100°C) (above T_g of PS) up to 120 bar; see Figure 6. These conditions coincided with those for which literature data for diffusivity and Henry's law constants were available. These experiments were carried out in the view cell and the photographs of specimens during different stages of the experiment were analysed using image analysis software to provide data for bubble radius versus time.

These measurements were performed at a higher temperature condition of 100°C and pressure ranges of 24 to 120 bar. Thus, all samples are initially in the rubbery state (above T_g). Bubbles appeared in all the experiments without exception. The nucleation appeared to occur randomly leading to subsequent bubble growth from these sites. With time, more bubbles nucleate and grow, and the bubbles initially appear to be circular. As more bubbles form and grow in close proximity to each other, their shape is influenced by the impingement of surrounding neighbours leading to polygon shaped bubbles towards the end, with each bubble separated from the neighbouring bubble by a thin film of polymer.

3.4. Incidence of Bubble Formation. At low temperature, there were no signs of bubble formation which is corresponding to the glassy state of PS. However, at high pressures and temperatures above T_g, nucleation occurred. Figure 7 illustrates these incidents of bubble formation for all the operating conditions. The results also compared with the reported values of plasticization of PS by high-pressure CO_2 [20]. It can be clearly seen that bubbles form only when the polymer has been in the rubbery state.

3.5. Possible Influence of the Glass Transition on Bubble Growth. It has been noted from the previous investigation in this work that T_g has the biggest influence on bubble growth. In Figure 8 (at the experimental condition of 24 bar

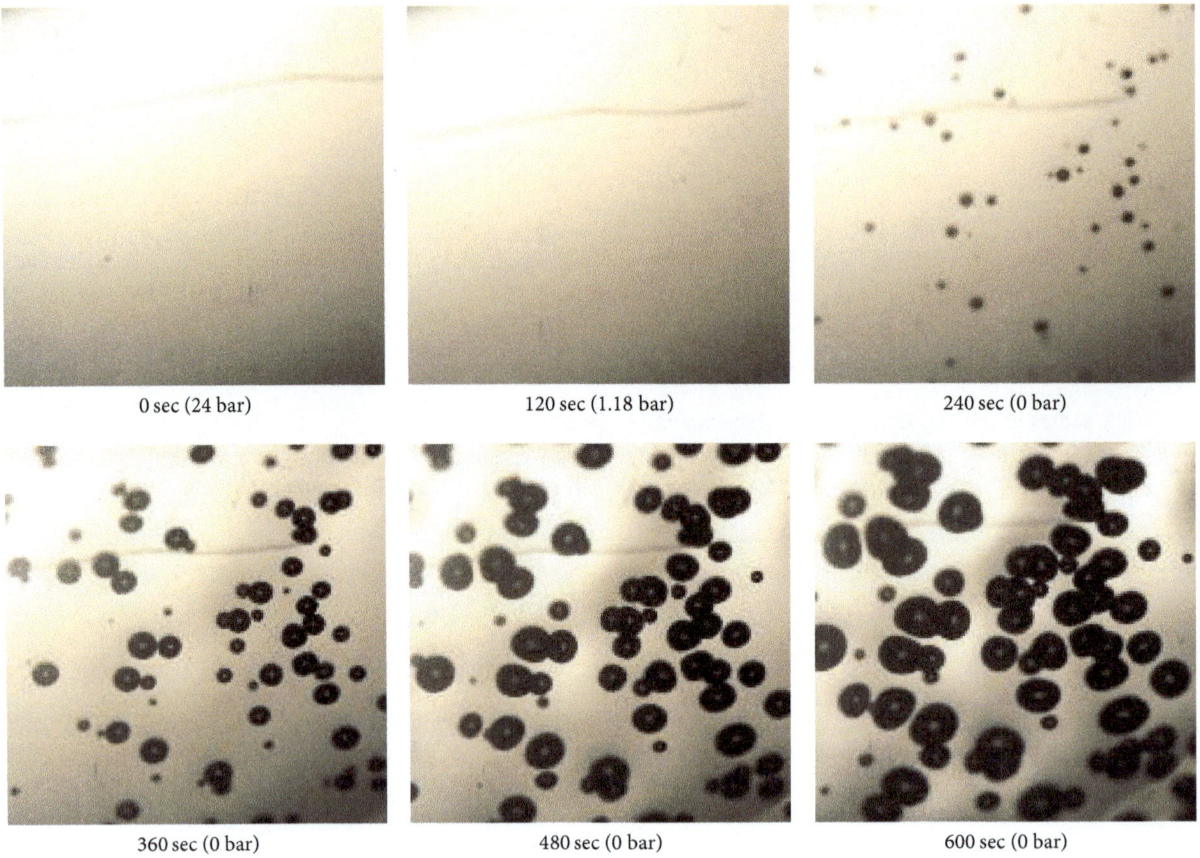

FIGURE 8: Photographs of a 1 mm thick PS sample at various times after the onset of depressurization after holding at 24 bar (in CO_2) and 100°C for 2 hours (scale of images is 6 × 6 mm).

and 100°C), it can be seen that the bubbles grow very freely without impinging but that the growth stops after 480 seconds before impingement occurred, which indicate that the growth here is not limited by impingement. Another experiment also showed that the sorption and diffusion of CO_2 cannot be used as a predictor of bubble growth. At the experimental conditions at 83 bar and the two different temperatures of 70°C (Figure 9) and 100°C (Figure 10), it was found that the bubble diameter at 70°C is much smaller than at 100°C. As the temperature rises, the viscosity falls which aids foaming. When CO_2 diffuses into a polymer, the polymer swells and the density of the polymer/CO_2 system reduces. Generally diffusion of carbon dioxide results in a strong plasticization effect for a number of polymers. It is also obvious that bubbles do not expand freely such that the CO_2 bubble pressure equilibrates with the cell pressure as this would result in much larger foam volumes than what are observed. It is thus likely that bubble growth reduces as the polymer (at some point) passes from the rubbery to the glassy state.

3.6. Pressure Depressurization Curves. Figure 11 illustrates the depressurization curves for all the experiments performed at 100°C. All the experiments reached 0 bar before 300 sec of depressurization. With low pressure (24 to 44 bar), the time to reach 0 bar was between 175 and 200 sec and for the high pressure (85 to 120 bar) it was around 275 to 300 sec. This is because the low pressure experiments have less gas to evacuate from the cell. We can see from all the previous Figures 4–6 that the bubbles keep growing even when the cell pressure has reached 0 bar, and some bubbles start to appear even after this time in the interstices between the bubbles. This trend happens at all pressures but it can be seen more clearly with the higher pressures. This shows that the CO_2 in the polymer, in the bubbles, and in the surrounding gas is not in equilibrium. In 100°C experiments the temperature is above the glass transition; therefore, the polymer is in the rubbery phase and the bubbles can grow very easily which made the bubbles very big compared with the earlier investigations in this work.

3.7. Bubble Growth Modelling

3.7.1. Model Framework. This model is to be used to fit to experimental data of bubble growth. The general model framework uses a finite difference technique as illustrated in Figure 12. The bubble, assumed to contain pure CO_2, is surrounded by a polymer sphere which is divided into 20 concentric shells of equal volume.

The polymer density is assumed to be constant, and hence the volumes associated with each shell remain constant. Therefore, knowledge of the current bubble radius at any time allows all the nodal radii to be evaluated.

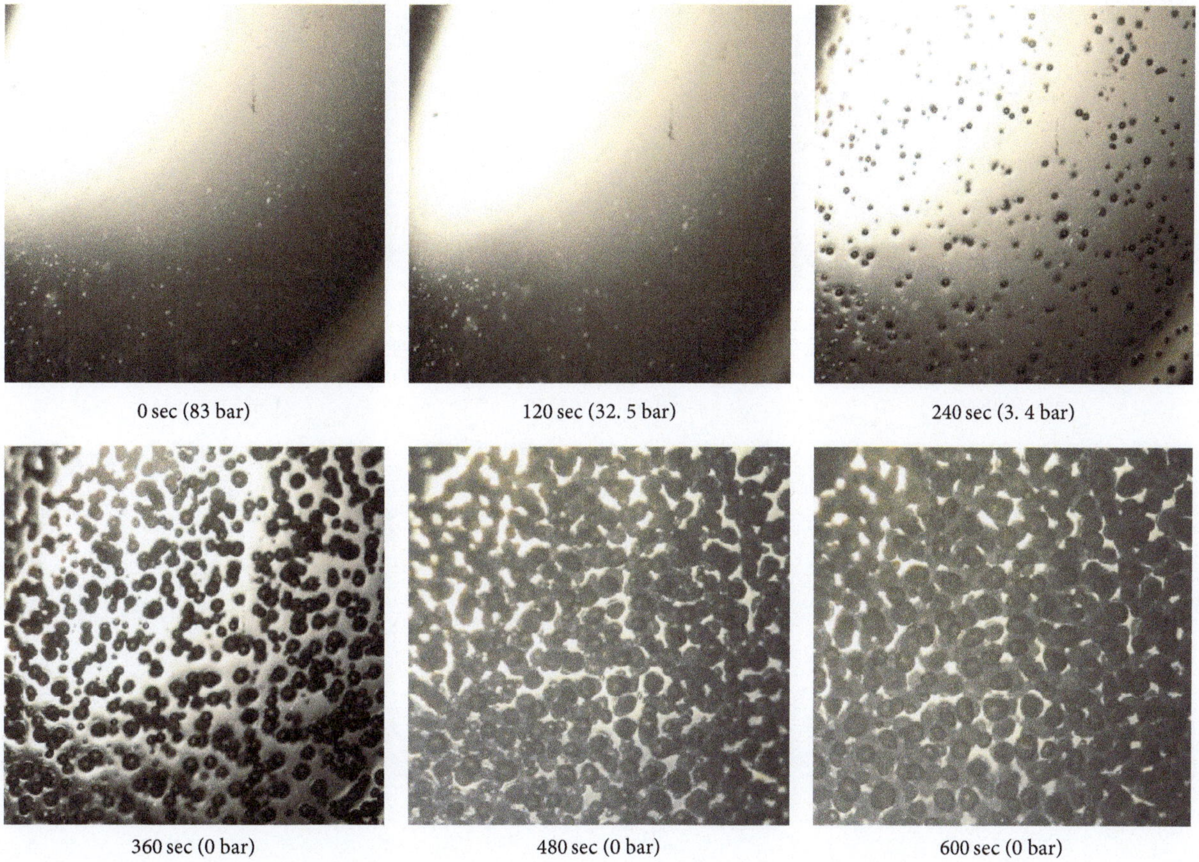

FIGURE 9: Photographs of a 1 mm thick PS sample at various times after the onset of depressurization after holding at 83 bar (in CO_2) and 70°C for 2 hours (scale of images is 6 × 6 mm).

$$\frac{4}{3}\pi r_a^3 + fV_{\text{polymer}} = \frac{4}{3}\pi r^3, \tag{1}$$

where V_{polymer} is the volume of polymer, f is the fraction of polymer volume inside the nodal radius, r_a is radius at node "a," and r is the radius at the node.

3.7.2. Diffusion Equations. Different equations are required to model concentrations at the bubble interface (node "a") and exterior (node "u") and within the polymer (nodes "b" to "t").

Main Body (Nodes "b" to "t"). Here node "q" is used as an example node for explanatory purposes with adjacent nodes "r" (outside) and "p" (inside). Equations for other nodes are essentially identical.

The volume associated with node "q" is a shell with outer radius $(1/2)(r_q + r_r)$ and inner radius $(1/2)(r_q + r_p)$.

$$\text{Flux In} = D_{qr}\frac{(c_r - c_q)}{(r_r - r_q)} \quad \text{(From node "}r\text{")},$$
$$\text{Area} = 4\pi\left(\frac{r_q + r_r}{2}\right)^2 \quad \text{of "}q, r\text{" boundary}, \tag{2}$$

$$\text{Flux Out} = D_{qp}\frac{(c_q - c_p)}{(r_q - r_p)} \quad \text{(to node "}p\text{")},$$
$$\text{Area} = 4\pi\left(\frac{r_q + r_p}{2}\right)^2 \quad \text{of "}q, p\text{" boundary}, \tag{3}$$

$$\text{Accumulation} = \frac{dc_q}{dt} \times \text{Volume}, \tag{4}$$

$$\text{Volume} = \frac{4}{3}\pi\left(\frac{r_q + r_r}{2}\right)^3 - \frac{4}{3}\pi\left(\frac{r_q + r_p}{2}\right)^3, \tag{5}$$

$$\text{Accumulation} = \text{Flux in} \times \text{Area} - \text{Flux out} \times \text{Area},$$

$$\frac{dc_q}{dt}\frac{4}{3}\pi\left[\left(\frac{r_q + r_r}{2}\right)^3 - \left(\frac{r_q + r_p}{2}\right)^3\right] = 4\pi\left(\frac{r_q + r_r}{2}\right)^2 D_{qr}\frac{(c_r - c_q)}{(r_r - r_q)} - 4\pi\left(\frac{r_q + r_p}{2}\right)^2 D_{qp}\frac{(c_q - c_p)}{(r_q - r_p)}. \tag{6}$$

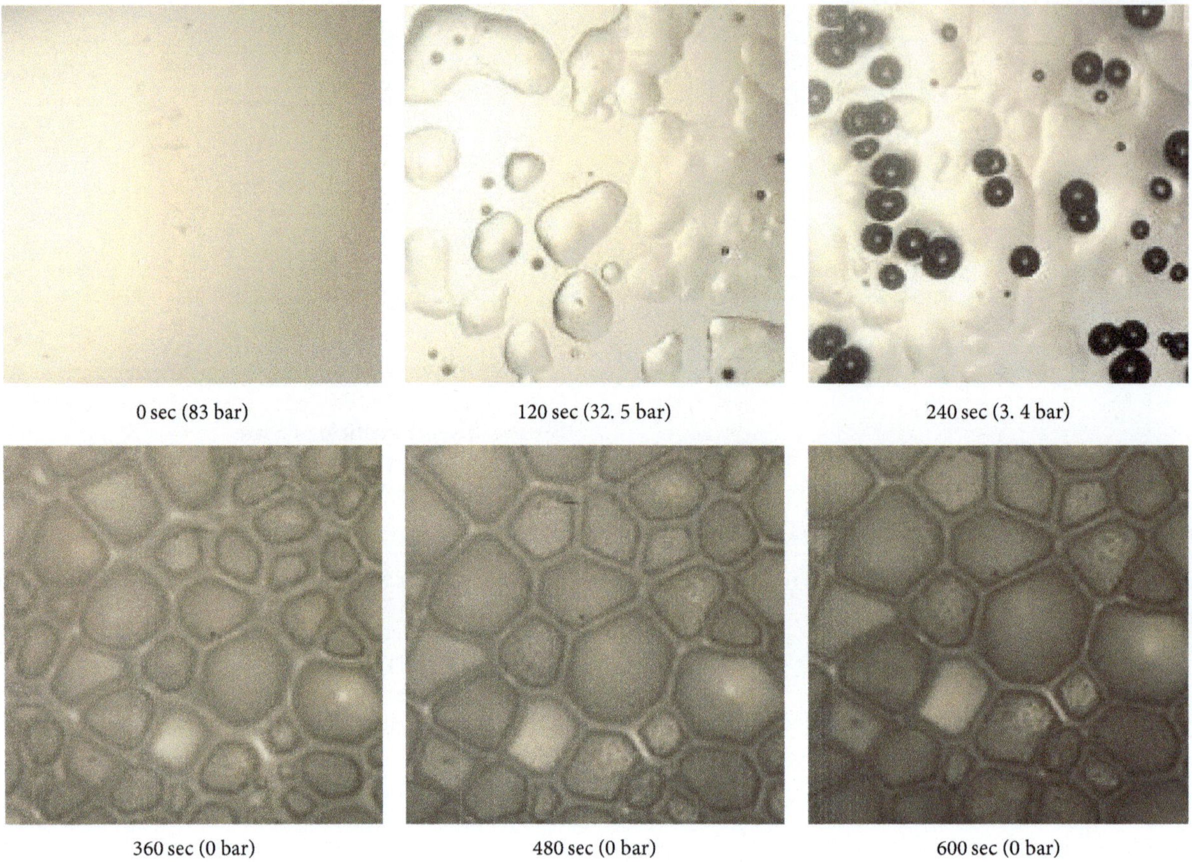

Figure 10: Photographs of a 1 mm thick PS sample at various times after the onset of depressurization after holding at 83 bar (in CO_2) and 100°C for 2 hours (scale of images is 6 × 6 mm).

Cancelling π,

$$\frac{dc_q}{dt}\frac{1}{6}\left[\left(r_q+r_r\right)^3-\left(r_q+r_p\right)^3\right] = \left(r_q+r_r\right)^2 D_{qr}\left(\frac{c_r-c_q}{r_r-r_q}\right) - \left(r_q+r_p\right)^2 D_{qp}\left(\frac{c_q-c_p}{r_q-r_p}\right). \quad (7)$$

Exterior Boundary (No Flux): Node "u." One has

$$\text{Flux Out} = D_{tu}\frac{(c_u-c_t)}{(r_u-r_t)} \quad (\text{to node "}t\text{"}),$$

$$\text{Area} = 4\pi\left(\frac{r_t+r_u}{2}\right)^2 \quad \text{of "}q, r\text{" boundary},$$

$$\text{Accumulation} = \frac{dc_u}{dt}\times\text{Volume},$$

$$\text{Volume} = \frac{4}{3}\pi\left(r_u\right)^3-\frac{4}{3}\pi\left(\frac{r_t+r_u}{2}\right)^3,$$

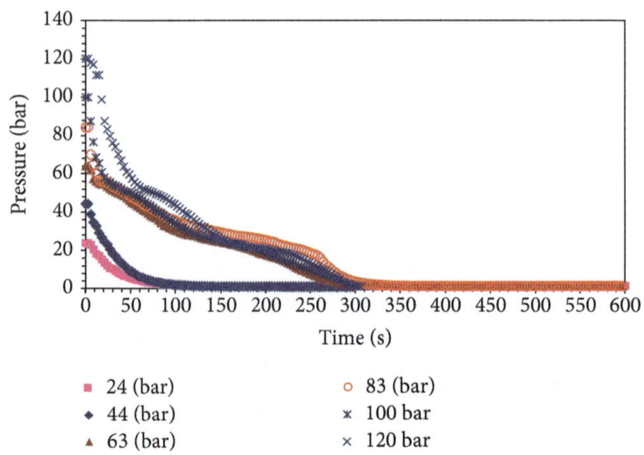

Figure 11: Pressure versus time plot for the depressurization experiments at 100°C.

$$\text{Accumulation} = -\text{Flux Out}\times\text{Area},$$

$$\frac{dc_u}{dt}\left[\frac{4}{3}\pi\left(r_u\right)^3-\frac{4}{3}\pi\left(\frac{r_t+r_u}{2}\right)^3\right] = -D_{tu}\frac{(c_u-c_t)}{(r_u-r_t)}4\pi\left(\frac{r_t+r_u}{2}\right)^2. \quad (8)$$

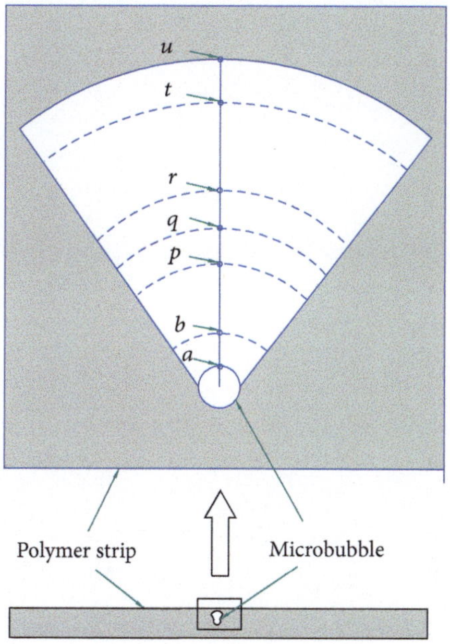

FIGURE 12: General bubble growth model sketch.

Cancelling π,

$$\frac{dc_u}{dt}\frac{1}{3}\left[(2r_u)^3 - (r_t + r_u)^3\right] = -D_{tu}\left(\frac{c_u - c_t}{r_u - r_t}\right)(r_t + r_u)^2. \quad (9)$$

Bubble Surface (Node "a"): Equilibrium Corresponding to Bubble Pressure. The bubble pressure is in equilibrium with the concentration of CO_2 in the polymer at the bubble surface (represented by the concentration at node "a").

$$P = Hc_A, \quad P = \rho\mathfrak{R}T \Longrightarrow \quad Hc_A = \rho\mathfrak{R}T. \quad (10)$$

As P and c_A are linked, the mass balance relating to node "a" should also include the CO_2 in the bubble as well as that in node "a."

The amount (moles, M) of CO_2 associated with node "a" and the bubble is

$$M = V_{\text{node}(a)} \times c_A + V_{\text{bubble}} \times \rho. \quad (11)$$

Hence, the accumulation term for the rate of change of M is

$$\frac{dM}{dt} = V_{\text{node}(a)}\frac{dc_A}{dt} + \rho\frac{dV_{\text{bubble}}}{dt} + V_{\text{bubble}}\frac{d\rho}{dt}. \quad (12)$$

This is equal to the rate at which CO_2 is transported cross the boundary between nodes "a" and "b."

$$\text{Flux} \times \text{Area} = \frac{dM}{dt}, \quad (13)$$

where

$$\text{Flux Into Bubble} = D_{ab}\frac{(c_b - c_a)}{(r_b - r_a)}, \quad \text{Area} = 4\pi\left(\frac{r_a + r_b}{2}\right)^2, \quad \frac{dM}{dt} = D_{ab}\frac{(c_b - c_a)}{(r_b - r_a)}4\pi\left(\frac{r_a + r_b}{2}\right)^2. \quad (14)$$

The molar density of CO_2 in the bubble is given by an Equation of State. At low pressure one can use the Ideal Gas Law $\rho = P/\mathfrak{R}T$, but in general we will use the Soave Redlich Kwong (SRK) Equation of State

$$P = \frac{\mathfrak{R}T}{V - b} - \frac{a}{V(V + b)}. \quad (15)$$

For CO_2, $a = 270485.7$, $b = 0.029683$, $\mathfrak{R} = 8314$ J/kmol-K, P (Pa), T (K), v (m^3/kmol) [21].

An iterative equation for calculating the molar volume for given values of pressure and temperature can be defined as follows:

$$V = b + \frac{\mathfrak{R}T}{P + a/V(V + b)}. \quad (16)$$

The value of molar volume can then be converted into a molar density value.

(a) Model A: Constant Diffusivity, Zero Viscosity. The simplest model (Model A) assumes that bubble growth is only limited by diffusion of gas into the bubble and is not limited by the viscosity of the polymer. The pressure of CO_2 inside the bubble is assumed to be the same as the exterior pressure.

If the exterior (and bubble) pressure varies with time then (12) is used; that is,

$$D_{ab}\frac{(c_b - c_a)}{(r_b - r_a)}4\pi\left(\frac{r_a + r_b}{2}\right)^2 = \left[\frac{4}{3}\pi\left(\frac{r_a + r_b}{2}\right)^3 - \frac{4}{3}\pi r_a^3\right]\frac{dc_a}{dt} + \rho 4\pi r_a^2\frac{dr_a}{dt} + \frac{4}{3}\pi r_a^3\frac{d\rho}{dt} \Longrightarrow$$

$$dr_a = \frac{1}{\rho r_a^2}\left[D_{ab}\frac{(c_b - c_a)}{(r_b - r_a)}\left(\frac{r_a + r_b}{2}\right)^2 dt - \left(\frac{1}{3}\left(\frac{r_a + r_b}{2}\right)^3 - \frac{r_a^3}{3}\right)dc_a - \frac{1}{3}r_a d\rho\right]. \quad (17)$$

(b) Model B: Constant Diffusivity and Viscosity. This model now assumes that the polymer viscosity has an effect on the bubble growth. Diffusion of gas into the bubble now leads to an increase of gas pressure in the bubble; the pressure difference between the inside and outside of bubble then leads to the inflation of the bubble.

This model will first consider the case of constant diffusivity and constant viscosity. First of all, an expression relating the rate of bubble growth to the pressure difference is derived.

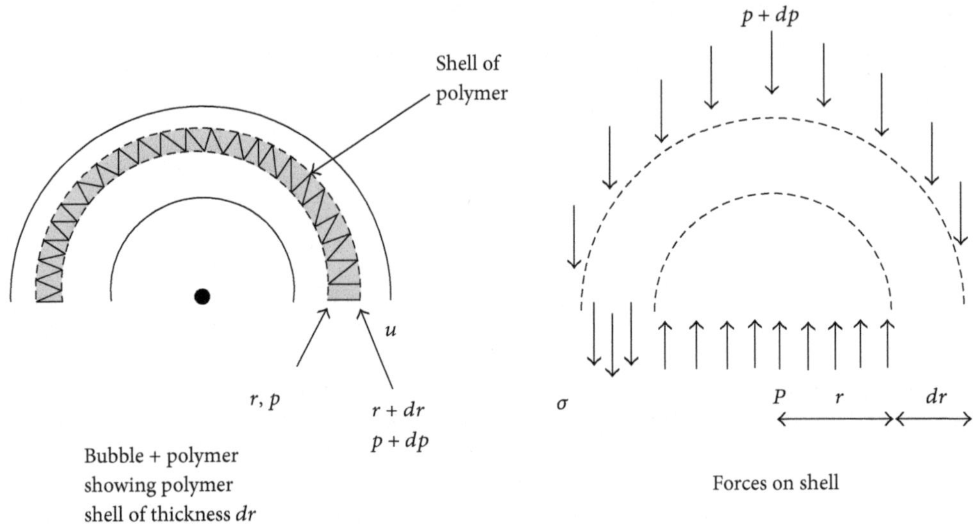

FIGURE 13: Force balance diagram.

A surface tension term is also included to model the pressure drop at the gas/polymer interface due to surface tension. Force balance on a spherical shell element of thickness dr (see Figure 13) is

Inside pressure force

= extensional force in polymer element

+ outside pressure force,

$$P\pi r^2 = \sigma 2\pi r dr + (p + dp)\pi (r + dr)^2, \quad (18)$$

$$Pr^2 = \sigma 2rdr + (p + dp)(r + dr)^2,$$

$$Pr^2 = \sigma 2rdr + (p + dp)(r^2 + 2rdr + dr^2),$$

$$Pr^2 = 2\sigma rdr + pr^2 + 2prdr + r^2 dp + 2rdrdp.$$

If $p \ll \sigma$, then $2prdr$ is a negligible term:

$$2\sigma rdr = -r^2 dp \Longrightarrow$$
$$2\sigma dr = -rdp. \quad (19)$$

Extensional stress σ is related to extensional strain rate ε' by

$$\sigma = \mu_{\text{ext}}\varepsilon', \quad \left(\varepsilon' = \frac{\dot{r}}{r}\right), \quad (20)$$

where μ_{ext} is the extensional viscosity.

Therefore,

$$2\mu_{\text{ext}}\frac{\dot{r}}{r}dr = -rdp. \quad (21)$$

However, $\dot{r}$ varies with r. Consider a bubble of radius R growing at $\dot{R}$ and a point at radius r moving at $\dot{r}$. The volume between r and R remains constant (see Figure 22).

$$\frac{4}{3}\pi r^3 = \frac{4}{3}\pi R^3 + c. \quad (22)$$

Differentiate with respect to t

$$4\pi r^2\frac{dr}{dt} = 4\pi R^2\frac{dR}{dt} + 0 \Longrightarrow$$
$$\dot{r} = \frac{R^2}{r^2}\dot{R}. \quad (23)$$

Insert (23) into (21):

$$2\mu_{\text{ext}}\frac{R^2}{r^2}\dot{R}\frac{1}{r}dr = -rdp \quad (24)$$

$$2\mu_{\text{ext}}\frac{R^2\dot{R}}{r^4}dr = -dp. \quad (25)$$

If μ_{ext} is a constant, we can integrate

$$2\mu_{\text{ext}}R^2\dot{R}\int_R^{R_{\text{ext}}}\frac{1}{r^4}dr = \int_{p_b}^{p_{\text{ext}}} -dp, \quad (26)$$

where R_{ext} is the external radius of the polymer sphere

$$2\mu_{\text{ext}}R^2\dot{R}\left[-\frac{1}{3}r^{-3}\right]_R^{R_{\text{ext}}} = \frac{2\mu_{\text{ext}}R^2\dot{R}}{3}\left[\frac{1}{R^3} - \frac{1}{R_{\text{ext}}^3}\right]$$
$$= p_b - p_{\text{ext}}. \quad (27)$$

Now $\mu_{\text{ext}} = 6\mu$, where μ is shear viscosity [17]

$$4\mu R^2\dot{R}\left[\frac{1}{R^3} - \frac{1}{R_{\text{ext}}^3}\right] = p_b - p_{\text{ext}}. \quad (28)$$

If surface tension forces are significant, then an extra term can be added to reflect this:

$$p_b - p_{\text{ext}} = 4\mu R^2\dot{R}\left[\frac{1}{R^3} - \frac{1}{R_{\text{ext}}^3}\right] + \frac{2\gamma}{R}, \quad (29)$$

where γ is the surface tension [22].

This final equation can be used to calculate the bubble growth rate $\dot{R}$, based on the pressure difference inside the bubble and outside the polymer.

$$\dot{R} = \frac{p_b - p_{\text{ext}} - (2\gamma/R)}{4\mu R^2 [1/R^3 - 1/R_{\text{ext}}^3]}. \quad (30)$$

In some studies, R_{ext} is considered large compared to R and thus the $1/R_{\text{ext}}^3$ term is regarded as negligible.

The bubble pressure is calculated based on the current size of the bubble and the number of nodes in the bubble and node a.

$$4\pi D_{AB} \frac{(c_b - c_a)}{(r_b - r_a)} \left(\frac{r_a + r_b}{2}\right)^2 = \frac{dM}{dt}, \quad (31)$$

$$(M = c_a V_a + \rho V_{\text{bub}}),$$

where V_{bub} is volume of the bubble

Then based on current values for V_{bub} (or R) and M, one can calculate ρ and c_A and p_B from

$$M = c_a V_a + \rho V_{\text{bub}} = \frac{V_a p}{H} + \frac{V_{\text{bub}}}{v}, \quad (32)$$

where

$$p_B = \left(\frac{\Re T}{V - b} - \frac{a}{V(V + b)}\right) \quad \text{(Equation of State)} \quad (33)$$

and v is the mole volume and p_B is the pressure of the bubble.

That is,

$$M = \frac{V_a}{H}\left(\frac{\Re T}{V - b} - \frac{a}{V(V + b)}\right) + \frac{V_{\text{bub}}}{V}. \quad (34)$$

Therefore, the correct value for v can be found numerically from ρ and c_a.

(c) Model C: Variable Diffusivity and Viscosity. The same general equation for bubble growth can be used as before (25).

$$2\mu_{\text{ext}} \frac{R^2 \dot{R}}{r^4} dr = -dp. \quad (35)$$

However, we now consider the case that viscosity is not a constant.

$$-\int_{p_b}^{p_{\text{ext}}} dp = 2R^2 \dot{R} \int_R^{R_{\text{ext}}} \frac{\mu_{\text{ext}}}{r^4} dr = 12R^2 \dot{R} \int_R^{R_{\text{ext}}} \frac{\mu}{r^4} dr. \quad (36)$$

In our analysis, we will consider viscosity and diffusivity to vary according to the Williams–Landel–Ferry (WLF) equation, where the local concentration will affect the local glass transition temperature. A similar approach was adopted by Goel and Bechman [8] for modelling viscosity (but not diffusivity) and by Chen et al. [16] who varied both the viscosity and diffusion according to the WLF equation.

The WLF equation is [Williams et al. [23]]

$$\log \frac{\mu}{\mu_o} = \frac{-c_1 (T - T_g)}{c_2 + (T - T_g)},$$

$$\frac{\mu}{\mu_o} = 10^{-c_1 (T - T_g)/(c_2 + (T - T_g))} = F, \quad (37)$$

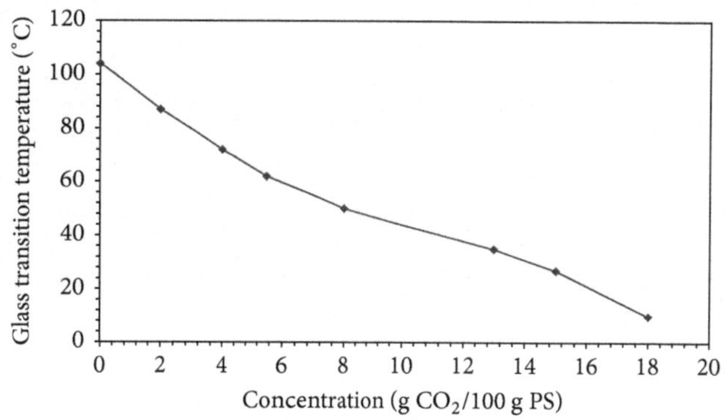

FIGURE 14: Variation of glass transition temperature of PS with the concentration of CO_2.

where $c_1 = 17.44$, $c_2 = 51.6\,\text{k}$, and F is a scaling factor arising from the WLF equation.

$$\text{As } D \sim \frac{1}{\mu} \quad (38)$$

[Bird et al. [24]]

$$\frac{D}{D_o} = 10^{(c_1 (T - T_g)/(c_2 + (T - T_g)))} = \frac{1}{F}. \quad (39)$$

Now we can say $T_g = T_{g0} - \beta c$ [Condo & Johnston [25]].

β is the slope of the plot of T_g versus concentration

$$\left(T_g = T_{g0} - mp, p = Hc \Longrightarrow \beta = mH\right). \quad (40)$$

These concentration values are plotted against temperature (the glass transition temperature) (see Figure 14). It is clear that T_g decreases with CO_2 concentration.

For each concentration, one can calculate

$$F = 10^{(-c_1 (T - T_{g0} + \beta c)/(c_2 + (T - T_{g0} + \beta c)))},$$

$$D = \frac{D_0}{F}, \quad (41)$$

$$\mu = \mu_0 F.$$

For the bubble growth relationship

$$-\int_{p_b}^{p_{\text{ext}}} dp = 2R^2 R' \int_R^{R_{\text{ext}}} \frac{\mu_{\text{ext}}}{r^4} dr = 12R^2 R' \int_R^{R_{\text{ext}}} \frac{\mu}{r^4} dr \quad (42)$$

as $1/r^4$ varies so dramatically with r it is better to change the variable for integration as per Tuladhar and Mackley [22] paper, $z = 1/r^3$ (m^{-3}).

$p_b - p_{\text{ext}} = 4R^2 \dot{R} \int_{z_R}^{z_{\text{ext}}} \mu\, dz = 4R^2 \dot{R} \int_{z_R}^{z_{\text{ext}}} F\, dz$, integrated numerically by trapezium rule

$$\dot{R} = \frac{p_b - p_{\text{ext}} - (2\gamma/R)}{4R^2 \mu_0 \int_{z_R}^{z_{\text{ext}}} F\, dz}. \quad (43)$$

Table 1: Models general parameters values.

Parameters	Values	Source
Initial radius of polymer sphere, $R_{polymer}$ (mm)	1	Assumed
Initial radius of gas bubble, R_{bubble} (mm)	0.01	Measured
Initial pressure, P_i (bar)	24 to 83	Measured
Final pressure, P_f (bar)	1	Measured
Temperature, T (°C)	100	Measured
Density of CO_2 in bubble, ρ (kmol/m^3)	0.032	Calculated
Surface tension, γ (mNm^{-1}), at 1 bar and 200°C	27.7	17
Henry's constant, H (bar), at 70°C	1140.0	Extrapolated
Henry's constant, H (bar), at 100°C	1628.2	30

Table 2: Model B best fit for shear viscosity values and diffusivity values [19].

P_i (bar)	Diffusion coefficient (10^{-10} m^{2}s^{-1})		Best fit for shear viscosity (108 Pa·s)	
	70°C	100°C	70°C	100°C
24	—	0.81	—	0.9
44	—	1.14	—	0.7
63	1.02	1.46	5	0.8
83	1.17	1.67	3.5	0.8

3.8. Discussion of the Model. The set of experiments used to test these models were those with pressure of 24, 44, 63, and 83 bar at a temperature of 70 and 100°C. All these experimental data were logged for 600 sec during depressurization. Each bubble was assumed to be spherical with an initial radius of 0.02 mm and surrounded by a finite volume of polymer. Three bubble growth models were used based on diffusion control only (Model A); constant viscosity and diffusivity (Model B); and varying diffusivity and viscosity according to the WLF equation (Model C). In the models, time zero is taken at the onset of nucleation, which is often sometime after depressurization has commenced and so the starting cell pressure for the model is usually lower than the equilibration pressure (which determines the initial concentration of CO_2 in the polymer). Most of the parameter values such as concentration (c_o), density (ρ), surface tension (γ), viscosity (μ), diffusivity (D), and temperature (T) are either provided or calculated. The general parameter values used in models are shown in Table 1. Each pressure has different diffusivity and shear viscosity as shown in Table 2.

Figures 15 and 16 compare Model A predications with the experimental data of 70 and 100°C for different pressure (24 to 83 bar). The model fit, however, significantly overpredicts bubble growth rates. As we know, Model A assumes that the growth is controlled by diffusion only which lets the growth happen very quickly without any viscous resistance. Therefore, Model B was applied to assess whether a viscous model with surface tension can predict bubble the growth with better accuracy. As we can see in Figures 17 and 18, Model B is better than Model A. This is because this model controls the growth by including viscosity and surface tension, but still the model fit is not good as it predicts a continuously increasing rate of bubble growth. Model C was then applied to determine whether a better prediction could be achieved using a varying viscosity and diffusivity model based on WLF equation.

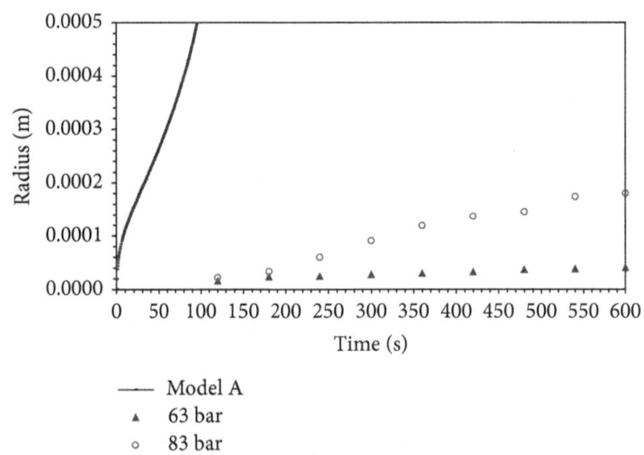

Figure 15: Model A versus experimental data for bubble growth at 70°C.

It is known that variable diffusivity and viscosity model (Model C) can provide realistic estimates of parameters which suggest that diffusion is major influence aspect of the softening of these polymer samples. Model C shows a very reasonable fit with our data by using realistic parameter values; see Table 3. Figures 19 and 20 compare Model C prediction with the experimental data of 70 and 100°C and different pressures (24, 44, 63, and 83 bar), where $c_1 = 0.95$, $c_2 = 32.1$ k, Sato et al. [19] and Nikitin et al. [26]. Figure 21 also depicts Model C experimental data for bubble growth at 100°C using WLF equation. This model shows a very reasonable fit with our data by using realistic parameter values and compares Model C prediction with the experimental

TABLE 3: Model C best fit diffusion coefficient at 100°C.

Pressure (bar)	Concentration of CO_2 (kmol/m^3)	Diffusion coefficient: Sato et al. [19] ($\times 10^{-10}$ m^2 /s)	Diffusion coefficient: fit ($\times 10^{-10}$ m^2/s)	Fitting error (error sq)
24	0.37123073	0.81	0.804970412	$2.52968E-05$
44	0.68058968	1.14	1.158677989	0.000348867
63	0.97448068	1.46	1.437437358	0.000509073
83	1.28383962	1.67	1.678843001	$7.81987E-05$

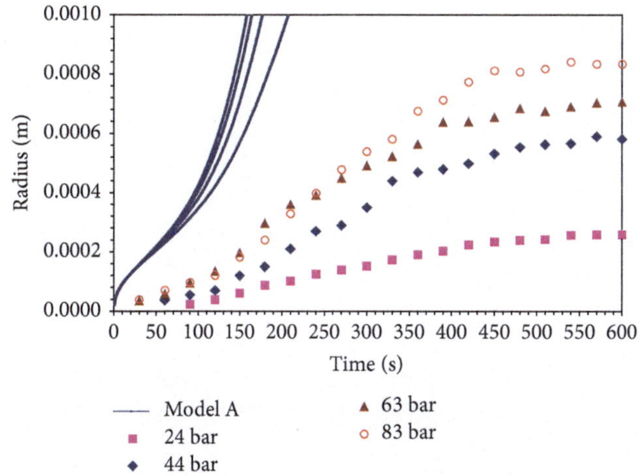

FIGURE 16: Model A versus experimental data for bubble growth at 100°C.

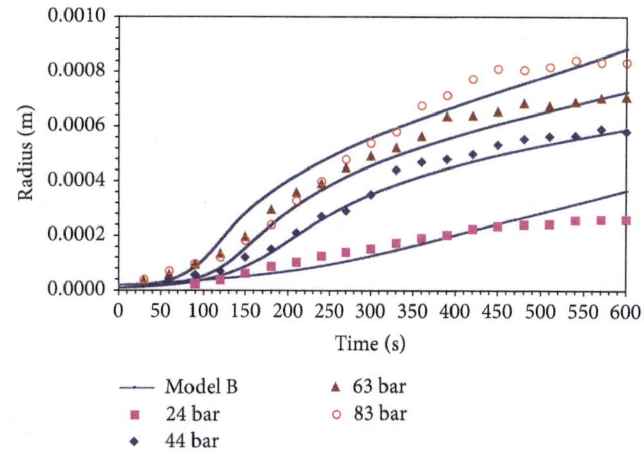

FIGURE 18: Model B versus experimental data for bubble growth at 100°C.

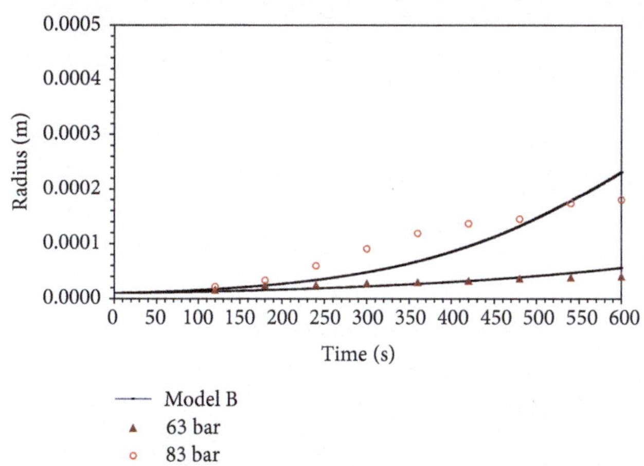

FIGURE 17: Model B versus experimental data for bubble growth at 70°C.

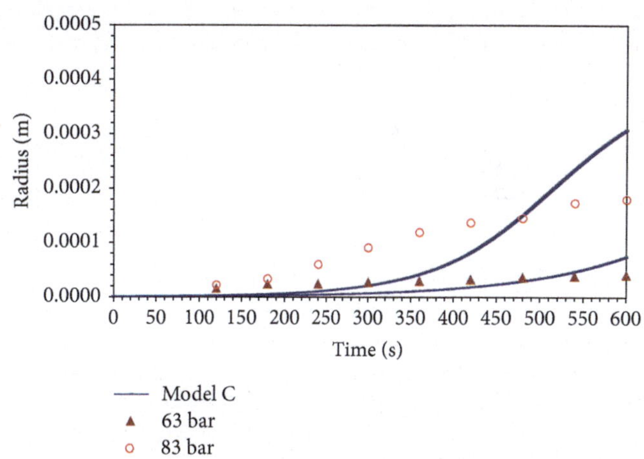

FIGURE 19: Model C versus experimental data for bubble growth at 70°C.

data of 100°C and different pressures. In general, foaming behavior is controlled by both diffusion and viscosity but these coefficients are varied according to the nature of the polymer.

4. Conclusion

Three sets of experiments were carried out at operating temperatures of 50, 70, and 100°C each over a range of pressures from 24 to 120 bar. For the 100°C experiments, the polymer will be in the rubbery state at the onset of depressurization and much of the subsequent time period. However, most of the 50°C experiments (below 100 bar) will be with the polymer entirely within the glassy state. Images during depressurization at 50°C and with initial pressure up to 83 bar and at 70°C at initial pressure up to 44 bar show no sign of any bubble formation. These correspond to conditions below T_g of PS. The third set of experiments was performed at a higher temperature (100°C) (above T_g of PS) up to 120 bar. The nucleation appeared to occur

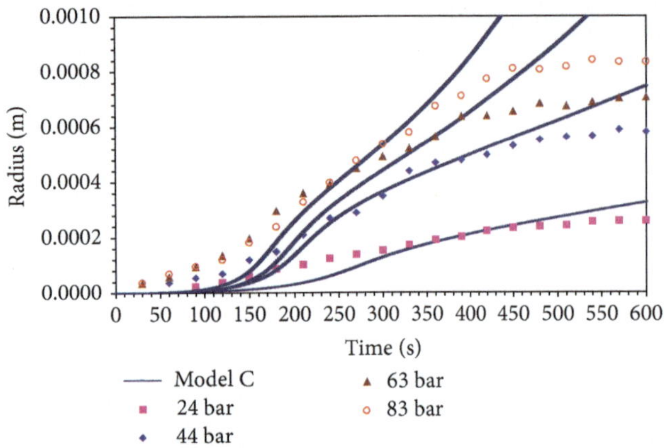

FIGURE 20: Model C versus experimental data for bubble growth at 100°C.

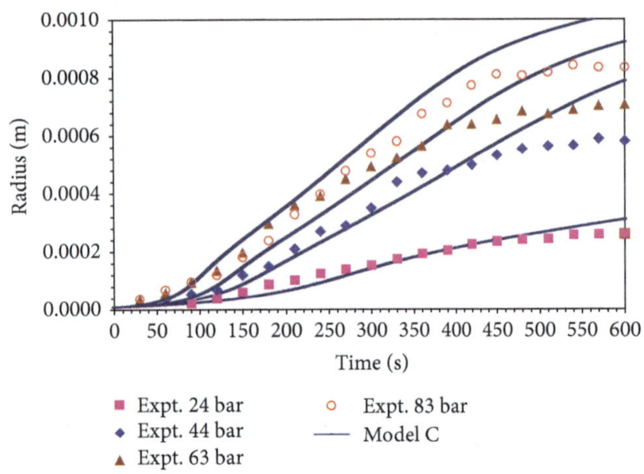

FIGURE 21: Model C versus experimental data for bubble growth at 100°C using WLF equation.

randomly leading to subsequent bubble growth from these sites, with maximum radius of 0.02–0.83 mm. Bubble growth has been theoretically addressed by number of researchers and most of the models are similar. The expansion occurs by diffusion of a dissolved gas from the supersaturated polymer envelope into the bubble. The main parameters that play the potentially significant role in bubble foaming are viscosity, diffusivity, and surface tension. Three models were applied on the foaming experimental data. Model C was applied to assess the variation of viscosity and diffusivity model with the WLF equation. The model shows very good agreement by using realistic parameter values.

Nomenclature

c: Concentration of CO_2 in the slab (g/ml)
D: Diffusion coefficient (m^2/s)
D_o: Coefficient in WLF equation for diffusivity (m^2/s)
Mw: Molecular weight of the polymer (g/mol)
P: Pressure (bar)

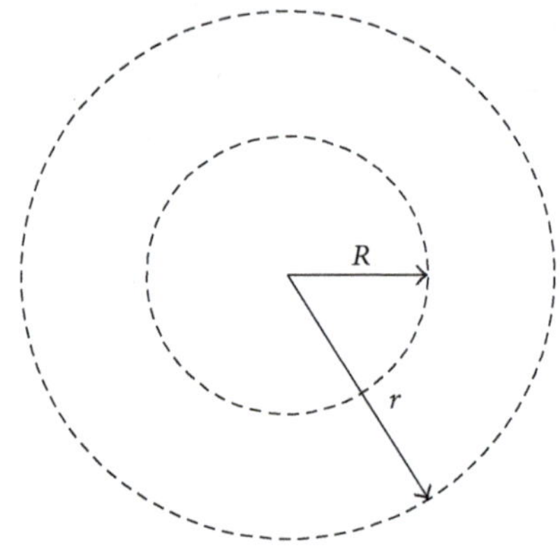

FIGURE 22

P_i: Initial pressure (bar)
P_f: Final pressure (bar)
H: Henry's constant (bar)
m: Mass of the specimen
R_{polyme}: Initial radius of polymer sphere (mm)
R_{bubble}: Initial radius of gas bubble (mm)
$\mathfrak{R}$: Universal gas constant = 8314 J/kmol-K
r: Radius of a representative bubble (m)
t: Time (sec)
ρ: Polymer density (kg/m^3, k mol/m^3)
μ: Viscosity (shear viscosity) (Pa·s)
μ_o: Viscosity in WLF equation for viscosity (Pa·s)
μ_{ext}: Extensional viscosity (Pa·s)
γ: Surface tension (external stress) (N/m)
σ: Extensional stress
ε': Extensional strain rate
$\varepsilon_{\max}$: Strain
β: Side group transition
V_{polymer}: Volume of polymer
f: Fraction of polymer volume inside the nodal radius
T: Temperature (K)
b: Width of the sample strips (m)
c: Concentration of CO_2 at time t and distance x into the slab (g/ml)
c_e: Equilibrium concentration in slab (corresponding to infinite time) (g/ml)
D: Diffusion coefficient (m^2/s)
D_o: Coefficient in WLF equation for diffusivity (m^2/s)
M: Moles of CO_2
Mw: Molecular weight of the polymer (kg/kmol)
Mw_{CO2}: Molecular weight of the carbon dioxide (mol/m^3)
P: Pressure (bar)
P_c: Absolute pressure (Mpa)
P_i: Initial pressure (bar)
P_f: Final pressure, (bar)
$\mathfrak{R}$: Gas constant, (8.3145 J/mol K)
r: Radius of a representative bubble (m)

S: Swelling ratio
t: Time (sec)
T_g: Glass transition temperature (°C)
V_{polymer}: Polymer volume (V_{Polymer}) (m^3)
V_{bub}: Volume of the bubble (m^3)
V_{Rod}: Rod volume (m^3)
V_{Sphere}: Volume of a representative bubble (m^3)
V_i: Volume of the specimen before foaming (m^3)
V_f: Volume of the specimen after foaming (m^3)
ΔV: The difference in the volume
V: Molar volume (m^3)
z: $1/r^3$ (m^{-3}).

Acknowledgments

The author acknowledges the Kuwait Institute for Scientific Research (KISR) for the financial support through Project no. PC 012K.

References

[1] D. L. Tomasko, H. Li, D. Liu et al., "A Review of CO_2applications in the processing of polymers," *Industrial & Engineering Chemistry Research*, vol. 42, no. 25, pp. 6431–6456, 2003.

[2] J. R. Royer, Y. J. Gay, J. M. Desimone, and S. A. Khan, "High-pressure rheology of polystyrene melts plasticized with CO2: Experimental measurement and predictive scaling relationships," *Journal of Polymer Science Part B: Polymer Physics*, vol. 38, no. 23, pp. 3168–3180, 2000.

[3] A. I. Cooper, "Polymer synthesis and processing using supercritical carbon dioxide," *Journal of Materials Chemistry*, vol. 10, no. 2, pp. 207–234, 2000.

[4] S. Areerat, E. Funami, Y. Hayata, D. Nakagawa, and M. Ohshima, "Measurement and prediction of diffusion coefficients of supercritical CO2 in molten polymers," *Polymer Engineering & Science*, vol. 44, no. 10, pp. 1915–1924, 2004.

[5] C. Drohmann and E. J. Beckman, "Phase behavior of polymers containing ether groups in carbon dioxide," *The Journal of Supercritical Fluids*, vol. 22, no. 2, pp. 103–110, 2002.

[6] H. Tai, C. E. Upton, L. J. White et al., "Studies on the interactions of CO2 with biodegradable poly(dl-lactic acid) and poly(lactic acid-co-glycolic acid) copolymers using high pressure ATR-IR and high pressure rheology," *Polymer Journal*, vol. 51, no. 6, pp. 1425–1431, 2010.

[7] J. J. A. Barry, H. S. Gidda, C. A. Scotchford, and S. M. Howdle, "Porous methacrylate scaffolds: Supercritical fluid fabrication and in vitro chondrocyte responses," *Biomaterials*, vol. 25, no. 17, pp. 3559–3568, 2004.

[8] S. K. Goel and E. J. Beckman, "Nucleation and growth in microcellular materials: supercritical CO2 as foaming agent," *AIChE Journal*, vol. 41, no. 2, pp. 357–367, 1995.

[9] S. K. Goel and E. J. Beckman, "Generation of microcellular polymeric foams using supercritical carbon dioxide. I: effect of pressure and temperature on nucleation," *Polymer Engineering & Science*, vol. 34, no. 14, pp. 1137–1147, 1994.

[10] I. Tsivintzelis, A. G. Angelopoulou, and C. Panayiotou, "Foaming of polymers with supercritical CO_2: an experimental and theoretical study," *Polymer Journal*, vol. 48, no. 20, pp. 5928–5939, 2007.

[11] I. Khan, D. Adrian, and S. Costeux, "A model to predict the cell density and cell size distribution in nano-cellular foams," *Chemical Engineering Science*, vol. 138, pp. 634–645, 2015.

[12] S. Costeux, I. Khan, S. P. Bunker, and H. K. Jeon, "Experimental study and modeling of nanofoams formation from single phase acrylic copolymers," *Journal of Cellular Plastics*, vol. 51, no. 2, pp. 197–221, 2014.

[13] Q. Guo, J. Wang, C. B. Park, and M. Ohshima, "A microcellular foaming simulation system with a high pressure-drop rate," *Industrial & Engineering Chemistry Research*, vol. 45, no. 18, pp. 6153–6161, 2006.

[14] K. Taki, "Experimental and numerical studies on the effects of pressure release rate on number density of bubbles and bubble growth in a polymeric foaming process," *Chemical Engineering Science*, vol. 63, no. 14, pp. 3643–3653, 2008.

[15] J. J. Feng and C. A. Bertelo, "Prediction of bubble growth and size distribution in polymer foaming based on a new heterogeneous nucleation model," *Journal of Rheology*, vol. 48, no. 2, pp. 439–462, 2004.

[16] X. Chen, J. J. Feng, and C. A. Bertelo, "Plasticization effects on bubble growth during polymer foaming," *Polymer Engineering & Science*, vol. 46, no. 1, pp. 97–107, 2006.

[17] M. Amon and C. D. Denson, "A study of the dynamics of foam growth: analysis of the growth of closely spaced spherical bubbles," *Polymer Engineering & Science*, vol. 24, no. 13, pp. 1026–1034, 1984.

[18] T. R. Tuladhar and M. R. Mackley, "The development of polymer foam microstructure: Experimental observations and matching modelling for polystyrene foams using different blowing agents," in *Proceedings of the 7th World Congress of Chemical Engineering (GLASGOW '05)*, pp. 10–14, Institution of Chemical Engineers (IChemE), Glasgow, Scotland, July 2005.

[19] Y. Sato, T. Takikawa, S. Takishima, and H. Masuoka, "Solubilities and diffusion coefficients of carbon dioxide in poly(vinyl acetate) and polystyrene," *The Journal of Supercritical Fluids*, vol. 19, no. 2, pp. 187–198, 2001.

[20] S. Al-Enezi, K. Hellgardt, and A. G. F. Stapley, "Mechanical measurement of the plasticization of polymers by high-pressure carbon dioxide," *International Journal of Polymer Analysis and Characterization*, vol. 12, no. 3, pp. 171–183, 2007.

[21] J. R. Calvert and R. A. Farrar, *An Engineering Data Book*, Palgrave, New York, NY, USA, 1999.

[22] T. R. Tuladhar and M. R. Mackley, "Experimental observations and modelling relating to foaming and bubble growth from pentane loaded polystyrene melts," *Chemical Engineering Science*, vol. 59, no. 24, pp. 5997–6014, 2004.

[23] M. L. Williams, R. F. Landel, and J. D. Ferry, "The temperature dependence of relaxation mechanisms in amorphous polymers and other glass-forming liquids," *Journal of the American Chemical Society*, vol. 77, no. 14, pp. 3701–3707, 1955.

[24] R. B. Bird, W. E. Stewart, and E. N. and Lighrfoot, *Transport phenomena*, Wiley, New York, NY, USA, 1960.

[25] P. D. Condo and K. P. Johnston, "Retrograde vitrification of polymers with compressed fluid diluents: experimental confirmation," *Macromolecules* , vol. 25, no. 24, pp. 6730–6732, 1992.

[26] L. N. Nikitin, M. O. Gallyamov, R. A. Vinokur et al., "Swelling and impregnation of polystyrene using supercritical carbon dioxide," *The Journal of Supercritical Fluids*, vol. 26, no. 3, pp. 263–273, 2003.

8

Effect of Polyvinyl Acetate Stabilization on the Swelling-Shrinkage Properties of Expansive Soil

Jin Liu,[1] Yong Wang,[1] Yi Lu,[2] Qiao Feng,[1] Faming Zhang,[1] Changqing Qi,[1] Jihong Wei,[1] and Debi Prasanna Kanungo[3]

[1]*School of Earth Sciences and Engineering, Hohai University, Nanjing 210098, China*
[2]*Key Laboratory of Earth Fissures Geological Disaster, Ministry of Land and Resource, Geological Survey of Jiangsu Province, Nanjing 210049, China*
[3]*CSIR-Central Building Research Institute (CBRI), Roorkee 247667, India*

Correspondence should be addressed to Jin Liu; jinliu920@163.com

Academic Editor: Ulrich Maschke

Polyvinyl acetate constitutes a class of polymers that can entirely dissolve in water to form a solution. In this study, polyvinyl acetate as a nontraditional chemical stabilizer was used in soil improvement. Laboratory tests were carried out to evaluate the effect of polyvinyl acetate on swelling-shrinkage properties of expansive soil. A series of shrink/swell tests were performed with adding polyvinyl acetate as amendment at a concentration 3 g/cm^3 to four aggregate sizes in the range of 0–0.5 mm, 0.5–1 mm, 1-2 mm, and 2–5 mm and five concentrations 1.5 g/cm^3, 3 g/cm^3, 4.5 g/cm^3, 6 g/cm^3, and 9 g/cm^3 to soils with aggregate size in the range of 0.5–1 mm for comparison of results with those of untreated soils. The results show that all the linear swelling ratio (LSWR) and linear shrinkage ratio (LSHR) values of the treated specimens decrease. SEM images and the test results indicate the achieved reduction in volume change of the soil tested using soil pore filling and particle encapsulation.

1. Introduction

Swelling-shrinkage of clayey soils due to the water moisture is a problem in many climatic and pedological zones that are characterized by alternate wet and dry seasons [1, 2]. Normally, the mineralogical composition, the volume fraction of such soils, the dimensions of soil aggregates, and the soil physicochemical environment are the main factors that affect the shrinkage-swelling potential of soil. The hydration characteristics of clay minerals in the soils and the surface force interactions including interparticle attractive and repulsive forces at the colloidal level influence volume change in clays. The presence of the more inert soil particles such as sand and silt in real soil is acknowledged but it is the clay function that undergoes volumes change reactions. Clayey soil stabilization methods must address the interactions among clay particles. The fundamental target of swelling soil stabilization measures is to change the balance between the interaction forces through the introduction of ions/molecules of the soil stabilization agent or alteration of the environmental conditions. Theng [3] introduced a clay swelling theory that recognizes the existence of a diffuse double layer of ions around the clay particles in an aqueous condition. Another feature is the access of liquids and substances to the interlayer spaces in clay during crystalline swelling. This theory often is used as the basis for phenomenological and quantitative interpretations [4–7]. Thus, soil stabilization substances should be able to inhibit clay swelling and shrinkage by reducing the expansion and shrinkage of the interlayer space through bonding the clay particles together on their external surfaces.

Recently, aqueous polymers which can provide these functions under the suitable physicochemical conditions have been analyzed as stabilizers for the control of the swelling-shrinkage of clayey soils [8–11]. Various kinds of polymers in aqueous solutions have been investigated [12–19]. These analyses often assess polymer potential for use as dust suppressants, antierosion agents, and crack inhibitors

TABLE 1: Important physicochemical properties of polyvinyl acetate.

Physical state	Solvability in water	Boiling point	pH	Viscosity (Mpa·s)	Density (g/cm^3)	Solid content (%)
Liquid-white colour	Solution	100°	6~7	400	1.05	41

in waste containment barrier materials. The sorption and interaction between aqueous molecules and clay surfaces have also been studied. The interaction between surfactant and surface active polymer was analogous to mix micelle formation. Jonsson et al. [20] explained that polymer sorption on solids was mostly irreversible owing to the high energy level required to debond the molecules from solids. Bae and Inyang [21] explained the effects of aqueous polymers on clay desiccation where polyethylenimine solution can retard liquid loss from montmorillonite with possible desirable effects on dust generation from exposed clayey soils. Liu et al. [22] found that the water stability of clay aggregates was improved by bonding the clay surfaces with aqueous polymers, and the ability of the polymer soil stabilizer depended on the membrane structure of the soil aggregate surface. Azzam [23] used a polymer nanocomposite material as a partial soil stabilizer to reduce the shrinkage-swelling potential of expansive soil by modifying the microstructure of soil and altering the texture of clay. Mousavi et al. [24] found that soil samples treated with RPP showed significantly lower swelling potential and swelling pressure in comparison with control samples. Based on the above stated advances, it is well known that aqueous polymers have a good potential for use as soil stabilizers in the control of the swelling-shrinkage of expansive clayey soils.

A new aqueous polymer soil stabilizer, named as polyvinyl acetate (PVA), is introduced in this research paper. In order to evaluate the effect of the use of PVA as a soil stabilizer in controlling the swelling-shrinkage of expansive clayey soils, the swelling test and shrinkage test on the untreated and treated expansive clayey soil with different aggregates were performed. Also, the soil structure modification mechanism is herein analyzed.

2. Experimental

2.1. Materials. PVA is a type of organic aqueous polymer soil stabilizer. Its main component is acetic-ethylene-ester polymer. It contains a large number of -$OOCCH_3$ functional groups. Important physicochemical properties of polyvinyl acetate are given in Table 1. It has a pH of 6-7, a solid content of 41%, a specific gravity of 1.05 g/cm^3, and a viscosity of 400 Mpa·s. As a new kind of soil stabilization material used in controlling the swelling-shrinkage of clayey soil, PVA has the following primary advantages: (i) it is a water-soluble material which can be diluted to different concentrations; (ii) the presence of PVA soil stabilizer can form an elastic and viscous membrane on soil surface at the natural situation; and (iii) it is an environment-friendly product and easy to be produced at low cost.

Expansive clayey soils used in this study were obtained from Nanjing area, China. The soil sample has a liquid limit of 52.6%, a plasticity index of 19.7, a specific gravity of 2.73, an optimum moisture content of 15.4% by weight, and a maximum dry density of 1.71 g/cm^3. The grain size distribution curve of the expansive soil is shown in Figure 1. The mineralogical compositions of the soil samples were tested by the X-ray diffraction (XRD) and the percentages of montmorillonite, kaolinite, and illite were 75.4, 13.5, and 10%, respectively. It has a specific surface of 41.66 m^2/g for the aggregate size smaller than 0.5 mm. Its free swelling ratio is 53% and is considered as a low expansive soil. The free swelling ratio of expansive soil was measured with the free swelling test. The aggregate size in this test is smaller than 0.5 mm. The ratio of the increased volume and original volume of sample was defined as the free swelling ratio.

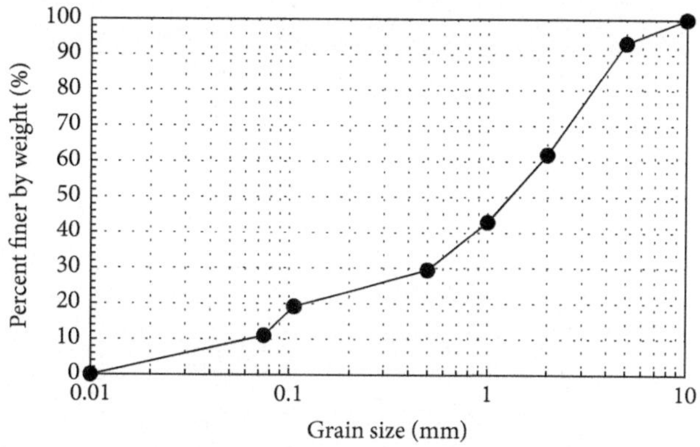

FIGURE 1: The grain size distribution curve of the expansive soil.

2.2. Preparation of Soil Samples. Soil samples were first oven-dried and four groups of aggregates were obtained through four different sieves. The room temperature was about 25°C in the test process. The four aggregate sizes were as follows: (i) less than 0.5 mm; (ii) 0.5–1 mm; (iii) 1-2 mm; and (iv) 2–5 mm. PVA solutions were prepared in concentrations of 1.5 g/cm^3, 3 g/cm^3, 4.5 g/cm^3, 6 g/cm^3, and 9 g/cm^3 by dissolving the appropriate quantity in the distilled water completely. Subsequently, the PVA solution with concentration 3 g/cm^3 was selected to study the soil aggregate size effect on swelling and shrinkage behavior. Each group of aggregates was mixed with the 16.7% by weight of the parent soil to prepare different soil aggregate samples. The aggregate with gain sizes between 0.5 and 1 mm was selected to study the effect of polymer concentrations on the soil swelling-shrinkage. The soil (0.5–1 mm) was mixed with each solution in a proportion of 16.7% of the weight of soil to make the different admixtures for the swelling and shrinkage tests. The soil mixed with the same weight of water was used as the control for all test groups.

FIGURE 2: WZ-2 dilatometer used in swelling test.

FIGURE 3: SS-1 contractometer used in shrinkage test.

The samples were prepared with the static compaction method. The tests complied with the national criterion for geotechnical tests in China which was set based on ASTM standards. The dry density is 1.3 g/cm^3. It should be pointed out that the two-layered compaction was adopted to keep the soil density relatively uniform within the diameter of 61.8 mm and height of 20 mm. Thereafter, specimens were air-dried at about 25°C for 24 hours for use in the swelling and shrinkage tests.

2.3. Swelling and Shrinkage Test. The equipment types used in the swelling and shrinkage tests in this study were WZ-2 dilatometer (Figure 2) and SS-1 contractometer (Figure 3), respectively. WZ-2 dilatometer is used to determine the expansion of clay during soaking and moisture content after expansion and stability. It is mainly composed of ring knife, regulating adjusting screw and inflation proof guide ring. SS-1 contractometer is used to determine the shrinkage limit moisture content, shrinkage, and body contraction coefficient of determination of clay in the dehydration process. It consists of the main body and bulldozing part. The main body is composed of bottom plate, watch rod, table clamp, and perforated plate. Bulldozing part is composed of guide ring, ring knife, positioning ring, and push head.

In the swelling test, each prepared specimen was filled to the rim in the sample box of 61.8 mm inner diameter and 20 mm height. Then, the dilatometer was immersed in the water, and the water level was kept 5 mm higher than the soil specimen. During immersing in the water, the specimen continued to expand for several days and the swelling height was recorded by the height indicator at specific time intervals. When there was no further change in height of the soil specimen, a final reading was taken. Then, the ratio of the final swelling height to the initial height 20 mm was calculated for each specimen. The average time to attain maximum swelling was about 7 days. Soil swelling is evaluated using linear swelling ratio (δ_{sw}), which is calculated according to the following formula:

$$\delta_{sw} = \left(\frac{Z_t - Z_0}{Z_0}\right) \times 100. \quad (1)$$

δ_{sw} is defined as linear swelling ratio (LSWR) at an immersion time (%), Z_t is the reading of indicator at the immersion time of T (mm), and Z_0 is the initial reading of indicator (mm).

In the shrinkage test, after specimen setting, the contractometer was subjected to a temperature around 25°C. The height of the specimen was observed to decrease with the water evaporation and the magnitude of this decrease was recorded by the height indicator at specific time intervals. The final reading with no further change in height was recorded to calculate the linear shrinkage ratio, δ_{sh}, using

$$\delta_{sh} = \left(\frac{Z_i - Z_f}{Z_i}\right) \times 100. \quad (2)$$

δ_{sh} is defined as linear shrinkage ratio (LSHR), Z_i is the initial height reading of indicator (mm), and Z_f is the final height reading (mm).

3. Results and Discussion

3.1. Effect of Aggregate Size on Swelling-Shrinkage Properties. The results of the swelling and shrinkage tests performed on expansive soils containing different aggregate sizes are presented in Tables 2 and 3 and Figures 4–7. Swelling and shrinkage are expressed in terms of the linear swelling ratio (LSWR) and the linear shrinkage ratio (LSHR), respectively, as previously defined. LSWR is the ratio of the increment of specimen height to the initial height and LSHR is the ratio of the decrement height to the initial height. The variation of LSWR with immersion time is shown in Table 2 and Figure 4. As seen in Table 2, all the LSWR values of specimens treated with concentration 3 g/cm^3 of polymer are smaller than the untreated specimens with the same time. It is clearly shown in Figure 4 that, in all cases, the greatest jump in LSWR occurs within the first 30 min and the LSWR values show a relative stable value after 5 hours. The LSWR at 7 days is selected as the final value to analyze the effect of aggregate size on swelling ratio.

Figures 5 and 6 illustrate the effect of aggregate size on LSWR and LSHR, respectively. As observed in Figure 5, LSWR of the treated and untreated soils increases with increase of the aggregate sizes beyond 1 mm. LSHR data exhibits the opposite tend with respect to aggregate size. For the results of shrinkage tests shown in Table 3, the LSHR of PVA treated specimens are smaller than the untreated ones. The LSHR of treated specimens with aggregate sizes 0–0.5 mm, 0.5–1 mm, 1-2 mm, and 2–5 mm are 56.48%,

TABLE 2: The linear swelling ratio of specimens with different aggregate sizes.

Time	3 (g/cm^3)				Untreated			
	0–0.5 mm	0.5–1 mm	1-2 mm	2–5 mm	0–0.5 mm	0.5–1 mm	1-2 mm	2–5 mm
0	0.00%	0.00%	0.00%	0.00%	0.00%	0.00%	0.00%	0.00%
5 min	2.06%	1.46%	3.00%	3.11%	4.00%	3.90%	4.90%	5.10%
10 min	3.23%	2.53%	5.09%	5.19%	6.53%	6.30%	8.50%	9.00%
15 min	4.05%	3.35%	5.88%	6.00%	7.90%	7.80%	10.90%	11.30%
20 min	4.98%	4.22%	6.72%	6.85%	9.50%	9.30%	12.90%	13.30%
30 min	5.47%	4.71%	7.27%	7.40%	11.07%	11.01%	14.50%	14.90%
1 hour	5.95%	4.96%	7.50%	7.93%	11.91%	11.86%	14.60%	15.10%
3 hours	6.40%	5.12%	7.67%	8.32%	12.60%	12.35%	14.65%	15.24%
5 hours	6.53%	5.45%	7.75%	8.45%	12.81%	12.43%	14.68%	15.28%
7 hours	6.60%	5.50%	7.77%	8.54%	12.94%	12.46%	14.69%	15.31%
9 hours	6.65%	5.54%	7.80%	8.60%	13.04%	12.50%	14.71%	15.33%
12 hours	6.70%	5.58%	7.83%	8.64%	13.07%	12.53%	14.75%	15.42%
1 day	6.78%	5.63%	7.86%	8.68%	13.10%	12.56%	14.80%	15.55%
2 days	6.83%	5.64%	7.88%	8.70%	13.12%	12.57%	14.82%	15.59%
3 days	6.87%	5.65%	7.89%	8.71%	13.14%	12.58%	14.85%	15.62%
4 days	6.88%	5.65%	7.90%	8.72%	13.15%	12.59%	14.87%	15.65%
5 days	6.89%	5.66%	7.91%	8.73%	13.16%	12.60%	14.88%	15.68%
6 days	6.90%	5.66%	7.92%	8.74%	13.17%	12.60%	14.90%	15.74%
7 days	6.92%	5.66%	7.92%	8.74%	13.17%	12.61%	14.91%	15.75%

TABLE 3: The linear shrinkage ratio of specimens with different aggregate sizes.

Aggregate sizes (mm)	Linear shrinkage ratio (%)	
	Untreated	PVA (3 g/cm^3)
0–0.5 mm	3.187	1.800
0.5–1 mm	2.895	1.279
1-2 mm	2.316	1.150
2–5 mm	2.044	0.930

44.18%, 49.65%, and 45.50% of the untreated ones. As seen in Figure 6, the LSHR of both untreated and treated specimens decreased with increase of aggregate size. Additionally, the relative decreases of LSWR and LSHR of the different aggregate size specimens are presented in Figure 7. It is observed that the relative decreases of both LSWR and LSHR are larger than 40% in all soils. Their respective maximum values (55.11% for LSWR and 55.82% for LSHR) are observed in soil specimens that have the aggregate sizes in the range of 0.5–1 mm.

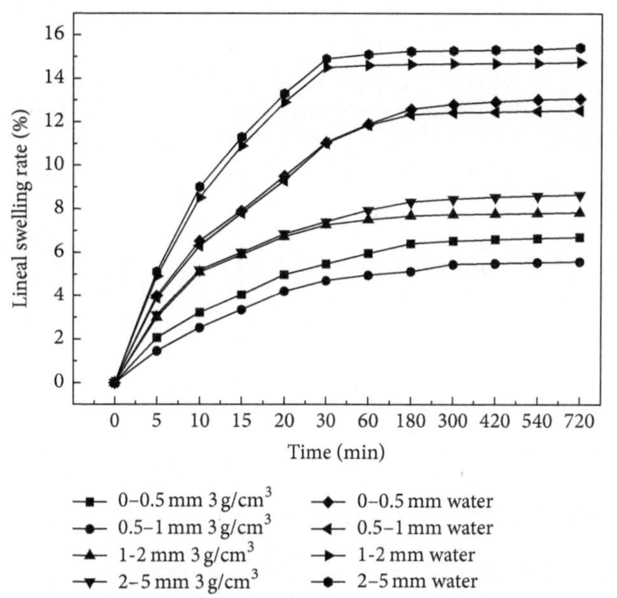

FIGURE 4: Variation of linear swelling ratio with immersed time.

3.2. Effect of Polymer Concentration on Swelling-Shrinkage Properties. The effects of polymer concentration on swelling and shrinkage properties of soils with aggregate size 0.5–1 mm are given in Tables 4 and 5 and Figures 8–10. As shown, with an increase in polymer concentration, the values of both LSWR and LSHR decrease. The variations of LSWR and LSHR with polymer concentration are presented in Figures 8 and 9. They show that polymer concentration plays an important role in controlling the swelling-shrinkage properties of soils. Compared with the values between the untreated and treated specimens with concentration 9 g/cm^3, LSWR decreases from 12.61% to 1.85% and LSHR decreases from 2.9% to 0.87%, respectively.

Figure 10 shows the relative decrease of LSWR and LSHR of specimens with the different polymer concentrations. It is clearly seen that LSWR and LSHR of all tested specimens decrease; the relative decrease ratios of both LSWR and LSHR increase with the increasing of polymer concentration. The relative decrease of LSWR of specimens treated with

TABLE 4: The linear swelling ratio of specimens with different concentration of PVA.

Concentration (g/cm^3)	0	1.5	3	4.5	6	9
Aggregate sizes	0.5–1 mm	0.5–1 mm	0.5–1 mm	0.5–1 mm	0.5–1 mm	0.5–1 mm
Swelling ratio of 7 d (%)	12.61	7.81	5.66	3.65	2.33	1.85

TABLE 5: The linear shrinkage ratio of specimens with different concentration of PVA.

Concentration (g/cm^3)	0	1.5	3	4.5	6	9
Aggregate sizes	0.5–1 mm	0.5–1 mm	0.5–1 mm	0.5–1 mm	0.5–1 mm	0.5–1 mm
Linear shrinkage ratio (%)	2.895	1.697	1.279	1.067	0.913	0.872

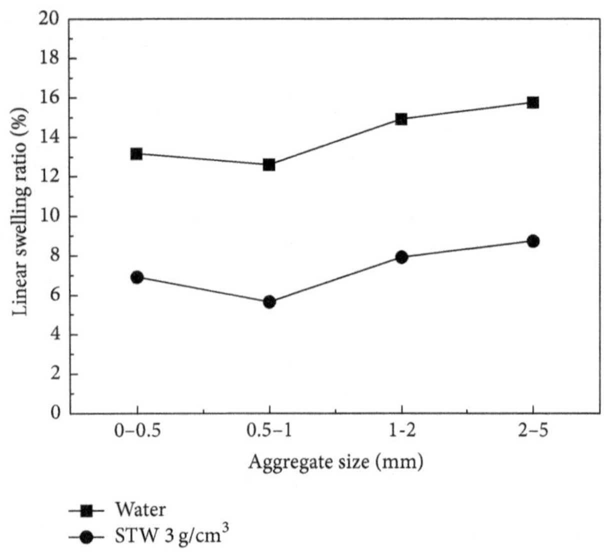

FIGURE 5: Effect of aggregate size on linear swelling ratio of soils.

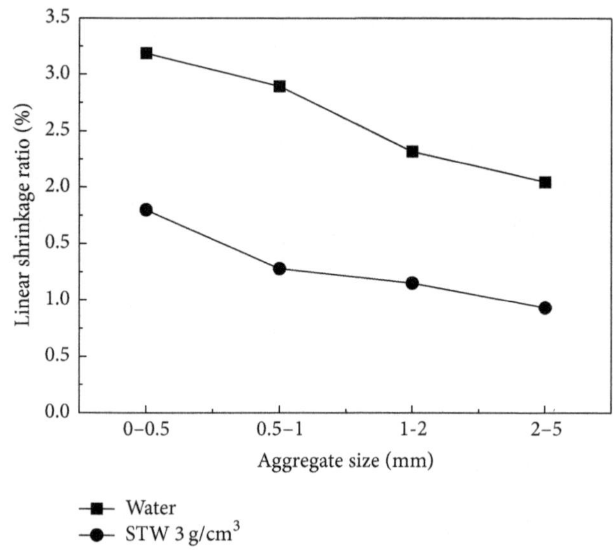

FIGURE 6: Effect of aggregate size on linear shrinkage ratio of soils.

the polymer concentrations 1.5 g/cm^3, 3 g/cm^3, 4.5 g/cm^3, 6 g/cm^3, and 9 g/cm^3 reaches 38.07%, 55.11%, 71.05%, 81.52%, and 85.33%, respectively. The relative decreases of LSHR of specimens with the different concentrations reach 41.38%, 55.82%, 63.14%, 68.46%, and 69.88%, respectively. These results indicate that PVA can effectively suppress the swelling and shrinkage of expansive soils.

4. Mechanism Analysis

The results of this investigation show that both the soil aggregate and polymer concentration influence swelling and shrinkage of soils to different extents. The aqueous polymer stabilization mechanism can be described as pore filling, physicochemical reaction, and enwrapping. The SEM photograph of the untreated soil at ×450 magnification is shown in Figure 11. It shows the presence of many voids and aggregates in the untreated soils. When PVA solution is applied to clayey soil, a part of the polymeric material fills up the voids of soil as shown in Figure 12, and part is sorted onto soil aggregate surface as shown in Figure 13. The hydrophilic groups (-$OOCCH_3$) in its molecular structure react chemically with the adsorbed ions of clay particles and create physicochemical bonds between polymer molecules and soil aggregates. The bonds are of the ionic, hydrogen, or Van der Waals varieties. Through these bonds, long-chain macromolecules of polymers encapsulate the aggregate surfaces and interlink around them to form elastic and viscous membrane structures that decrease the swelling-shrinkage capacity of soil.

Photographs that enable the comparison of untreated and treated soil aggregates are shown in Figure 14. As shown, when the soil aggregate is immersed in water, the untreated soil aggregate collapses, but the treated soil maintains the structure reasonably well. For a soil that collapses, expansion occurs because of the diffuse double layer of ions of inner clay mineral and the access of liquids and substances to interlayer spaces in clay. With higher polymer concentration, the soil has a greater bonding and stability to keep the structure relatively intact as more polymer molecules penetrate the intra-aggregate pores and encapsulate the soil surface. Thus, aqueous PVA polymer decreases the clayey soil swelling potential. The decrease in swelling ratio increases with the increase of polymer concentration.

As evident in the test results for soils of different aggregate sizes, the soil swelling decrease may not depend on the polymer effect alone but on the existing soil skeleton. When the soil aggregate size is smaller than 1 mm, the soil swelling

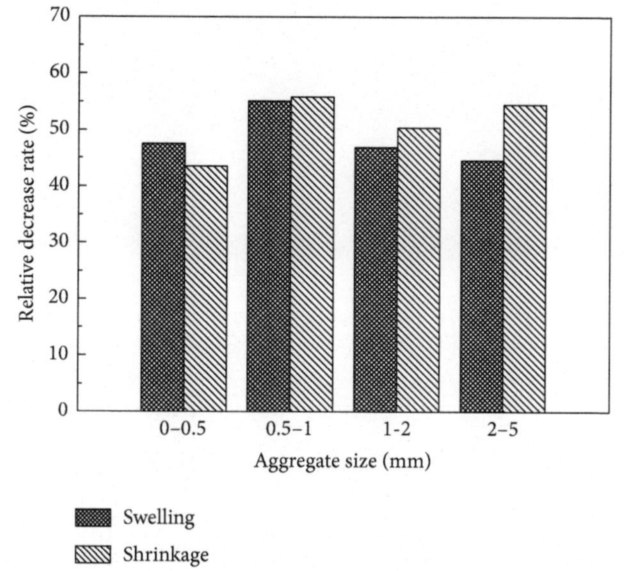

FIGURE 7: Relative decrease of LSWR and LSHR of different aggregate size specimens.

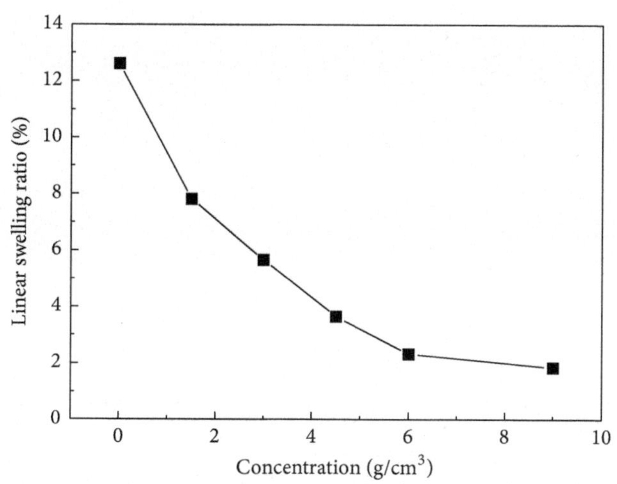

FIGURE 8: Effect of aqueous polymer concentration on linear swelling ratio of soils.

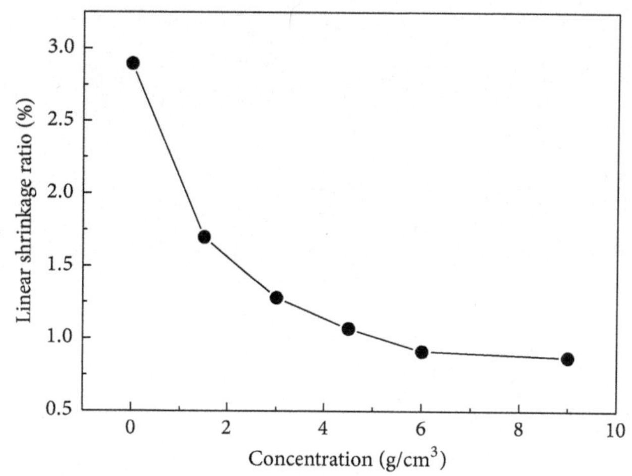

FIGURE 9: Effect of aqueous polymer concentration on linear shrinkage ratio of soils.

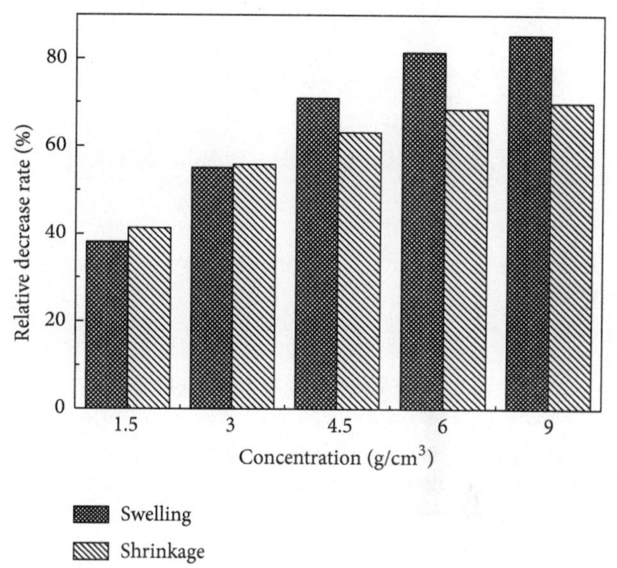

FIGURE 10: Relative decrease of LSWR and LSHR in different polymer concentration specimens.

is mainly due to the diffuse double layer of ions. In this case, the larger the specific surface area of the smaller aggregate, the greater the effect of double layer of ions. Conversely, the result of the soil aggregate that is larger than 1 mm in average particle size has the opposite trend. The structure of the soil skeleton affects the swelling of soil with the larger aggregate. The liquids and other substances that gain access into large voids and interlayer spaces in clay during swelling may lead to an expansion of the whole soil skeleton. This effect should become more significant with the increase in aggregate size, so that swelling increases with increase in aggregate size in both untreated and treated soils.

With the decrease in moisture content of soils, the double layer of ions becomes thinner, resulting in dry shrinkage of soil. The presence of polymer creates a reticular membrane structure among soil particles. This structure inhibits the movement of soil particles. Meanwhile, since the PVA concentration is increasing, the probability of winding and bonding during the interaction of polymer molecules with soil aggregates also increases, aiding the formation of a more complete membrane structure in soil. So, the soil shrinkage decreases because of the presence of PVA, and the shrinkage ratio also decreases with increase in PVA concentration. Furthermore, considering that the specific surface area decreases with increase in aggregate size, the already thin or negligible double layer of ions becomes thinner with increase in aggregate size. This is the reason why the shrinkage ratios of both untreated and treated soils decrease with increase in aggregate size.

It should be noted that this paper is only limited to select PVA 3 g/cm^3 to evaluate the effect of soil aggregate size and

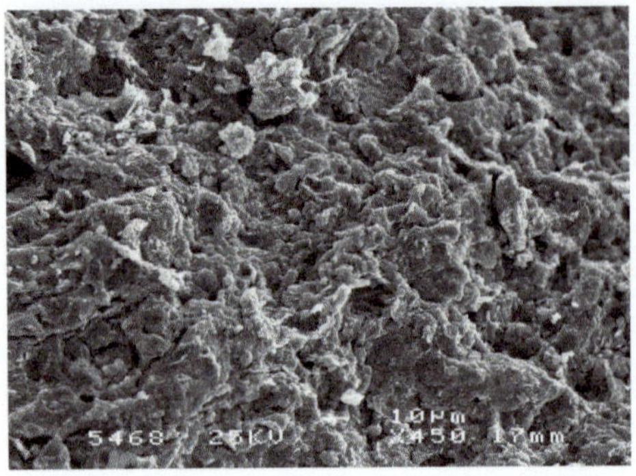

FIGURE 11: SEM image of untreated soil at ×450.

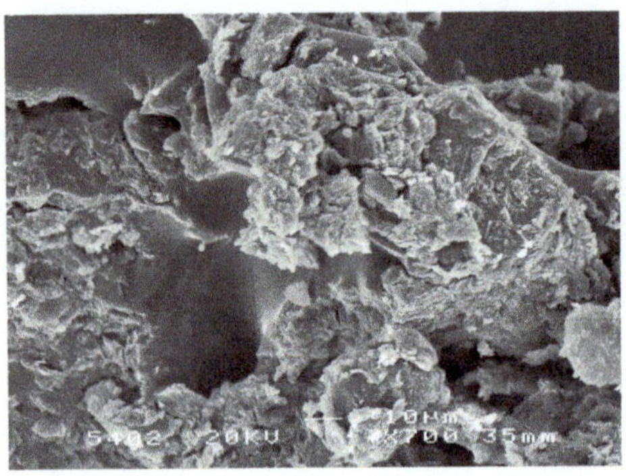

FIGURE 12: SEM image of soil void filled by STW at ×700 magnification.

the soil aggregate size 0.5–1 mm to evaluate the effect of PVA concentration. Furthermore, the PVA concentration should be adjusted to meet the different engineering requirements in practical application.

5. Conclusions

The results of this investigation indicate that PVA is effective in controlling the swelling and shrinkage of soils with size smaller than 5 mm. Both soil aggregate classes and PVA concentration play important roles in determining the magnitude of the effect on the swelling-shrinkage characteristics of the clayey soil investigated.

It is observed from swelling and shrinkage tests that soil swelling mainly occurs at after 5 hours of immersion in water. For aggregate size smaller than 1 mm, the linear swelling ratios (LSWR) of both untreated and treated soil decrease with the increase of aggregate size. However, when the aggregate size is greater than 1 mm, the trend is reversed. The values of both LSWR and LSHR increase with increase in PVA concentration. SEM photos indicate that PVA produces a reticular membrane structure as soil through pore through filling, physicochemical reaction, and encapsulation to inhibit the swelling and shrinkage of soil.

The findings of this research indicate that it is possible to use PVA as soil stabilizer to control volume change of expansive soil in practical soil stabilization projects.

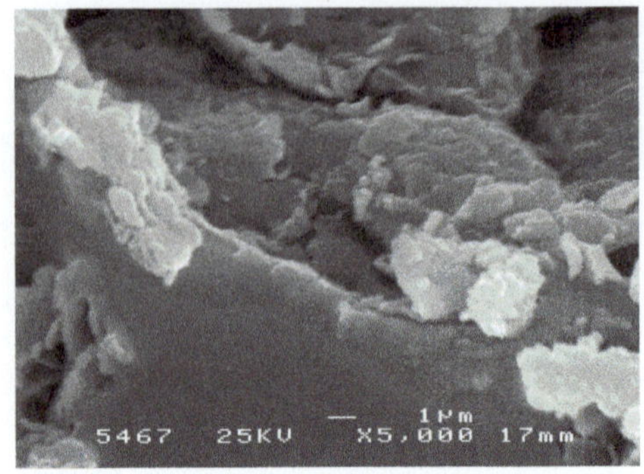

FIGURE 13: SEM image of treated soil surface at ×5,000 magnification.

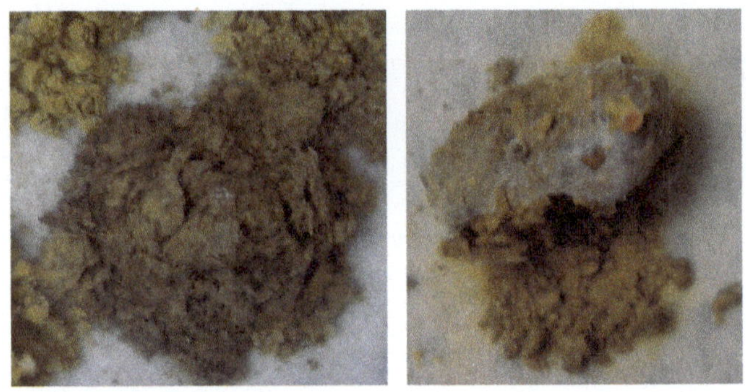

FIGURE 14: The photo of comparison between untreated and treated soil aggregate.

Acknowledgments

This research was financially supported by the National Natural Science Foundation of China (Grant no. 41472241) and Natural Science Foundation of Jiangsu Province, China (Grant no. BK20141415 and no. BK20151011).

References

[1] S. M. Rao, B. V. V. Reddy, and M. Muttharam, "The impact of cyclic wetting and drying on the swelling behaviour of stabilized expansive soils," *Engineering Geology*, vol. 60, no. 1–4, pp. 223–233, 2001.

[2] H. I. Inyang, S. Bae, G. Mbamalu, and S.-W. Park, "Aqueous polymer effects on volumetric swelling of Na-montmorillonite," *Journal of Materials in Civil Engineering*, vol. 19, no. 1, pp. 84–90, 2007.

[3] B. K. G. Theng, *Formation and Properties of Clay-Polymer Complexes*, Elsevier Scientific, Amsterdam, The Netherlands, 1979.

[4] A. C. D. Newman, "The specific surface of soils determined by water sorption," *European Journal of Soil Science*, vol. 34, no. 1, pp. 23–32, 1983.

[5] D. E. Ulberg, N. V. Churaev, V. V. Ilyin, and G. L. Malashenko, "Molecular dynamics simulation of the aggregation of colloidal particles," *Colloids & Surfaces A: Physicochemical & Engineering Aspects*, vol. 80, no. 2-3, pp. 93–102, 1993.

[6] D. A. Laird, "Layer charge influences on the hydration of expandable 2 : 1 phyllosilicates," *Clays and Clay Minerals*, vol. 47, no. 5, pp. 630–636, 1999.

[7] J. Chen, A. Anandarajah, and H. I. Inyang, "Pore fluid properties and compressibility of kaolinite," *Journal of Geotechnical & Geoenvironmental Engineering*, vol. 126, no. 9, pp. 798–807, 2000.

[8] F. Yazdandoust and S. S. Yasrobi, "Effect of cyclic wetting and drying on swelling behavior of polymer-stabilized expansive clays," *Applied Clay Science*, vol. 50, no. 4, pp. 461–468, 2010.

[9] A. S. Mohammed and C. Vipulanandan, *Behavior of Polymer Modified Expansive CH Soil*, CIGMAT, Houston, Tex, USA, 2012.

[10] W. R. Azzam, "Durability of expansive soil using advanced nanocomposite stabilization," *International Journal of Geomate*, vol. 7, no. 1, pp. 927–937, 2014.

[11] S. R. Bekkouche and G. Boukhatem, "Experimental characterization of clay soils behavior stabilized by polymers," *Journal of Fundamental and Applied Sciences*, vol. 8, no. 3, pp. 1193–1205, 2016.

[12] P. V. Barry, D. E. Stott, R. F. Turco, and J. M. Bradford, "Organic polymers' effect on soil shear strength and detachment by single raindrops," *Soil Science Society of America Journal*, vol. 55, no. 3, pp. 799–804, 1991.

[13] F. G. Bell, "An assessment of cement-PFA and lime-PFA used to stabilize clay-size materials," *Bulletin of Engineering Geology and the Environment*, vol. 49, no. 1, pp. 25–32, 1994.

[14] A. Nadler, E. Perfect, and B. D. Kay, "Effect of polyacrylamide application on the stability of dry and wet aggregates," *Soil Science Society of America Journal*, vol. 60, no. 2, pp. 555–561, 1996.

[15] K. N. Nwankwo, "Polyacrylamide as a soil stabilizer for erosion control," Tech. Rep., Wisconsin Department of Transportation, Madison, Wis, USA, 2001.

[16] H. I. Inyang and S. Bae, "Polyacrylamide sorption opportunity on interlayer and external pore surfaces of contaminant barrier clays," *Chemosphere*, vol. 58, no. 1, pp. 19–31, 2005.

[17] J. L. Daniels and H. I. Inyang, "Contaminant barrier material textural response to interaction with aqueous polymers," *Journal of Materials in Civil Engineering*, vol. 16, no. 3, pp. 265–275, 2004.

[18] S. Bae and H. I. Inyang, "Confirmation of aqueous polymer sorption on contaminant barrier clay using TGA," *Journal of Materials in Civil Engineering*, vol. 18, no. 2, pp. 307–311, 2006.

[19] M. A. Mohsin and N. F. Attia, "Inverse emulsion polymerization for the synthesis of high molecular weight polyacrylamide and its application as sand stabilizer," *International Journal of Polymer Science*, vol. 2015, Article ID 436583, 10 pages, 2015.

[20] B. Jonsson, B. Lindman, K. Holmberg, and B. Kronberg, *Surfactants and Polymers in Aqueous Solution*, Wiley, New York, NY, USA, 1998.

[21] S. Bae and H. I. Inyang, "Effects of various polyethylenimine solutions on desiccation of Na-montmorillonite," *Soil and Sediment Contamination*, vol. 10, no. 6, pp. 675–685, 2001.

[22] J. Liu, B. Shi, H. Jiang, S. Bae, and H. Huang, "Improvement of water-stability of clay aggregates admixed with aqueous polymer soil stabilizers," *Catena*, vol. 77, no. 3, pp. 175–179, 2009.

[23] W. R. Azzam, "Reduction of the shrinkage-swelling potential with polymer nanocomposite stabilization," *Journal of Applied Polymer Science*, vol. 123, no. 1, pp. 299–306, 2012.

[24] F. Mousavi, E. Abdi, and H. Rahimi, "Effect of polymer stabilizer on swelling potential and CBR of forest road material," *KSCE Journal of Civil Engineering*, vol. 18, no. 7, pp. 2064–2071, 2014.

Thermal Synthesis of Polypeptides from *N*-Butyloxycarbonyl Oligopeptides Containing Aspartyl Residue at *C*-Terminus

Toratane Munegumi[1] and Takafumi Yamada[2]

[1]*Department of Science Education, Naruto University of Education, Naruto, Tokushima 772-8502, Japan*
[2]*Department of Chemistry, University of Tsukuba, Tsukuba, Ibaraki 305-857, Japan*

Correspondence should be addressed to Toratane Munegumi; tmunegumi@naruto-u.ac.jp

Academic Editor: Peng He

The thermal reactions of amino acids have been investigated for pure organic synthesis, materials preparation in industry, and prebiotic chemistry. *N*-t-Butyloxycarbonyl aspartic acid (Boc-Asp) releases 2-butene and carbon dioxide upon heating without solvents. The resulting mixture of the free molten aspartic acid was dehydrated to give peptide bonds. This study describes the thermal reactions of *N*-t-butyloxycarbonyl peptides (Boc-Gly-L-Asp, Boc-L-Ala-L-Asp, Boc-L-Val-L-Asp, and Boc-Gly-Gly-L-Asp) having an aspartic residue at the carboxyl terminus. The peptides were deprotected upon heating at a constant temperature between 110 and 170°C for 1 to 24 h to afford polypeptides in which the average molecular weight reached 7800.

1. Introduction

Polypeptides [1] have been well investigated as protein model compounds [1–8]. Numerous reports on the methodology for the synthesis of polypeptides have been published [1–14]. The *N*-carboxyl-α-amino acid anhydride (NCA) method (1) [2, 3, 9, 10], polymerization of amino acids using active esters (2) [1, 4–7], solid-phase peptide synthesis (3) [8, 12], and the heating of amino acids (4) [13, 14] are typical examples. The NCA method (1) is suitable for making homopolypeptides and random copolypeptides but is not suitable for the synthesis of sequential copolyamino acid, which is more important for the build-up of functional polypeptides. Sequential polyamino acids have repetitive amino acid residues, in which the amino acid residues can be like -(Gly-Gly-Asp)*n*-. The active ester method (2) [1, 4–7] and solid phase synthesis (3) [8] are more suitable for the synthesis of sequential copolyamino acid. However, the problems of methods (2) and (3) are a long reaction time and the use of much solvent. In contrast, the synthesis of polyamino acid by heating a derivative of the amino acid (4) [13, 14] requires neither long reaction time nor solvents.

In previous papers [15, 16], we reported the synthesis of homopolypeptides [15] and random copolypeptides [16] upon heating of *N*-t-butyloxycarbonyl aspartic acid anhydride (Boc-Asp anhydride) and mixtures of Boc-L-Asp, anhydride, and Boc-Gly, Boc-L-Ala, or Boc-Val. In this paper, we report a trial for the synthesis of sequential copolypeptides by the heating of Boc-peptides instead of these anhydrides. As shown in Figures 1 and 2, Boc-peptides (**5a–d**) and Boc-L-Asp (**5e**) were prepared for heating under a stream of N_2.

2. Materials and Methods

2.1. Instrumentation. A nuclear magnetic resonance (NMR) (JEOL FX-100 NMR system (JEOL, Tokyo, Japan)) was used for the collection of ^{1}H-NMR spectra. A Hitachi model 260-50 infrared (IR) spectrophotometer (Hitachi, Tokyo, Japan) was used for the collection of IR spectra. A Hitachi 200-10 spectrophotometer was used for spectrophotometry measurements.

A Jasco DIP-181 digital polarimeter (Jasco, Tokyo, Japan) was used for the measurement of the optical rotation of the peptide derivatives. A Hitachi 163 gas chromatograph

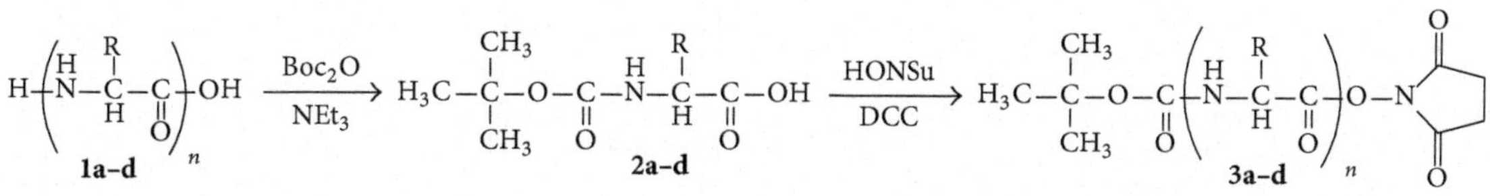

1a: n = 1, R = -H (Gly)
1b: n = 1, R = $-CH_3$ (L-Ala)
1c: n = 1, R = $-CH(CH_3)_2$ (L-Val)
1d: n = 2, R = -H (Gly-Gly)

2a: n = 1, R = -H
2b: n = 1, R = $-CH_3$
2c: n = 1, R = $-CH(CH_3)_2$
2d: n = 2, R = -H

3a: n = 1, R = -H
3b: n = 1, R = $-CH_3$
3c: n = 1, R = $-CH(CH_3)_2$
3d: n = 2, R = -H

FIGURE 1: Preparation of N-t-butyloxycarbonyl-amino acid and peptide active esters. Boc_2O: di-tert-butyl dicarbonate; NEt_3: triethylamine; HONSu: N-hydroxysuccinimide; DCC: N,N'-dicylohexylcarbodiimide.

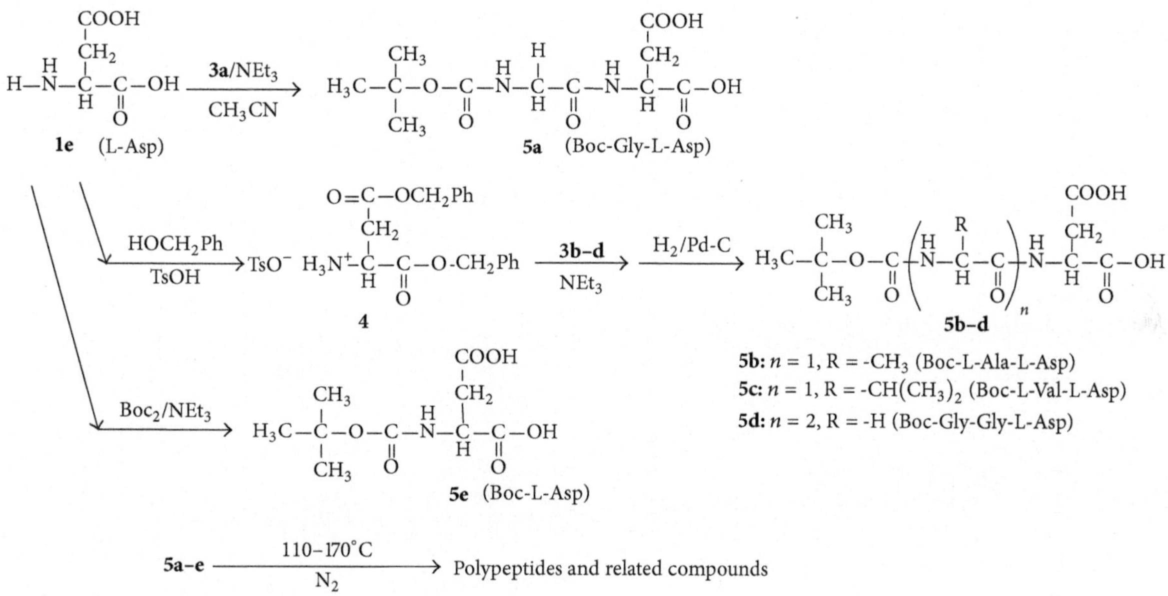

FIGURE 2: Preparation of N-t-butyloxycarbonyl peptides containing aspartyl residue at C-terminus. TsOH: p-toluenesulfonic acid; TsO-: p-toluenesulfonate; Boc_2: di-t-butyl dicarbonate.

equipped with a chiral glass capillary column Chirasil-Val [17, 18] was used for the separation of the enantiomeric derivatives of amino acids. For thermal analysis, we used a Shimadzu DT-40 thermal analyzer (Shimadzu, Kyoto, Japan). A Jasco Trirotar-V as the flow pump and a Jasco UVIDEC-100-IV spectrophotometer as the detector were used for the HPLC system equipped with a gel permeation column G-3000 PW (TSK, Yamaguchi, Japan). Analysis of the evolved gases from the thermal analyzer was performed with a Shimadzu GCMS-QP1000A.

2.2. Materials

2.2.1. Starting Materials and Reagents for the Preparation of Substrate Peptide Derivatives. Glycine (**1a**), L-alanine (**1b**), and L-valine (**1c**) were supplied by Nippon Rika Co., Ltd. (Tokyo, Japan). Glycylglycine (**1d**) and di-t-butyl dicarbonate (Boc_2O) were purchased from Peptide Institute, Inc. (Minoh-shi, Osaka, Japan). L-Aspartic acid (**1e**) and N-hydroxysuccinimide (HONSu) were purchased from Nacalai Tesque (Kyoto, Japan). N,N'-Dicyclohexylcarbodiimide (DCC) was supplied by Watanabe Chemical Industries, Ltd. (Hiroshima, Japan). Palladium on charcoal was purchased from Nippon Engelhard Ltd. (now N.E. CHEMCAT, Tokyo, Japan). Trifluoroacetic anhydride and triethylamine (NEt_3) were purchased from Tokyo Chemical Industries Co., Ltd. (Tokyo, Japan). Hydrochloric acid (6 M) for the hydrolysis of peptides and acetic acid were purchased from Wako Pure Chemical Industries, Ltd. (Osaka, Japan). The 2-propanol solution containing 2.0 M hydrogen chloride was prepared by bubbling HCl into 2-propanol.

2.2.2. Preparation of Substrates

*(1) N-t-Butyloxycarbonyl Amino Acids (**2a–d**)*

*Boc-Gly (**2a**). Glycine* (7.51 g, 0.100 mol) was dissolved in an aqueous solution (150 mL) containing 0.100 mol NaOH, to which a 1,4-dioxane solution (150 mL) including di-t-butyl dicarbonate (24.0 g, 0.11) was added dropwise. After 2 h stirring, the reaction solution was evaporated in vacuo to an oily product, which was dissolved in 10% potassium hydrogen sulfate to reach a pH of 2.5. The resulting solution was extracted with ethyl acetate (100 mL, three times). The extracts were combined, washed with brine, and dried with anhydrous $MgSO_4$. The resulting ethyl acetate solution was evaporated in vacuo. The obtained precipitate was recrystallized with ethyl acetate and petroleum ether to give 10.1 g (yield 57%); the melting point (mp) was 91°C (in lit. [19], it is 94-95°C). ^{1}H-NMR (DMSO-D_6): δ 7.04 (t, J = 6.0 Hz, H, NH), 3.57 (d, J = 7.0 Hz, 2H, CH_2), 1.38 ppm (s, 9H, CH_3); IR (KBr, cm^{-1}): 3410, 3350, 3116 (NH), 1750 (COOH), 1671 (amide I), 1539 (amide II); elementary analysis: Calcd for $C_7H_{13}NO_4$: C, 47.99; H, 7.48; N, 8.00%. Found: C, 48.09; H, 7.58; N, 7.91%.

Other N-t-butyloxycarbonyl-amino acids were prepared in a similar manner. Their physical properties are detailed below.

*Boc-L-Ala (**2b**)*. Yield 76%, mp 87-88°C (lit. [19] 83–8°C). ^{1}H-NMR (DMSO-D_6): δ 7.07 (d, J = 4 Hz, 1H, NH), 3.97 (q, J = 10 Hz, 1H, CH), 1.22 (d, J = 8.0 Hz, 3H, CH_3), 1.38 ppm (s, 9H, CH_3). IR (KBr, cm^{-1}): 3386 (NH), 1742 (COOH), 1690 (amide I), 1520 (amide II). Elementary analysis: Calcd for $C_8H_{15}NO_4$: C, 50.78; H, 7.99; N, 7.40%. Found: C, 50.78; H, 8.10; N, 7.27%. $[\alpha]^{15}{}_D$ −24.2 (c 1.30, acetic acid) (lit. $[\alpha]^{20}{}_D$ −27 (c 2.26, acetic acid) [19]).

*Boc-L-Val (**2c**)*. Yield 93%, mp 78-79°C (lit. 80°C [19]). ^{1}H-NMR (DMSO-D_6): δ 6.90 (d, J = 9.0 Hz, 1H, NH), 3.80 (m, 1H, CH), 1.84–2.11 (m, 1H, CH_3), 1.39 (s, 9H, CH_3), 0.87 ppm (d, J = 4.0 Hz, 6H, CH_3). IR (KBr, cm^{-1}): 3312 (NH), 1740 (COOH), 1649 (amide I), 1500 (amide II). Elementary analysis: Calcd for $C_{10}H_{19}NO_4$: C, 55.28; H, 8.81; N, 6.45%. Found: C, 55.42; H, 8.86; N, 6.45%. $[\alpha]^{15}{}_D$ −5.1 (c 0.97, acetic acid) (lit. $[\alpha]^{20}{}_D$ −5.8 (c 1.2, acetic acid) [19]).

*Boc-Gly-Gly (**2d**)*. Yield 73%, mp 138-139°C. ^{1}H-NMR (DMSO-D_6): δ 8.02 (t, J = 6.0 Hz, 1H, NH), 6.97 (q, J = 6.0 Hz, 1H, NH), 3.77 (d, J = 5.0 Hz, 3H, CH_2), 3.57 (d, J = 6.0 Hz, 2H, CH_2), 1.38 ppm (s, 9H, CH_3). IR (KBr, cm^{-1}): 3362 (NH), 1742, 1690 (COOH), 1618 (amide I), 1526 (amide II). Elementary analysis: Calcd for $C_9H_{16}N_2O_5$: C, 46.55; H, 6.94; N, 12.06%. Found: C, 46.70; H, 7.37; N, 12.02%.

*(2) N-t-Butyloxycarbonyl-Amino Acid N-Hydroxysuccinimide (HONSu) Esters (**3a–d**)*

*Boc-Gly-ONSu (**3a**)*. *N*-t-Butyloxycarbonyl-glycine (23.73 g, 0.135 mol) and *N*-hydroxysuccinimide (HONSu) (17.26 g, 0.15 mol) were dissolved in ethyl acetate (410 mL). *N, N'*-dicyclohexylcarbodiimide (DCC) (34.05 g, 0.165 mol) dissolved in ethyl acetate (60 mL) was added to the cooled ethyl acetate solution at 0°C. The reaction mixture was stirred for about 36 h at 0°C. After the precipitate was filtered off, the obtained filtrate was evaporated in vacuo to give a colorless crystal, which was recrystallized with 2-propanol to give a colorless crystal (18.45 g, 50%), mp 155°C (lit. 168–170°C [19]). ^{1}H-NMR (DMSO-D_6): δ 7.48 (t, J = 6.5 Hz, 1H, NH), 4.09 (d, J = 6.0 Hz, 2H, CH_2), 2.81 (s, 4H, CH_2), 1.39 ppm (s, 9H, CH_3). IR (KBr, cm^{-1}): 3302 (NH), 1824, 1792 (ONSu), 1740 (ester). Elementary analysis: Calcd for $C_{11}H_{16}N_2O_6$: C, 48.53; H, 5.92; N, 10.29%. Found: C, 48.05; H, 5.90; N, 10.32%. By the similar manner, the other succinimide esters (**3b–d**) were prepared as follows.

*Boc-L-Ala-ONSu (**3b**)*. 91%, mp 165–167°C (lit. 167°C [19]). ^{1}H-NMR (DMSO-D_6): δ 7.63 (d, J = 7.0 Hz, 1H, NH), 4.38 (q, J = 7.3 Hz, 1H, CH), 3.08 (s, 4H, CH_2), 1.40 (d, J = 6.0 Hz, 3H, CH_3), 1.39 ppm (s, 9H, CH_3). IR (KBr, cm^{-1}): 3296 (NH), 1827, 1792 (ONSu), 1742 (ester). Elementary analysis: Calcd for $C_{12}H_{18}N_2O_6$: C, 50.35; H, 6.34; N, 9.79%. Found: C, 50.52; H, 6.72; N, 9.73%. $[\alpha]^{15}{}_D$ −48.2 (c 2.16, 1,4-dioxane) (lit $[\alpha]^{20}{}_D$ −49 (c 2, 1,4-dioxane) [19]).

*Boc-L-Val-ONSu (**3c**)*. 85%, mp 131-132°C (lit. [19] 128-129°C). ^{1}H-NMR (DMSO-D_6): δ 7.57 (d, J = 7.0 Hz, 1H, NH), 4.23 (d, J = 5.0 Hz, 1H, CH), 2.81 (s, 4H, CH_2), 2.05–2.26 (m, 1H, CH), 1.41 (s, 9H, C CH_3), 1.00 ppm (d, J = 7.0 Hz, 6H, CH_3). IR (KBr, cm^{-1}): 3354 (NH), 1810, 1783 (ONSu), 1742 (ester), 1671 (amide I), 1539 (amide II). Elementary analysis: Calcd for $C_{14}H_{22}N_2O_6$: C, 53.49; H, 7.05; N, 8.91%. Found: C, 53.56; H, 7.19; N, 8.84%. $[\alpha]^{15}{}_D$ −23.4 (c 1.81, 1,4-dioxane) (lit. $[\alpha]^{20}{}_D$ −37.0 (c 2, 1,4-dioxane) [19]).

*Boc-Gly-Gly-ONSu (**3d**)*. 71%, mp 164-165°C. ^{1}H-NMR (DMSO-D_6): δ 8.45 (t, J = 6.0 Hz, 1H, NH), 7.19 (t, J = 10 Hz, 1H, NH), 4.26 (d, J = 6.0 Hz, 2H, CH_2), 3.60 (d, J = 7.0 Hz, 2H, CH_2), 2.82 (s, 4H, CH_2), 1.39 ppm (s, 9H, CH_3). IR (KBr, cm^{-1}): 3416, 3300 (NH), 1825, 1792 (ONSu), 1742 (ester), 1702, 1665 (amide I), 1576, 1516 (amide II). Elementary analysis: Calcd for $C_{13}H_{19}N_3O_7$: C, 47.42; H, 5.82; N, 12.76%. Found: C, 47.50; H, 5.88; N, 12.63%.

*(3) N-t-Butyloxycarbonyl-Peptides (**5a–d**)*

*TsOH L-Asp(OBzl)$_2$ (**4**)*. L-Aspartic acid (**1e**) (19.97 g, 0.150 mol), *p*-toluenesulfonic acid monohydrate 1 (29.10 g, 0.153 mol), benzyl alcohol (300 mL), and benzene (300 mL) were mixed in a three-necked flask that was connected to a Dean-Stark apparatus. The esterification reaction was carried out by refluxing the reaction mixture and removing water in the flask for 7 days. The cooled reaction solution was evaporated in vacuo to give a colorless solid, which was recrystallized with chloroform and petroleum ether. Yield 49.8 g (68%), mp 157-158°C (lit. 159-160°C [19]). ^{1}H-NMR (DMSO-D_6): δ 8.54 (s, 1H, SO_3H), 7.51 (d, J = 10 Hz, 2H, ArH), 7.36 (s, 10H, ArH), 7.11 (d, J = 9.0 Hz, 2H, ArH), 5.14 (d, J = 8.0 Hz, 2H, NH_2), 4.50 (t, J = 5.5 Hz, 1H, CH), 3.40 (s, 4H, CH_2), 3.05 (d, J = 8.0 Hz, 2H, CH_2), 2.28 ppm (s, 3H, CH_3). IR (KBr, cm^{-1}): 3042 (NH_3), 1758, 1738, 1127 (ester), 1185 (-SO3), 814, 739, 685. Elementary analysis: Calcd for $C_{25}H_{27}NO_7S$: C, 61.84; H, 5.61; N, 2.89%. Found: C, 61.94; H, 5.63; N, 2.85%. $[\alpha]^{15}{}_D$ −1.04 (c 2.21, methanol) (lit. $[\alpha]^{25}{}_D$ +1.0 (c 1, methanol) [19]).

*Boc-Gly-L-Asp (**5a**)*. L-Aspartic acid (**1e**) (7.98 g, 0.060 mol) and triethylamine (NEt_3) (12.14 g, 0.12 mol) were dissolved in distilled water (250 mL), and Boc-Gly-ONSu (13.61 g, 0.050 mol) dissolved in 1,4-dioxane (250 mL) for 4 h was added to the solution. The resulting solution, which was acidified with 10% $KHSO_4$ to give a pH of 2.5, was extracted with ethyl acetate (100 mL, three times). The combined extracts were washed with brine and dried with anhydrous $MgSO_4$. The filtrated solution was evaporated in vacuo to give a colorless crystal that was recrystallized with ethyl acetate and petroleum ether to yield 10.17 g (70%); mp 130°C; ^1H-NMR (DMSO-D_6): δ 8.02 (d, J = 8.0 Hz, 1H, NH), 6.96 (t, J = 4.0 Hz, 1H, NH), 4.56 (t, J = 10 Hz, 1H, CH), 3.56 (q, J = 2.0 Hz, 2H, CH_2), 2.66 (d, J = 6.0 Hz, 2H, CH_2), 1.38 ppm (s, 9H, CH_3). IR (KBr, cm^{-1}): 3432, 3340 (NH), 1756, 1711 (COOH), 1659 (amide I), 1547 (amide II). Elementary analysis: Calcd for $C_{11}H_{18}N_2O_7$: C, 45.52; H, 6.25; N, 9.65%. Found: C, 45.43; H, 6.35; N, 9.48%. $[\alpha]^{15}_D$ +20.7 (c 1.06, acetic acid).

Boc-L-Ala-L-Asp (5b). Boc-L-Ala-ONSu (**3b**) (0.020 mol) was coupled with *TsOH L-Asp (OBzl)$_2$* (**4**) in dichloromethane to yield Boc-L-Ala-L-Asp (OBzl)$_2$ of 84%, which was hydrogenolyzed over 5% Pd on charcoal to yield *Boc-L-Ala-L-Asp*. Yield: 3.21 g, 70%, 85°C (decomposed). ^1H-NMR (DMSO-D_6): δ 7.97 (d, J = 8.0 Hz, 1H, NH), 6.94 (d, J = 7.0 Hz, 1H, NH), 4.50 (t, J = 10 Hz, 1H, CH), 3.95 (q, J = 7.0 Hz, 2H, CH), 2.63 (d, J = 6.0 Hz, 2H, CH_2), 1.37 (s, 9H, CH_3), 1.17 ppm (d, J = 7.0 Hz, 3H, CH_3). IR (KBr, cm^{-1}): 3432, 3340 (NH), 1756, 1711 (COOH), 1659 (amide I), 1547 (amide II). Elementary analysis: Calcd for $C_{12}H_{20}N_2O_7$ 0.45THF: C, 49.22; H, 7.06; N, 8.32%. Found: C, 48.98; H, 7.09; N, 8.07%. $[\alpha]^{15}_D$ +3.81 (c 0.972, acetic acid). Other substrates (**5c**, **5d**) were prepared in a similar manner.

Boc-L-Val-L-Asp (5c). 72%, mp 85°C (decompose). ^1H-NMR (DMSO-D_6): δ 8.10 (d, J = 7.0 Hz, 1H, NH), 6.65 (d, J = 9.0 Hz, 1H, NH), 4.53 (t, J = 10 Hz, 1H, CH), 3.83 (d, J = 2.0 Hz, 1H, CH), 2.65 (d, J = 9.0 Hz, 2H, CH_2), 1.70–1.83 (m, 1H, CH), 1.38 (s, 9H, CH_3), 0.86 (d, J = 6.0 Hz, 3H, CH_3), 0.80 ppm (d, J = 6.0 Hz, 3H, CH_3). IR (KBr, cm^{-1}): 3432, 3340 (NH), 1756, 1711 (COOH), 1659 (amide I), 1547 (amide II). Elementary analysis: Calcd for $C_{14}H_{24}N_2O_7$ 0.45THF: C, 52.02; H, 7.63; N, 7.68%. Found: C, 52.03; H, 7.60; N, 7.41%. $[\alpha]^{15}_D$ −5.67 (c 1.04, acetic acid).

Boc-Gly-Gly-L-Asp (5d). 97%, mp 82–92°C. ^1H-NMR (DMSO-D_6): δ 8.20 (d, J = 8.0 Hz, 1H, NH), 8.02 (t, J = 2.5 Hz, 1H, NH), 7.10 (t, J = 11 Hz, 1H, NH), 4.55 (t, J = 10 Hz, 1H, CH), 3.75 (d, J = 5.0 Hz, 2H, CH_2), 3.57 (d, J = 6.0 Hz, 2H, CH_2), 2.64 (d, J = 0.2 Hz, 2H, CH_2), 1.39 ppm (s, 9H, CH_3). IR (KBr, cm^{-1}): 3354 (NH), 1725 (COOH), 1657 (amide I), 1535 (amide II). Elementary analysis: Calcd for $C_{13}H_{21}N_3O_8$ 0.50THF: C, 46.99; H, 6.57; N, 10.96%. Found: C, 47.05; H, 6.56; N, 10.85%. $[\alpha]^{15}_D$ +16.2 (c 1.04, acetic acid).

Boc-L-Asp (5e). 60%, mp 116–118°C (lit. [19] 118-119°C). ^1H-NMR (DMSO-D_6): δ 7.02 (d, J = 8.0 Hz, 1H, NH), 4.26 (q, J = 7.3, 15 Hz, 1H, CH), 2.63 (d, J = 4.0 Hz, 2H, CH_2), 1.38 ppm (s, 9H, C CH_3). IR (KBr, cm^{-1}): 3358 (NH), 1720 (COOH), 1694 (amide I), 1539 (amide II). Elementary analysis: Calcd for $C_9H_{15}NO_6$: C, 46.35; H, 6.48; N, 6.01%. Found: C, 46.37; H, 6.59; N, 5.85%. $[\alpha]^{15}_D$ −4.5 (c 1.15, methanol) (lit. $[\alpha]^{20}_D$ −6.4 (c 1, methanol) [19]).

2.3. Thermal Analysis. A small scale of the reaction was performed in the thermal analyzer using the substrate under an N_2 stream to monitor the thermal gravimetry (TG) and differential thermal analysis (DTA); Boc-Gly-L-Asp (3.30 mg) was heated from 50°C to 175°C at a rate of 4.7°C per min. The gas mixture that formed during the heating reaction was directly derived into a mass spectrometer (Shimadzu QP-1000A). The gases were determined by selective ion monitoring as CO_2 (m/z = 44), isobutene (m/z = 56), and water (m/z = 18).

2.4. Thermal Reaction. N-Boc-dipeptides (**5a–5c**, 0.5 mmol) and N-Boc-tripeptide (**5d**, 0.3 mmol), which were put into different Pyrex test tubes (165 mm × 18 internal diameter (i.d.)), were kept for 5 min under N_2 flow and then heated in an oil bath controlled at a constant temperature under N_2 flow. After the heating reactions, the reaction mixtures in the test tubes were kept under vacuum for 24 h at room temperature. The weight decrease and IR spectra of the resulting samples were measured.

2.5. Gel Filtration. The overall amount of each reaction mixture after the heating reaction was dissolved in 5 mL of 0.5 M acetic acid and the resulting solution was loaded onto a gel permeation chromatograph (910 mm × 15 mm i.d.). During the elution with 0.5 M acetic acid, the eluate was collected by 3 mL fractions in 95 test tubes. The ultraviolet (UV) absorption of each collected fraction was recorded at 230 nm. 3 M acetic acid was used only for the reaction mixture using Boc-L-Val-L-Asp (**5c**) instead of 0.5 M acetic acid.

2.6. Molecular Weight Estimation. The collected fractions were tested for ninhydrin using thin-layer chromatography in a developing solvent: 1-butanol–acetic acid–water (4 : 1 : 2 (v/v)). Ninhydrin-negative fractions were combined as higher molecular weight fractions, and ninhydrin-positive fractions were combined as lower molecular weight fractions, which showed not a spot but a tailed area with *Rf* values in the range from 0 to ca. 0.4. The higher molecular weight fractions were lyophilized to afford an amorphous peptide powder, of which a part (1 to 2 mg) was dissolved in 0.1 M sodium phosphate buffer (NaH_2PO_4-Na_2HPO_4, pH 6.9). A part (20 μL) of the sample solution was mixed with 20 μL blue dextran solution (1.5 mg/mL 0.1 M sodium phosphate buffer pH 6.9) and was injected into the TSK gel G-3000 PW in an HPLC system. The retention of the sample on the chromatogram was compared with the calibration line that was prepared with several retention times of proteins of known molecular weight.

2.7. Hydrolysis of Polypeptides and Amino Acid Analysis. About 2 mg of each higher molecular weight fraction described in Section 2.5 was mixed with 2 mL of 6 M-HCl in a glass test tube that was sealed under vacuum. The sealed glass was heated to 110°C and held for 8 h, which was enough for the complete hydrolysis of the sample peptide. The resulting hydrolysate was analyzed by means of the automatic amino acid analyzer.

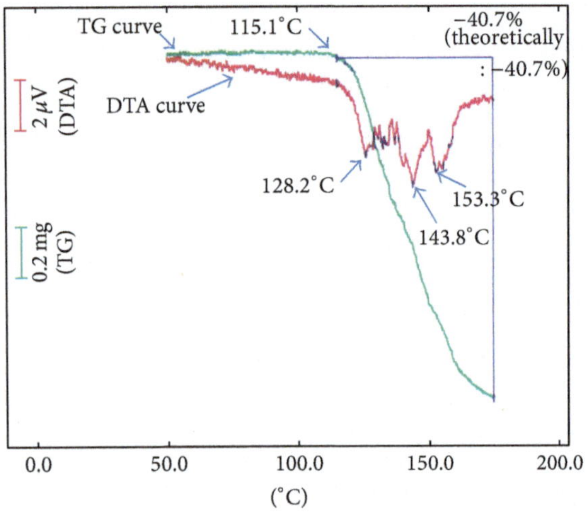

FIGURE 3: Thermal analysis of Boc-Gly-L-Asp (**5a**). Heating rate: 4.7°C/min; carrier gas: He, 50 mL/min; AMP range: DTA: 200 μV; TG: 10 mg on a Pt cell (2.65 mm × 5.40 i.d.).

2.8. D/L Ratio Analysis. 1 mL of each acid-hydrolysate described above was put in a glass vial and kept under a vacuum. The obtained constant weight of the sample was esterified with 2 mL of 2-propanol solution containing 1.5 M HCl and then with trifluoroacetic anhydride to give *N*-trifluoroamino acid 2-propyl ester (*N*-TFA-AA-O-2-Pr).

3. Results and Discussion

3.1. Weight Decrease and Ion Monitoring during the Heating Reactions of Boc-Peptides and Boc-L-Asp. The charts of DTA and TG for Boc-Gly-L-Asp (**5a**) are shown in Figure 3.

Although the DTA line curved gently down during the time the TA line was flat, both lines go rapidly down after the temperature of 115°C was reached, which corresponded with a temperature a little lower than the mp (130°C). The thermal absorption on the DTA line continued and became almost flat at 175°C. The TG line also became flat and the total decrease in weight was 40.7% at 175°C. This ratio corresponds to the theoretical weight decrease after the decomposition of the Boc group (34.5%) and dehydration (6.2%) between free peptides from the initial weight of the Boc-Gly-L-Asp compound. To monitor the evolved gases during the heating reaction of Boc-Gly-L-Asp, the thermal analyzer was directly connected to a mass spectrometer (MS). The MS records showed 2-butene (m/z = 56), CO_2 (m/z = 44), and water (m/z = 18) at temperature conditions higher than about 110°C (Figure 4). Monitoring water generation on the line, there were two peaks of water. The peak at the lower temperatures seems to be the water generation during the peptide bond formation between deprotected dipeptides. Another peak at the higher temperatures is supposed to be the water generation during the imide formation.

As a control experiment, the thermal analysis of Boc-L-Asp was conducted during monitoring the generated gases by mass spectrometry as shown in Figures 5 and 6.

The curves of TA and DTA of Boc-L-Asp in Figure 5 show almost flat lines (the first flat) from room temperature to 100°C and then a rapid decrease of weight on TA plus an energy absorption peak on DTA. In the range of higher temperatures, the curve of DTA became flat again (the second flat) at 180°C and rapidly falls down and then is flat (the third flat) at 216°C. The weight decrease from the first to the second flat corresponds with 49.2% of the initial weight of Boc-Gly-L-Asp. The theoretical weight decrease is caused almost totally (50.7%) by 2-butene (24.1%), CO_2 (18.9%), and water (7.7%). These gases were monitored in the lower temperature range up to 180°C (Figure 6). The weight decrease (11.5%) from the second to the third flat may include generation of water (7.7%) by imide formation and a little decomposition. The temperature range of the weight decrease (11.5%) corresponded to the energy absorption peak on the DTA curve (Figure 5) and the second peak of water generation in Figure 6.

Compared with the thermal analyses of Boc-Gly-L-Asp and Boc-L-Asp, the dehydration in the thermal reaction of Boc-Gly-L-Asp might proceed at temperatures lower than those of the thermal reaction of Boc-L-Asp [15, 20], which did not produce peptides but amino acid at lower temperatures. A similar feature was observed for Boc-DL-Asp [20] and Boc-L-asparagine (Boc-L-Asn) [21]. The latter case released ammonia as well as 2-butene, CO_2, and water.

*3.2. Thermal Reactions of N-Boc-Peptides (**5a–5d**).* Table 1 shows the weight decrease during the heating reactions of *N*-Boc-peptides under different conditions. The heating reaction of Boc-Gly-L-Asp under these conditions gave about 37% to 46% weight decrease, almost equal to the total release (41%) of 2-butene, CO_2, and water. Boc-L-Ala-L-Asp, Boc-L-Val-L-Asp, and Boc-Gly-Gly-L-Asp theoretically release 2-butene, CO_2, and water at the ratio of 39%, 34%, and 29%, respectively. At temperatures below 130°C, these compounds showed a lower decrease than the theoretical decrease. The reactions of these substrates at 130°C for 4 h are enough for the complete release of Boc and water.

3.3. Gel Filtration of Reaction Mixtures. The thermal reaction products were loaded onto a gel permeation chromatograph. A constant volume (ca. 3 mL) of eluate was added to the test tubes. Figures 7–9 show the absorbance of each collected eluate at 230 nm plotted against the test tube number, which was proportional to the total volume of elution after the start of loading.

Figure 7 shows chromatograms of the reaction mixtures of Boc-Gly-L-Asp at 130°C for varying reaction times. Two peaks appear in each chromatogram. The first and the second peaks included the higher and lower molecular weight fraction, respectively. The first peak is revealed faster with the reaction time proceeded until 8 h reaction (d) and is revealed later after 8 h reaction. The results suggested that the polymerization reaction mainly proceeded during 8 h and degradation proceeded mainly after 8 h.

Figure 8 shows the chromatograms of the heating reaction mixtures of Boc-Gly-L-Asp (**5a**) at varying temperatures. Comparing the first peaks of chromatograms ((a) and (b)) at

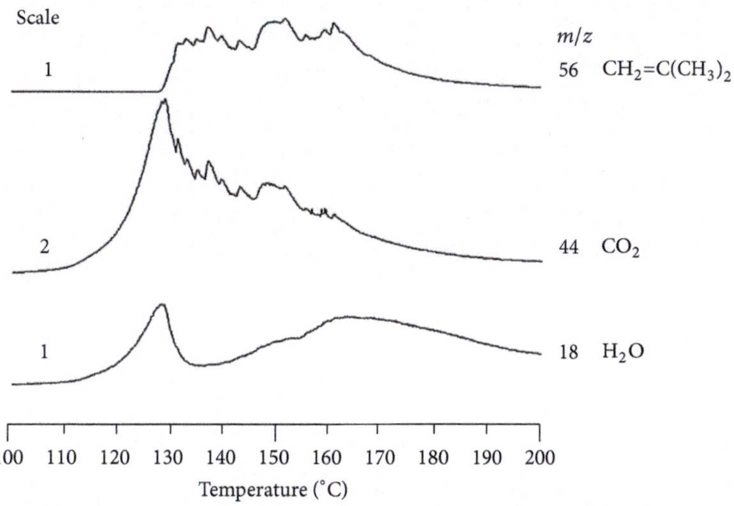

FIGURE 4: Thermal release patterns of ions for compounds from Boc-Gly-L-Asp (**5a**) by DTA/TG-MS analysis. Ion source: EI, 70 eV; temperature: 250°C. Detection: EI range, m/z = 17 to 100; interval, 1.5 s; gain, 2.5; pipe heater: 280°C. GC oven: 150°C; injection: 200°C.

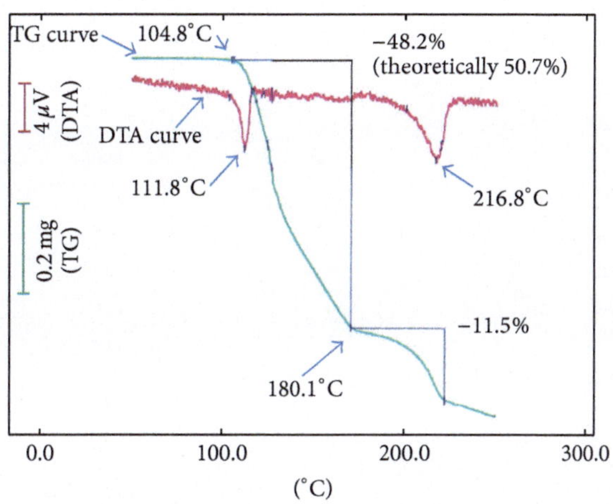

FIGURE 5: Thermal analysis of Boc-L-Asp (**5e**). Heating rate: 4.7°C/min; carrier gas: He, 50 mL/min; AMP range: DTA (differential thermal analysis): 200 μV, TG (thermogravimetry): 10 mg on a platinum cell (2.65 mm × 5.40 i.d.).

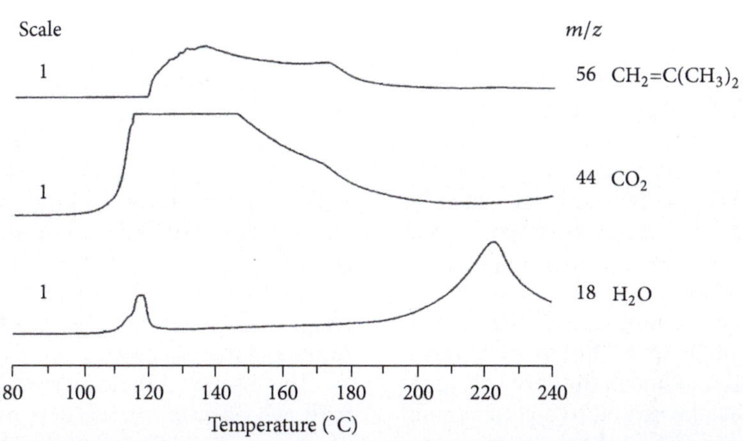

FIGURE 6: Thermal release patterns of ions for compounds from Boc-L-Asp (**5a**) by DTA/TG-MS analysis. Ion source: EI, 70 eV; temperature: 250°C. Detection: EI range, m/z = 17 to 100; interval, 1.5 sec; gain, 2.5. Pipe heater: 280°C. GC oven: 150°C; injection: 200°C.

Table 1: Weight decrease in the heating reaction of substrate peptides.

Substrate	Temperature/°C	Reaction time/h	Weight/mg		Weight decrease	
			Initial	Later	mg	%
Boc-Gly-L-Asp (**5a**)	110	24	145	88	57	39
	120	24	145	89	56	39
	130	1	145	92	53	37
	130	2	143	86	57	40
	130	4	146	86	60	41
	130	8	146	83	63	43
	130	16	148	84	64	43
	130	24	145	84	61	42
	150	2	146	81	65	45
	170	2	147	80	67	46
Boc-L-Ala-L-Asp (**5b**)	120	1	161	120	41	25
	120	2	151	117	34	23
	120	4	162	121	41	25
	130	1	146	92	54	37
	130	2	174	100	74	43
	130	4	179	99	80	45
Boc-L-Val-L-Asp (**5c**)	110	1	164	160	4	2
	110	2	160	132	28	18
	110	4	163	126	37	23
	120	1	161	134	27	17
	120	2	186	160	26	14
	120	4	169	115	54	32
	130	1	170	141	29	17
	130	2	165	116	49	30
	130	4	166	109	57	34
Boc-Gly-Gly-L-Asp (**5d**)	120	1	113	107	6	5
	120	2	107	99	8	7
	120	4	110	74	36	33
	130	1	109	80	29	27
	130	2	104	69	35	34
	130	4	104	71	33	32

different temperatures, the first peak on (b) (120°C, 24 h) was revealed faster than (a) (110°C, 24 h). The result suggested that the condition at 120°C, 24 h, was better for polymerization than 110°C, 24 h. Comparing the first peaks of chromatograms ((c) and (d)) at different temperatures, the first peak of (c) (150°C, 2 h) was revealed faster than (a) (170°C, 2 h). The result suggested that the condition at 150°C, 2 h, was better for polymerization than 170°C, 2 h. A high temperature, such as 170°C, fostered the degradation of polymers generated from the substrate Boc-Gly-L-Asp (**5a**).

The chromatograms of the reaction mixture using other substrates (**5b–d**) are shown in Figure 9. The chromatograms of **5b** and **5c** showed two peaks, although the chromatogram of the reaction mixture of **5c** shows a not-so-clear second peak. This might have been caused by the use of an eluate with a higher concentration of acetic acid (3 M) which had stronger absorption than 0.5 M acetic acid at 230 nm but could enable the reaction mixture to dissolve in it.

The IR spectrum (Figure 10) of the higher molecular weight products shows the absorption of typical acidic polypeptides: amide I, amide II, -NH, and -COOH groups.

3.4. Estimation of the Molecular Weight of the Higher Molecular Weight Fraction. The molecular weight of the higher molecular weight fractions was estimated with the calibration curve (Figure 11) that shows the relationship between the common logarithm and the ratio V_e/V_0 between the elution volume and the void volume. Table 2 shows the values V_e/V_0 for the higher molecular weight fractions at varying reaction times and temperatures.

The estimated molecular weight was almost comparable with the position of the first peak on the chromatograms of Figures 7–9. The samples from the reaction of Boc-Gly-L-Asp for 24 h gave the following values: 4400, 7000, and 4600 Da for 110, 120, and 130°C, respectively. A proper combination of the reaction conditions gave the highest

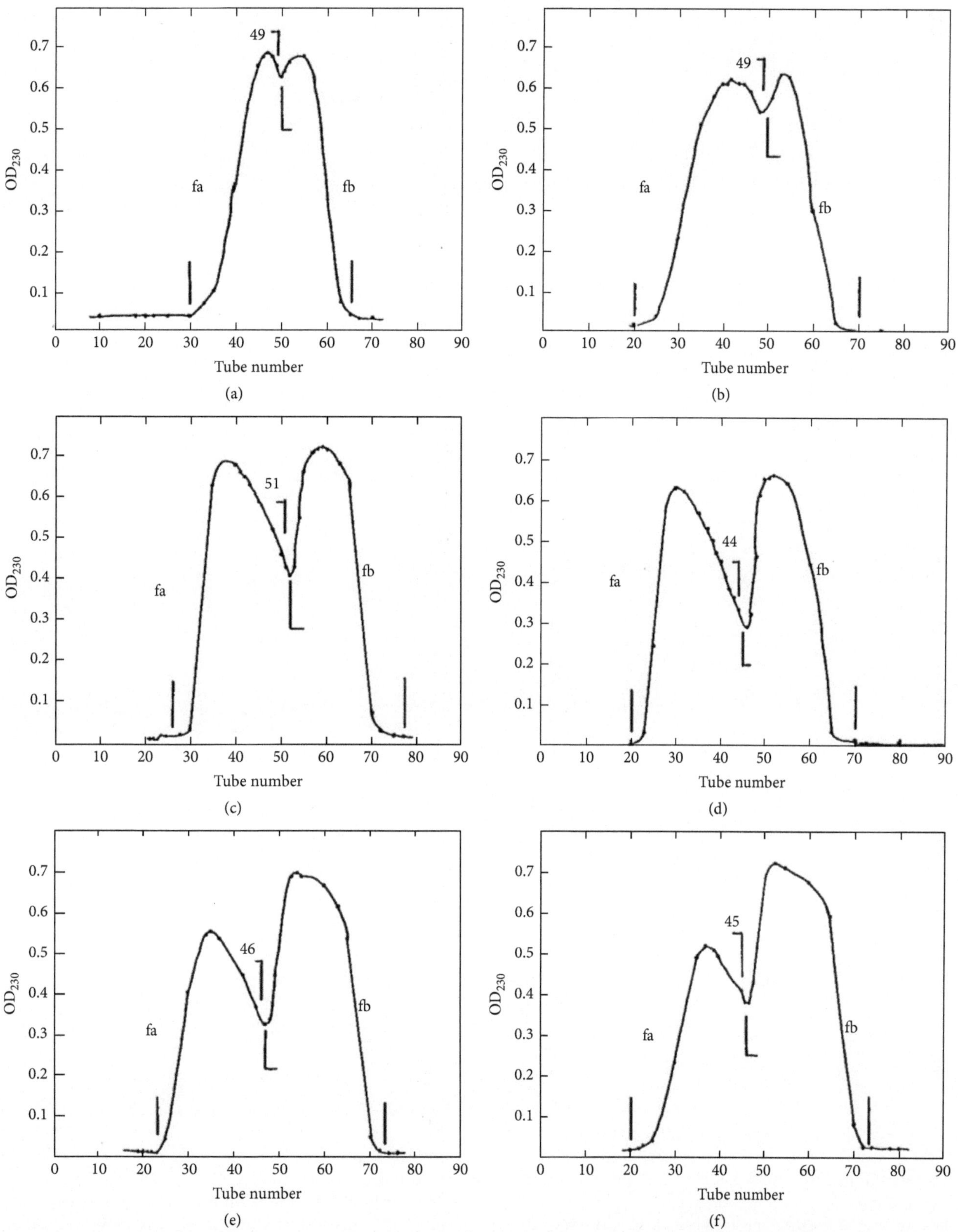

FIGURE 7: Gel permeation chromatograms of the reaction mixtures of Boc-Gly-L-Asp at 130°C for varying reaction times: (a) 1 h; (b) 2 h; (c) 4 h; (d) 8 h; (e) 16 h; and (f) 24 h. Solid phase: Sephadex G-25F. Eluate: 0.5 M acetic acid. Volume per test tube: 3 mL. fa: ninhydrin-negative, higher molecular weight fractions; fb: ninhydrin-positive lower molecular weight fractions.

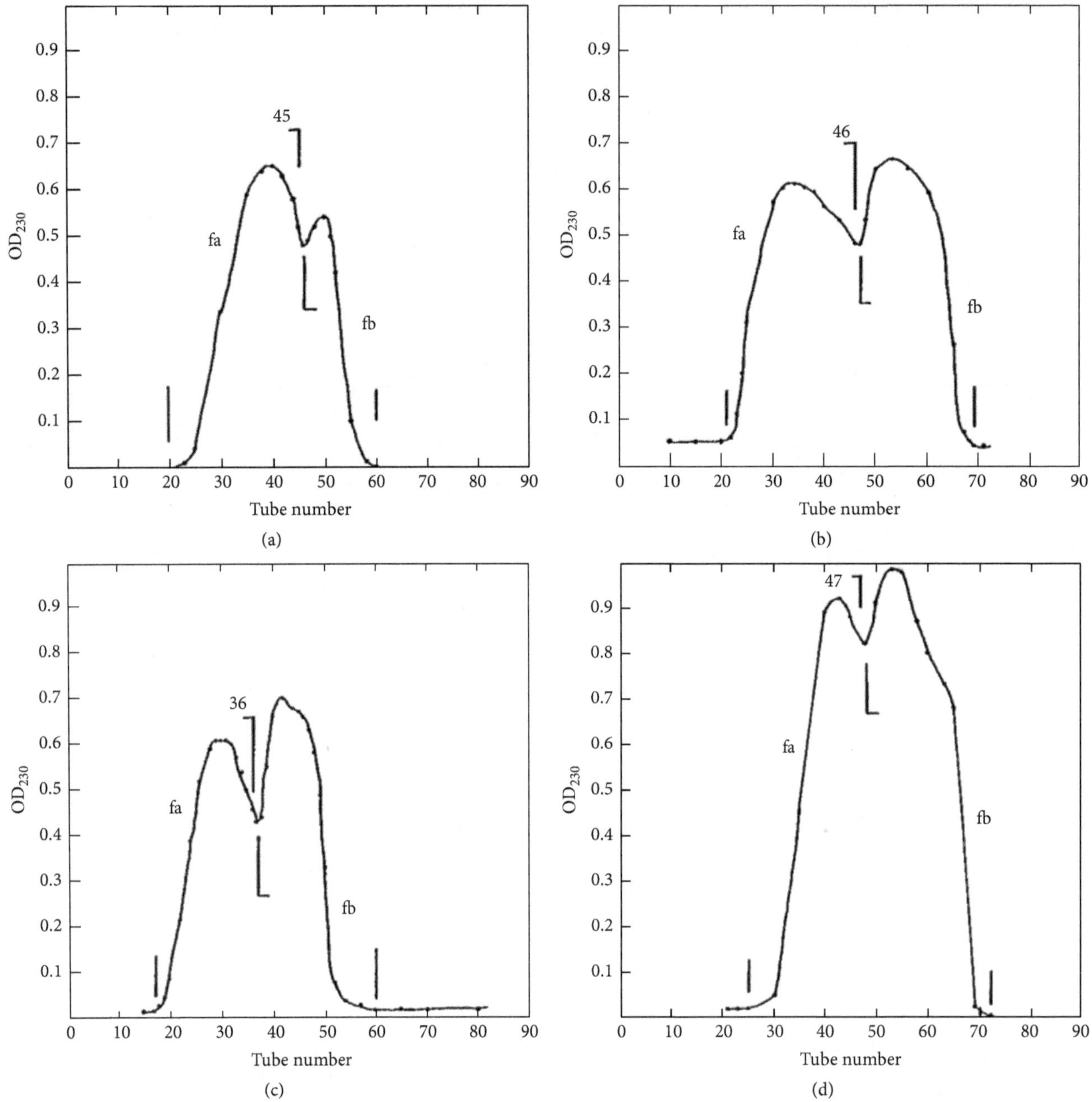

FIGURE 8: Gel permeation chromatograms of the reaction mixtures of Boc-Gly-L-Asp (**5a**) at varying temperatures and times: (a) 110°C, 24 h; (b) 120°C, 24 h; (c) 150°C, 2 h; and (d) 170°C, 2 h. Solid phase: Sephadex G-25F. Eluate: 0.5 M acetic acid. Volume per a test tube: 3 mL. fa: ninhydrin-negative, higher molecular weight fractions; fb: ninhydrin-positive lower molecular weight fractions.

molecular weight, and a temperature that was too high gave a lower molecular weight because of enhanced degradation. The effect of reaction time at the same temperature could be seen in detail in the reactions at 130°C. The shorter reaction times gave a lower molecular weight but it was higher with reaction time (2800, 4100, and 6900 for 1, 2, and 4 h, resp.); the medium reaction time gave the highest molecular weight: 7800 for 8 h; the longer reaction time gave a lower molecular weight (5400 and 4600 Da for 16 and 24 h, resp.) again. The other substrates resulted in a similar tendency as for substrate **5a**.

3.5. Amino Acid Composition and D/L Ratio of the Residue in the Higher Molecular Weight Fraction. The amino acid composition and the D/L ratio in the acid-hydrolysate of the higher molecular weight fractions are listed in Table 3.

The amino acid compositions were almost even at the level of the theoretical value in the hydrolysates obtained with the higher molecular weight fractions of the Boc-dipeptides at lower temperatures and for shorter reaction times. However, the hydrolysates of the obtained higher molecular weight fractions at the higher temperature and longer reaction times gave higher compositions for amino

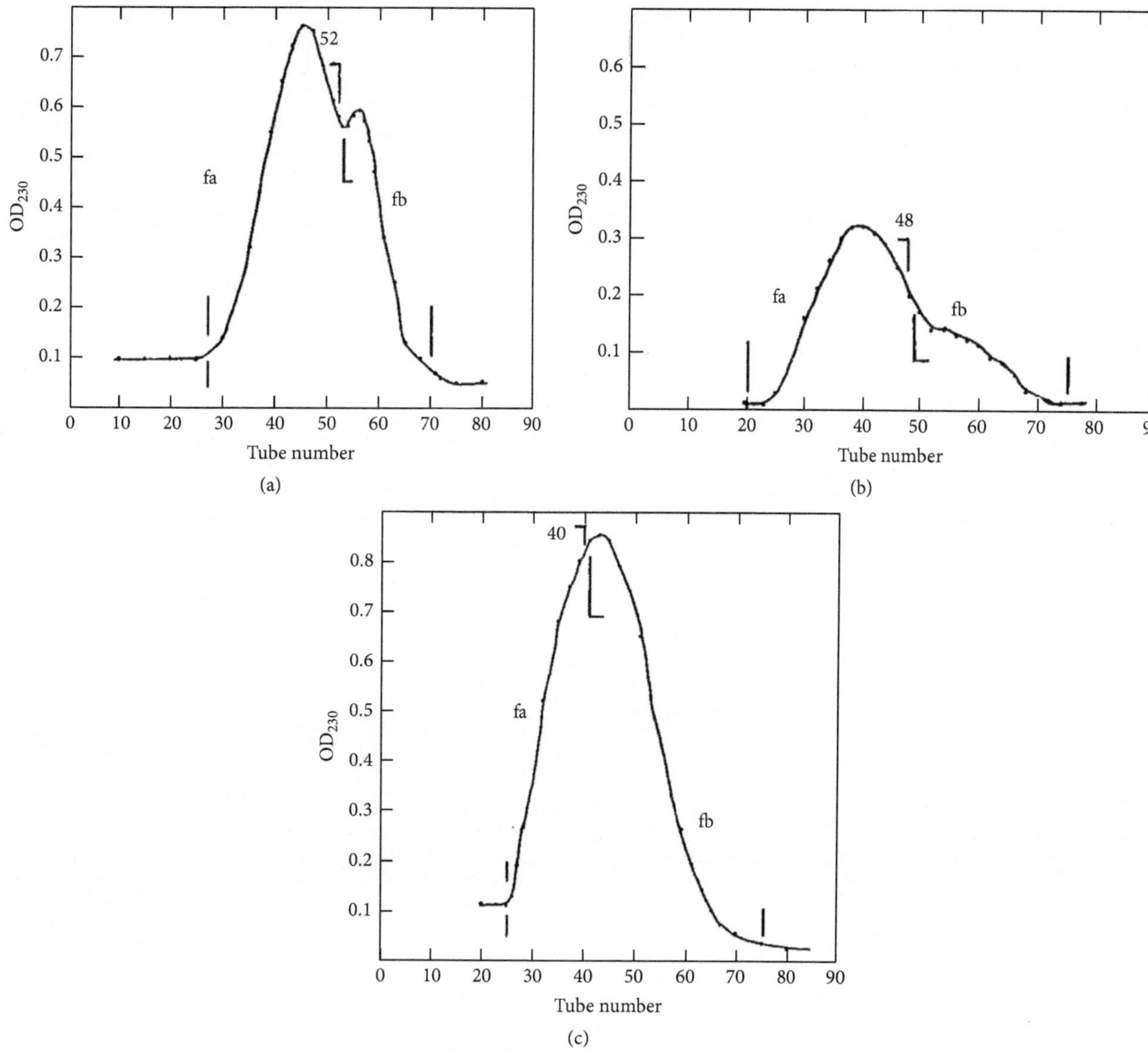

FIGURE 9: Gel permeation chromatograms of the reaction mixtures of peptide derivatives under different conditions. (a) Boc-L-Ala-L-Asp, 130°C, 1 h; (b) Boc-L-Val-L-Asp, 130°C, 2 h; and (c) 120°C, 4 h. Solid phase: Sephadex G-25F. Eluate: 0.5 M acetic acid. Volume per a test tube: 3 mL. fa: ninhydrin-negative, higher molecular weight fractions; fb: ninhydrin-positive lower molecular weight fractions.

acids other than aspartic acid, because the latter is more labile to decomposition at higher temperatures than other amino acids. The hydrolysates of the product formed a tripeptide derivative: Boc-Gly-Gly-L-Asp gave Gly and Asp with the ratio of 2 : 1. The amino acid recovery after hydrolysis was from 76% to almost the quantitative amount.

Regarding the D/L ratio, the molar ratio of the D-isomer to L-isomer of aspartic acid was higher in the hydrolysates of the products with the higher temperature and the longer reaction. The D/L ratio from the products was 0.142 at 130°C for 24 h, 0.189 at 150°C for 2 h, and 0.596 at 170°C for 2 h. The D/L ratios of other amino acids were lower than that for aspartic acid. The existence of D-isomers may be explained by racemization [22] of L-aspartic acid or epimerization of L-aspartyl residues in the polypeptides. L-Aspartic acid undergoes the fastest racemization and epimerization of the all proteinous amino acids, because particularly aspartic residue forms imide structure which fosters the epimerization [23–26]. However, if some better reaction conditions were selected, such racemization and epimerization could be minimized to be lower than 10%. In contrast, racemization and epimerization of Asp residue might proceed in the range of 1% to 2% during 8 h hydrolysis at 110°C. The actual D/L ratio in the polypeptides must be 1% to 2% lower than the results in Table 3.

3.6. Analysis of the Lower Molecular Weight Fractions. HPLC results of the heating reaction mixtures of Boc-dipeptides (**5a–c**) gave two peaks on the chromatogram by means of G-3000PM, whereas the Boc-tripeptide (Boc-Gly-Gly-L-Asp:

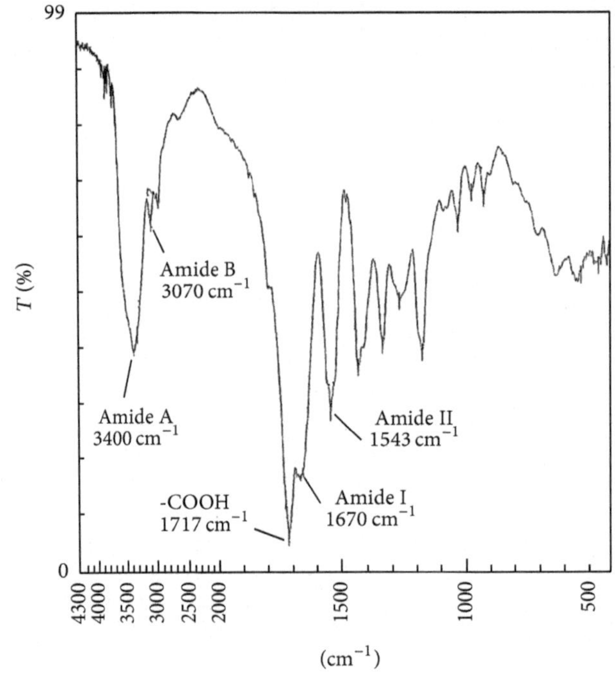

FIGURE 10: IR spectrum of the higher molecular weight fraction of the reaction mixture obtained from Boc-Gly-L-Asp at 130°C for 8 h.

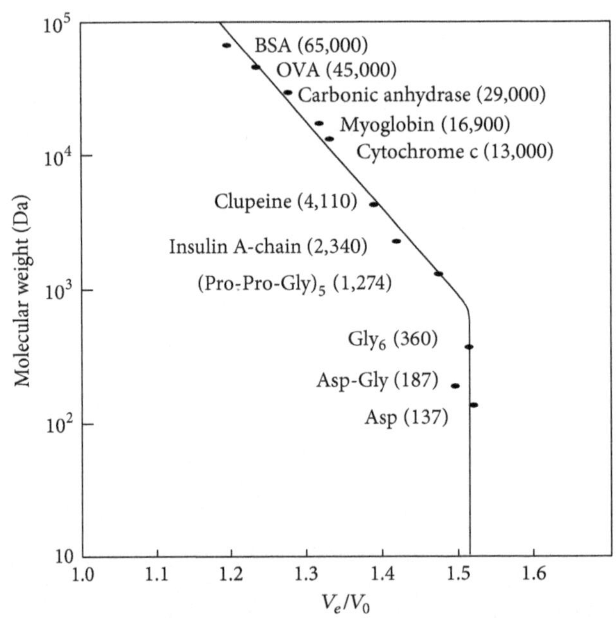

FIGURE 11: Calibration curve of the molecular weight of peptides using HPLC with a TSK-GEL G-3000PW (300 mm × 7.5 mm i.d.). Eluent: 0.1 M sodium phosphate buffer, pH 6.9; flow rate: 0.7 mL/min; detection: UV 230 nm. V_e: volume (mL) of elution for each sample; V_0: elution volume of blue dextran as void volume (mL).

5d) gave only one (Figure 12: (a) Boc-L-Val-L-Asp (**5c**) and (b) Boc-Gly-Gly-L-Asp (**5d**)).

Chromatogram (a) gave only one peak, but (b) gave two peaks, in which the longer retention time peak was caused by the lower molecular weight products. The IR spectrum (Figure 12) of the lower molecular weight fraction from Boc-L-Val-L-Asp showed absorption for the groups of Amide I (1663 cm^{-1}) and -COOH (1717 cm^{-1}) or imide and the IR spectrum of Boc-Gly-Gly-L-Asp showed Amide II, as well as -COOH, and Amide I (1545, 1717, and 1663 cm^{-1}, resp.). As the lower molecular weight fraction from the reaction mixture of Boc-L-Val-L-Asp (**5c**) has -COOH and amide groups but

TABLE 2: Estimated molecular weight of the higher molecular weight fraction.

Substrate	Temperature/°C	Reaction time/h	V_e/V_0[a]	Molecular weight/Da	Number (n) of residues (AA-Asp)n
Boc-Gly-L-Asap (**5a**)	110	24	1.395	4400	26
	120	24	1.362	7000	41
	130	1	1.423	2800	16
	130	2	1.397	4100	24
	130	4	1.363	6900	40
	130	8	1.354	7800	45
	130	16	1.381	5400	31
	130	24	1.389	4600	23
	150	2	1.430	2500	15
	170	2	1.468	1400	8
Boc-L-Ala-L-Asp (**5b**)	130	1	1.466	1500	8
	130	2[b]	1.421	2900	16
	130	4[b]	1.382	5200	28
Boc-L-Val-L-Asp (**5c**)	120	4[b]	1.447	2000	9
	130	2	1.441	2200	10
	130	4[b]	1.393	4200	20
Boc-Gly-Gly-L-Asp (**5d**)	120	4	1.354	7800	34
	120	4[b]	1.404	3700	16
	130	2[b]	1.402	4000	17
	130	4[b]	1.385	5000	21

[a]V_e: volume of elution; V_0: void volume; [b]intact reaction mixture without gel filtration.

TABLE 3: Amino acid composition and D/L ratio of hydrolysates of the higher molecular weight fractions.

Substrate	Temperature/°C	Reaction time/h	Mw/Da	AA composition		D/L ratio		AA recovery/%
				AA	Asp	AA	Asp	
Boc-Gly-L-Asp (**5a**)	110	24	4400	1.11	1.00	—	0.064	95
	120	24	7000	0.98	1.00	—	0.061	104
	130	1	2800	1.10	1.00	—	0.058	87
	130	2	4100	1.08	1.00	—	0.072	76
	130	4	6900	1.10	1.00	—	0.087	93
	130	8	7800	1.10	1.00	—	0.098	88
	130	16	5400	1.12	1.00	—	0.122	84
	130	24	4600	1.09	1.00	—	0.142	83
	150	2	2500	1.11	1.00	—	0.189	100
	170	2	1400	1.13	1.00	—	0.596	100
Boc-L-Ala-L-Asp (**5b**)	130	1	1500	1.00	1.00	0.042	0.067	88
Boc-L-Val-L-Asp (**5c**)	130	2	2200	0.94	1.00	0.017	0.099	78
Boc-Gly-Gly-L-Asp (**5d**)	120	4	7800	1.95	1.00	—	0.059	102

not the linear peptide, the fractions are suggested to include the structure of 2,5-diketopiperazine (DKP) [27] or a kind of imide [20] structure.

3.7. Postulated Mechanism of the Thermal Reaction of Boc-Peptides. From the many features of the thermal reactions of substrate Boc-peptides and the products described above, we propose the reaction mechanism of the thermal reaction depicted in Figure 14.

There might be three reaction intermediates, **6**, **7**, and **8**, from substrates **5b–d** to polypeptide **9**. Intermediate **6** may be produced by releasing 2-butene and carbon dioxide first. Intermediate **7** may be produced by releasing water first. Intermediate **8** may be produced from intermediates **6** and **7**

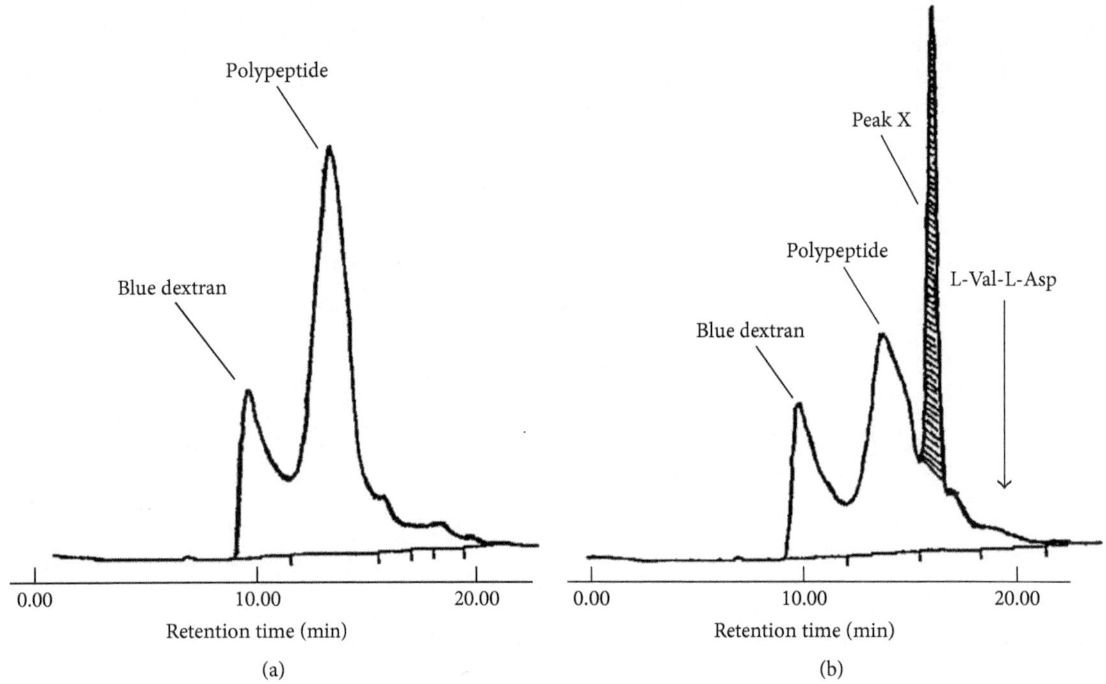

FIGURE 12: Chromatograms of the crude products obtained from Boc-Gly-Gly-L-Asp (a) at 120°C for 4 h and Boc-L-Val-L-Asp (b) at 130°C for 2 h in the heating reactions using an HPLC equipped with a TSK-GEL G-3000PW (300 mm × 7.5 mm i.d.). Eluent: 0.1 M sodium phosphate buffer, pH 6.9; flow rate: 0.7 mL/min; detection: UV 230 nm.

or from substrate **5a–d** directly. The actual reaction mixture can include the three intermediates (**6**, **7**, and **8**), because the analysis of the gases directly injected from thermal reaction to the mass spectrometer indicated that the three gases continuously evolved at the same temperature, although the order is CO_2 plus water and then 2-butene (Figures 3 and 4). Therefore, intermediates **6** and **8** can yield polypeptide **9**. Intermediate **6** would directly dehydrate between an amino group and a carboxy group to yield a peptide bond. Intermediate **8** would be attacked by an amino group to open its anhydride group and form two kinds of peptide bond that are linked using the α- and β-carboxy groups; and then complete dehydration makes imide (**10**) bonds in polypeptides (**9**).

A part of dipeptide derivatives (**5a–c**) form six-membered ring compounds (2,5-diketopiperazine: DKP (**11**) [27]), which may not polymerize anymore. The existence of DKP in the lower molecular weight fractions can be supported by the IR spectrum (B) (Figure 13), which has the absorbance of DKP (1663 cm^{-1}) and carboxyl group of -COOH (1710 cm^{-1}). Another estimated structure showing absorbance at 1710 cm^{-1} may be a tetrapeptide imide structure like compound **10** ($n = 1$; $q = 1$), which could form by coupling of two dipeptides, for instance, Val-Asp from **5c** in Figure 14. However, the anhydride structure of compound **10** would have partly opened by hydrolysis during chromatography. If the tetrapeptide imide structure cyclized from head to tail, the resulting product would have two five-membered cyclic imides and a twelve-membered cyclic

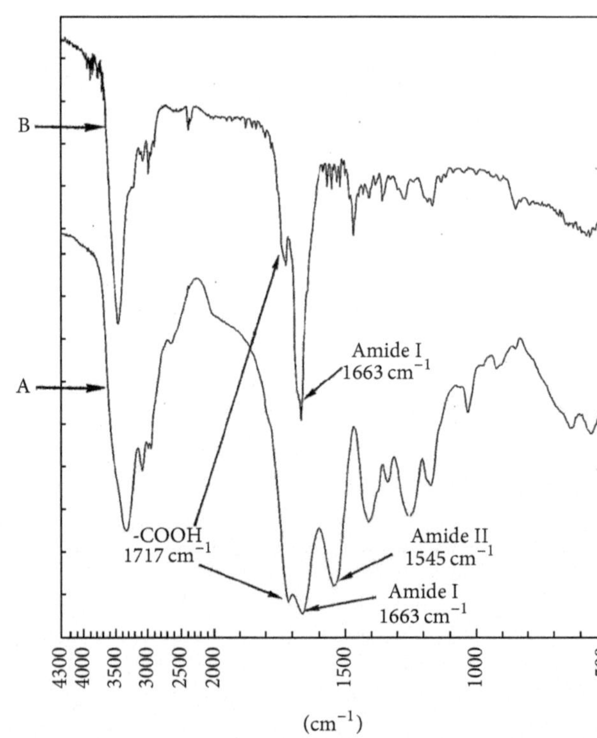

FIGURE 13: IR spectra of the lower molecular weight fractions obtained from Boc-Gly-Gly-L-Asp (A) at 120°C for 4 h and Boc-L-Val-L-Asp (B) at 130°C for 2 h in the reaction heating reactions.

peptide. Therefore, the lower molecular weight fraction may include DKP and the tetrapeptide imide structure. Further

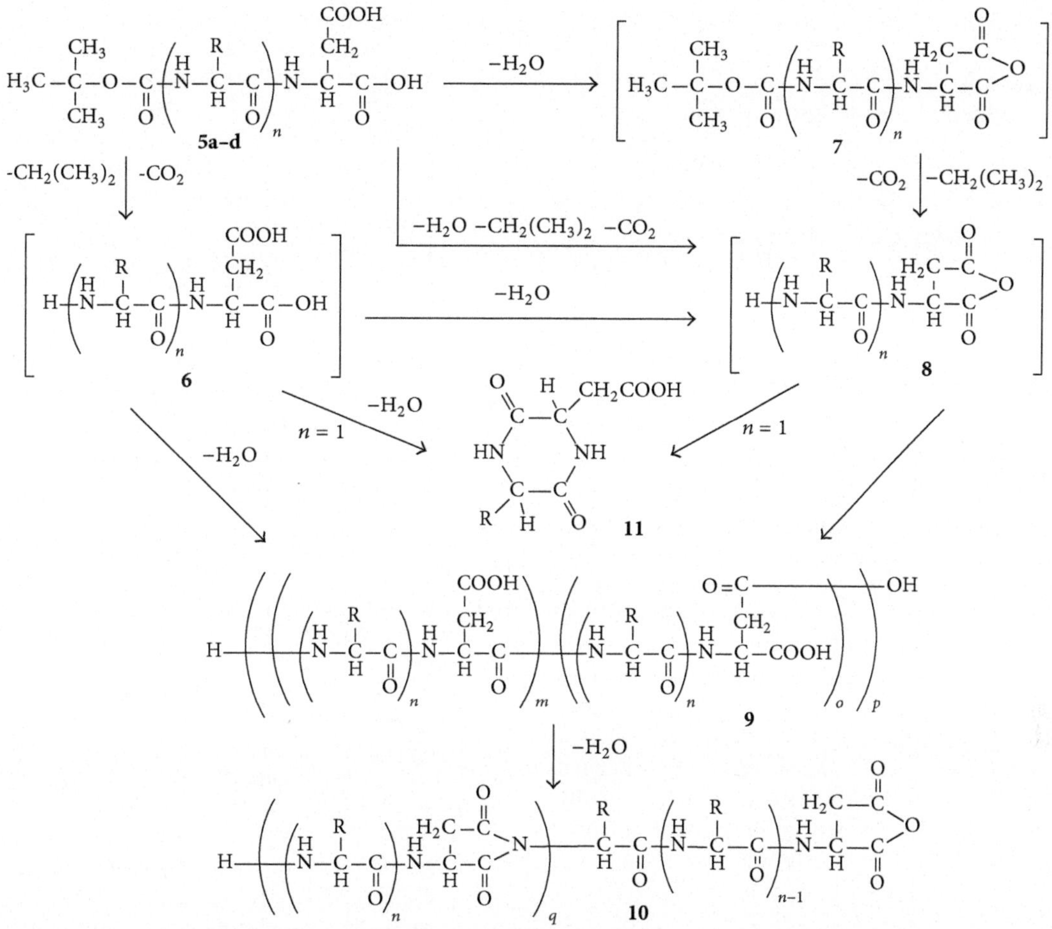

FIGURE 14: Postulated mechanism of the thermal reaction of Boc-peptides (**5a–d**). $n = 1$ for **5a–c**; $n = 2$ for **5d**. m, n, o, p, and q: natural number (0, 1, 2, 3, . . .).

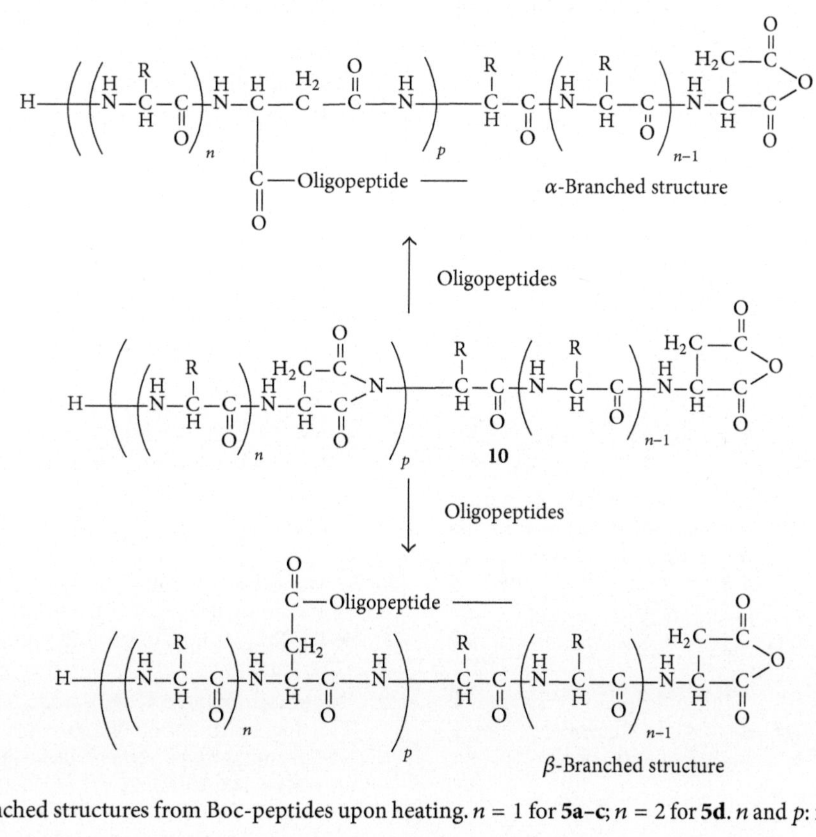

FIGURE 15: Postulated branched structures from Boc-peptides upon heating. $n = 1$ for **5a–c**; $n = 2$ for **5d**. n and p: natural number (1, 2, 3, . . .).

study of isolation of these compounds will give proof of the hypothesis.

The results in this paper reveal that the heating reactions of N-Boc-oligopeptides gave higher molecular peptides, which have almost the same amino acid compositions as their starting substrates.

The results suggest that the polypeptide products have a sequential structure but it is not established clearly. There may be some cases to yield branched polypeptides, which can be made during the reactions of the imide structure with the deprotected oligopeptides as shown in Figure 15. Since the chemical structure is very complexed, further research may be needed to clarify the details in the future research.

4. Conclusions

This investigation has demonstrated the first simple successive heating synthesis of copolypeptides by using peptide derivatives. In the thermal reactions, Boc-peptides melted, released a protecting group, and dehydrated to polypeptides. The mechanism was supported by thermal analysis accompanied by mass spectrometry. The holding amino acid residues in the polyamino acid structure were supported by amino acid analysis. The D/L ratio was suppressed to be below 10% at lower temperatures. These results suggested that the thermal reaction using Boc-peptides might be useful for producing sequential polypeptides. The sequential structure that is not proven should be clarified in the future research.

Disclosure

Takafumi Yamada is currently affiliated Center for Basic Technology Research, Tokyo Metropolitan Institute of Medical Science, Setagaya-ku, Tokyo 156-8506, Japan.

Acknowledgments

The authors thank the late professor emeritus Kaoru Harada of the University of Tsukuba for fruitful discussions.

References

[1] J. H. Johns, *Chemistry and Biochemistry of Amino Acids, Peptides*, vol. 4, Marcel Decker Inc., New York, NY, USA, 1977.

[2] Y. Imanishi, "Polymerization Of α-amino acid N-carboxyanhydride in the presence of preformed poly(α-Amino Acid)—from chain effect to stereoselective polymerization," *Pure and Applied Chemistry*, vol. 53, no. 3, pp. 715–727, 1981.

[3] H. R. Kricheldorf, "Polypeptides and 100 years of chemistry of α-amino acid N-carboxyanhydrides," *Angewandte Chemie*, vol. 45, no. 35, pp. 5752–5784, 2006.

[4] P. C. Brown and L. E. Glynn, "The antigenicity of sequential polypeptides. II. The antigenicity of some sequential polymers including several related to collagen," *Immunology*, vol. 25, no. 2, pp. 251–260, 1973.

[5] J. Ramachandran, A. Berger, and E. Katchalski, "Synthesis and physicochemical properties in aqueous solution of the sequential polypeptide poly(Tyr-Ala-Glu)," *Biopolymers*, vol. 10, no. 10, pp. 1829–1851, 1971.

[6] R. S. Rapaka and D. W. Urry, "Coacervation of sequential polypeptide models of tropoelastin. Synthesis of H-(Val-Ala-Pro-Gly)n-Val-OMe and H-(Val-Pro-Gly-Gly)n-Val-OMe," *International Journal of Peptide and Protein Research*, vol. 11, no. 2, pp. 97–108, 1978.

[7] R. S. Rapaka, K. Okamoto, and D. W. Urry, "Non-elastomeric polypeptide models of elastin. Synthesis of polyhexapeptides and a cross-linked polyhexapeptide," *International Journal of Peptide and Protein Research*, vol. 11, no. 2, pp. 109–127, 1978.

[8] S. Sakakibara, Y. Kishida, Y. Kikuchi, R. Sakai, and K. Kakiuchi, "Synthesis of poly-(L-prolyl-L-prolylglycyl) of defined molecular weights," *Bulletin of the Chemical Society of Japan*, vol. 41, no. 5, pp. 1273-1273, 1968.

[9] H. Leuchs, "Ueber die Glycin-carbonsäure," *Berichte der Deutschen Chemischen Gesellschaft*, vol. 39, no. 1, pp. 857–861, 1906.

[10] R. B. Woodward and C. H. Schramm, "Synthesis of protein analogs," *Journal of the American Chemical Society*, vol. 69, no. 6, pp. 1551-1552, 1947.

[11] A. B. Hughes, *Amino Acids, Peptides and Proteins in Organic Chemistry, Building Blocks, Catalysis and Coupling Chemistry (Amino Acids, Peptides and Proteins in Organic Chemistry (VCH))*, Wiley & Sons, New York, NY, USA, 2011.

[12] Humana Press, "Peptide Synthesis and Applications," in *Methods in Molecular Biology*, K. J. Jensen, A. P. Tofteng, and S. L. Pedersen, Eds., Springer, New York, NY, USA, 2nd edition, 2013.

[13] A. Vegotsky, K. Harada, and S. W. Fox, "The characterization of polyaspartic acid and some related compounds," *Journal of the American Chemical Society*, vol. 80, no. 13, pp. 3361–3366, 1958.

[14] J. Kovacs, H. N. Kovacs, I. Könyves, J. Császár, T. Vajda, and H. Mix, "Chemical studies of polyaspartic acids," *Journal of Organic Chemistry*, vol. 26, no. 4, pp. 1084–1091, 1961.

[15] T. Munegumi, Y.-Q. Meng, and K. Harada, "Thermal Syntheses of Polypeptides from N-Boc-amino acid (Aspartic acid, β-aminoglutaric acid) anhydrides," *Chemistry Letters*, vol. 17, no. 10, pp. 1643–1646, 1988.

[16] T. Munegumi and K. Harada, "Synthesis of polypeptides from amino acid derivatives melted upon heating," *Peptide Chemistry 1990*, pp. 75–78, 1991.

[17] W. Parr and P. Howard, "Separation of amino acid enantiomers by gas chromatography with an optically active stationary phase (*N*-TFA-L-valyl-L-leucine cyclohexyl ester)," *Chromatographia*, vol. 4, no. 4, pp. 162–166, 1971.

[18] H. Frank, G. J. Nicholson, and E. Bayer, "Rapid gas chromatographic separation of amino acid enantiomers with a novel chiral stationary phase," *Journal of Chromatographic Science*, vol. 15, no. 5, pp. 174–176, 1977.

[19] J. S. Davis, *Amino Acid and Peptides*, Chapman and Hall, London, UK, 1985.

[20] T. Munegumi, Y. Qing Meng, and K. Harada, "Polypeptide formation by heating *N*-t-butyloxycarbonyl acidic amino acid derivatives," *Asian Journal of Chemistry*, vol. 26, no. 15, pp. 4716–4722, 2014.

[21] T. Munegumi, K. Akao, Y. Kawatu, T. Yamada, and K. Harada, "Heating reactions of N-t-Butyloxycarbonyl-Asparagine and related compounds," *Asian Journal of Chemistry*, vol. 26, no. 19, pp. 6541–6548, 2014.

[22] G. G. Smith and B. S. De Sol, "Racemization of amino acids in dipeptides shows COOH > NH_2 for non-sterically hindered residues," *Science*, vol. 207, no. 4432, pp. 765–767, 1980.

[23] K. Harada, M. Matsuyama, and E. Kokufuta, "The aqueous thermal polycondensation of asparagine and isoasparagine and the structure of polyaspartic acid," *Polymer Bulletin*, vol. 1, no. 3, pp. 177–180, 1978.

[24] J. L. Radkiewicz, H. Zipse, S. Clarke, and K. N. Houk, "Accelerated racemization of aspartic acid and asparagine residues via succinimide intermediates: an ab initio theoretical exploration of mechanism," *Journal of the American Chemical Society*, vol. 118, no. 38, pp. 9148–9155, 1996.

[25] J. L. Radkiewicz, H. Zipse, S. Clarke, and K. N. Houk, "Neighboring side chain effects on asparaginyl and aspartyl degradation: an Ab initio study of the relationship between peptide conformation and backbone NH acidity," *Journal of the American Chemical Society*, vol. 123, no. 15, pp. 3499–3506, 2001.

[26] C. Brückner, S.-C. Bunz, D. Imhof, C. Neusüß, and G. K. E. Scriba, "Isomerization and epimerization of the aspartyl tetrapeptide Ala-Phe-Asp-GlyOH at pH 10-A CE study," *Electrophoresis*, vol. 34, no. 18, pp. 2666–2673, 2013.

[27] A. D. Borthwick, "2,5-diketopiperazines: synthesis, reactions, medicinal chemistry, and bioactive natural products," *Chemical Reviews*, vol. 112, no. 7, pp. 3641–3716, 2012.

10

(Bio)degradable Ionomeric Polyurethanes Based on Xanthan: Synthesis, Properties, and Structure

T. V. Travinskaya, A. N. Brykova, Yu. V. Savelyev, N. V. Babkina, and V. I. Shtompel

Institute of Macromolecular Chemistry, NAS of Ukraine, Kharkovskoe Shosse 48, Kiev 02160, Ukraine

Correspondence should be addressed to T. V. Travinskaya; travinskaya-tamara@rambler.ru

Academic Editor: Shida Miao

New (bio)degradable environmentally friendly film-forming ionomeric polyurethanes (IPU) based on renewable biotechnological polysaccharide xanthan (Xa) have been obtained. The influence of the component composition on the colloidal-chemical and physic-mechanical properties of IPU/Xa and based films, as well as the change of their properties under the influence of environmental factors, have been studied. The results of IR-, PMS-, DMA-, and X-ray scattering study indicate that incorporation of Xa into the polyurethane chain initiates the formation of a new polymer structure different from the structure of the pure IPU (matrix): an amorphous polymer-polymer microdomain has occurred as a result of the chemical interaction of Xa and IPU. It predetermines the degradation of the IPU/Xa films as a whole, unlike the mixed polymer systems, and plays a key role in the improvement of material performance. The results of acid, alkaline hydrolysis, and incubation into the soil indicate the increase of the intensity of degradation processes occurring in the IPU/Xa in comparison with the pure IPU. It has been shown that the introduction of Xa not only imparts the biodegradability property to polyurethane, but also improves the mechanical properties.

1. Introduction

Problem of environmental pollution with polymeric waste has shown a clear need for ecologically friendly materials based on renewable resources: vegetable oils and natural polymers, in particular, polysaccharides, which results in the development of a new trend in polymer chemistry, creation of sustainable biodegradable polymers having a wide range of applications [1–4]. Several methods of incorporating polysaccharides into the polyurethanes are discussed depending on the desired final properties of the polyurethane [5]. Due to the success of biotechnologies, a particular attention in modern chemistry of biodegradable polymers is paid to microbial polysaccharides (exopolysaccharides), which are climate and season independent. One of the leading places in this line belongs to xanthan, an extracellular polysaccharide of the bacteria *Xanthomonas campestris*. Xanthan molecules are prone to the self-association in aqueous solutions; a rise of solution ionic strength or polysaccharide concentration results in gel formation [6, 7]. Xanthan provides the synthetic polymers, in particular, polyurethanes, with ability to degrade under the influence of natural factors. It was shown that the chemical bond between synthetic and natural components played a crucial role in giving to the synthetic polymer biodegradability, opposed to mechanical mixtures, where only the natural component decomposed with time [8, 9]. Aqueous film-forming polyurethane dispersions have found diverse applications as finishing, impregnating, stiffener, protective materials, and adhesives in leather, textile, and furniture industry [10, 11]. The synthesis and characterization of novel aqueous polyurethane dispersions derived from plant oil and starch with the use of dimethylolpropionic acid as internal emulsifier which allows us to incorporate hydrophilic groups into the polymer chain and achieve stable self-emulsifying dispersions are discussed in [12–14]. It should be noted that the synthesis of degradable polyurethanes based on renewable feedstock is one of developing branches of green polyurethane chemistry. Such materials may be used as packaging, finishing, impregnating, stiffener, protective materials, and adhesives in leather, textile, and furniture industry, biologically active substances in agriculture. A distinctive feature of degradable polymers is to preserve the necessary operational parameters during the time of their use in combination with accelerated degradation under the influence of various natural factors after the expiration. Microorganisms are the main biological systems which destroy the polymeric compounds. The ability of polymers to degrade and be assimilated by microorganisms depends on a number of their characteristics: the chemical structure of the polymer, molecular weight, branching of macrochain, and supramolecular structure [15]. The aim of the latest developments in the field of degradable polymers is to establish the common principles in selecting the components for synthesis of polymer which combine a high level of performance with the (bio)degradation ability and knack of adjusting the (bio)degradation processes to ensure fast and safe (bio)degradation of polymer materials at the expiration of their working time. The paper considers the synthesis and complex study of "composition, structure, properties, and (bio)degradation ability" relationship of structurally modified polymeric materials based on polyurethane ionomer and microbial exopolysaccharide xanthan.

2. Experimental

2.1. Materials. Hexamethylene diisocyanate (HMDI) (Merck), polyoxytetramethylene glycol (POTMG 1000), dimethylolpropionic acid (DMPA), triethylamine (TEA), and Xa in the form of powder (Brookfield viscosity of 1% solution, according to Certificate of analysis, amounts 959 cps) were purchased from Aldrich and used as received; acetone was purchased from Fluka.

2.2. Preparation of Xanthan Containing Ionomer Polyurethane. Xanthan containing ionomer polyurethanes (IPU/Xa) in the form of aqueous dispersions was prepared by the reaction of an isocyanate precursor based on POTMG and HMDI with DMPA as ion centers carrier, taken in a molar ratio of 1 : 2 : 0.6. The reaction time for the isocyanate precursor formation was 2 h at 80°C until the NCO content reached the theoretical value to produce NCO-terminated prepolymer. The content of NCO groups was controlled according to [16]. Then, DMPA was added and reacted with prepolymer at 80°C for 1 h. Suitable amount of acetone was added to the system to decrease the viscosity of reaction mixture. Xa $(C_{35}H_{49}O_{29})_n$, was added to the reaction mixture in a form of dry powder (the reaction time was 1 h at 56°C). Neutralization of the DMPA fragments' carboxyl groups of obtained IPU/Xa hard block was performed with TEA (the molar ratio of TEA : DMPA = 1 : 1). Next, a simultaneous elongation and water dispersion were carried out, followed by removal of acetone. Film-forming opalescent dispersions were then obtained by phase inversion. Polymer films were formed at room temperature on a Teflon disk, followed by drying in an oven at 65°C and in a vacuum oven at 55°C to constant weight.

IPU/Xa systems with different Xa weight concentrations, 5%, 10%, and 30% (IPU/5Xa, IPU/10Xa IPU/30Xa), were prepared. IPU matrix was synthesized similar to IPU/Xa (using water as chain extender, without addition of Xa) and has been selected as an object of comparison.

2.3. Characterization

2.3.1. pH Values. pH values were determined using pH-meter "pH-150 M" (Russia).

2.3.2. Particle Size. Particle size measurements were determined from the turbidity spectrum using FEK-56M, according to [17].

2.3.3. Mechanical Properties. Measurements were performed using a tensile testing machine FU-1000 (VEB MWK "Fritz Heckert," Germany) at a tensile speed of 100 mm/min and temperature 25°C. The number of samples used in each measurement was three. Samples were prepared in a form of strips (width: 4 mm, operating length: 2 mm). Measurements were carried out in accordance with standard 14236-81; allowed error is 3%.

2.3.4. Water Absorption. Preliminary weighed dry films of IPU matrix, IPU/Xa, were immersed in water for 24 h, whereupon the excess water was removed with filter paper and samples were weighed. Water absorption (W_{H2O}, wt.%) was calculated according to $W_{H2O}(\%) = [(W_w - W_d)/W_d] \cdot 100\%$, where W_w and W_d are weight of the films in a wet and dry state, correspondingly.

2.3.5. Degree of Hydrolysis in Acid and Alkali Medium. Degree of hydrolysis in acid and alkali medium was determined by evaluation of the weight change and physic-mechanical characteristics of the samples after hydrolysis. Preweighed samples were immersed in 0.1 N solution of KOH and HCl and kept in a thermostat for 30 days at $T = 25°C$; afterwards the films were dried to constant weight with following weight control and tensile test.

2.3.6. IR Measurements. IR measurements were performed on Bruker "Tensor-37" Fourier transform infrared spectrometer in the region of wave numbers of 4500 cm^{-1}–500 cm^{-1}. Samples' surfaces were studied by attenuated total internal reflection infrared spectroscopy.

2.3.7. Thermodestruction of Xa-Containing IPU. Thermodestruction of Xa-containing IPU was studied by pyrolytic mass spectrometry (PMS). A device consisting of a mass spectrometer MX-1321 for determination of the components of gas mixtures in the range of mass numbers of 1–4000 and of cells for linearly programmable pyrolysis in the temperature range 25–400°C was used. The sample was placed into the cell which was evacuated ($1.33 \cdot 10^{-4}$ Pa) for 30 minutes at 25°C. The same pressure was maintained during the experiment. The heating rate was (6 ± 1)°C/min. The accuracy of determining the temperature of the sample was ±1°C. The ionization energy in the chamber of the mass spectrometer was 70 eV.

2.3.8. The X-Ray Diffraction Patterns. The wide-angle X-ray diffraction patterns (WAXS) were recorded with

Table 1: Properties of aqueous IPU/Xa dispersions and based films.

Xa content, %	Dispersion properties r_{av*}, nm	pH	Film characteristics Water absorption, 24 hours, %	Contact angle, degree	Tensile strength σ, MPa	Elongation at a break point ε, %
0	71	7,84	2,6	68	7,3	1100
5,0	176	7,55	23	50	8,1	742
10,0	272	7,17	65	45	16,9	530
30,0	504	6,82	405	29	36,4	360
100,0	—	—	The film was dissolved	—	**	**

r_{av*}: average particle radius, ** not applicable (the film is brittle).

X-ray diffractometer DRON-4-07 with roentgen schema made according to Debye-Scherrer method (transmission). Supramolecular structure was studied by small-angle X-ray diffraction (SAXS), using small-angle camera KRM-1 with flat-filled collimator, made according to Kratky method. All X-ray measurements were carried out in CuK_α radiation, monochromatic with Ni-filter at $T = 22 \pm 2^\circ C$.

2.3.9. Dynamic Mechanical Analysis (DMA). Dynamic mechanical measurements were carried out on a dynamic mechanical analyzer Q 800, TA Instruments in the tension mode at frequency of 10 Hz and heating rate of 2.0°C/min. The samples for DMA study were cut from the polymer films and had the following dimensions: length: 12.75 mm, width: 4 mm, and thickness: 0.3 mm. The viscoelastic properties, that is, mechanical loss factor ($\tan\delta$) and storage modulus (E'), were recorded as function of temperature. The glass transition temperature (T_g) was determined by the position of the loss factor maximum.

3. Results and Discussion

3.1. Colloid-Chemical Properties. The composition, colloid-chemical, and physic-mechanical properties of synthesized IPU/Xa dispersions and films are presented in Table 1.

The average particle size reports the degree of physical interaction of the constituent macromolecules and the dispersion medium. The introduction of Xa results in increasing of r_{av} owing to growth of the specific content of three-dimensional Xa fragments in macromolecules (as previously was shown using the small Xa concentrations [18]), which, however, does not lead to a loss of dispersions' stability [19]. With increase of Xa content, the pH of dispersions is systematically reduced as a result of the presence in Xa of a large number of acid pyruvic cycles.

3.2. Water Absorption. The degree of swelling of polymer components and their affinity to water is one of the indices of decomposition rate of polymeric materials under the influence of environmental factors. The investigation of the hydrolytic stability of IPU/Xa has shown the increase of the water absorption up to a maximum value of 405% at 30% Xa content. Such increase may be due to the presence of free polar fragments (carboxyl and OH groups), which determine the hydrophilic properties of the polymer. Probably, a part of Xa hydrophilic hydroxyls involved in the formation of intermolecular bonds with the polar groups of IPUs remains unengaged, following an increase of their concentration in the surface layers and contributing to the elevation of the hydrophilicity degree and, normally, to decreasing of contact angle values (Table 1).

3.3. Physic-Mechanical Properties. The tensile strength of IPU/Xa films significantly increases as the Xa content increases and exceeds this index of the IPU matrix in 1–5 times (Table 1). The increase of tensile strength for the IPU/Xa films indicates the occurrence of the intermolecular interaction between IPU and Xa. The loss of elasticity is specified by decrease of the specific weight of the flexible IPU segments in polymer composition, which is in agreement with the IR spectroscopic data indicating the hydrogen association of ether oxygen of POTMG with OH groups of Xa and steric hinders effected by volume molecules of Xa.

3.4. IR Spectroscopy. IR spectra of IPU matrix (1), IPU/10Xa (2), and Xa (3) are presented in Figure 1(a). The IR spectrum of the initial Xa is characterized by intense absorption bands of stretching vibrations of O-H bonds in the region of 3236–3613 cm^{-1}, 1031 cm^{-1}, by peak of medium intensity of stretching vibrations of C-H bonds (2926 cm^{-1}), an intense band with a frequency of 1736 cm^{-1}, apparently corresponding to the vibrations of ester groups. IPU matrix shows all characteristic bands of polyurethanes: $\nu(NH)_{assoc}$: 3314 cm^{-1}; $\nu(CH_2)$ and $\nu(CH_3)$ intermolecular hydrogen bonds: 2939 cm^{-1} and 2850 cm^{-1}, respectively; $\nu(C{=}O)_{free.}$: 1720 cm^{-1} and $\nu(C{=}O)_{assoc.}$: 1705 cm^{-1} of urethane groups; $\delta(NH)_{free}$: 1540 cm^{-1}; ν(CO-N): 1415 cm^{-1}; ν(COO) (ester fragments of urethane group) 1250 cm^{-1}; and ν(C-O-C) (of flexible segments of the matrix): 1105 cm^{-1}. The high-frequency shoulder at the peak of the stretching vibrations of NH groups in the IPU spectrum (3375 cm^{-1}) indicates the presence of free NH groups. In the spectrum of IPU/10Xa (Figure 1(a)), this shoulder disappears, and the intensity of hydrogen-bonded NH groups (3314 cm^{-1}) increases. A redistribution of the intensities of free 1720 cm^{-1} and bonded 1705 cm^{-1} CO groups is observed. The appearance of low-frequency shoulder (1031 cm^{-1}), which refers to the stretching

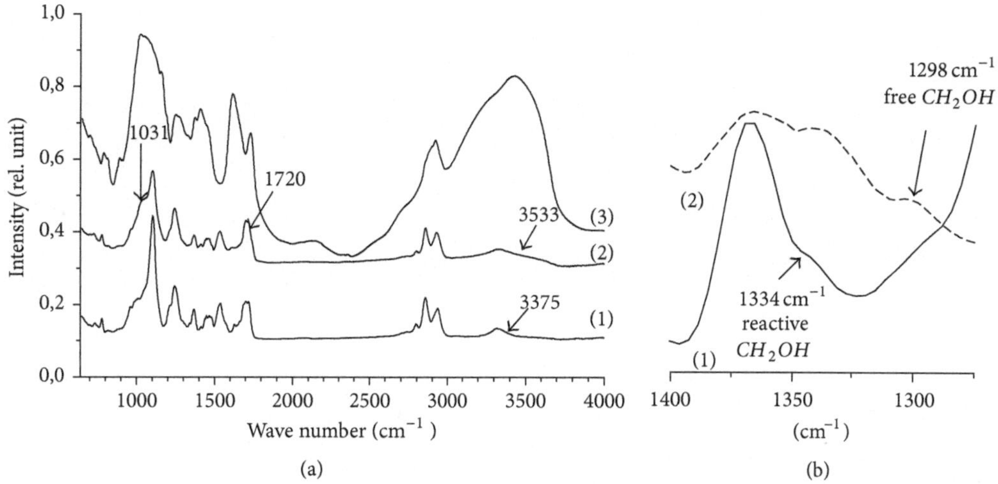

FIGURE 1: IR spectra: (a) IPU (1), IPU/10Xa (2), and Xa (3); (b) IPU /10Xa (1) and Xa (2).

vibrations of C-O and decrease of intensity of the valence symmetric vibrations of C-O-C 1105 cm^{-1} (Figures 1(a) and 2), indicates the hydrogen association of ether oxygen of polyurethane with OH -groups of Xa. The appearance of a broad weak band at 3533 cm^{-1} is associated with the presence of free OH groups of Xa.

At the same time, the appearance in the spectrum of the IPU/Xa (Figure 1(b), curve (1)) of the band at 1334 cm^{-1}, assigned to the C-H bond of CH_2OH group, with intensity lower than that for the native Xa and absence of the band 1298 cm^{-1} (Figure 1(b), curve (2)), indicates the participation of these groups in the formation of a chemical bond with the NCO groups of ionomeric oligourethane.

3.5. DMA Results. The temperature dependencies of $\tan\delta$ и E' and viscoelastic characteristics for the IPU and IPU/Xa films are shown in Figure 2 and Table 2. The dependence $\tan\delta$-T (Figure 2(a)) for the IPU is typical for segmented polyurethanes: there are two relaxation processes, corresponding to the soft and hard blocks. The maximum in the temperature range of −70 to 10°C (T_g = −30°C) corresponds to the IPU soft blocks' relaxation transition. In the same temperature range, a sharp drop of E' (Figure 2(b)) is observed. Sharp peak of $\tan\delta$ at 90°C and sharp drop of E' higher that 70°C indicate the existence of hard block in IPU.

Such viscoelastic behavior is inherent to many segmented IPU. This is due to the segmental mobility in the hard blocks' microregions (hard domains) and their destruction. The relaxation transition typical for the soft block of IPU is also observed for all IPU/Xa films along with the fixed beginning of the relaxation transition for the hard block (Figure 2(a)). However, the viscoelastic behavior of the IPU/Xa is significantly affected by incorporation of Xa. The increase of Xa content results in significant lowering of the relaxation peak corresponding to the soft block and in T_g reduction (Table 2). Thus, 30% of Xa content (IPU/30Xa) leads to lowering of T_g by 20°C in regard to T_g of IPU.

TABLE 2: Viscoelastic characteristics of xanthan containing IPU.

Xa content, %	T_g, °C (according to $\tan\delta_{max}$)	$\tan\delta_{max}$	E', MPa (25°C)
0	−30	0,54	6,5
5	−42	0,21	136
10	−45	0,18	247
30	−50	0,09	1230

Such changes in relaxation behavior may be caused by three-dimensional Xa molecules which form steric hindrances during the IPU/Xa soft block formation. It leads to its partial destruction and T_g lowering. At the same time, the decrease of specific weight of IPU/Xa soft-segmented part and interaction between the polar groups of the IPU and the OH groups of Xa results in the blocking of flexible polymer chains mobility and, consequently, decreasing $\tan\delta_{max}$. The relaxation transition corresponding to the hard block of IPU/Xa begins at higher temperatures compared to IPU matrix and is characterized by a smoother increase of mechanical loss factor (Figure 2(a)). Possibly the destruction of hard domains in IPU/Xa is preceded by the gradual destruction of intermolecular hydrogen bonds between the OH groups of Xa and urethane and urea groups of IPU. When Xa content amounts to 30%, the growth of mechanical loss factor is not observed up to 240°C. Perhaps a significant increase in the proportion of bound urethane groups at such Xa concentration leads to a sharp restriction or complete blocking of segmental mobility in the hard block. The low-intensity relaxation maxima on the temperature dependence of $\tan\delta$ indicate the heterogeneity of the IPU/Xa systems and the presence of amorphous microregions with different compositions. Incorporation of Xa results in substantial increase of storage modulus (Figure 2(b)). Thus, at T = 25°C, the value of the storage modulus for IPU/5Xa is more than 20 times higher than that for the pure IPU. When the content of Xa reaches 30% (IPU/30Xa), the value of storage modulus increases almost in 200 times (Table 2). Such significant

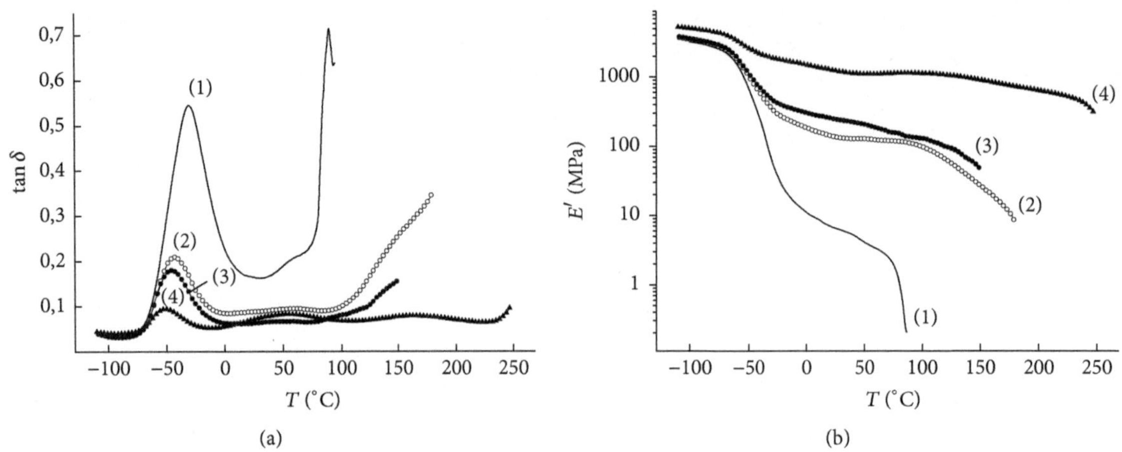

FIGURE 2: Temperature dependence of tan δ (a) and E' (b) of IPU matrix (1), IPU/5Xa (2), IPU/10Xa (3), and IPU/30Xa (4).

increase of the storage modulus of Xa-comprising IPUs confirms the presence of chemical and hydrogen bonding between the components.

Thus, viscoelastic properties of IPU/30Xa are determined by Xa content and intermolecular interactions between the components.

3.6. Acid and Alkaline Hydrolysis. The presence of Xa in the IPU chain determines the nature of hydrolysis, one of the main factors of the materials' degradation under environmental conditions. The higher the Xa content, the greater the mass loss and the lower the strength and elasticity of the films after hydrolysis (Table 3); that is, IPU/Xa films are more susceptible to hydrolytic destruction in comparison with the IPU matrix.

The IPU/Xa (bio)degradation ability was studied by a technique that allows us to simulate the processes taking place under the natural conditions [20]. Samples were incubated in containers with soil of medium biological activity (pH = 6.82, relative humidity: 60%, T = 14–25°C) for a period of 1–4 months [21]. The analysis of the soil's bacterial population has shown the presence of fungi of the following genera: *Rhizopus, Aspergillus,* and *Penicillium*.

The rate of degradation was controlled by weight loss of incubated samples through regular intervals. The higher the Xa content, the greater the mass loss of the samples (Table 4), and within 4 months it reaches 10.2% (IPU/5Xa) and 38% (IPU/30X a), which exceeds the actual Xa content and the specified matrix characteristic. This indicates the possibility of control of destruction rate by changing the component composition.

Soil-born microorganisms (MO) affect the films' properties. They provoke a decrease of physic-mechanical parameters (σ/ε) after remaining 4 months in the soil: for IPU matrix by 2.8/1.4%, for IPU/5Xa by 50.1/18, 0%, respectively, and for IPU/10Xa and IPU/30Xa; these indices are not available, because the films have lost their integrity. Thus, the presence of Xa promotes biodegradation of polymer materials. Visual assessment of the films after testing in the soil also indicates a sufficiently high degree of samples' damage with MO.

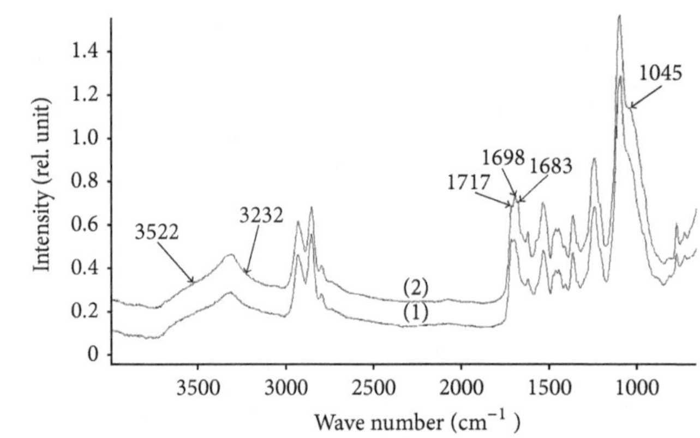

FIGURE 3: IR spectra of IPU/10Xa before (1) and after (2) incubation into the soil for 4 months.

The degradation of the samples was confirmed by IR spectroscopy on the example of IPU/10Xa. Figure 3 shows IR spectra of IPU/10Xa before (1) and after (2) incubation into the soil. In the absorption region (1000–1800) cm^{-1} of the after-ground sample (spectrum 2), a redistribution of the intensities of the bonded 1698 cm^{-1} and free 1717 cm^{-1} C=O groups is observed. The appearance of a new band 1683 cm^{-1} is associated with the decomposition of COOH groups: the ester group turns into an ether one. In addition, the expanse of the OH- groups' band to the more (3522 cm^{-1}) and less (3232 cm^{-1}) frequency region is observed. The appearance of a new band 1045 cm^{-1} is a result of the ether bonds decomposition after sample incubation in the soil.

3.7. Thermodestruction. The processes of thermal destruction of IPU and IPU/Xa, a comparative analysis of their structure, and the depth of (bio)degradation in the soil were estimated by the PMS method.

Analysis of the temperature dependence of the total ion current of the emission of volatile degradation products of the IPU (Figure 4(a)) has shown its complete thermal decomposition in two stages with maximum decomposition rates at the

TABLE 3: Physic-mechanical characteristics of the IPU, IPU/Xa films after acid and alkaline hydrolysis.

Xa content, %	0,1 N solution of KOH			0,1 N solution of HCl		
	Weight loss, %	Tensile strength σ, MPa	Elongation at a break point, ε, %	Weight loss, %	Tensile strength σ, MPa	Elongation at a break point, ε, %
0	0,1	5,7	970	0,22	6,7	860
5,0	1,8	1,3	442	1,7	1,7	402
10	3,3	*	*	3,4	*	*
30	10,6	*	*	11	*	*

*Fragmentation of the film.

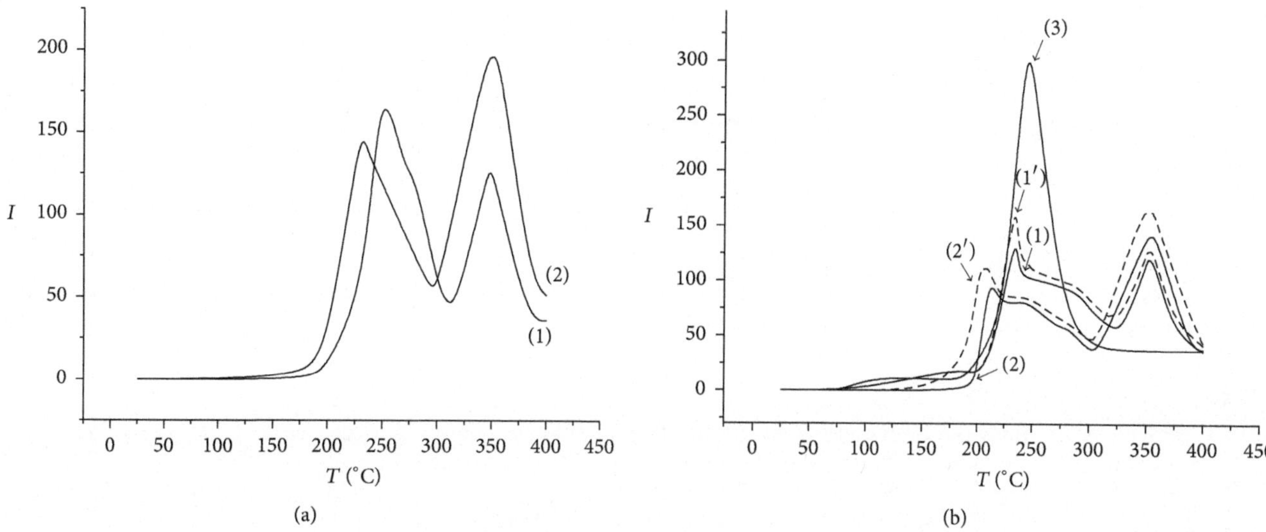

FIGURE 4: The temperature dependence of the ion current intensity for IPU (a), IPU/10Xa (1, 2), and IPU/30Xa (1′, 2′) (b): (1, 1′) initial sample; (2, 2′) sample after incubation in the soil (4 months); (3) initial Xa.

TABLE 4: Results of of IPU/Xa incubation in the soil.

Characteristic (4 months)	Xa content, %			
	0	5,0	10,0	30,0
Weight loss, %	1,3	10,2	13	38
Changing of physic-mechanical indices after the test (σ/ε), %	2,8/1,4	50,1/18,0	*	*

*Fragmentation of the film.

temperatures of 250°C and 350°C, which corresponds to the decomposition temperatures of hard and soft blocks. The first stage corresponds to the pyrolysis of urethane and urea bonds and the second to the pyrolysis of oligoether fragments [22]. The maximum of the peak of IPU kept in the soil shifts to the lower temperature (220°C) and is supplemented by a decrease of the intensity of the decomposition peak of hard blocks. The maximum decomposition peak of soft blocks remains at the same temperature of 350°C; however, the intensity of the total ion current of volatile products is significantly increased. Thus, keeping the IPU films in the soil results in the primary degradation of oligoether component, since it is known [23] that the oligoether fragments are predominantly situated in the IPU surface layers. In the initial IPU/10Xa (Figure 4(b)), the high intensity of the volatile products' release is observed at the hard blocks' decomposition and is accompanied by the release of water, as well as the following fragments: (*m/z*: 28 (CO, C_2H_4, N_2, CHNH), 31 (CH_3O, CH_2OH), 43 (C_2H_5N), C_3H_7), CH_3CO), 55 (C_4H_7), CH_2CHCO), 71 (CH_2CHCH_2CHO), and 73 ($OHCCH_2CHOH$)), which are the pyrolysis products of urethane and urea groups. With the increase of Xa content (sample IPU/30Xa), the intensity of release of the volatile products of the hard block decomposition increases (Figure 4(b) (1′)). After the incubation into the soil, similar to IPU, a greater intensity of volatile products' release is observed at the decomposition of the soft blocks for both IPU/10Xa and IPU/30Xa. The shift of the maximum (350°C) is not observed. For the incubated IPU/10Xa and IPU/30Xa samples, the temperature of the maximum intensity of the release of volatile decomposition products of the hard block is shifted towards the lower temperatures, from 230 to 210°C for IPU/10Xa and from 230 to 200°C for IPU/30Xa.

The dependence of the ion current on temperature for Xa has only one maximum at 250°C (Figure 4(b)). The fragments with *m/z* 18 and 44, corresponding to water and carbon dioxide, have the maximum specific content. It should be noted that there are practically no fragments typical for the degradation of native Xa (*m/z* 15 (CH_3), 17 (OH), 32 (O_2),

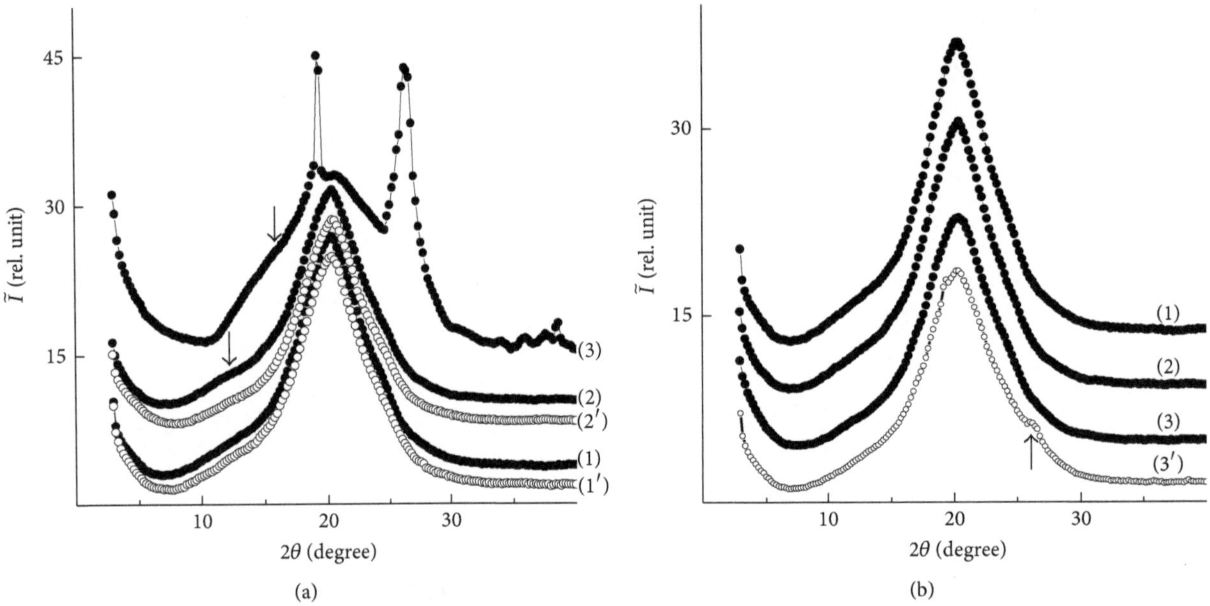

FIGURE 5: WAXS patterns of IPU and Xa-containing IPU: (a) initial IPU (1), IPU/5Xa (2), Xa (3), and being exposed in a soil for 4 months: IPU (1′) and IPU/5Xa (2′); (b) initial IPU (1), IPU/5Xa (2), and IPU/10Xa (3, 3′): (3) experimental and (3′) additive.

and 60 (CH_2CO_2H)) among the degradation products of IPU/10Xa and IPU/30Xa, which indicates the chemical binding of Xa. As it follows from the temperature dependence, the destruction of the IPU hard block occurs with higher intensity in comparison with the IPU/10Xa and IPU/30Xa, which may be a result of intra- and intermolecular hydrogen bonding between OH groups of Xa and urethane and urea groups of IPU. Therefore, the destruction of intermolecular hydrogen bonds precedes the destruction of the hard blocks and results in the decrease of the IPU/Xa hard blocks' decomposition intensity. However, the higher the Xa content, the lower the degradation temperature.

3.8. X-Ray Analysis. The presence of a single diffuse diffraction maximum with angular position $2\theta_m = 20{,}3^\circ$ in WAXS patterns of IPU and IPU/5Xa (Figure 5(a), curves (1, 2)) shows that these polymers are characterized by short-range order at the translation in expanse of their macrochain fragments.

The average distance (d) between the centers of macromolecular chain layers of IPU and IPU/5Xa, according to the Bragg equation $d = \lambda\,(2\sin\theta) - 1$ (where λ is the wavelength of the CuK_α radiation $\lambda = 0{,}154$ nm), amounts to 0.437 nm.

Two discrete diffraction maxima, singlet and multiplet at $2\theta_m = 19{,}3^\circ$ and $26{,}2^\circ$, respectively, appeared against a background of evident asymmetric diffraction peak $2\theta_m \approx 20{,}6^\circ$ in WAXS pattern of Xa (Figure 5(a), curve (3)), testifying to the amorphous-crystalline structure of Xa.

Nonlinear change of scattering intensity in the range of $2\theta \sim 11.2^\circ$–17.0° is indicated by poorly detected diffraction maximum at $2\theta_m \sim 15{,}8^\circ$ (arrow). This maximum describes the short-range order at the translation in the Xa volume of its side branch fragments.

We evaluated the relative level of crystallinity X_{cr} of polysaccharide Xa in accordance with Matthews's method [24], $X_{cr} \approx 20\%$, and determined effective size L of Xa crystallites using Scherer method [25], $L \approx 18$ nm.

The amorphous-crystalline structure of Xa was not detected on the X-ray diffraction pattern of IPU/5Xa and IPU/10Xa (Figure 5(b), curve (2, 3)), due to intermolecular interaction between components. Comparison of experimental and calculated additive (when interaction between components is absent) X-ray diffraction patterns of IPU/10Xa (Figure 5(b), curves (3, 3′)) has served as evidence that the absence of Xa crystalline structure phenomenon in IPU/Xa composition is caused by intermolecular interactions between IPU and Xa components.

Calculated additive X-ray diffraction pattern of IPU/10Xa (Figure 5(b), curve (3′)) has shown that in case of the absence of components' interaction there is a weak expression of the most intense diffraction peaks (at $2\theta_m = 19{,}3^\circ$ and $26{,}2^\circ$) that characterize the crystalline structure of Xa. This is a conformation of intermolecular interactions between IPU and Xa which results in suppression of Xa capacity for crystallization.

The invariable intensity and angular position of the amorphous halo ($2\theta_m \approx 20{,}3^\circ$) of the initial and aged in the soil IPU and IPU/5Xa samples (Figure 5(a), curves (1, 1′) and (2, 2′)) indicate that there is no change in their amorphous structure.

For more complete structural characterization of the initial 13 and aged for 4 months in the soil IPU and IPU/5Xa samples, we study their microheterogeneous structure. SAXS results (Figures 6(a) and 6(b)) have shown that all studied polymers have microheterogeneous structure. There are the areas of microheterogenity in their volume, the electron density ($\Delta\rho$) between which is different from zero, $\Delta\rho =$

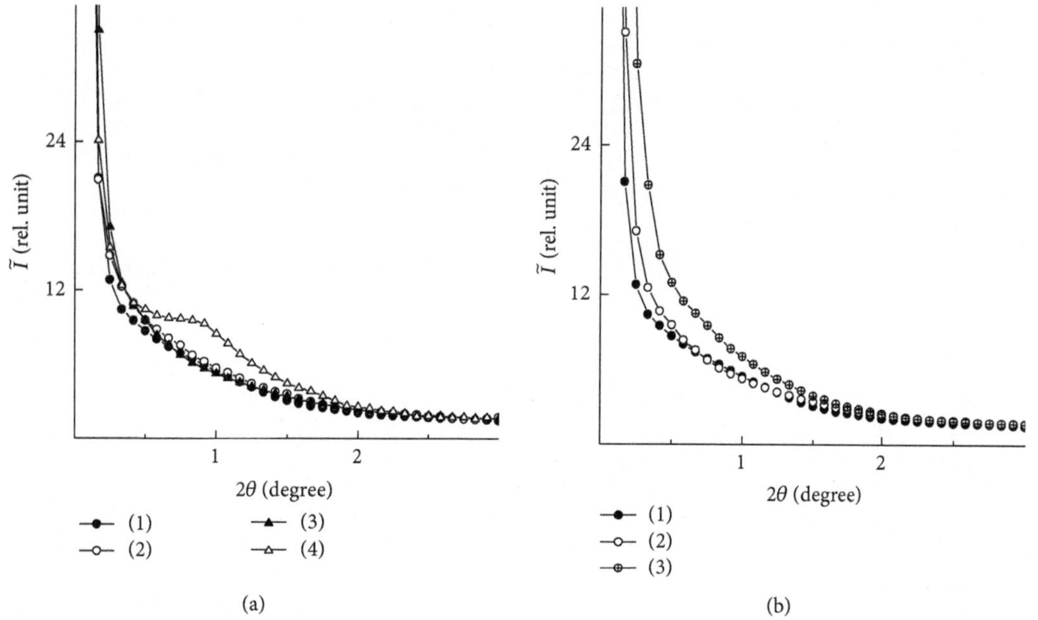

FIGURE 6: SAXS patterns of IPU and Xa-containing IPU: (a) initial IPU (1) and IPU/5Xa (2), and being exposed in soil for 4 months: IPU (3) and IPU/5Xa (4); and (b) initial IPU (1), IPU/5Xa (2), and IPU/10Xa (3).

$\rho - \langle\rho\rangle$, where ρ and $\langle\rho\rangle$ are the local and average value of the electron density in two-phase system.

The comparison of the profiles of the initial and aged for 4 months in soil IPU and IPU/5Xa samples has shown that the initial IPU has the lowest scattering intensity as a result of thermodynamic incompatibility between soft and hard IPU units [26]. The sample IPU/5Xa has a slightly higher scattering intensity than the IPU. Even higher scattering intensity and, correspondingly, the value of the electron density contrast have the sample IPU/10Xa. At that the lack of interference maximum on the intensity profiles indicates a disordered placement of microareas of heterogeneity in polymer volume (Figure 6(b)). Attention is drawn to the fact that exposure of samples to soil for 4 months results in a growth of scattering intensity of both IPU and, particularly, IPU/5Xa. The appearance of the interference maximum $2\theta_m \approx 0{,}92^\circ$ in a form of "shoulder" (Figure 6(a), curves (3, 4)) indicates the existence of periodicity in distribution of microareas of heterogeneity with different size of local electron density in IPU/5Xa volume. According to the above Bragg equation, the value of period D of alternation in the volume of monotype ρ-sized microareas of heterogeneity is 9.6 nm. It should be noted that the increase of the SAXS intensity at the transition from IPU to IPU/5Xa and IPU/10Xa, as well as the result of exposure of the first two samples for 4 months in soil, characterizes the variations of the level of heterogeneity of their structure. To quantify the relative level of heterogeneity of the structure, we calculated the structural parameter "Porod invariant" Q' [27], the value of which is independent (invariant) of the form of microareas of heterogeneity:

$$Q' = \int_0^\infty q\tilde{I}(q)\,dq, \qquad (1)$$

where q is directed magnitude of wave vector s ($q = 2\pi s$).

TABLE 5: Parameters of microheterogeneous structure of initial IPU, IPU/5Xa, IPU/10Xa and IPU, IPU/5Xa after 4 months of exposure in soil.

Sample	Q', rel. unit	l_p, nm
IPU	5,9	6,5
IPU/5Xa	6,4	5,4
IPU/10Xa	7,8	6,1
IPU, 4 months in soil	6,5	5,8
IPU/5Xa, 4 months in soil	7,8	5,3

This parameter characterizes the integral intensity of X-ray scattering by two-phase system and has a direct connection with the quadratic fluctuation of electron density in its volume.

According to calculated values of Q' (Table 5), IPU possesses the least level of structure heterogeneity, while IPU/10Xa and IPU//5Xa, which are kept for 4 months in soil, have the largest level of structure heterogeneity.

Another characteristic of microheterogeneous structure of studied systems is the average size of microareas of heterogeneity existing in their volume. The range of heterogeneity l_p was determined by the Ruland method [28]. This parameter is directly related to the average diameter of the microareas of heterogeneity in the two-phase system. It was determined that presence of Xa results in a decrease of the effective size of microareas of heterogeneity both in the initial samples and after 4 months of exposure in soil (Table 4). Unlike the level of heterogeneity Q', the transition from IPU to IPU/5Xa and IPU/10Xa causes, in general, the reduction of the range of heterogeneity l_p (Table 4).

Thus, as a result of X-ray study, Xa has been found to miss its ability to crystallization due to the intermolecular interactions between components in IPU/Xa systems. The increase of Xa content in IPU/Xa systems reduces the size of microareas of heterogeneity. The disappearance of the diffraction maximum ($2\theta_m \approx 12{,}2^{\circ}$) after exposure of the sample IPU/5Xa in the soil indicates a change of its amorphous structure as a result of (bio)degradation following a consecutive increase of structure heterogeneity on nanosized level.

4. Conclusions

New ecologically friendly IPUs were prepared on the basis of the renewable exopolysaccharide Xa. Introduction of Xa allows partially replacing exhaustible oil row materials and improving the strength properties of pure IPU matrix: the tensile strength of IPU/Xa systems is 1–5 times higher compared with IPU. Along with retention of other functional characteristics of the IPU, Xa imparts it a property of (bio)degradation after the end of lifetime that leads to the deep chemical transformations occurring in the IPU/Xa systems. The proven covalent and hydrogen bonding between components ensures the occurrence of destructive processes of the entire system as a whole. With an increase of Xa content, the mass loss of IPU/Xa systems as a result of hydrolytic splitting and degradation in the soil increases and exceeds the actual content of Xa and the value of mass loss of the IPU matrix. The results of PMS, DMA, and X-ray scattering indicate that the presence of Xa in polymer macrochain leads to the formation of a new structural organization different from the structure of the IPU matrix due to the chemical bonding between the exopolysaccharide and diisocyanate. The structural and operational properties and degradability of studied polymers are determined by the structure and content of the natural component. Film-forming aqueous polyurethane dispersions on the basis of exopolysaccharide Xa are perspective as biologically active substances in agriculture: immunostimulants and protective coating for seeds and plants; antitranspirants for reducing water scarcity and optimization of the production process of crops in drought conditions; and binders for biologically active substances granulation. The advantages of such material lie in environmentally friendly production technology due to the absence of organic solvent, economy through the use of cheap renewable raw materials, and reducing the harmful impact on the environment through the regulated level of (bio)degradation after the expiration of life time.

References

[1] S. Rogovina, K. Aleksanyan, E. Prut, and A. Gorenberg, "Biodegradable blends of cellulose with synthetic polymers and some other polysaccharides," *European Polymer Journal*, vol. 49, no. 1, pp. 194–202, 2013.

[2] S. A. Ashter, "Overview of biodegradable polymers," in *Introduction to Bioplastics Engineering*, pp. 19–30, Elsevier, Amsterdam, Netherlands, 2016.

[3] T. Travinskaya and Y. Savelyev, "Aqueous polyurethane dispersions—sodium alginate based blends and hydrogels," *Frontiers in Heterocyclic Chemistry*, vol. 2, no. 1, pp. 20–25, 2016.

[4] P. Alagi, Y. J. Choi, and S. C. Hong, "Preparation of vegetable oil-based polyols with controlled hydroxyl functionalities for thermoplastic polyurethane," *European Polymer Journal*, vol. 78, pp. 46–60, 2016.

[5] M. J. Donnelly, J. L. Stanford, and R. H. Still, "The conversion of polysaccharides into polyurethanes: A review," *Carbohydrate Polymers*, vol. 14, no. 3, pp. 221–240, 1991.

[6] J. G. Southwick, H. Lee, A. M. Jamieson, and J. Blackwell, "Self-association of xanthan in aqueous solvent-systems," *Carbohydrate Research*, vol. 84, no. 2, pp. 287–295, 1980.

[7] S. C. Moldovenau, *Analytical Pyrolysis of Natural Organic Polymers*, vol. 20, Brown & Williamson Tobacco Corp., Macon, Ga, USA, 1998, p. 510.

[8] T. V. Travinskaya, A. N. Brykova, I. K. Kurdish, A. V. Chevychalova, and Y. V. Savelyev, "Degradable ionomer polyurethane on the basis of xanthan," *Reports of the Academy of Sciences*, vol. 7, pp. 132–139, 2014.

[9] Y. V. Savelyev, T. V. Travinskaya, L. A. Markovskaya, and A. N. Brykova, "The method of obtain of degradable polymer composition," Pat. No 93372 Ukraine, Publ. 25.09.2014. Bull. no. 18, 2014.

[10] Q. B. Meng, S.-I. Lee, C. Nah, and Y.-S. Lee, "Preparation of waterborne polyurethanes using an amphiphilic diol for breathable waterproof textile coatings," *Progress in Organic Coatings*, vol. 66, no. 4, pp. 382–386, 2009.

[11] V. Sriram, S. Sundar, A. Dattathereyan, and G. Radhakrishnan, "Synthesis and characterization of cationomeric AB crosslinked polyurethane polymers based on different chain extenders," *Reactive and Functional Polymers*, vol. 64, no. 1, pp. 25–34, 2005.

[12] J. Bullermann, S. Friebel, T. Salthammer, and R. Spohnholz, "Novel polyurethane dispersions based on renewable raw materials—Stability studies by variations of DMPA content and degree of neutralisation," *Progress in Organic Coatings*, vol. 76, no. 4, pp. 609–615, 2013.

[13] T. Travinskaya, Y. Savelyev, and E. Mishchuk, "Waterborne polyurethane based starch containing materials: preparation, properties and study of degradability," *Polymer Degradation and Stability*, vol. 101, no. 1, pp. 102–108, 2014.

[14] S. J. Lee and B. K. Kim, "Covalent incorporation of starch derivative into waterborne polyurethane for biodegradability," *Carbohydrate Polymers*, vol. 87, no. 2, pp. 1803–1809, 2012.

[15] Y. V. Savelyev, T. V. Travinskaya, L. P. Robota et al., "Biodegradable polyurethane materials of different origin based on natural components," *Austin Journal of Biomedical Engineering*, vol. 2, no. 1, article 1030, 2015, http://www.austinpublishinggroup.com/.

[16] ASTM D2572-03, "Standard test method for isocyanate groups in urethane materials or prepolymers," ASTM, West Conshohocken, Pa, USA, 2003.

[17] S. Y. Shegolev and V. I. Klenin, "Determination of parameters of complicated disperse polymer system from turbidity spectrum," *Vysokomolekulyarnye Soedineniya B*, vol. 13, no. 12, pp. 2809–2815, 1971.

[18] T. V. Travinskaya, A. N. Brykova, V. I. Bortnitskiy, and Yu. V. Savelyev, "Preparation and Properties of (bio)degradable ionomer polyurethanes based on xanthan," *Polymernyj Journal*, vol. 36, no. 4, pp. 393–400, 2014.

[19] N. I. Levchenko, S. A. Sukhorukova, and T. V. Travinskaya,

"Aqueous anionactive polyurethanes for high—quality coatings," in *Proceedings of the Partnership in Polymers, the Cambridge Polymer Conference*, pp. 195–200, Cambridge, UK, 1996, Special conference issue of full papers.

[20] B. S. Lee, M. Vert, and E. Holler, *Water-Soluble Aliphatic Polyesters: Poly(malic acid)s*, Wiley-VCH Verlag Gmbh, Weinheim, Germany, Polyester, 1st edition, 2002.

[21] I. P. Babaeva and G. M. Zenova, *Biology of Soils*, Moscow University, Moscow, Russia, 1989.

[22] V. A. Zaikin, *Mass Spectroscopy of Synthetic Polymers*, All-Russian Mass Spectrometric Society, Moscow, Russia, 2009.

[23] V. V Boyko, L. V. Kobrina, S. V. Riabov, and R. L. Gaiduk, "Investigation of biodegradable properties of polyurethane compositions filled by chitosan," *Polymernyj Journal*, vol. 26, no. 4, pp. 235–238, 2004.

[24] J. L. Matthews, H. S. Peiser, and R. B. Richards, "The X-ray measurement of the amorphous content of polythene samples," *Acta Crystallographica*, vol. 2, no. 2, pp. 85–90, 1949.

[25] A. Guiner, *Radiography of Crystals. Theory And Practice*, Nauka, Moscow, Russia, 1961, p. 604.

[26] V. I. Shtompel and Y. Y. Kercha, *Structure of Linear Polyurethanes*, Nauka, Moscow, Russia, 2008, Kiev, p. 248.

[27] G. Porod, in *General Theory, Small-Angle X-Ray Scattering*, O. Glatter and O. Kratky, Eds., pp. 17–51, Academic Press, Cambridge, Mass, USA, 1982, London.

[28] R. Perret and W. Ruland, "Eine verbesserte Auswertungsmethode für die Röntgenkleinwinkelstreuung von Hochpolymeren," *Kolloid-Zeitschrift & Zeitschrift für Polymere*, vol. 247, no. 1-2, pp. 835–843, 1971.

Effect of Bolt-Hole Clearance on Bolted Connection Behavior for Pultruded Fiber-Reinforced Polymer Structural Plastic Members

Sang-Pyuk Woo,[1] Sun-Hee Kim,[2] Soon-Jong Yoon,[1] and Wonchang Choi[2]

[1]Department of Civil Engineering, Hongik University, Seoul 04066, Republic of Korea
[2]Department of Architectural Engineering, Gachon University, 1342 Seongnam-daero, Sujeong-gu, Seongnam-si, Gyeonggi-do 13120, Republic of Korea

Correspondence should be addressed to Wonchang Choi; wonchang.choi@gmail.com

Academic Editor: Reza Aliha

Bolt-hole clearance affects the failure mode on the bolted connection system of pultruded fiber-reinforced polymer plastic (PFRP) members. The various geometric parameters, such as the shape and cross-sectional area of the structural members, commonly reported in many references were used to validate the bolt-hole clearance. This study investigates the effects of the bolt-hole clearance in single-bolt connections of PFRP structural members. Single-bolt connection tests were planned using different bolt-hole clearances (e.g., tight-fit and clearances of 0.5 mm to 3.0 mm with 0.5 mm intervals) and uniaxial tension is applied on the test specimens. Most of the specimens failed in two sequential failure modes: bearing failure occurred and the shear-out failure followed. Test results on the bolt-hole clearances are compared with results in the previous research.

1. Introduction

Until the 1990s, the use of fiber-reinforced polymer plastic (PFRP) composites was limited to aerospace and military applications [1]. However, the fiber-reinforced plastic (FRP) composites have many advantageous mechanical properties, such as an excellent strength-to-weight ratio and stiffness-to-weight ratio, which make them highly desirable also as a building material for civil engineering applications [2, 3]. Therefore, efforts to include FRP materials in civil engineering have been increased markedly over recent decades. In order to use FRP materials in the construction field, the pultruded structural member must be connected. Several types of connections are currently used for this purpose, including bolted, bonded, a combination of bolted and bonded, and interlocking connections. For civil engineering applications, bolted connections are preferred and deemed the most practical because they are easy to assemble and disassemble, easy to maintain, and are usually cost-effective when compared with the other types of connections [4].

In general, there are four possible failure modes for pultruded fiber-reinforced polymer plastic (PFRP) single-bolted connections that are subjected to tensile forces [5–12], as shown in Figure 1. Figure 1(a) illustrates net-tension failure, which is attributable to the reduced cross-sectional area of a FRP member that is due to the bolt hole. Figure 1(b) shows cleavage-tension failure where the cross-sectional area of the bolt is not resistant to tensile loading and breaks away from the contact point. Figures 1(c) and 1(d) show bearing failure and shear-out failure, respectively.

Conventionally, FRP materials are failed in a brittle mode [12], whereas bearing failure allows ductile behavior unlike the other failure modes (net-tension failure, cleavage-tension failure, and shear-out failure). Bearing failure is the most likely to occur when FRP material is used in a structural member. Various parameters need to be taken into account to ensure safety in the design with regard to the bolt connection in order to induce bearing failure. Researchers and manufacturers who have studied bolt connections of composite materials have suggested a geometric recommendation index based on experiments and experience [16–24]. Geometric

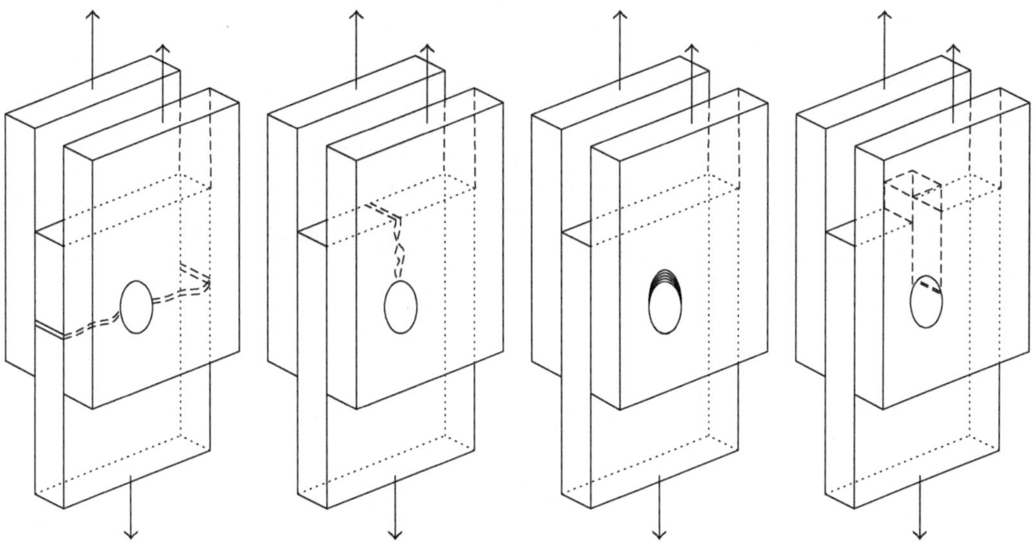

FIGURE 1: Typical failure modes of bolted connections.

coefficients for a bolt's width-to-diameter ratio of 6 (i.e., $w/d = 6$) and end-to-diameter ratio of 3 (i.e., $e/d = 3$) were proposed to induce bearing failure according to ASTM Standard D5961/D5961 M-96 [21]. Rosner and Rizkalla [6] reported that bearing failure mode occurs predominantly with an increase in the use of high geometric coefficient values for both the width-to-diameter ratio and end-to-diameter ratio. They recommended that the geometric coefficients (w/d and e/d) should be greater than 5.0. Therefore, the geometric coefficients used in this study are $w/d = 5$ and $e/d = 5$.

The clearance between the bolt diameter (d_b) and the hole diameter (d_h) is one of the parameters that allows constructability and ductile failure in the bolt connection. However, very few experiments have been conducted with regard to the necessary clearance, and existing codes recommend various values to determine the appropriate bolt-hole clearance. Thus, this study aims to address this informational deficit and determine the adequate bolt-hole clearance to prevent brittle failure in a single-bolted connection system for PFRP materials. A total of 98 single-bolt connection specimens with various bolt-hole clearances were tested under tensile loading. Also, the existing available code values found in the Eurocomp Design Code [13] and the Italian National Research Council (CNR) standards [14] and the bolt-hole clearance reported by Mottram [15] were compared with the experimental results obtained.

2. Experimental Program

2.1. Mechanical Properties of PFRP Structural Members. Two shapes, angle, and I-shape, of PFRP structural members were investigated. The PFRP structural members were fabricated in Korea using the pultrusion process and manufactured using E-glass fiber and polyester resin with the fiber volume fraction of 0.578. In the pultruded structural shapes, E-glass fiber bundles are placed along the longitudinal direction of the member.

Figure 2 presents the dimensions of two types of PFRP specimens.

The material properties of the PFRP members were determined from tensile tests, compression tests, and shear tests. The tensile test specimens were taken in the longitudinal direction (i.e., the member axis direction that coincided with the reinforcing fiber direction) and prepared, with slight modification, according to ASTM D3039/D3039 M-08 [26]. Figure 3 shows the prismatic tensile test specimens and test setup. Fifteen specimens were prepared, including five specimens from the angle (specimen dimension: 250.00 × 24.78 × 9.82 mm) and ten specimens from the I-shape (specimen dimension: 250.00 × 25.44 × 9.62 mm). Each specimen was loaded up to failure with a loading speed of 3 mm/min in accordance with the displacement control method. In these tensile tests, all the specimens are failed in a brittle manner within the gage length.

In order to determine the modulus of elasticity along the transverse direction of each specimen, tensile strength test specimens should be taken along the transverse direction of the member to comply with ASTM D3410/D3410 M-03. However, in this study, the compressive strength test, instead of tensile test, suggested by Yoon [27] was adopted because taking a tension test specimen in the transverse direction of the member [28] is not possible because the width of flange and web are too small. Figure 4 shows the compression specimens and test setup used in this study. Fifteen specimens were prepared, including five specimens from the angle of structural member (specimen dimension: 80.00 × 19.79 × 9.83 mm) and ten specimens from the I-shape (specimen dimension: 80.00 × 18.29 × 9.50 mm). Each specimen was loaded up to failure with a loading speed of 3 mm/min

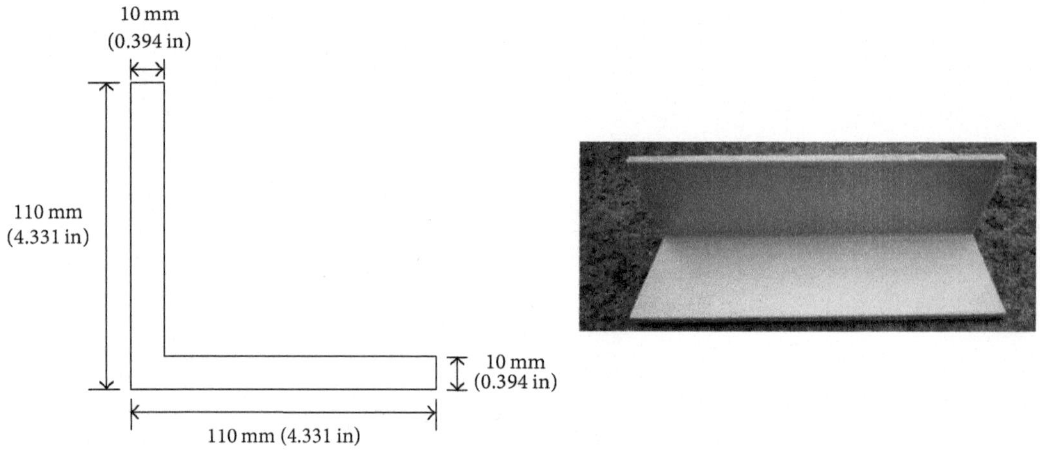

(a) Angle

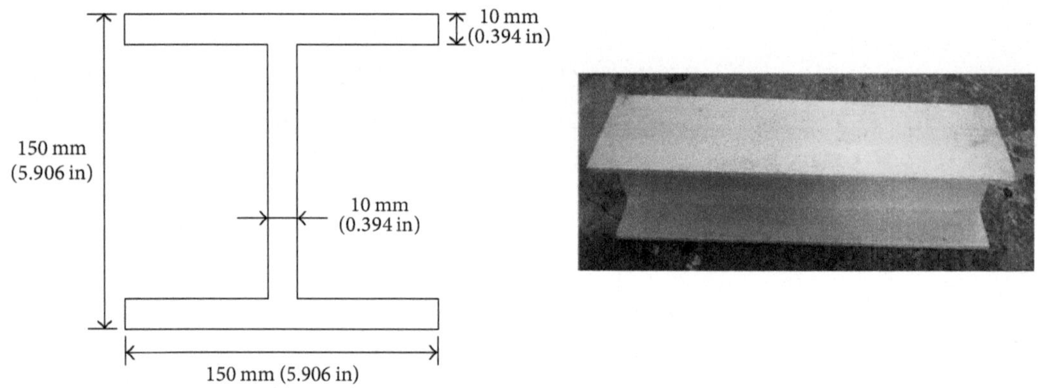

(b) I-shape

Figure 2: Structural shape and its cross-section dimension.

Table 1: Mechanical properties of PFRP members.

Test	PFRP member		Average strength (MPa)	Average modulus of elasticity (GPa)	Number of specimens
Tensile	Angle		524.32 ± 25.10	35.12 ± 8.87	5
	I-shape		415.31 ± 86.45	32.65 ± 3.33	10
Compression	Angle		147.32 ± 7.38	13.24 ± 1.67	5
	I-shape		161.91 ± 9.01	12.31 ± 2.84	10
Shear	Angle	Axial	87.31 ± 3.80	6.30 ± 0.97	5
		Transverse	61.28 ± 4.83	6.01 ± 0.58	5
	I-shape	Axial	76.30 ± 4.17	5.40 ± 0.68	10
		Transverse	41.41 ± 6.77	6.33 ± 1.97	10

according to the displacement control method. In these compressive tests, all the specimens are failed in a brittle manner within the gage length.

Finally, shear tests for the PFRP specimens were also conducted according to the method found in ASTM D5379/D5379 M-12 [29]. Figure 5 shows a shear test specimen and test setup. Thirty specimens were prepared, including 10 specimens from the angle type (specimen dimension: 76.27 × 12.06 × 9.22 mm) of structural member in both the longitudinal direction and transverse directions and 20 specimens from the I-shape type (specimen dimension: 76.01 × 12.04 × 12.20 mm) also in the longitudinal and transverse directions. Load was applied with a speed of 1.27 mm/min according to the displacement control method.

Table 1 presents a summary of the average test results for the three types of tests performed in terms of the specimens' mechanical properties.

2.2. Single-Bolted Connections. The connection specimens tested in this investigation consisted of rectangular plates that

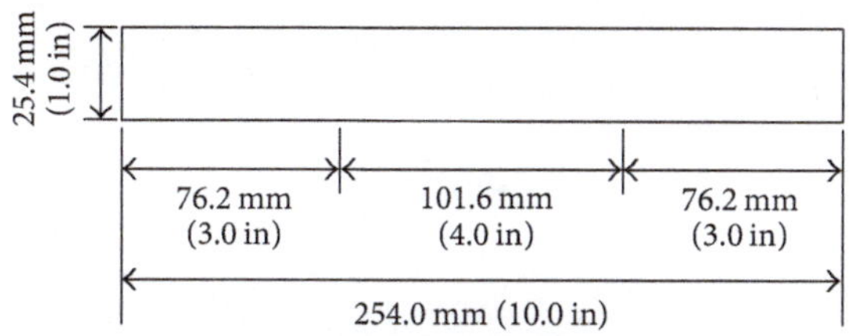

(a) Specimen dimension

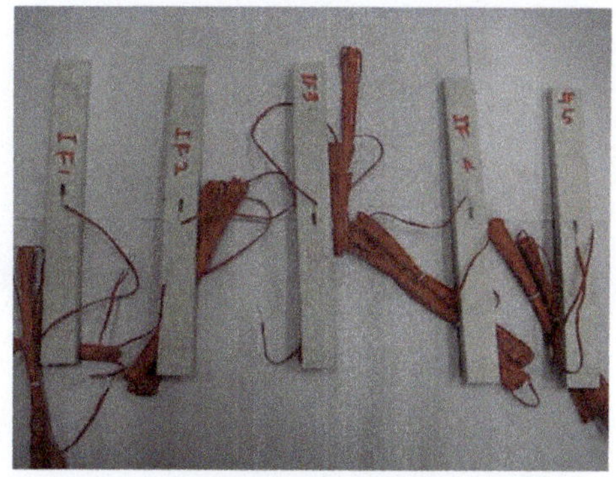

(b) Tensile test specimens

(c) Test setup

(d) Failed specimen

FIGURE 3: Tensile test specimens and test setup.

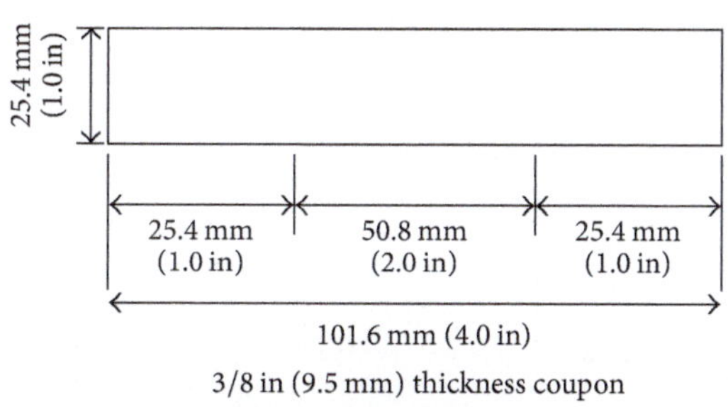

(a) Specimen dimension

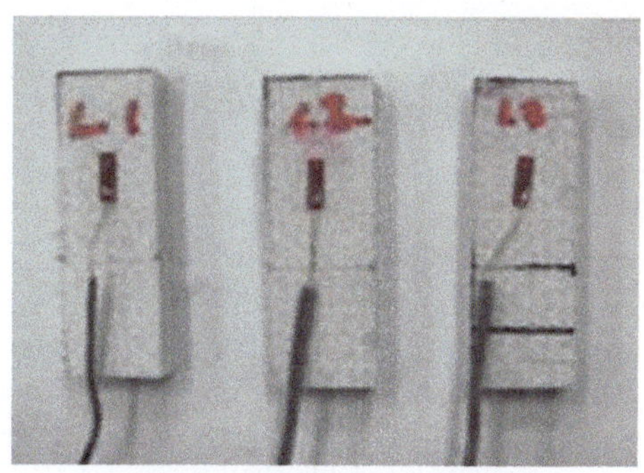

(b) Compressive test specimens

(c) Test setup

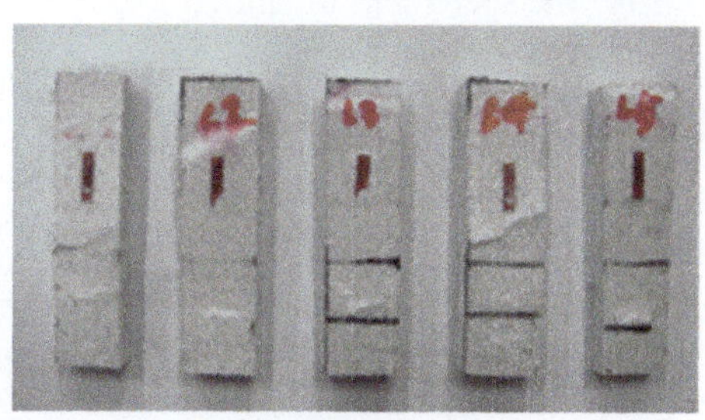

(d) Failed specimen

FIGURE 4: Compression test specimens and test setup.

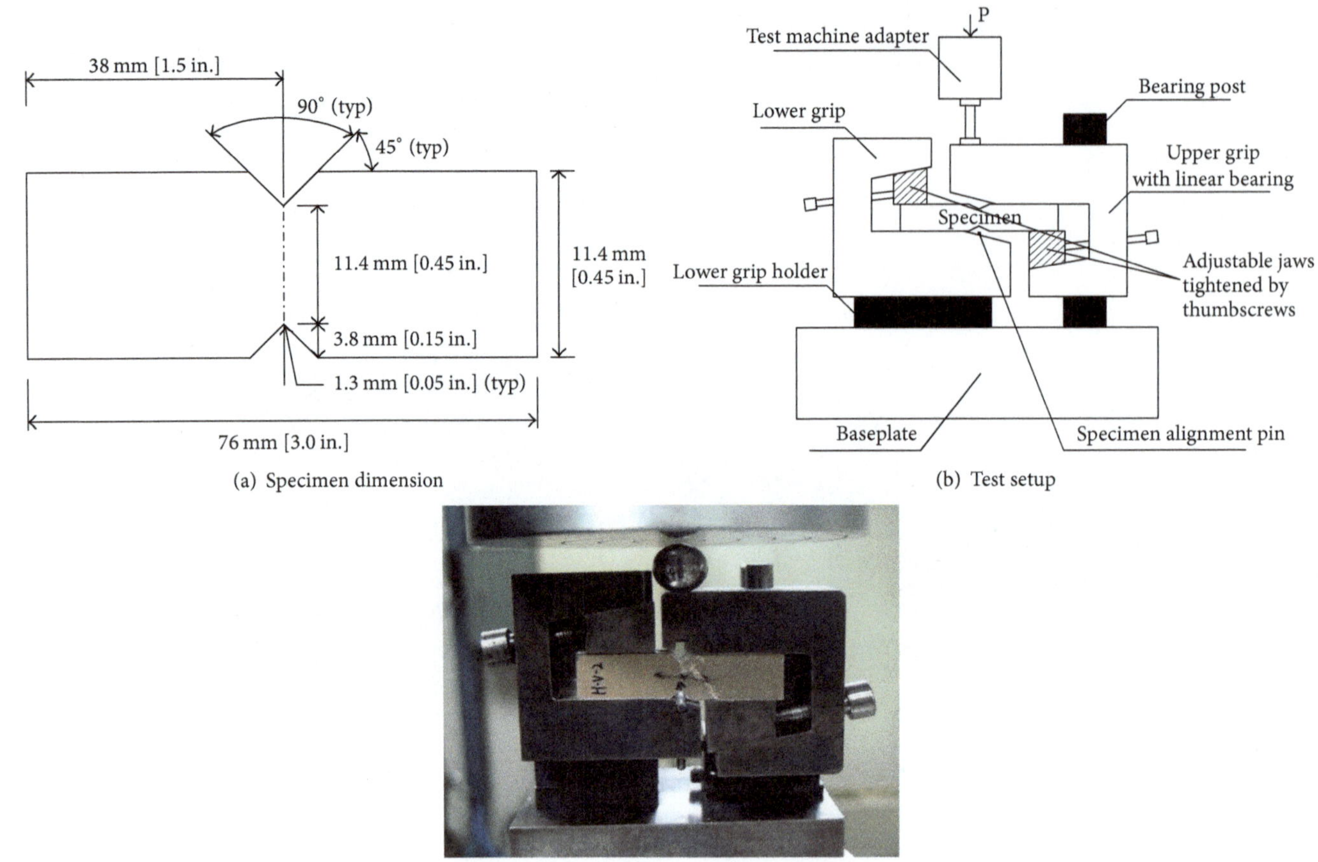

(a) Specimen dimension

(b) Test setup

(c) Failed specimen

Figure 5: Shear test specimen and test setup.

were cut from two types of PFRP structural members (angle and I-shape) using a table-saw with a diamond-tipped blade. At one end of each plate, holes were marked and drilled the desired end distance according to the specified pattern for the connection. Rectangular plates for I-shape-1, I-shape-2, and I-shape-3 were cut from the flange and web of I-shape PFRP member. Table 2 provides detailed dimensions of the single-bolted test specimens. A total of 98 bolted connection specimens (i.e., angle and I-shape-3 specimens: three for each test variation, I-shape-1, and I-shape-2 specimens: four for each test variation) were prepared by drilling holes with different bolt-hole clearances. The specimens identified with a member shape, bolt diameter, and bolt-hole clearance.

Stainless steel hexagon head screws (M10) [30], stainless steel hexagon nuts (M10) [31], and stainless steel plain washers [32] were used in the fabrication of the connection test specimens that were taken from the angle and I-shape-1 structural members. Steel hexagon head bolts (M10, M12) [30], steel hexagon head nuts (M10, M12) [31], and stainless steel plain washers [32] were also used in the fabrication of the connection test specimens taken from the I-shape-2 and I-shape-3 specimens. Figures 6 and 7 describe the geometric parameters for the single-bolted connections and the three types of bolts (including nuts and washers) used in the test, respectively.

Four basic geometric parameters that may affect the strength and the failure mode of single-bolt connections were investigated in the experimental program: the width of the member (w), the end distance (e), the hole clearance ($d_h - d_b$), and the reinforcing fiber direction of the PFRP structural shapes (refer to Figure 6). Rosner and Rizkalla [6] reported that the thickness (t) of the specimen barely affects the experimental results in overlapped spliced joint tests; thus, the dimensions specified by the manufacturer were used without any further process.

2.3. Tension Tests for Single-Bolted Connections. Tension tests using single-bolted connections were conducted using a 1000-kN Universal Testing Machine. Figure 8 presents the tension test for a double-lapped joint and the test setup. The grip plates were made using a stainless steel plate. Also, a stainless steel hexagon head screw (M10) and steel hexagon head bolt (M10, M12) were used to fabricate the connection test system. Tensile loading was applied with a displacement rate of 1 mm/min (0.167 mm/sec) in accordance with ASTM D953-10 [33].

3. Test Results and Discussion

3.1. Tension Test Results for Single-Bolted Connections. As shown in Figure 9, local failure load is defined as the load

TABLE 2: Dimensions of test specimens connected with a single-bolt.

Shape	Specimen designation	Geometric parameter (mm)					e/d_b	w/d_b	d_b/t	$d_h - d_b$	Number of specimens
		e	w	t	d_b	d_h					
Angle	A-C0.0	50	50	10	10	10	5	5	1	0	3
	A-C0.5					10.5				0.5	3
	A-C1.0					11				1.0	3
	A-C1.5					11.5				1.5	3
	A-C2.0					12				2.0	3
	A-C2.5					12.5				2.5	3
	A-C3.0					13				3.0	3
I-shape-1	I-C0.0	50	50	10	10	10	5	5	1	0	4
	I-C0.5					10.5				0.5	4
	I-C1.0					11				1.0	4
	I-C1.5					11.5				1.5	4
	I-C2.0					12				2.0	4
	I-C2.5					12.5				2.5	4
	I-C3.0					13				3.0	4
I-shape-2	I-B10-C0.0	50	50	10	10	10	5	5	1	0	4
	I-B10-C0.5					10.5				0.5	4
	I-B10-C1.0					11.0				1.0	4
	I-B10-C1.5					11.5				1.5	4
	I-B10-C2.0					12.0				2.0	4
	I-B10-C2.5					12.5				2.5	4
	I-B10-C3.0					13				3.0	4
I-shape-3	I-B12-C0.0	50	50	10	12	12	4.17	4.17	1.2	0	3
	I-B12-C0.5					12.5				0.5	3
	I-B12-C1.0					13				1.0	3
	I-B12-C1.5					13.5				1.5	3
	I-B12-C2.0					14				2.0	3
	I-B12-C2.5					14.5				2.5	3
	I-B12-C3.0					15				3.0	3

at which the bearing failure mode changes to the shear-out failure mode. The load-displacement curves for the specimens indicate a local failure load after which the load increases and then decreases slightly and continuously up to the point where the structural fracture load is reached.

Figures 10 and 11 show the failure modes at each stage of load increment. Table 3 presents a summary of the local failure loads, structural fracture loads, and corresponding modes of failure for all the tested connection specimens for the two structural profile sections used in this investigation. In Table 3, the failure modes are described as CT (cleavage-tension), B (bearing), and S (shear-out). There is no net-tension failure (NT) in this study.

3.2. Effects of Bolt-Hole Clearance. For civil engineering applications, maintaining a uniform and precise size for the bolt-hole clearance is important for constructability. Table 4 shows the bolt-hole clearance suggested in the Eurocomp Design Code [13], the Italian National Research Council (CNR) standard [14], and previous research [15].

To investigate the effects of variation in the bolt-hole clearances, the failure loads were plotted with respect to the bolt-hole clearances ($d_h - d_b$), as shown in Figures 12–15.

Figures 12–14 shows no noticeable trend in the changes in the structural fracture load with respect to the bolt-hole clearance. In contrast, a significant decreasing trend is evident in the results for the local failure load. Therefore, the difference between the local failure load and the structural fracture load is larger if the bolt-hole clearance has increased. However, Figure 15 shows that the structural fracture load for I-shape-3 with respect to the bolt-hole clearance has a decreasing trend.

In previous research [34], coupon tests were conducted using the whole cross-section of pultruded I-shape specimens. A total of 18 coupons were cut from the cross-section: six coupons from the upper flange, six coupons from the

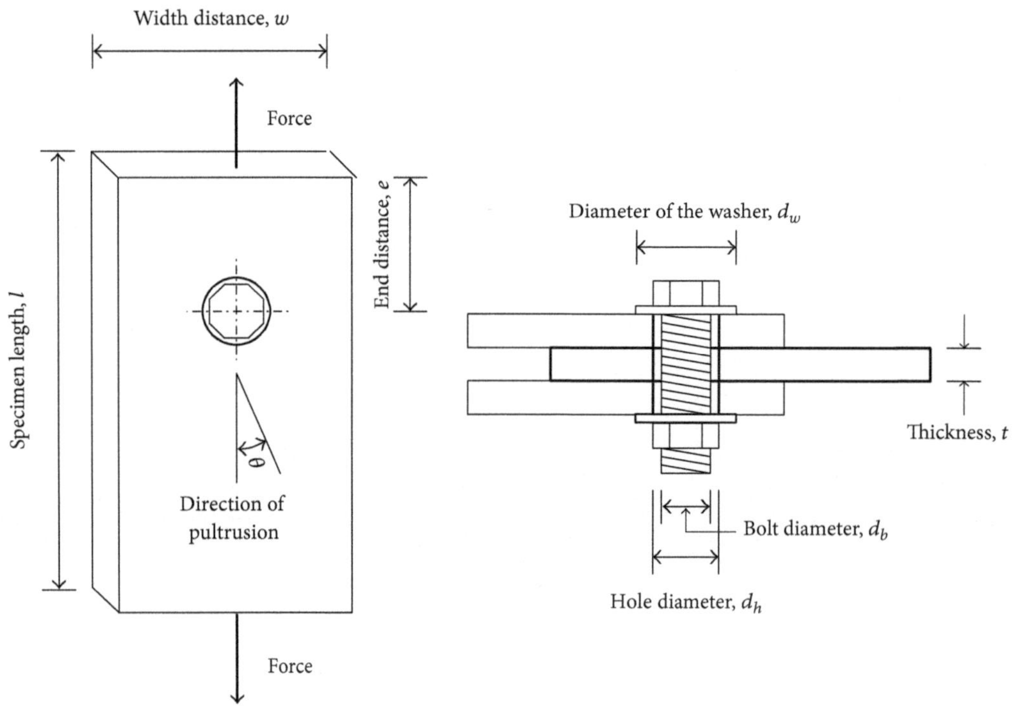

FIGURE 6: Geometric parameters for single-bolted connections.

(a) Stainless steel hexagon head screw (M10) (b) Steel hexagon head bolt (M10) (c) Steel hexagon head bolt (M12)

FIGURE 7: Type of bolt, nut, and washer used in the test.

lower flange, and six coupons from the web. The results for the material property variation tests show that the longitudinal elastic modulus is different in the range of 9 percent to 23 percent. This variation of the elastic modulus may affect the failure load of the specimen.

The reinforcing fiber direction and sampling location in the structural shape produced by the pultrusion process may result in differences between the elastic moduli of the web and flange of the PFRP structural shapes. Therefore, the structural fracture load decreased slightly. Similarly, the discrepancy between the local failure load and structural fracture load increased with an increase in the bolt-hole clearance.

A relatively large bolt-hole clearance is preferred for easy of fabrication of the structure to secure constructability. The recommended value suggested in the Eurocomp Design Code [13] seems to be too small to use in practice. In addition, the application of the Eurocomp simplified method for the design of bolted joints for PFRP materials is questionable because of its reliance on single-curve normalized stress distributions. The CNR standard [14] proposes a bolt-hole clearance of 1 mm. However, because the time required for the structural fracture load to occur after the local failure load test is too short, the test specimen is not suitable for use as a design basis in consideration to safety. Therefore, the suggested bolt-hole clearance of 1.6 mm (1/16 in.) found in the Mottram [15] is preferable because this clearance dimension accounts for the span between the local failure point (which is the ductile failure mode) and the structural failure point (which is the fracture point), even if the FRP is a brittle material. The ASCE Design Guide for FRP Composite Connections [24]

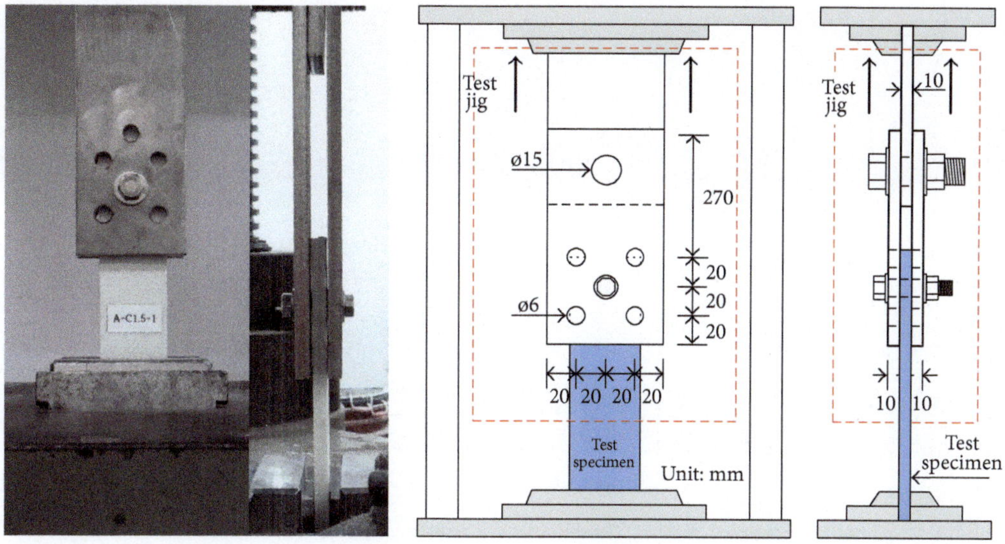

Figure 8: Tension test setup for a single-bolted connection [25].

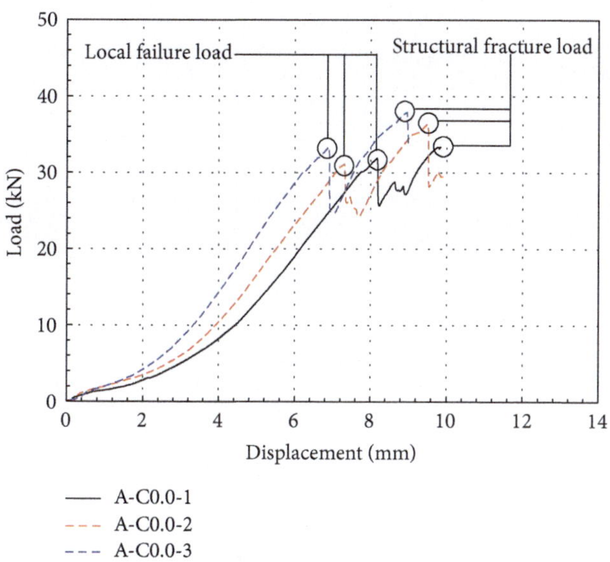

(a) Three specimens for A-C0.0

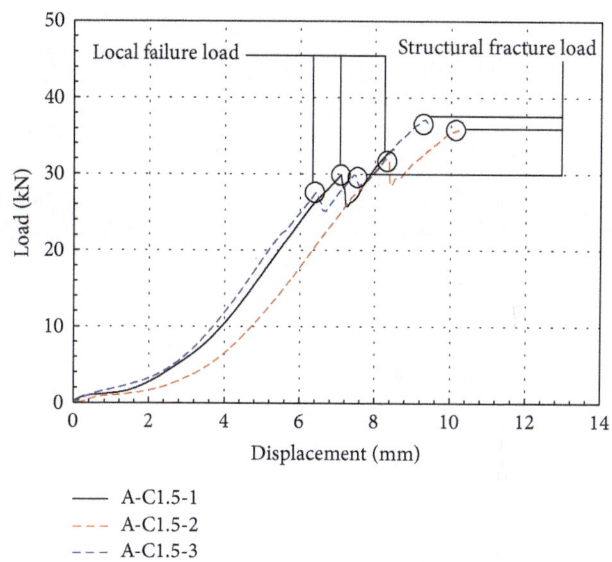

(b) Three specimens for A-C1.5

Figure 9: Relationship between load and displacement.

(a) Failed specimen

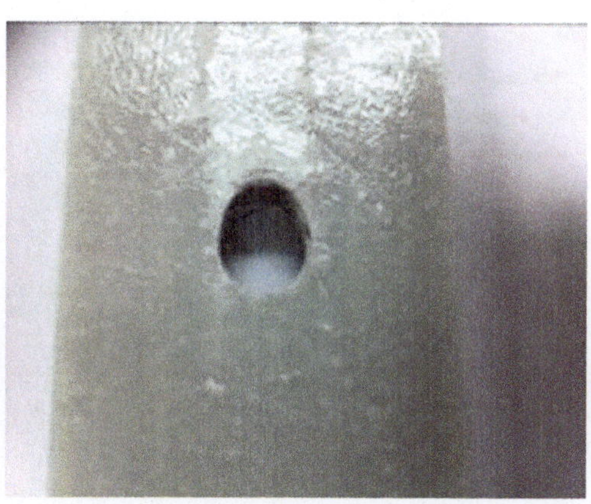

(b) Local failure mode (bearing failure)

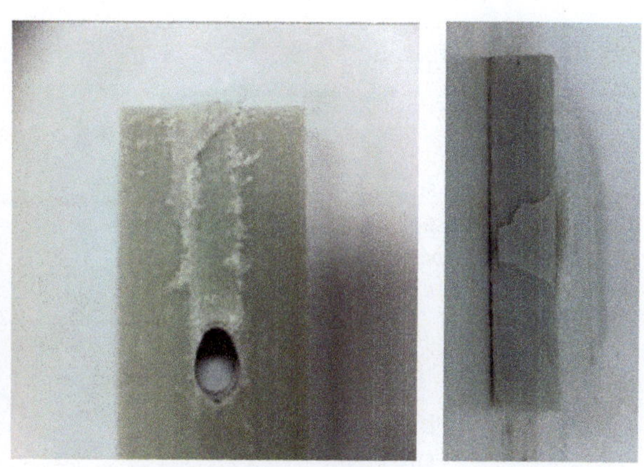

(c) Structural fracture mode (shear-out failure)

Figure 10: Examples of local failure and structural fracture modes (I-B10-C1.5 specimen).

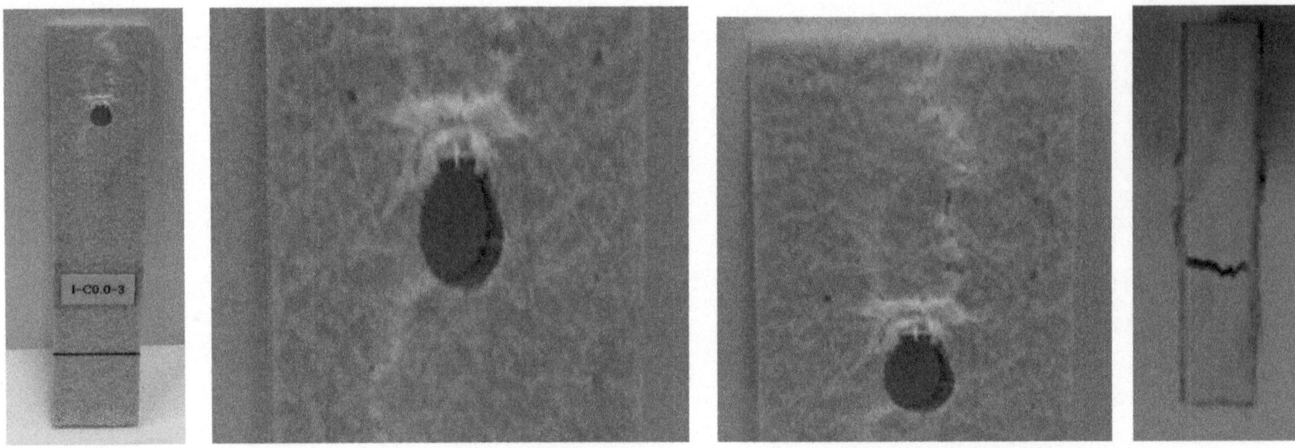

(a) Failed specimen (b) Local failure mode (bearing failure) (c) Structural fracture mode (cleavage-tension failure)

FIGURE 11: Examples of local failure and structural fracture modes (I-C0.0 specimen).

TABLE 3: Experimental results for PFRP bolted connections.

Shape	ID	e/d_b	w/d_b	d_b/t	$d_h - d_b$	Average local failure load (kN)	Local failure mode	Average structural fracture load (kN)	Structural fracture mode
Angle	A-C0.0	5	5	1	0	32.25 ± 1.00	B	35.97 ± 2.43	S
	A-C0.5				0.5	28.59 ± 2.74	B	38.26 ± 1.02	S
	A-C1.0				1.0	29.00 ± 3.15	B	38.56 ± 3.68	S
	A-C1.5				1.5	29.81 ± 2.19	B	35.44 ± 1.98	S
	A-C2.0				2.0	29.64 ± 1.81	B	35.95 ± 2.77	S
	A-C2.5				2.5	25.37 ± 3.38	B	38.65 ± 2.28	S
	A-C3.0				3.0	24.63 ± 1.05	B	37.19 ± 2.28	S
I-shape-1	I-C0.0	5	5	1	0	30.63 ± 3.64	B	30.77 ± 3.93	CT
	I-C0.5				0.5	27.81 ± 1.83	B	30.45 ± 3.32	S
	I-C1.0				1.0	27.79 ± 2.79	B	28.96 ± 3.25	S
	I-C1.5				1.5	27.66 ± 3.44	B	29.54 ± 2.04	S
	I-C2.0				2.0	25.66 ± 1.78	B	27.96 ± 2.62	S
	I-C2.5				2.5	25.37 ± 4.34	B	30.69 ± 2.28	S
	I-C3.0				3.0	27.19 ± 2.87	B	30.62 ± 3.56	S
I-shape-2	I-B10-C0.0	5	5	1	0	29.73 ± 0.57	B	30.07 ± 0.92	S
	I-B10-C0.5				0.5	23.83 ± 5.13	B	30.47 ± 1.51	S
	I-B10-C1.0				1.0	23.68 ± 4.70	B	29.69 ± 0.69	S
	I-B10-C1.5				1.5	20.95 ± 7.73	B	26.65 ± 3.55	S
	I-B10-C2.0				2.0	18.43 ± 4.45	B	29.06 ± 4.27	S
	I-B10-C2.5				2.5	22.41 ± 5.36	B	29.31 ± 1.77	S
	I-B10-C3.0				3.0	24.12 ± 2.27	B	28.72 ± 3.57	S
I-shape-3	I-B12-C0.0	4.17	4.17	1.2	0	34.13 ± 0.12	B	34.40 ± 0.53	CT
	I-B12-C0.5				0.5	25.73 ± 8.70	B	32.40 ± 2.96	S
	I-B12-C1.0				1.0	33.13 ± 0.61	B	34.87 ± 2.95	S
	I-B12-C1.5				1.5	31.13 ± 3.44	B	32.93 ± 1.36	S
	I-B12-C2.0				2.0	24.20 ± 0.92	B	32.07 ± 3.51	S
	I-B12-C2.5				2.5	19.00 ± 6.94	B	26.67 ± 6.80	S
	I-B12-C3.0				3.0	25.67 ± 3.25	B	30.07 ± 2.20	S

Note. Although all of the specimens were assumed to have exact sizes for bolt-hole clearance, however, in practice some error in clearance size was unavoidable during the specimen preparation and testing, which may affect the test results.

Table 4: Recommended bolt-hole clearance.

Reference	Eurocomp [13]	Italian CNR [14]	Mottram [15]
Bolt hole clearance	Tight fit ($0.05d_b$)	≤1 mm	1/16 in (1.6 mm)

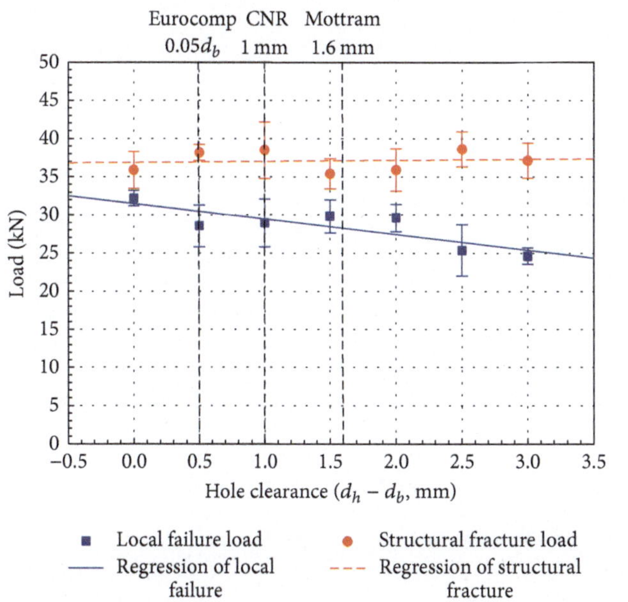

Figure 12: Relationship between load and bolt-hole clearance (specimen from angle).

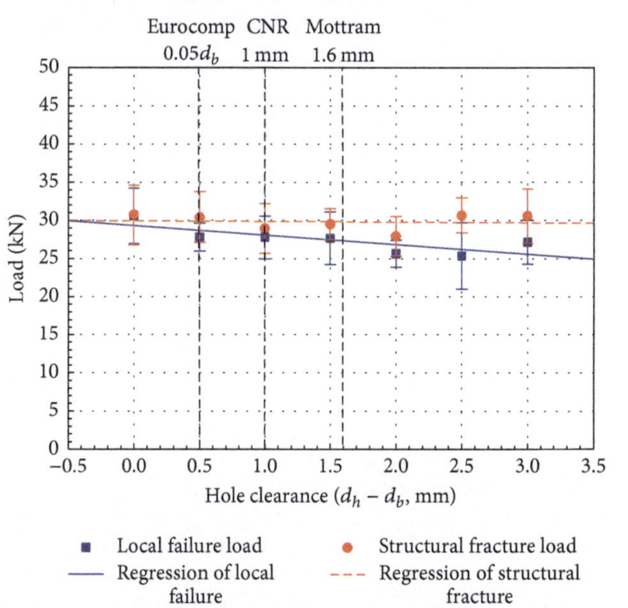

Figure 13: Relationship between load and bolt-hole clearance (specimen from I-shape-1).

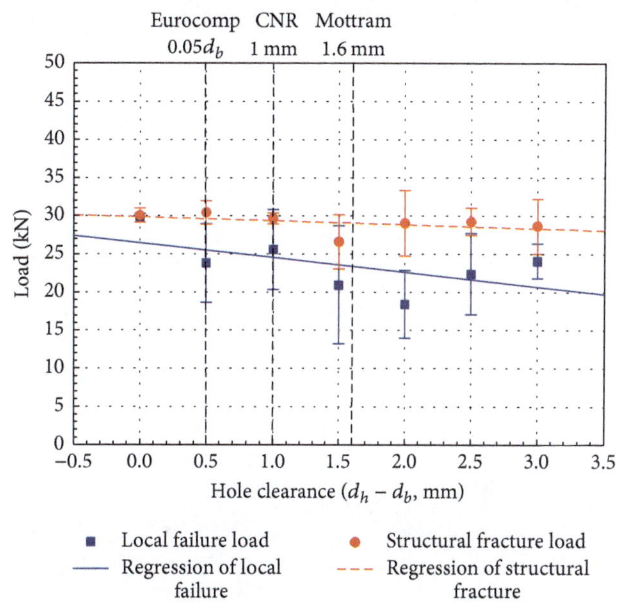

Figure 14: Relationship between load and bolt-hole clearance (specimen from I-shape-2).

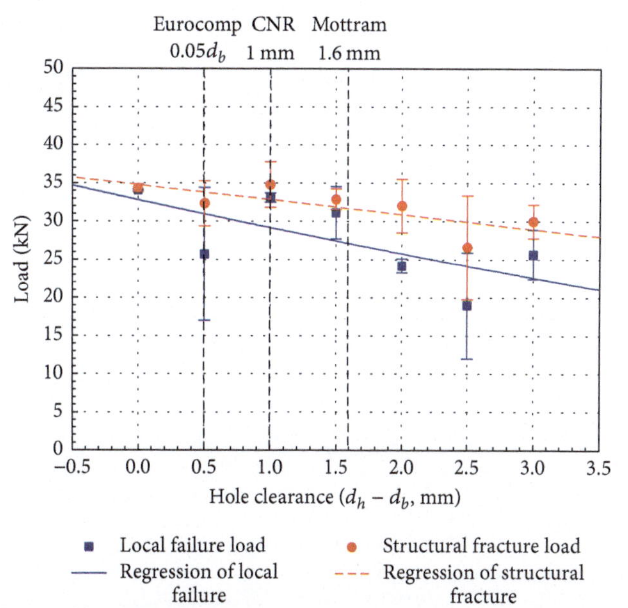

Figure 15: Relationship between load and bolt-hole clearance (specimen from I-shape-3).

was discussed on the effect of bolt fit (hole clearance). The hole diameter, d, equals the bolt diameter, d_b, plus 5/8 in. (15.875 mm) which may acceptable in practice. However, it is felt that 15.875 mm hole clearance is too large to be efficient to induce bearing failure mode.

4. Conclusions

In this paper we investigated the effects of bolt-hole clearance in single-bolted connections in the PFRP structural members. Different sizes of bolt-hole clearance from tight-fit to 3.0 mm with 0.5 mm intervals were investigated. The experimental results in terms of local failure load, structural fracture load, and failure mode were analyzed with respect to the geometric parameters (i.e., the bolt-hole clearances). The following results were found.

(1) The specimens, in general, failed with two sequential failure modes. The bearing failure mode appeared first and the shear-out failure mode followed. Therefore, the geometric parameters of the specimens, that is, the e/d and w/d, were needed to maintain with sufficient values regardless of the bolt-hole clearance. For each case, bearing failure was the predominant failure mode.

(2) When the bolt-hole clearances were in the range of 0 mm to 3 mm, no significant trend was evident with regard to the structural fracture loads. However, the local failure load decreased if the bolt-hole clearance was increased. Differences between the structural fracture loads and the local failure loads were greater when the bolt-hole clearance was increased.

(3) Constructability can be ensured by maintaining a minimum bolt-hole clearance. The Eurocomp Design Code recommends a bolt-hole clearance that is $(d_h - d_b)$ $0.05d_b$ (5%) of the bolt diameter, but this recommended clearance may be not efficient in practice.

(4) The recommended bolt-hole clearances found in both the Eurocomp and CNR standards are not suitable in terms of safety due to the small interval from the local failure load to the structural fracture load. The bolt-hole clearance of 1.6 mm (1/16 in.) found in the previous research [15] is appropriate and allows ductile failure.

Acknowledgments

This work was supported by the Energy Research & Development of the Korea Institute of Energy Technology Evaluation and Planning (KETEP) grant funded by the Korean Government's Ministry of Trade, Industry & Energy (no. 20161120200190).

References

[1] L. Feo, G. Marra, and A. S. Mosallam, "Stress analysis of multi-bolted joints for FRP pultruded composite structures," *Composite Structures*, vol. 94, no. 12, pp. 3769–3780, 2012.

[2] M. C. Wang, *Connection capacity of pultruded gfrp channels in multidirectional loading [M.S. Thesis]*, University of Colorado, 2013.

[3] M. Shokrieh and M. Heidari-rarani, "Experimental and analytical studies on one-way concrete slabs reinforced with GFRP molded gratings," *Steel and Composite Structures*, vol. 9, no. 6, pp. 569–584, 2009.

[4] CN. Rosner, *Single-bolted connections for orthotropic fibre-reinforced composite structural members [M. S.Thesis]*, University of Manitoba, 1992.

[5] B. Vangrimde and R. Boukhili, "Descriptive relationships between bearing response and macroscopic damage in GRP bolted joints," *Composites Part B: Engineering*, vol. 34, no. 7, pp. 593–605, 2003.

[6] C. N. Rosner and S. H. Rizkalla, "Bolted connections for fiber-reinforced composite structuralmembers: Experimental program," *Journal of Materials in Civil Engineering*, vol. 7, no. 4, pp. 223–231, 1995.

[7] S. F. M. Abd-El-Naby and L. Hollaway, "The experimental behaviour of bolted joints in pultruded glass/ polyester material. Part 1: Single-bolt joints," *Composites*, vol. 24, no. 7, pp. 531–538, 1993.

[8] C. Cooper and G. J. Turvey, "Effects of joint geometry and bolt torque on the structural performance of single bolt tension joints in pultruded GRP sheet material," *Composite Structures*, vol. 32, no. 1-4, pp. 217–226, 1995.

[9] G. J. Turvey, "Single-bolt tension joint tests on pultruded GRP plate - effects of tension direction relative to pultrusion direction," *Composite Structures*, vol. 42, no. 4, pp. 341–351, 1998.

[10] Ü. Esendemir, "An experimental study of mechanically fastened composite joints with clearance," *International Journal of Damage Mechanics*, vol. 20, no. 3, pp. 464–480, 2011.

[11] F. Sen, M. Pakdil, O. Sayman, and S. Benli, "Experimental failure analysis of mechanically fastened joints with clearance in composite laminates under preload," *Materials & Design*, vol. 29, no. 6, pp. 1159–1169, 2008.

[12] J. T. Mottram and G. J. Turvey, "Physical test data for the appraisal of design procedures for bolted joints in pultruded FRP structural shapes and systems," *Progress in Structural Engineering and Materials*, vol. 5, no. 4, pp. 195–222, 2003.

[13] *Eurocomp Design Code and Handbook. Structural Design of Polymer Composites*, The European Structural Polymeric Composites Group, 1996.

[14] Technical Document, "Guide for the Design and Construction of Structures Made of FRP Pultruded Elements," Tech. Rep. CNR-DT 205/2007, Italian National Research Council (CNR), Rome, Italy, 2008.

[15] J. T. Mottram, "Design guidance for bolted connections in structures of pultruded shapes: gaps in knowledge," in *Proceedings of the 4th International Conference on Advanced Composites in Construction (ACIC 2009)*, pp. 483–495, NetComposites Ltd, Chesterfield, UK, 2009.

[16] V. P. Lawlor, M. A. McCarthy, and W. F. Stanley, "An experimental study of bolt-hole clearance effects in double-lap, multi-bolt composite joints," *Composite Structures*, vol. 71, no. 2, pp. 176–190, 2005.

[17] T. A. D. Tajeuna, F. Légeron, S. Langlois, P. Labossière, and M. Demers, "Effect of geometric parameters on the behavior of bolted GFRP pultruded plates," *Journal of Composite Materials*, vol. 50, no. 26, pp. 3731–3749, 2016.

[18] Y. Xiao and T. Ishikawa, "Bearing strength and failure behavior of bolted composite joints (part I: Experimental investigation)," *Composites Science and Technology*, vol. 65, no. 7-8, pp. 1022–1031, 2005.

[19] G. Kelly and S. Hallström, "Bearing strength of carbon fibre/epoxy laminates: Effects of bolt-hole clearance," *Composites Part B: Engineering*, vol. 35, no. 4, pp. 331–343, 2004.

[20] M. A. McCarthy, V. P. Lawlor, W. F. Stanley, and C. T. McCarthy, "Bolt-hole clearance effects and strength criteria in single-bolt, single-lap, composite bolted joints," *Composites Science and Technology*, vol. 62, no. 10-11, pp. 1415–1431, 2002.

[21] ASTM, "Standard test method for bearing response of polymer matrix composite laminates," ASTM International 5961/D 5961 M-05, West Conshohocken, PA, USA, 2005.

[22] F. Ascione, L. Feo, and F. Maceri, "An experimental investigation on the bearing failure load of glass fibre/epoxy laminates," *Composites Part B: Engineering*, vol. 40, no. 3, pp. 197–205, 2009.

[23] A. A. Pisano, P. Fuschi, and D. De Domenico, "A layered limit analysis of pinned-joints composite laminates: Numerical versus experimental findings," *Composites Part B: Engineering*, vol. 43, no. 3, pp. 940–952, 2012.

[24] A. S. Mosallam, "Design for FRP composite connections," *ASCE Manuals and Reports on Engineering Practice*, 2011.

[25] Y.-G. Lee, E. Choi, and S.-J. Yoon, "Effect of geometric parameters on the mechanical behavior of PFRP single bolted connection," *Composites Part B: Engineering*, vol. 75, pp. 1–10, 2015.

[26] ASTM, "Standard test method for tensile properties of polymer matrix composite materials," ASTM International 3039/D 3039 M-08, West Conshohocken, PA, USA, 2008.

[27] S. J. Yoon, *Local buckling of pultruded I-shape columns [Ph.D. thesis]*, Georgia Institute of Techonlogy, Atlanta, GA, USA.

[28] Y. G. Lee, J. S. Park, J. H. Nam, D. J. An, and S-J. Yoon, "Structural behavior of PFRP connection with single bolt," in *Proceedings of the 18th International Conference on Composite Materials (ICCM' 18)*, pp. 1–5, Jeju Island, Korea, August 2011.

[29] ASTM, "Standard test method for shear properties of composite materials by the V-notched beam method," ASTM International D 5379/D 5379 M-12, West Conshohocken, PA, USA, 2012.

[30] KS B 1002. Hexagon Head Bolts and Hexagon Head Screws. Korea: Korean Standards Association, 2001.

[31] KS B 1012. Hexagon Nuts and Hexagon Thin Nuts. Korea: Korean Standards Association, 2001.

[32] KS B 1326. Plain Washers. Korea: Korean Standards Association, 2009.

[33] ASTM, "Standard test method for bearing strength of plastics," ASTM International D 953-10, West Conshohocken, PA, USA, 2010.

[34] H. Choi, J. W. Choi, H. J. Joo, D. J. An, and S-J. Yoon, "Effects of material properties variations on the local buckling loads of pultruded structural shapes," *Journal of Computational Design and Engineering*, vol. 1, pp. 645–648, 2009.

The Influence of Nanohydroxyapatite on the Thermal, Mechanical, and Tribological Properties of Polyoxymethylene Nanocomposites

Kinga Pielichowska,[1] Dariusz Bieliński,[2] Michał Dworak,[1,3] Ewelina Kilian,[1] Beata Macherzyńska,[1] and Stanisław Błażewicz[1]

[1]*Department of Biomaterials, Faculty of Materials Science and Ceramics, AGH University of Science and Technology, Al. Mickiewicza 30, 30-059 Kraków, Poland*
[2]*Polymer Institute, Technical University of Lodz, ul. Zwirki 36, 90-924 Łódź, Poland*
[3]*Institute of Materials Science, Faculty of Computer Science and Materials Science, University of Silesia, ul. 75 Pułku Piechoty 1A, 41-500 Chorzów, Poland*

Correspondence should be addressed to Kinga Pielichowska; kingapie@agh.edu.pl

Academic Editor: Hossein Roghani-Mamaqani

The influence of nanohydroxyapatite on the glass transition region and its activation energy, as well as on the tribological and mechanical properties of polyoxymethylene nanocomposites, was investigated using DMA, TOPEM DSC, nanoindentation, and nondestructive ultrasonic methods. It was found that the glass transition for unmodified POM was in the lower temperature range than in POM/HAp nanocomposites. Moreover, ΔC_p and activation energy were larger for POM/HAp nanocomposites. Friction coefficient was higher for POM/HAp nanocomposites in comparison to both POM homopolymer and POM copolymer. Simultaneously, the indentation test results show that microhardness is also higher for POM/HAp nanocomposites than for POM. From ultrasonic investigations it was found that the highest values of both longitudinal and transverse propagation waves and Young's and shear modulus for POM homopolymer (DH) and POM copolymer T2H and their nanocomposites can be attributed to their higher degree of crystallinity in comparison to UH copolymer. Moreover, for POM/HAp nanocomposites with 5% of HAp, ultrasonic longitudinal wave velocity was almost constant even after 1000000 mechanical loading cycles, evidencing an enhancement of mechanical properties by HAp nanoparticles.

1. Introduction

Polyoxymethylene (POM) is one of the major thermoplastic engineering polymers and it is mainly applied in mechanical parts due to excellent performance against wear, high strength, stiffness, and excellent chemical resistance [1–3]. POM is also attractive material for use in biomedical applications such as bone tissue replacement especially in hip prosthesis [4], dentistry [5], and cardiology [6]. POM copolymer has been used for the spiked washer to reattach together with a bone screw ruptured ligaments to the bone. In the joint replacement applications, different POM implants like artificial finger and shoulder joints are well-known products [4]. In all these applications mechanical properties of POM are of crucial importance. POM exhibits good fatigue resistance, high impact strength, creep resistance, low friction, and good wear resistance due to its high degree of crystallinity and high bond energy [7]. As POM is a semicrystalline polymer with a relatively high degree of crystallinity, the bigger spherulites may act as the sites of stress concentration and, in consequence, the polymer has high notch sensitivity [8].

Various methods of improvement of POM mechanical properties have been proposed including incorporation of inorganic nanoparticles in the powder form. Hence, Xu et al. [9] found that the addition of reactable nano-SiO_2

can enhance the tensile strength and Young's modulus of composites, and the impact strength of composites has a significant improvement at low contents of reactable nano-SiO_2, while, with the further increase of nanofiller content, the impact strength of composites shows a gradual decrease. Romero-Ibarra and coworkers [10] studied the influence of a radiopaque reinforcement-spherical barium sulphate ($BaSO_4$) in nanometric size, on the mechanical properties of POM. They observed that the introduction of 1 phr $BaSO_4$ nanoparticles does not substantially modify the mechanical properties of the matrix, whereas upon increasing $BaSO_4$ content to 3 phr mechanical properties diminished.

POM/Al_2O_3 nanocomposites were prepared by melt processing method and studied by Wacharawichanant et al. [11]. They revealed that the tensile strength and Young's modulus of the nanocomposites increased by adding Al_2O_3 particles in a range of 0.5–1 wt.% and then decreased. With increasing filler content, the impact strength of the nanocomposites decreased.

Wang et al. investigated [8] POM matrix composites with different contents of nano-ZrO_2 particles and they found that the addition of ZrO_2 nanoparticles can improve the tensile strength and impact strength by about 8% and 20%, respectively, when the nano-ZrO_2 content was 1.0 wt.% and the lowest wear appeared at 1.0 wt.% content of ZrO_2 nanoparticles.

To improve the bioactivity and biocompatibility of POM implants, Pielichowska et al. introduced to POM matrix hydroxyapatite (HAp, $[Ca_{10}(PO_4)_6(OH)_2]$), which is a well-known bioceramics, chemically similar to human inorganic hard tissue [12, 13]. Bone tissue is built from mineralized collagen (Col) fibrils, reinforced with 50–60 wt.% of apatite crystals. HAp exhibits excellent biocompatibility, bioactivity, bone bonding properties, and osteoinduction, as shown by in vivo studies [14]. Due to the chemical similarity, HAp facilitates the formation of new bone without resorption and interacts with the living system [13]. On the other hand, incorporation of nanoparticles into POM matrix causes the effect of heterogeneous nucleation leading to thinner spherulites and lower degree of crystallinity, which can help to improve impact strength. The decrease of the crystallinity degree reduced the tensile strength [8, 15, 16]. We found in an in vitro evaluation that addition of HAp nanoparticles to POM matrix enhances bioactivity of the nanocomposites, and an excellent stability of POM/HAp nanomaterials under these conditions was confirmed [16].

Since for load bearing orthopaedic applications mechanical properties are of crucial importance, the aim of this study was to investigate in detail the influence of HAp nanoparticles on the tribological and mechanical (static and fatigue dynamic) properties of POM nanocomposites with hydroxyapatite. Please note that in our previous papers we investigated only tensile behaviour of POM/HAp nanocomposites [15–17]. In this study three POM types (homopolymer and two copolymers) with similar melt flow index and processability but with different mass-average molar mass have been used to find the best formulation for long-term orthopaedic implants.

Table 1: Description of the samples.

POM	HAp content [wt.%]	Sample code
Delrin® 111P	0.0	DH0
	1.0	DH1
	2.5	DH2.5
	5.0	DH5
Ultraform® H2320 004	0.0	UH0
	1.0	UH1
	2.5	UH2.5
	5.0	UH5
Tarnoform® T200	0.0	T2H0
	1.0	T2H1
	2.5	T2H2.5
	5.0	T2H5

2. Experimental Part

2.1. Materials and Preparation. Three commercial grades of POM were investigated: POM homopolymer (Delrin® 111P) (DH) with M_w = 96323 Da was supplied by Du Pont; POM copolymer (Ultraform® H2320 004) (UH) with M_w = 78,565 Da was obtained from BASF (Ludwigshafen, Germany), and POM T200 copolymer (Tarnoform®) (T2H) with M_w = 100,768 Da was produced by Z.A. Tarnów-Mościce. The melt flow index was 2.4, 2.6 and 2.5 g/10 min, respectively, according to MFR 190/2.16, ISO 1133. Hydroxyapatite $Ca_{10}(PO_4)_6OH_2$ nanopowder (99% of nanoparticle with size below 100 nm) was supplied by nGimat Co. (Atlanta, USA). POM and POM/HAp nanocomposites were obtained using melt processing methods. Firstly, POM was dried, and then the POM and HAp were mechanically mixed (0, 0.5, 1.0, 2.5, 5, and 10.0% w/w of HAp) (Table 1), extruded in a twin-screw extruder ThermoHaake PolyLab PTW 16/25 (rotary speed of 50 rpm). Compositions were then shaped by using DSM laboratory injection molding machine. The proper dispersion of HAp nanoparticles in POM matrix was confirmed using TEM and SEM techniques. FTIR analysis results show that there is no chemical bonding between POM matrix and HAp nanoparticles, but HAp influenced POM crystallinity and mechanical properties [15–17].

2.2. Techniques. For the DSC measurements DSC1 (Mettler Toledo) operating in dynamic mode with heating rate of 10 K/min and cooling rate of 10 K/min was employed. Sample of ca. 6 mg weight was placed in sealed and pierced aluminum pan. Nitrogen was used as an inert gas. For TOPEM DSC measurements the underlying heating rate was 0.5 K/min, the amplitude of the temperature pulse was ±0.5 K, and the switching time range to limit the duration of the pulses had the minimum of 15 s and the maximum of 50 s. Sample of ca. 15 mg weight was placed in sealed aluminum pans. Nitrogen with flow rate 30 mL/min was used as the inert gas and the intracooler was used for cooling. DSC was calibrated using indium and mercury standards; an empty aluminum pan was used as a reference.

Dynamic mechanical analysis was performed using a DMA 242C Netzsch in three-point bending mode (20 mm sample holder) on bars with diameters of 10 mm × 26 mm, thickness 1 mm (obtained by injection molding method) in the temperature range within −160–150°C in dry air. Measurements were performed at the heating rate of 2 deg/min and perturbation frequencies of 0,1, 1, 5, and 10 Hz, under maximal dynamic force of 6 N, at constant part of static force of 0.3 N and static force 10% higher than dynamic one. The maximal amplitude was set to 50 μm. As a cooling medium liquid nitrogen was used.

The friction force of the stainless steel spherical tip of r = 5 μm sliding over the film surface was studied using a NanoTest 600 instrument (Micro Materials, UK), equipped with a friction attachment. Measurements were carried out at a normal load of 30 mN and a sliding speed of 10 μm/s over a distance of 5 mm, under ambient conditions. Coefficient of friction was calculated using well-known Amontons' formula—as an average value of friction force after stabilization of friction divided by the load.

Mechanical properties of the surface layer of polymers were determined using the NanoTest 600 instrument. A Berkovitch indenter penetrated the surface layer of polymers with the loading/unloading rate of dP/dt = 0.10 mN/s, up to the maximum load of P = 5.00 mN. Mechanical parameters: hardness and elastic modulus were determined from unloading curves using the procedure of Oliver and Pharr [18]. The measurements were made at T = 20 ± 2°C and relative humidity of 60 ± 5%. For each sample two series of seven measurements in various places on sample surface were performed and the data averaged.

Ultrasonic measurements were performed using the Ultrasonic Measuring System UZP-1 (INCO-VERITAS) with enhanced accuracy of the time of propagation of ultrasonic waves (±0.01 μs). An ultrasonic pulse technique was used to determine the elastic properties (E, G, and μ). The 10 MHz transducer with a diameter of 10 mm connected to a sample by paraffin oil was used for longitudinal waves, and the same transducer was glued with the samples by the Canadian balm for transverse waves. Material constants values were calculated using the following formula [19]:

$$G = C_T^2 * \rho,$$

$$E = C_L^2 \cdot \rho \cdot \frac{(1+\mu)\cdot(1-2\mu)}{(1-\mu)}, \qquad (1)$$

$$\mu = \frac{C_L^2 - 2C_T^2}{2\left(C_L^2 - C_T^2\right)},$$

where E is ultrasonic Young's modulus, G is Kirchhoff's modulus, μ is Poisson's Ratio, ρ is apparent density, C_L is speed of a longitudinal wave, and C_T is speed of a transverse wave.

Mechanical dynamic loading was performed in the three-point bending mode under a constant force amplitude using specimens with dimensions 40 mm × 5 mm × 2 mm. The level of maximum force amplitude amounted to 50% of the sample strength determined in static mechanical test. After each 200000 cycles the samples were ultrasonically tested using Unipan-ultrasonic CT3 equipment with the 1 MHz transducer and ultrasonic wave velocities were determined.

3. Results and Discussion

DMA results in the form of storage modulus (E') and loss factor (tan δ) as a function of temperature (T) are demonstrated in Figures 1 and 2.

Earlier studies on the relaxation effects in pristine POM revealed three kinds of relaxations—α, β, and γ. The first one recorded at ca. −70°C as a broad peak was assigned to the glass to rubber γ transition because of its "cooperative character" [20]. The other two peaks, observed at ca. −10 and 100°C, correspond to β and α relaxations, respectively. β relaxation was assigned to the motions of long molecular segments in disordered phase [21], while α relaxation to movement of well-ordered crystalline phases. As it can be seen in Figures 1 and 2 also γ-mode at ca. −74°C, β-mode at −10°, and α at 90°C can be observed for POM copolymers (UH and T2H) while for POM homopolymer (DH) just α-mode at ca. −73°C and α at ca. 100–110°C can be seen. Based on the frequency dependence of tan δ, glass transition at different frequencies was determined and from Arrhenius frequency map lnf versus 1000/T_g (Figure 3) the activation energy of α-mode was calculated. Moreover, to get a deeper insight into the glass transition region of investigated POM nanocomposites TOPEM DSC experiments were performed. Using conventional DSC it was impossible to determine γ-mode, but based on the reversing heat flow from TOPEM DSC one can detect γ relaxation (Figure 4). Glass temperature, ΔC_p determined from TOPEM DSC, and activation energy (E_a) data from DMA data [22, 23] are presented in Table 2.

As it can be seen from both TOPEM DSC and DMA results, the glass temperature for unmodified POM was lower than T_g of POM/HAp nanocomposites. Moreover, ΔC_p and E_a were larger for POM/HAp nanocomposites. A similar trend was observed by other authors for different polymer nanocomposites. Hence, Wong et al. [24] reported the linear increase in the glass transition temperature of poly(methyl methacrylate)/ZnO nanocomposites versus ZnO quantum dots weight content. It was assumed that the confinement effect between nanoparticles affects T_g as a polymer thin film is sandwiched between two substrates [25, 26]. Ceramic nanofillers such as HAp do not directly contribute to the T_g value of the polymer matrix, but the interaction of the polymer chains with the surface of the particles can drastically alter the chain kinetics and mobility in the region surrounding the particle due to the presence of the interface. Due to the large surface area of nanofillers, for a well-dispersed system, even low filler volume fractions provide a substantial amount of interfacial area through which the bulk properties of the polymer can be altered [27]. Rittigstein et al. [25] found that, in nanocomposites with well-dispersed nanofiller, T_g can exhibit substantial deviations in relation to the bulk polymer, decreasing when polymer-nanofiller interfaces yield free surfaces [27, 28] and increasing when wetted interfaces with attractive interactions are present. When interactions, such as hydrogen bonds, are present

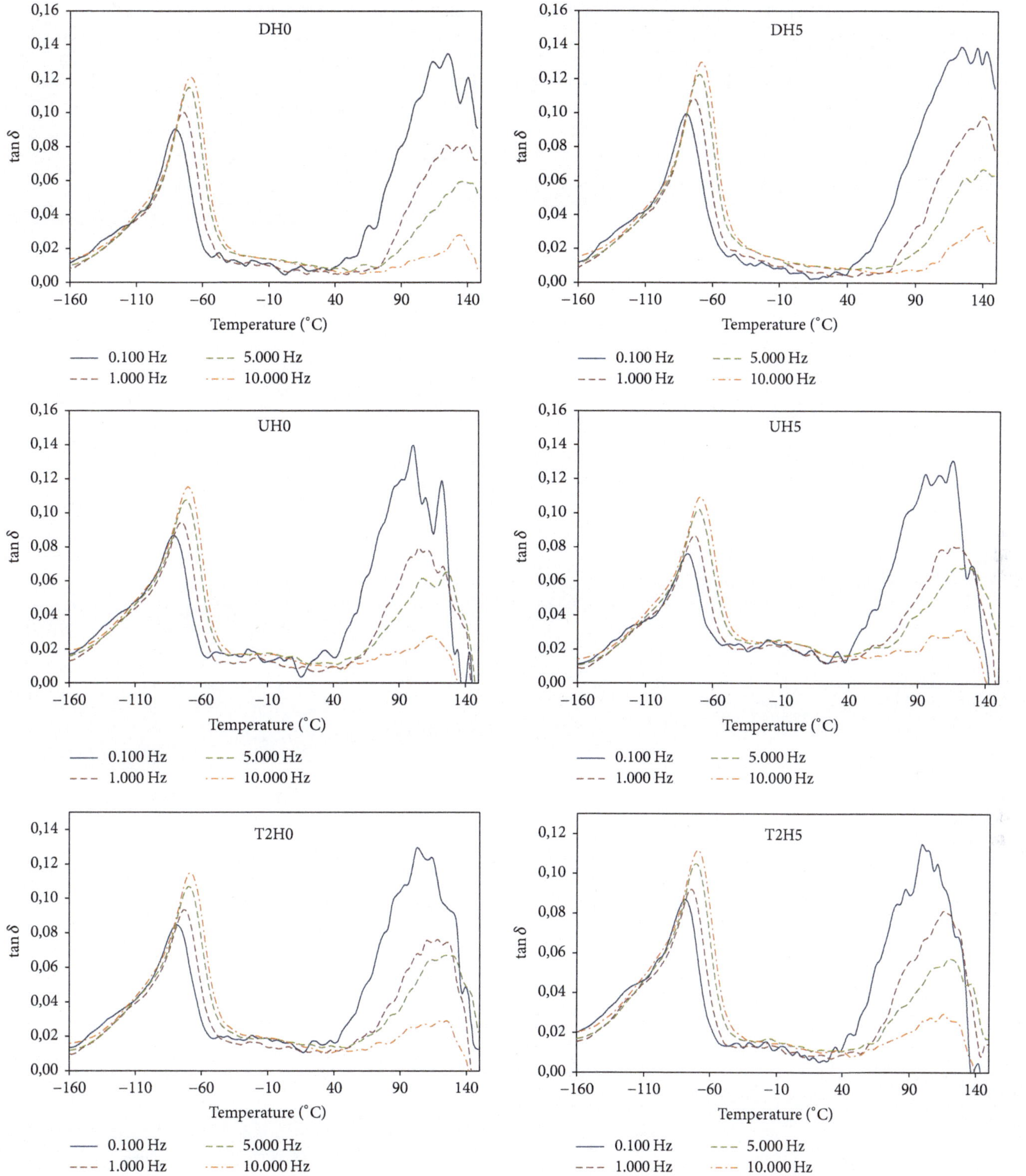

FIGURE 1: DMA curves for POM and POM/5% HAp nanocomposites: tan δ versus temperature.

at a polymer-substrate interface, confinement effects can lead to enhancements relative to bulk polymer properties [28]. The effects of confinement on T_g in polystyrene (PS), poly(methyl methacrylate) (PMMA), and poly(2-vinyl pyridine) (P2VP) nanocomposites containing silica and alumina nanospheres were investigated, too [28]. In the alumina-containing nanocomposites, T_g increases relative to $T_{g,\text{bulk}}$ by as much as 16 K in P2VP, decreases by 5 K in PMMA, and is invariant in PS. These results were explained by wetted P2VP-nanofiller interfaces with attractive interactions, nonwetted

TABLE 2: Glass temperature, ΔC_p, and activation energy of glass transition for POM and POM/HAp nanocomposites.

Sample	T_g [°C] (TOPEM DSC) by reversing heat flow	ΔC_p [J/molK] (TOPEM DSC)	T_g [°C] (DMA 1 Hz)	E_a [kJ/mol]
DH0	−74.5	0.070	−74.3	128.23
DH5	−71.3	0.076	−73.7	139.48
UH0	−77.0	0.052	−74.7	134.25
UH5	−73.1	0.108	−73.9	160.60
T2H0	−76.7	0.049	−74.4	140.91
T2H5	−67.0	0.066	−73.4	148.58

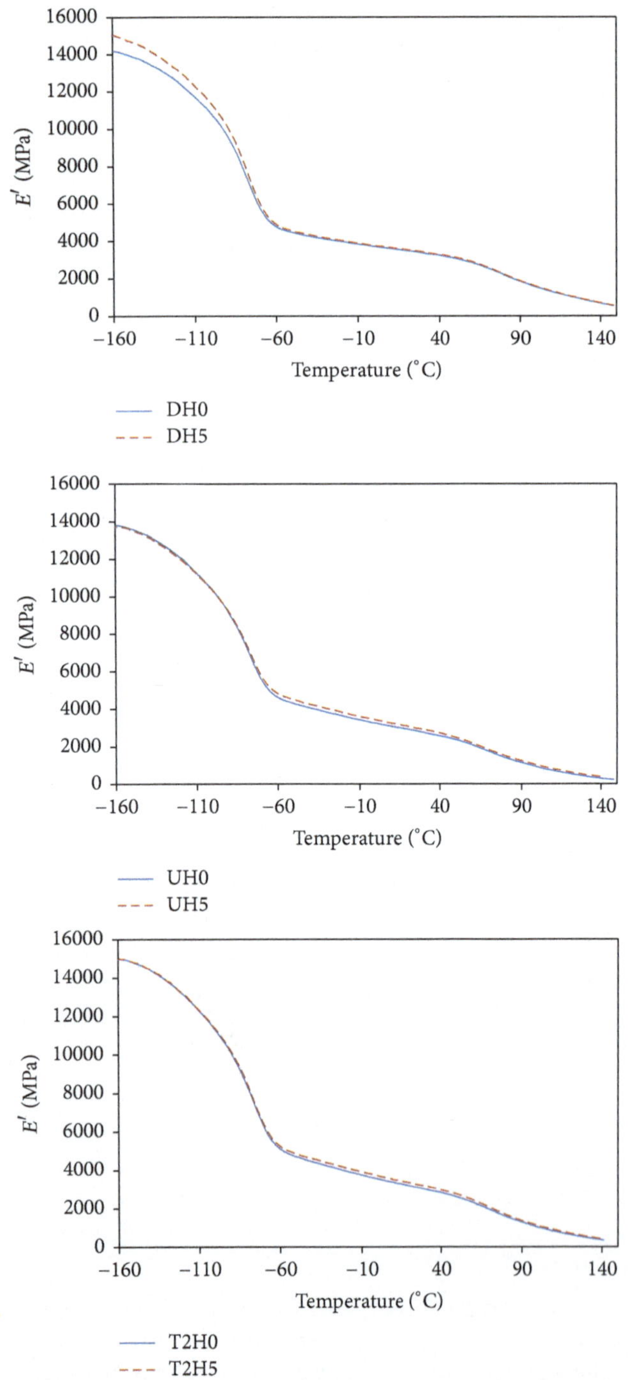

FIGURE 2: DMA curves for POM and POM/5% HAp nanocomposites: E' versus temperature.

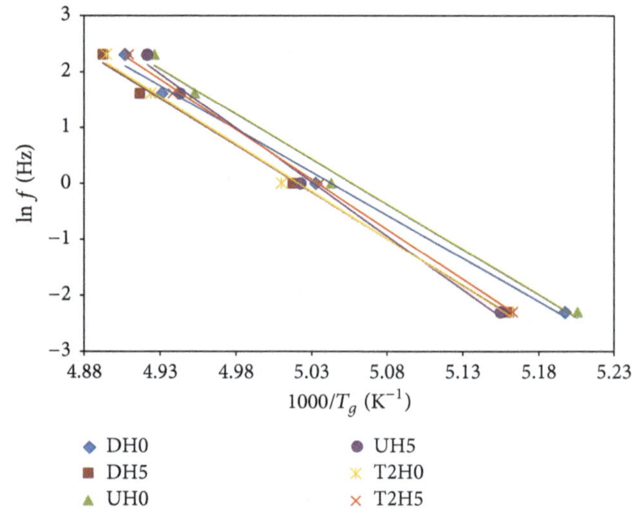

FIGURE 3: Arrhenius frequency map of POM and POM/HAp nanocomposites calculated from DMA results. DMA points correspond to the γ peak at tan δ.

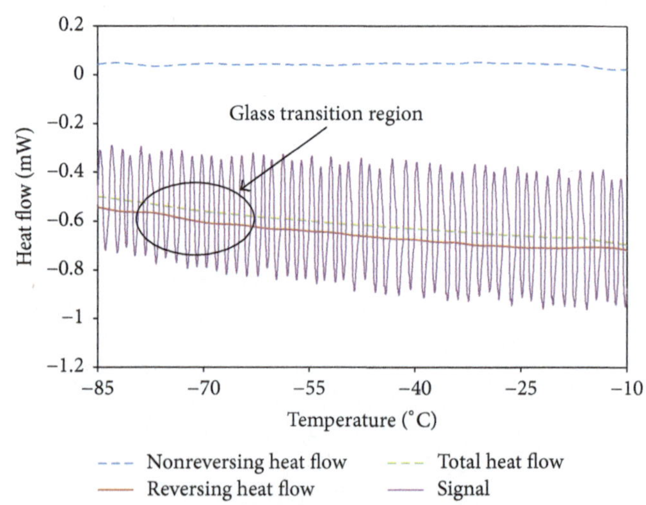

FIGURE 4: TOPEM DSC curves for DH0.

PMMA-nanofiller interfaces (free space at the interface), and wetted PS-nanofiller interfaces lacking attractive interactions, respectively. Additionally, the molecular dynamics simulations endorsed the hypothesis that T_g of polymer nanocomposites can be enhanced or depressed by tuning the polymer-nanofiller interactions [27–30]. For POM/HAp

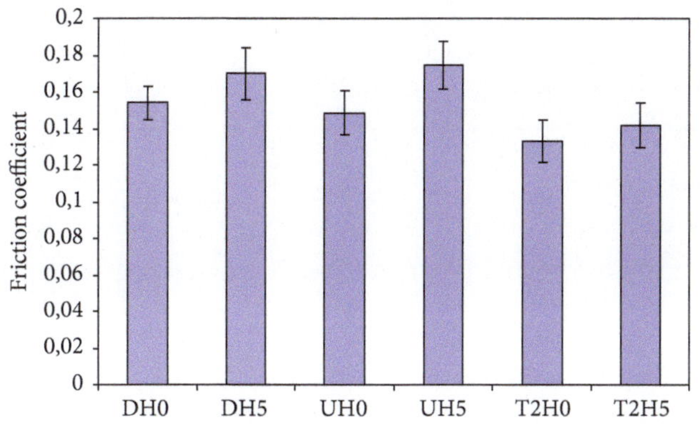

Figure 5: Friction coefficient for POM and POM/HAp nanocomposites.

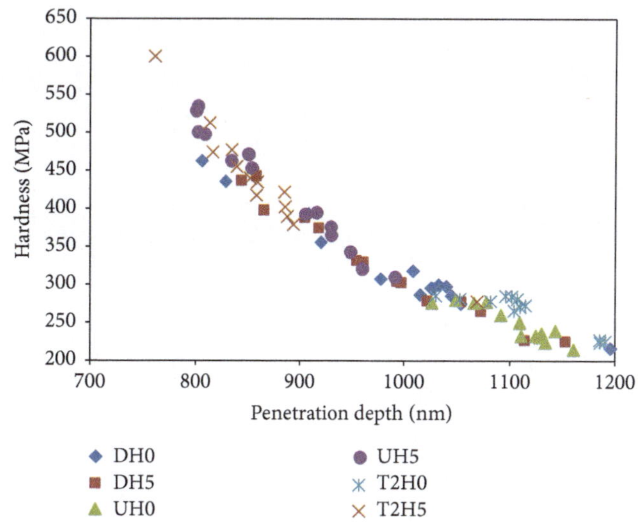

Figure 6: Hardness versus penetration depth for POM and POM/HAp nanocomposites.

nanocomposites, an increase in the glass temperature can be attributed to the interactions between HAp nanoparticles that contain polar groups—OH^- and PO_4^{3-}—and polymer chain with hydroxyl groups, as well as ether oxygen and acetic functionalities. The chemical structure of the components enables hydrogen bond formation between POM chains and HAp nanoparticles causing an increase of the glass temperature in POM/HAp nanocomposites. The increased activation energy of glass transition can be attributed to the confinement effects of polymer chains between HAp nanoparticles, leading to restricted chains mobility. Moreover, it can be observed that the lowest activation energy was found for POM homopolymer and higher for POM copolymers probably due to differences in macromolecular dynamics of POM chain between POM homo- and copolymer [31], while the glass temperature determined from DMA results was almost the same for all POM types.

In the next step, microtribological investigations were performed. Friction coefficient values of POM and POM/HAp nanocomposites are presented in Figure 5.

Generally, the friction coefficient is higher for POM/HAp nanocomposites in comparison to pure POM, both homopolymer and copolymer. POM resin has been widely used as a self-lubricating material in numerous fields, such as automobile, electronic appliances, and machine engineering, because it exhibits good fatigue and creep resistance and high impact strength. POM low friction coefficient is attributed to the flexibility of the linear molecular chains, and simultaneously its good wear resistance is connected with its high crystallinity and high bond energy [7, 32]. The influence of nanoadditives on the tribological properties depends on the kind of nanofiller. Sun et al. [7] investigated POM/3% Al_2O_3 nanocomposite and POM nanocomposites with polytetrafluoroethylene (PTFE), molybdenum disulfide, and Al_2O_3 (POM/PTFE/MoS_2/3% Al_2O_3) nanocomposite and reported that, under dry sliding conditions, POM/3% Al_2O_3 nanocomposite has higher friction coefficient and wear volume than neat POM, while the friction coefficient and wear volume of POM/PTFE/MoS_2/3% Al_2O_3 nanocomposite decreased compared to neat POM. They suggested that, for POM/3% Al_2O_3 nanocomposite, the peeling off and agglomeration of nanoparticles lead to a longer running-in period and also a higher friction coefficient. In contrast, for POM/PTFE/MoS_2/3% Al_2O_3 nanocomposite, nanoparticles can accelerate the formation of compact and uniform transfer films leading to a steady friction coefficient [7]. By analogy, in POM/HAp nanocomposites investigated in the course of this study, an increased friction coefficient was determined.

Results of nanoindentation investigations are presented in Figures 6 and 7.

All the materials studied exhibited the similar surface gradient of hardness. The indentation test results show that microhardness is generally higher for POM/HAp nanocomposites than for POM. Larger differences should be also noted in hardness of POM and POM/HAp nanocomposites for POM copolymer (UH and T2H) whereby for POM homopolymer hardness increases only slightly. The highest hardness at the lowest penetration depth was measured for UH5 and T2H5 nanocomposites. The obtained results stay in good agreement with crystallinity level of the materials studied. For unmodified POM the highest degree of crystallinity was found for DH0, next one was for T2H0, and the lowest one was for UH0 (50.66%, 49.26%, and 45.17%, resp.) while for POM HAp nanocomposites the highest degree of crystallinity was determined for T2H5, next one was for UH5, and the lowest one was for DH5 (52.94%, 46.98%, and 41.18%, resp.) [15–17]. The same trend was observed for the hardness results that confirm our previous results. Different hardness values (Figure 6) can be probably attributed to crystalline and amorphous regions as shown in SEM microphotographs [15–17]. This dependence was well documented in the literature for different macromolecular systems, and, generally, microhardness increases with increasing degree of crystallinity of polymer [33]. It was postulated that for polymers with crystallinity around 35% containing nanocrystals randomly distributed within the amorphous phase a network of physical crosslinks connected by bridging molecules is formed. It induces a significant reinforcement of the material that leads

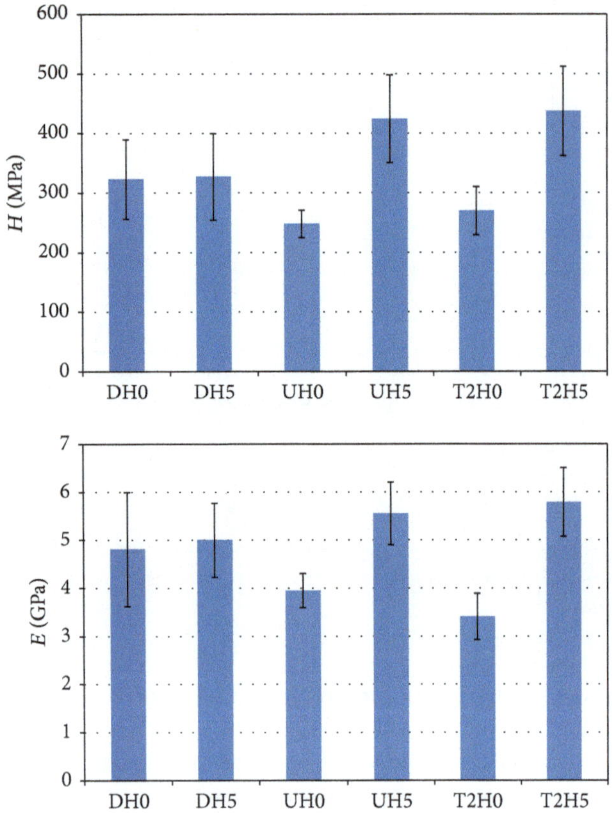

FIGURE 7: Hardness and elastic modulus of POM and POM nanocomposites.

to an increase of microhardness, and the higher the amount of crystals the larger the reinforcing effect [33].

A similar trend is observed for elasticity modulus—higher values of modulus were found for POM/HAp nanocomposites. Similar effects were observed by Shen et al. [34] for polyamide 66/organoclay nanocomposites. These authors additionally postulated that the mechanical properties observed are correlated with the inhomogeneous microstructures of the systems under investigation.

Because of the application of POM/HAp nanocomposites in the field of bone implants, determination of their mechanical properties is of primary importance. Ultrasonic nondestructive testing method is a useful tool for characterization of nanocomposites homogeneity and elasticity. At first, the propagation velocity of longitudinal C_L and transverse waves (C_T) was determined—Figure 8.

The highest propagation velocities of both longitudinal and transverse ultrasonic waves can be observed for POM homopolymer (DH) and POM copolymer T2H and their nanocomposites. This effect can be attributed to the higher degree of crystallinity of DH and T2H as well as their nanocomposites when comparing to the degree of crystallinity of UH copolymer and its nanocomposites [15–17]. The reason is that ordered crystalline fractions produce stronger intermolecular interactions, leading to increased mechanical properties and, consequently, to an increase of ultrasonic wave velocity [35]. Ultrasonic waves in polymeric materials can propagate through slight translocation of atoms, groups, and chain segments around their equilibrium positions. The forces operating along chain segments and between macromolecular chains lead to translocations into neighbouring macromolecules, thus involving stress waves through the material [19, 36, 37]. Next, based on the measurements of ultrasonic wave velocity, the Poisson ratio was calculated—Table 3.

TABLE 3: Apparent density and Poisson ratio for POM and POM/HAp nanocomposites.

Sample	ρ [g/cm^3]	μ
DH0	1.42	0.41
DH1	1.40	0.41
DH2.5	1.41	0.41
DH5	1.36	0.40
UH0	1.44	0.42
UH1	1.39	0.42
UH2.5	1.34	0.41
UH5	1.42	0.42
T2H0	1.57	0.42
T2H1	1.51	0.42
T2H2.5	1.45	0.40
T2H5	1.52	0.40

The Poisson ratio is in the range 0.40–0.42 that is a value typical for semicrystalline polymers [38, 39] and is generally lower for POM/HAp nanocomposites with higher HAp loading. During semicrystalline polymer stretching firstly the onset of isolated inter- and intralamellar slip processes is observed after the initial Hooke elastic range, next it changes into a collective activity of slip motions of crystal blocks at the point of maximum curvature of the true stress-strain curve, and, finally, the beginning of destruction of crystal blocks followed by recrystallization with formation of fibrils occurs. It is accompanied by the beginning of disentanglement of the amorphous network or strain hardening due to the stretching of the amorphous entangled network at high deformations [40]. For polymer composites, the stress concentration causes around the filler particles the effect of external load, whereby the actual stress distribution decides about the extent of local micromechanical deformation processes [41]. In polymer composites the prevalent deformation mechanism is debonding on the matrix/filler interface that leads to some volume increase during deformation [42, 43]. For POM/HAp nanocomposites we found in our previous study [15] that the strain-stress curves have very clear yielding points showing ductile characteristics, but, with an increase of HAp content, the ductile properties decrease and the materials become brittle. This also proves the excellent filler-matrix adhesion of POM/HAp nanocomposites—the better the adhesion between nanoparticles and polymer matrix, the more difficult the movement of macrochains that restricts deformation during tensile stress and leads to a decrease of Poisson ratio. It is noteworthy that for unmodified POM these restrictions are not present. For POM/HAp nanocomposites with lower HAp loading the restrictions in the chain mobility are too low to change the Poisson ratio.

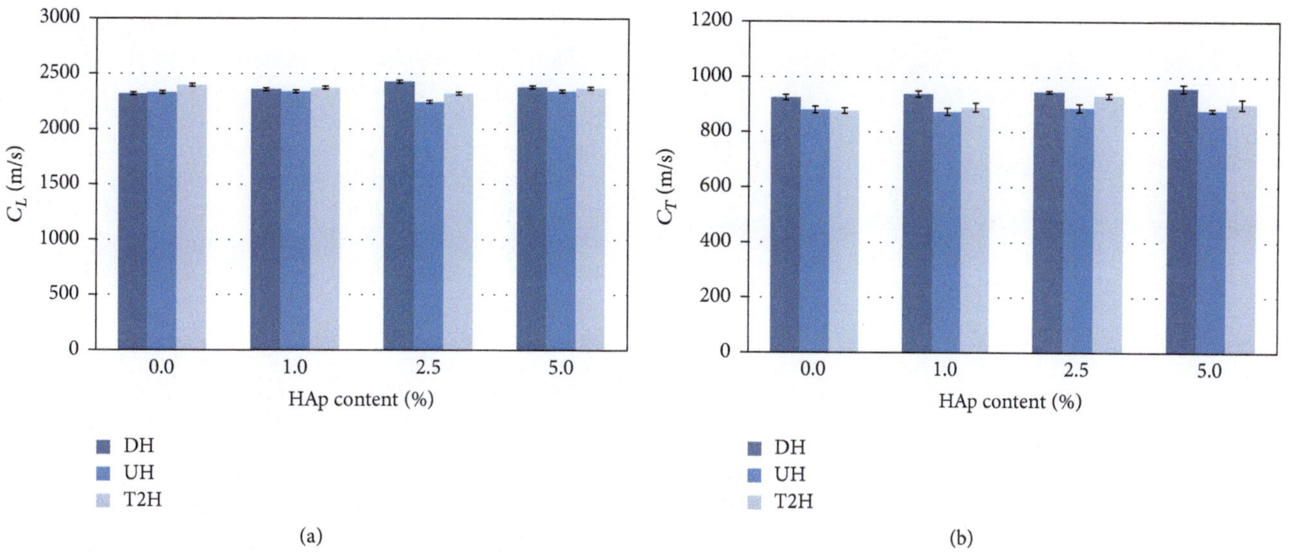

FIGURE 8: Longitudinal wave velocity (C_L) (a) and transverse wave velocity (C_T) (b) of POM and POM/HAp nanocomposites with different HAp content.

Based on these results, ultrasonic Young's modulus E [GPa] and Kirchhoff's modulus G [GPa] were calculated—Figure 9.

It can be seen that the elastic properties strongly depend on the HAp concentration and type of POM and their crystallinity. The highest values of Young's and shear moduli were found for polymers with the highest degree of crystallinity—higher moduli were obtained for DH and T2H systems in comparison to UH copolymer. To check the mechanical properties stability, the samples were tested up to 10^6 cycles under dynamical loading. Results of measurements of longitudinal ultrasonic wave velocity after these tests are presented in Figure 10.

Larger changes in ultrasonic longitudinal wave velocity were detected for unmodified POM while for POM/HAp nanocomposites with 5% of HAp longitudinal wave velocity was almost constant even after 10^6 cycles, proving an enhancement of the mechanical properties stability by HAp nanoparticles. This effect can probably be attributed to the restriction of POM chain mobility due to the presence of HAp nanoparticles and their nucleating effect on POM chains that leads to the formation of larger amounts of smaller and more defected spherulites.

4. Conclusions

In this work, the influence of nanohydroxyapatite on the glass transition region and its activation energy and tribological and mechanical properties of polyoxymethylene nanocomposites was investigated. It was found that glass transition for unmodified POM was in the lower temperature range than for POM/HAp nanocomposites. Moreover, ΔC_p and activation energy were larger for POM/HAp nanocomposites which can be attributed to confinement effect of polymer chains between HAp nanoparticles which restricted chains mobility. The lowest activation energy was found

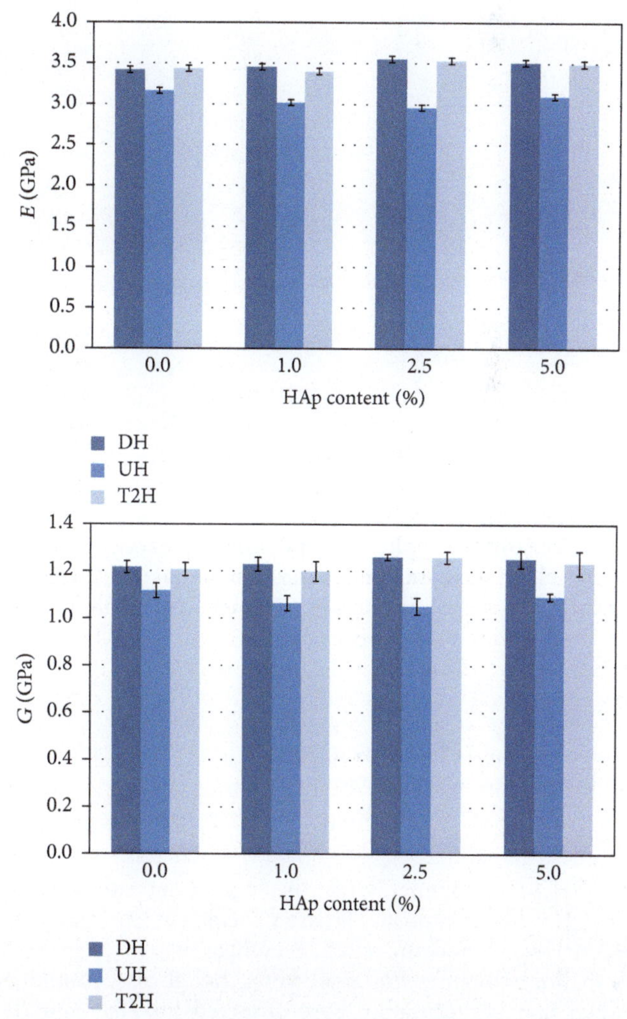

FIGURE 9: Ultrasonic Young's modulus E [GPa] and Kirchhoff's modulus G [GPa] calculated based on measurements of ultrasound velocities. Error bars represent standard deviation.

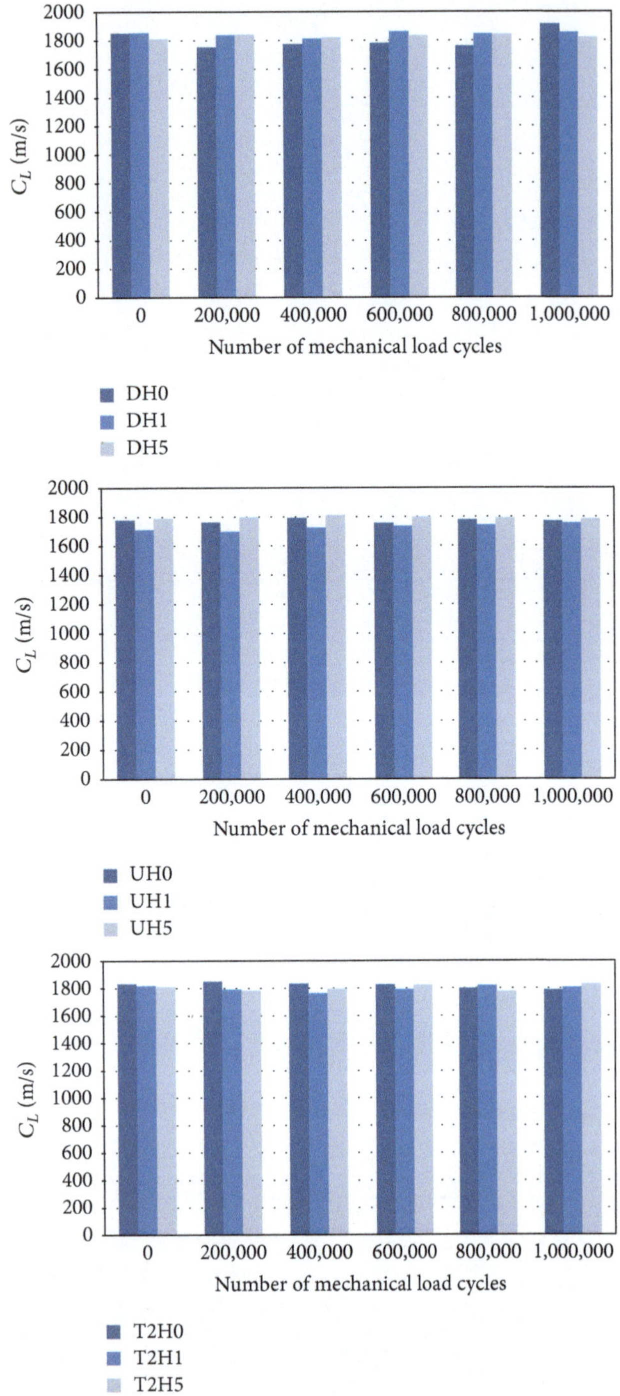

FIGURE 10: Longitudinal wave velocity (C_L) for POM and POM/HAp nanocomposites after dynamical loading.

for DH homopolymer, while the highest activation energy for UH5 nanocomposite. Friction coefficient was higher for POM/HAp nanocomposites in comparison to pure POM both homopolymer and copolymer, but it also should be noted that higher values were observed for DH. Simultaneously, the indentation test results show that microhardness is also higher for POM/HAp nanocomposites than for POM, but more significant changes were observed for POM copolymers. Based on the literature data this effect can be explained by inhomogeneous distributions of the crystalline morphology as well as by HAp concentration changes along the indentation direction. From ultrasonic investigations the highest propagation velocity was found of both longitudinal and transverse ultrasonic waves and ultrasonic Young's and shear modulus for POM homopolymer (DH) and POM copolymer T2H and their nanocomposites which can be attributed to their higher degree of crystallinity in comparison to UH copolymer. Poisson ratio was in the range 0.40–0.42 and was generally lower for POM/HAp nanocomposites with higher HAp loading due to good adhesion between POM and HAp that hinders the movements of macrochains and restricts deformation under force. Besides, for POM/HAp nanocomposites with 5% of HAp, ultrasonic longitudinal wave velocity was almost constant even after 10^6 mechanical loading cycles proving an enhancement of mechanical properties stability by HAp nanoparticles. This probably can be attributed to the restriction of POM chain mobility due to the presence of HAp nanoparticles showing good adhesion with POM matrix, as well as their nucleating effect on the POM macrochains.

Acknowledgments

One of the authors (Kinga Pielichowska) is grateful to the Polish Ministry of Science and Higher Education for financial support under Grant no. PBZ/MEiN/01/2006/15. This work was partially financed by statute fund of Faculty of Materials Science and Ceramics AGH-UST under Contract no. 11.11.160.616.

References

[1] S. Wacharawichanant, S. Thongyai, A. Phutthaphan, and C. Eiamsam-ang, "Effect of particle sizes of zinc oxide on mechanical, thermal and morphological properties of polyoxymethylene/zinc oxide nanocomposites," *Polymer Testing*, vol. 27, no. 8, pp. 971–976, 2008.

[2] K. Takasa, N. Miyashita, and K. Takeda, "Young modulus and degree of crystallization of highly-elongated polyoxymethylene," *Journal of Applied Polymer Science*, vol. 101, no. 2, pp. 1223–1227, 2006.

[3] K. Pielichowska, "Polyacetals," in *Handbook of Thermoplastics*, O. Olabisi and K. Adewale, Eds., CRC Press, Boca Raton, Florida, USA, 2nd edition, 2015.

[4] L. Eschbach, "Nonresorbable polymers in bone surgery," *Injury*, vol. 31, supplement 4, pp. D22–D27, 2000.

[5] T. Eliades, "Orthodontic materials research and applications: Part 2. Current status and projected future developments in materials and biocompatibility," *American Journal of Orthodontics and Dentofacial Orthopedics*, vol. 131, no. 2, pp. 253–262, 2007.

[6] SH. Teoh, "New insight into the fracture and wear problems of a mechanical heart valve — in vitro microstrain, creep rupture, and wear studies," in *Heart Replacement*, T. Akutsu and H. Koyanagi, Eds., pp. 348–352, Springer, Japan, 1998.

[7] L.-H. Sun, Z.-G. Yang, and X.-H. Li, "Mechanical and tribological properties of polyoxymethylene modified with nanoparticles and solid lubricants," *Polymer Engineering and Science*, vol. 48, no. 9, pp. 1824–1832, 2008.

[8] J. Wang, X. G. Hu, M. Tian, and R. Stengler, "Study on mechanical and tribological property of nanometer ZrO2-filled polyxymethylene composites," *Polymer - Plastics Technology and Engineering*, vol. 46, no. 5, pp. 469–473, 2007.

[9] X. Xu, L. Guo, Y. Zhang, and Z. Zhang, "Mechanical and thermal properties of reactable nano-SiO 2/polyoxymethylene composites," *Advanced Materials Research*, vol. 535-537, pp. 103–109, 2012.

[10] I. C. Romero-Ibarra, E. Bonilla-Blancas, A. Sánchez-Solís, and O. Manero, "Influence of X-ray opaque $BaSO_4$ nanoparticles on the mechanical, thermal and rheological properties of polyoxymethylene nanocomposites," *Journal of Polymer Engineering*, vol. 32, no. 4-5, pp. 319–326, 2012.

[11] S. Wacharawichanant, P. Sahapaibounkit, and U. Saeueng, "Study on Mechanical and Morphological Properties of Polyoxymethylene/Al_2O_3 Nanocomposites," in *Proceedings of the TIChE International Conference 2011*, Songkhla, Thailand, 2011.

[12] E. Bouyer, F. Gitzhofer, and M. I. Boulos, "Morphological study of hydroxyapatite nanocrystal suspension," *Journal of Materials Science: Materials in Medicine*, vol. 11, no. 8, pp. 523–531, 2000.

[13] K. Pielichowska and S. Blazewicz, "Bioactive polymer/hydroxyapatite (nano)composites for bone tissue regeneration," *Advances in Polymer Science*, vol. 232, no. 1, pp. 97–207, 2010.

[14] R. K. Roeder, G. L. Converse, R. J. Kane, and W. Yue, "Hydroxyapatite-reinforced polymer biocomposites for synthetic bone substitutes," *JOM*, vol. 60, no. 3, pp. 38–45, 2008.

[15] K. Pielichowska, A. Szczygielska, and E. Spasówka, "Preparation and characterization of polyoxymethylene-copolymer/hydroxyapatite nanocomposites for long-term bone implants," *Polymers for Advanced Technologies*, vol. 23, no. 8, pp. 1141–1150, 2012.

[16] K. Pielichowska, "Polyoxymethylene-homopolymer/hydroxyapatite nanocomposites for biomedical applications," *Journal of Applied Polymer Science*, vol. 123, no. 4, pp. 2234–2243, 2012.

[17] K. Pielichowska, "The influence of molecular weight on the properties of polyacetal/ hydroxyapatite nanocomposites. Part 1. Microstructural analysis and phase transition studies," *Journal of Polymer Research*, vol. 19, no. 2, article no. 9775, 2012.

[18] W. C. Oliver and G. M. Pharr, "Improved technique for determining hardness and elastic modulus using load and displacement sensing indentation experiments," *Journal of Materials Research*, vol. 7, no. 6, pp. 1564–1580, 1992.

[19] J. Krautkrämer and H. Krautkrämer, *Introduction. ultrasonic testing of materials*, Springer, Berlin, Heidelberg, 1990.

[20] G. C. Psarras, S. Siengchin, P. K. Karahaliou, S. N. Georga, C. A. Krontiras, and J. Karger-Kocsis, "Dielectric relaxation phenomena and dynamics in polyoxymethylene/polyurethane/alumina hybrid nanocomposites," *Polymer International*, vol. 60, no. 12, pp. 1715–1721, 2011.

[21] B. B. Sauer, P. Avakian, E. A. Flexman, M. Keatinc, B. S. Hsiao, and R. K. Verma, "A.C. Dielectric and TSC studies of constrained amorphous motions in flexible polymers including poly(OxyMethylene) and miscible blends," *Journal of Polymer Science, Part B: Polymer Physics*, vol. 35, no. 13, pp. 2121–2132, 1997.

[22] G. Rotter and H. Ishida, "Dynamic mechanical analysis of the glass transition. Curve resolving applied to polymers," *Macromolecules*, vol. 25, no. 8, pp. 2170–2176, 1992.

[23] G. Li, P. Lee-Sullivan, and R. W. Thring, "Determination of activation energy for glass transition of an epoxy adhesive using dynamic mechanical analysis," *Journal of Thermal Analysis and Calorimetry*, vol. 60, no. 2, pp. 377–390, 2000.

[24] M. Wong, R. Tsuji, S. Nutt, and H.-J. Sue, "Glass transition temperature changes of melt-blended polymer nanocomposites containing finely dispersed ZnO quantum dots," *Soft Matter*, vol. 6, no. 18, pp. 4482–4490, 2010.

[25] P. Rittigstein, R. D. Priestley, L. J. Broadbelt, and J. M. Torkelson, "Model polymer nanocomposites provide an understanding of confinement effects in real nanocomposites," *Nature Materials*, vol. 6, no. 4, pp. 278–282, 2007.

[26] A. Bansal, H. Yang, C. Li et al., "Quantitative equivalence between polymer nanocomposites and thin polymer films," *Nature Materials*, vol. 4, no. 9, pp. 693–698, 2005.

[27] B. J. Ash, L. S. Schadler, and R. W. Siegel, "Glass transition behavior of alumina/polymethylmethacrylate nanocomposites," *Materials Letters*, vol. 55, no. 1-2, pp. 83–87, 2002.

[28] P. Rittigstein and J. M. Torkelson, "Polymer-nanoparticle interfacial interactions in polymer nanocomposites: Confinement effects on glass transition temperature and suppression of physical aging," *Journal of Polymer Science, Part B: Polymer Physics*, vol. 44, no. 20, pp. 2935–2943, 2006.

[29] J. Koszkul and D. Kwiatkowsk, "The creep tests of polyoxymethylene composites reinforced with the microspheres from fly ashes," *Journal of Materials Processing Technology*, vol. 157-158, pp. 360–363, 2004.

[30] Y. Sun, Z. Zhang, K.-S. Moon, and C. P. Wong, "Glass transition and relaxation behavior of epoxy nanocomposites," *Journal of Polymer Science, Part B: Polymer Physics*, vol. 42, no. 21, pp. 3849–3858, 2004.

[31] E. G. Merino, S. Atlas, M. Raihane et al., "Molecular dynamics of poly(ATRIF) homopolymer and poly(AN-co-ATRIF) copolymer investigated by dielectric relaxation spectroscopy," *European Polymer Journal*, vol. 47, no. 7, pp. 1429–1446, 2011.

[32] J. Chen, Y. Cao, and H. Li, "Investigation of the friction and wear behaviors of polyoxymethylene/linear low-density polyethylene/ethylene-acrylic-acid blends," *Wear*, vol. 260, no. 11-12, pp. 1342–1348, 2006.

[33] A. Flores, F. Ania, and F. J. Baltá-Calleja, "From the glassy state to ordered polymer structures: a microhardness study," *Polymer*, vol. 50, no. 3, pp. 729–746, 2009.

[34] L. Shen, I. Y. Phang, T. Liu, and K. Zeng, "Nanoindentation and morphological studies on nylon 66/organoclay nanocomposites. II. Effect of strain rate," *Polymer*, vol. 45, no. 24, pp. 8221–8229, 2004.

[35] F. Lionetto, F. Montagna, and A. Maffezzoli, "Ultrasonic dynamic mechanical analysis of polymers," *Applied Rheology*, vol. 15, no. 5, pp. 326–335, 2005.

[36] Committee ASMH. Metals Handbook: Nondestructive evaluation and quality control: American Society for Metals.

[37] F. Lionetto and A. Maffezzoli, "Polymer characterization by ultrasonic wave propagation," *Advances in Polymer Technology*, vol. 27, no. 2, pp. 63–73, 2008.

[38] D. I. Bower, *An Introduction to Polymer Physics*, Cambridge University Press, 2002.

[39] G. N. Greaves, A. L. Greer, R. S. Lakes, and T. Rouxel, "Poisson's ratio and modern materials," *Nature Materials*, vol. 10, no. 11, pp. 823–837, 2011.

[40] C. De Rosa and F. Auriemma, "Stress-induced phase transitions in metallocene-made isotactic polypropylene," in *Progress in Understanding of Polymer Crystallization. Lecture Notes in Physics. 714*, G. Reiter and G. Strobl, Eds., pp. 345–371, Springer, Berlin Heidelberg, 2007.

[41] C. Bucknall, Toughened plastics. London: Applied Science Publishers, Ltd, 1977.

[42] B. Pukánszky, M. Van Es, F. H. J. Maurer, and G. Vörös, "Micromechanical deformations in particulate filled thermoplastics: volume strain measurements," *Journal of Materials Science*, vol. 29, no. 9, pp. 2350–2358, 1994.

[43] A. Sudar, J. Moczo, G. Voros, and B. Pukanszky, "The mechanism and kinetics of void formation and growth in particulate filled PE composites," *eXPRESS Polymer Letters*, vol. 1, pp. 763–772, 2007.

Effect of Polypropylene Modification by Impregnation with Oil on its Wear and Friction Coefficient at Variable Load and Various Friction Rates

Paweł Sędłak,[1] Beata Białobrzeska,[2] Tomasz Stawicki,[1] and Piotr Kostencki[1]

[1] *Department of Agrotechnical Systems Engineering, West Pomeranian University of Technology, Papieza Pawla VI Street 1, 71-459 Szczecin, Poland*

[2] *Department of Materials Science, Welding and Strength of Materials, Wrocław University of Technology, Smoluchowskiego Street 25, 50-370 Wrocław, Poland*

Correspondence should be addressed to Beata Białobrzeska; beata.bialobrzeska@pwr.edu.pl

Academic Editor: Saad Khan

Laboratorial two-body wear testing was carried out in order to assess effects of polypropylene modification by impregnating it with oils on friction coefficient and wear in comparison to those parameters of unmodified polypropylene, Teflon, and polyamide during operation under conditions of sliding friction without lubrication. Wear behaviour of the tested specimens was investigated using ASTM G77-98 standard wear test equipment. Recording program made it possible to visualise and record the following parameters: rotational speed and load, linear wear, friction coefficient, temperature of the specimen, and ambient temperature. In addition, wear mechanisms of the analysed materials were determined with use of scanning electron microscopy. In the case of the remaining tested polymers, the most important mechanism of wear was adhesion (PP, PTFE, PA 6.6, and PA MoS_2), microcutting (PTFE, PA 6.6, and PA MoS_2), fatigue wear (PTFE), forming "roll-shaped particles" combined with plastic deformation (PA 6.6 and PA MoS_2), and thermal wear (PP). Impregnation of polypropylene with engine oil, gear oil, or RME results in significant reduction of friction coefficient and thus of friction torque, in relation to not only unmodified polypropylene but also the examined polyamide and Teflon.

1. Introduction

Being a science engaged in the phenomena occurring in friction areas of mating parts, tribology considers difficult phenomena determining friction and wear of engineering materials. The factors decidedly affecting tribological processes in friction nodes include friction type, intensity and resistance, nature and value of applied loads, relative velocity, working temperature, and geometry of the friction node, as well as sliding direction with respect to injection flow of the polymer [1]. Wear processes are also affected by stereometry of the outside surface created in the manufacturing processes. The above-mentioned factors determine nature and intensity of destructive processes occurring in the cooperation areas of friction pairs composed of various materials: metal-metal, metal-nonmetal (e.g., a polymer), and polymer-polymer [2, 3]. A set of all factors influencing a friction node brings about a specific kind of wear and, more precisely, the dominating wear process. Besides the above-mentioned factors, type of wear is also affected by heat dissipation from the friction area, lubricant supply and properties, and mechanism and intensity of removing wear products, as well as shape and dimensions of the wearing surfaces.

Influence of individual factors on friction and wear processes in friction pairs based on traditional constructional materials can be evaluated, among others, thanks to well-recognised crystalline structure of these materials and to relatively comprehensive set of experimental data and the information coming from practice. However, traditional tribology developed originally for metals cannot be applied in the case of friction pairs containing polymers because of their rapid improvement [4], and so still new research works are required in order to recognise influence of the a.m. factors on friction and wear processes. It is very difficult to indicate the factors that generate the prevailing wear process and to attribute to them the mechanism responsible for wear. This results from various physicomechanical features of polymer materials in relation to those of the mating materials that are most often metals and from a significant influence of temperature on their properties. Significant amount of heat generated in friction pairs based on metals and metal alloys is effectively dissipated to the environment thanks to good thermal conductivity of metals combined with heat collection by the lubricating agent. However, in the pairs based on polymers (including polymer-based composites) characterised by thermal conductivity lower than that of metals, generated heat significantly affects their physicomechanical properties [4]. This restricts the possibility of using polymers in sliding friction nodes to those with low linear velocities and low loads [5]. In particular, polymers are used for friction nodes in machines and facilities in that traditional lubrication is difficult or impossible or building a lubrication system is nonviable. At dry friction, bearings made of polymers show features of self-lubricating (maintenance-free) bearings. When working conditions are known, properties of polymers used in sliding nodes are to some extent deliberately formed by selection of a specific chemical composition that determines their structure, as well as mechanical, thermal, and tribological properties [6, 7]. Frequently, in order to improve mechanical and tribological parameters of polymers, modification of chemical composition is applied. Most often, metals and alloys (tin, lead, copper, bearing alloys, and oxides of some metals) or nonmetallic additives (graphite, soot, molybdenum disulphide, and chalk of other polymers) are used as modifiers. Form of an addition influences the obtained mechanical and tribological effects. The additives can be applied in form of fine particles with various granularity, fibres with various length, mats, or fabrics, creating with the polymer a polymeric composite [8–10].

Another way of modifying physicomechanical features of pure polymers is their impregnation with liquid lubricants. In this process, porous structure of polymers is utilised, which makes it possible to accumulate the lubricant necessary to reduce frictional resistance and wear in the sliding node [11]. In this case, an issue is uniform and durable impregnation of polymer in its entire volume, as well as influence of the lubricant on physicomechanical properties of the polymer. Quantity of lubricant that can be gathered in the polymer volume depends on structure of its macromolecules and on its cross-linking degree (crystallisation). Here, an indication can be absorbability of water, which is one of the parameters determined by manufacturers of polymers. In addition, structure and size of lubricant particles can influence the impregnation process, since they determine mobility of molecules within the polymer structure. Attention should be also paid to easiness of impregnation of polymeric parts with liquid lubricants and wide selection of possible lubricating agents. However, the number of scientific studies concerning impregnation of polymers with various lubricants and their tribological properties is still insufficient.

2. Purpose of the Research

The research was aimed at determining influence of polypropylene modification by impregnation with oils on its friction coefficient and wear in comparison to these properties of unmodified polypropylene, Teflon, and polyamide during their operation at sliding friction without lubrication.

Good properties of polypropylene obtained by impregnation with oils indicate a possibility to use this inexpensive material in maintenance-free sliding nodes.

3. Materials and Methods

Specimens of polypropylene (PP) were made of a rod dia. 10.5 mm. Specimens of Teflon (PTFE) and polyamide (PA 6.6 and PA MoS_2) were made of rods dia. 10 mm. The rods were mechanically cut to 19 mm long pieces and then shortened to 17 mm and their edges were chamfered. After thorough cleaning with extraction naphtha and technical acetone, the specimens were weighed on laboratory scales WPS510/C/1 to ±0.001 g and their diameters were measured with a workshop minimeter to ±0.001 mm.

The polypropylene specimens were modified by impregnation with vegetable oil, that is, rapeseed oil methyl ester (RME), and with two mineral oils, that is, gear oil Hipol GL-4 80W/90 and engine oil HD Diesel Formula 15W/40. Polypropylene specimens were impregnated by submerging in the a.m. oils heated up to 105°C and holding for 24 hours. Impregnated specimens were then thoroughly cleaned, measured, and weighed. After the next 24 h, measurements of mass and dimensions were repeated in order to verify whether quantity of the absorbed oil did not change and geometry of the specimens was stable. Depending on the applied oil, various impregnation degrees were obtained (measured by relations of masses after and before impregnation): 15–17% for RME, 7–9% for gear oil, and 9–11% for engine oil. Increments of diameters of the specimens were 0.05–0.08 mm (0.62% in average) for RME, 0.03–0.05 mm (0.4% in average) for gear oil, and 0.04–0.06 mm (0.5% in average) for engine oil.

A counterspecimen for all the examined polymers was a ring dia. 35 mm and 8 mm wide, made of bearing steel ŁH 15 acc. to PN-H-84041:1974 with the following chemical composition: 0.95–1.1% C, 0.25–0.45% Mn, 1.3–1.65% Cr, 0.15–0.35% Si, 0.025% P_{max}, 0.025% S_{max}, 0.3% Ni_{max}, and 0.3% Cu_{max}, which is equivalent to steel 100Cr6 acc. to EN ISO 683-17:2015-01. The counterspecimen was hardened and tempered to 48 HRC and its surface was subjected to finishing grinding ($R_a = 0.19\,\mu$m).

Individual polymers were examined at linear velocities of 0.13, 0.26, 0.44, and 0.62 $m{\cdot}s^{-1}$. The velocities were selected on the grounds of preliminary examinations that showed clear reduction of friction torque at the velocities over 0.62 $m{\cdot}s^{-1}$ due to excessive increase of the polymer temperature, as was also found in [12]. In the examinations, force acting on the friction node increased at constant rate from 0 to 1030 N and the test time was 480 s. With these assumptions, load growth rate was 2.16 $N{\cdot}s^{-1}$ and friction distances were 62.4, 124.8, 211.2, and 297.6 m, respectively, for the accepted working velocities. For some material pairs, the assumed maximum force loading the friction node (1030 N) could not be reached. In these cases, friction torque reached the accepted maximum

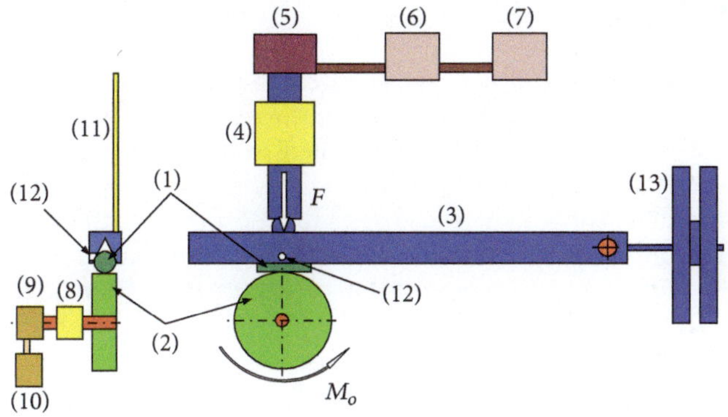

FIGURE 1: Layout of the tribometer used in the tests: (1) specimen, (2) counterspecimen, (3) specimen holder, (4) force sensor, (5) pneumatic cylinder, (6) pneumatic transducer and pressure regulator, (7) compressor, (8) torque meter, (9) electric motor, (10) frequency converter, (11) movement sensor, (12) thermocouple for temperature measurements of the specimen, and (13) counterweight.

permissible value of 350 Ncm at lower load values, and the test station was automatically stopped to avoid its possible failure.

For each material mating and working velocity of the tribological pair, measurements were made in 7 series.

Friction torque between the specimen and the counterspecimen was measured with use of a tribometer designed and manufactured in the Department of Agrotechnical Systems Engineering of West Pomeranian University of Technology, Szczecin [13]. Structure of the tribometer is shown in Figure 1. The test station was composed of a frequency converter FREQVAR 3000 (10) controlling the electric motor (9), connected via a torque meter (8) (operating ranges 500 Ncm and 3000 rpm, accuracy class 0.2%) to the counterspecimen (2) creating the examined tribological pair with the specimen (1). Load was applied by a pneumatic cylinder FMP Agromet-Mogilno (5) cooperating with a compressor (7), a pneumatic transducer A201-A 101 D2, and a pressure regulator AW 3000-03 (6). Load of the friction pair was controlled by a force sensor (4) (operating range 2 kN, reading error 0.1%) located between the cylinder and the specimen holder (3). Displacement of the specimen with respect to the counterspecimen was measured by an induction movement sensor LVDT5 (11) (reading error 0.015 mm). The tribometer was equipped with a thermocouple TP-202 type K (12) (reading error 0.1°C). For technical reasons, the thermocouple was located ca. 3 mm from the friction area (along with the specimen wear, the temperature measurement point came nearer to the friction area). This made it possible to measure general temperature of the specimen and not temperature at the contact point between the specimen and the counterspecimen.

Each time, the test station was prepared and the tests were started according to the guidelines in ASTM G77. The counterspecimen was cleaned with use of waterproof abrasive paper number 600 and then degreased together with the specimen. Next, specimens were once more weighed and fastened in the tribometer holder and the test run was started. Each run commenced from preliminary grinding-in the frictional parts at the set rotational speed and constant load 103 N. During grinding-in, the following phases occurred: accelerating the tribometer to the set rotational speed for 21 s, increasing the load from 0 to 103 N for 10 s, maintaining the constant rotational speed at 103 N for 120 s, decreasing the load from 103 N to 0 for 5 s, and maintaining the constant rotational speed for another 5 s. The last phase of this run was aimed at determining linear wear of the specimen. As a result of grinding-in, all components of the tribological system were properly mutually arranged, and surfaces of the specimen and the counterspecimen were matched. After grinding-in, the actual test run commenced without stopping the counterspecimen.

Operation of the tribometer was controlled by the interface MW-5 cooperating with the software STEDIT 101 installed in a PC. By means of this system, load of the friction pair with its planned increase rate, rotational speed, and rotational direction of the counterspecimen were set in a controlled way. The recording program permitted rotational speed of the counterspecimen, load, linear wear of the specimen, friction torque between the specimen and the counterspecimen, and temperature of the specimen to be recorded. After a completed test run, the specimen was again weighed in order to determine its mass wear.

On the grounds of averaged actual values of friction torque in the examined tribological pairs, actual values of friction coefficient between the specimen and the counterspecimen were calculated from the following formula:

$$\mu = \frac{M_t}{F \cdot r_1}, \tag{1}$$

where μ is actual value of friction coefficient, M_t is actual value of friction torque, Nm, F is actual value of force loading the tribological pair, N, and r_1 is diameter of the counterspecimen, m.

The calculated values were used for graphical presentation of friction coefficient in a function of force loading the tribological pair and friction distance.

Values of actual surface pressure at the contact point between the specimen and the counterspecimen were also determined. Calculations were made with formula (2), commonly used for determining pressures in slide bearings:

$$p = \frac{F}{A}, \tag{2}$$

where p is actual value of surface pressure, MPa, F is actual value of force loading the tribological pair, N, and A is actual value of contact area between the specimen and the counterspecimen projected on the plane perpendicular to vertical axis of the counterspecimen, m^2.

A result of wear of the specimen during its mating with the counterspecimen is trace of wear. Projection of this trace on the plane perpendicular to vertical axis of the counterspecimen is close to an ellipse (Figure 2), whose surface area can be determined when lengths of its axes are known.

Actual value of the parameter A was determined from formula (3) in that actual lengths of the ellipse axes were

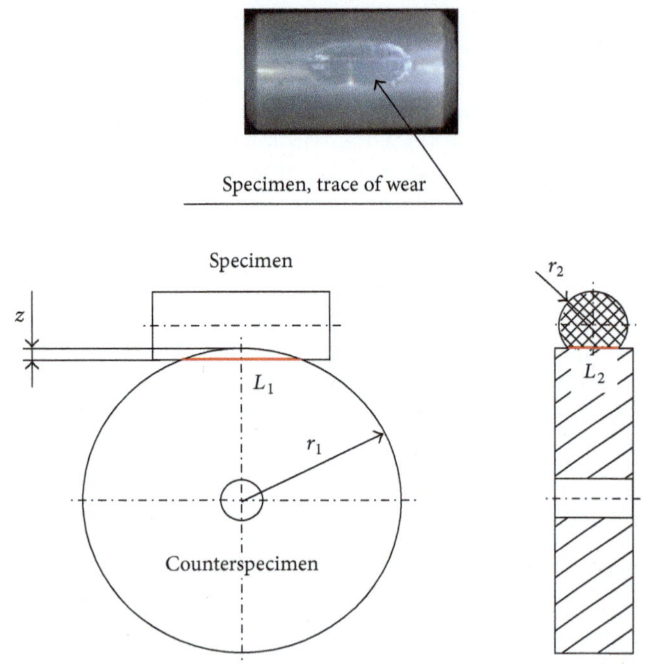

FIGURE 2: Specimen with trace of wear and axes of the ellipse being projection of the trace on the plane perpendicular to vertical axis of the counterspecimen: z, linear wear of the specimen; axis L_1, projected length of trace of wear; axis L_2, width of trace of wear.

related to actual value of linear wear of the specimen and to diameters of the specimen and the counterspecimen:

$$A = \frac{\pi}{4} \cdot L_1 \cdot L_2 = \pi \cdot \sqrt{r_1^2 - (r_1 - z)^2} \cdot \sqrt{r_2^2 - (r_2 - z)^2}, \quad (3)$$

where A is actual value of contact area between the specimen and the counterspecimen projected on the plane perpendicular to vertical axis of the counterspecimen, m^2, L_1 and L_2 are lengths of the ellipse axes (projection of trace of wear), m, z is actual value of linear wear of the specimen, m, r_1 is diameter of the counterspecimen, m, and r_2 is diameter of the specimen, m.

Initial value of the area A was determined for linear wear of the specimen found after grinding-in.

It should be emphasised that, in the above calculations, a simplification concerning linear wear of the specimen z was applied. Namely, value z recorded during the test was applied in calculations and that value was composed of both material loss and deformation of the specimen caused by the force loading the friction pair.

After the tests, wear surfaces of the examined polymers were subjected to visual observations by means of a stereoscopic microscope and a scanning electron microscope in order to identify mechanisms of wear. In the case of polypropylene, SEM observations were carried out on an unmodified specimen, since otherwise there was a risk of damaging the microscope.

4. Results of the Research

4.1. Friction Coefficient. Wear rate is higher at a higher pv value, where p is contact pressure and v is sliding speed [14]. The wear process is controlled to a greater extent by material transfer, which is a very characteristic phenomenon in contacts of polymers and plays the most important role in friction and wear processes in polymeric tribosystems [14].

Figure 3 shows representative curves of friction coefficient for PTFE, PA 6.6, PA MoS_2, and unmodified and modified PP, tested at room temperature at sliding velocities of 0.13, 0.26, 0.44, and 0.62 $m{\cdot}s^{-1}$, depending on sliding distance and load. The friction coefficient curve is a very important characteristic of the friction process. It should be noted that friction coefficient is also affected by sliding velocity through its effect on frictional heating. For all values of sliding velocity, impregnated PP showed an improvement in comparison to unmodified PP, PTFE, PA 6.6, and PA MoS_2.

At the beginning it should be explained why, for polyamide PA 6.6 and PA MoS_2, the lines illustrating values of friction coefficient in some diagrams (Figures 3(b)–3(d)) (like the lines illustrating friction torque, see Figures 4(b) and 4(c)) show distinct folds, while such folds do not occur for the other materials. For PA 6.6 and PA MoS_2, the folds are present in most cases of the applied working velocities for higher load values (over 500 N). These folds in the diagrams were caused by different behaviour of these materials in individual tests. Within high loads, friction torque recorded at a part of repetitions rapidly increased (during 5 to 7 s), reaching the value of 350 Ncm, at which the stoppage of the test stand followed, and thus the given test was finished before the intended maximum load of the friction node could be reached (for polyamide PA 6.6 at 0.62 $m{\cdot}s^{-1}$ no specimen completed the test at the maximum permissible load). Of course, high values of friction torque were accompanied by high values of friction coefficient. However, such large growth of friction torque did not occur in other repetitions and measurements were continued till the planned maximum load of the friction node. The folds visible in the diagrams were caused by averaging actual values of friction coefficient obtained in individual tests, since large values of friction torque in the tests finished before reaching the maximum load were eliminated. Variability of friction coefficient, visible in the diagrams, indicates unstable operation of polyamide PA 6.6 and PA MoS_2 in the tribological pair at its high loads.

At the smallest linear velocity (0.13 $m{\cdot}s^{-1}$) in relation to friction distance and increasing load, it can be seen that shapes of curves presenting changes of friction coefficient are similar for all the examined materials. At the beginning, values of friction coefficient increase rapidly and next stabilise, while this stabilisation occurs for individual materials at various moments, that is, at various friction distances and increasing load.

The largest friction coefficient and the latest stabilisation of its values were recorded for unmodified polypropylene. Beyond the set friction distance of ca. 25 m and load of ca. 400 N, the friction coefficient is ca. 0.18.

(In the context of test results, a name of a polymer always means the friction pair steel-polymer. However, in view of

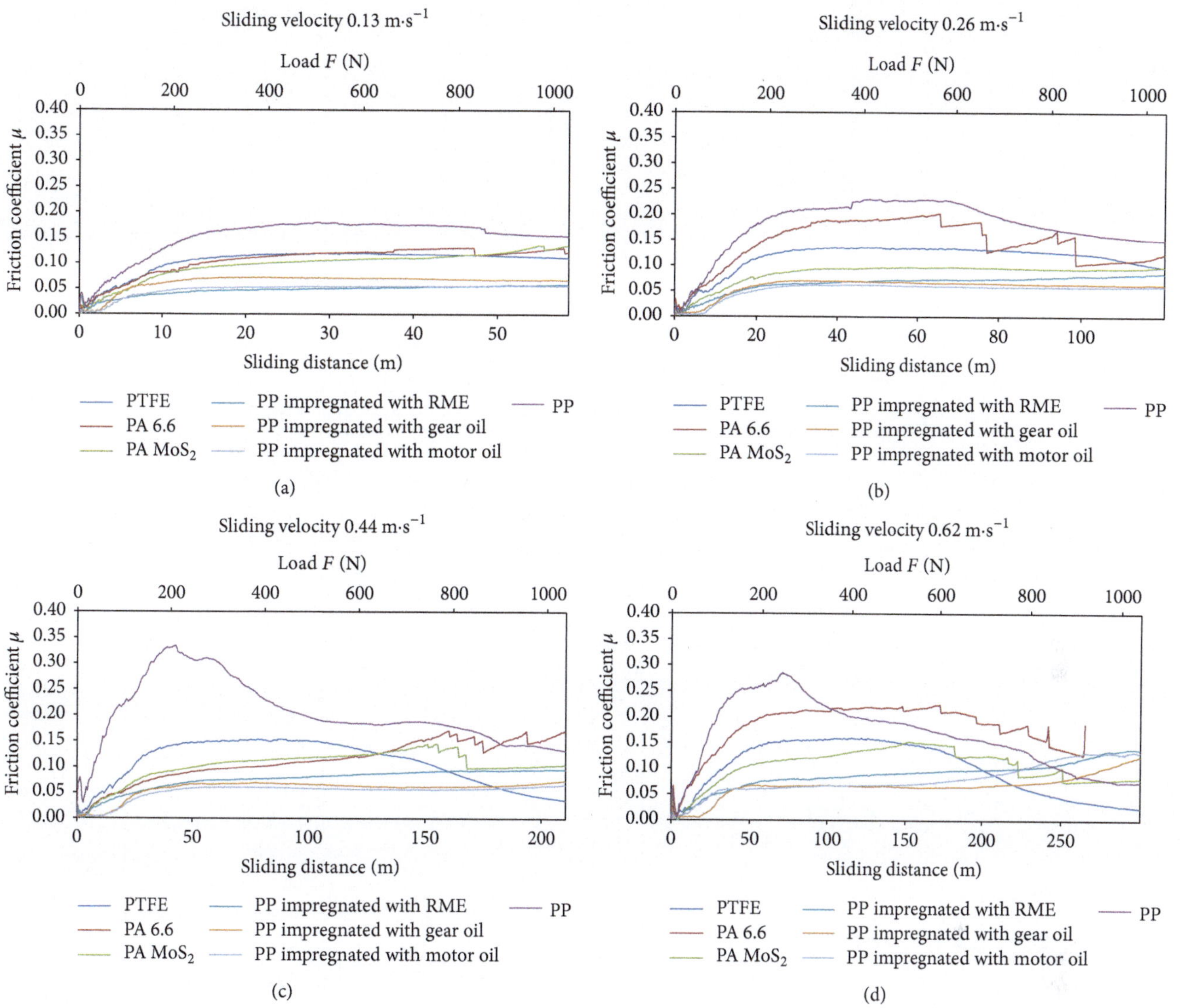

FIGURE 3: Relationship between sliding distance, load, and friction coefficient of the specimens at various sliding velocities: (a) 0.13, (b) 0.26, (c) 0.44, and (d) $0.62\,m{\cdot}s^{-1}$.

better readability of the paper, a polymer name alone is used in the following text.)

It is interesting that shapes of friction curves for PA 6.6, PA MoS_2, and PTFE at $0.13\,m{\cdot}s^{-1}$ are similar (see Figure 3(a)). After stabilisation, friction coefficient ranges from ca. 0.11 for PTFE to over 0.12 for polyamides. Along with increasing load or friction distance, friction coefficient continually slightly increases for polyamides and decreases for PTFE. During abrasion of PTFE balls on glass plates at linear velocity of $0.1\,m{\cdot}s^{-1}$, Makinson and Tabor [15] also found high values of friction coefficient between 0.1 and 0.16 at the beginning of the test and next, after a determined friction distance, these values decreased, which was explained by a PTFE "filter" created as a result of its adhesion to the counterspecimen.

Analysis of friction curves for polypropylene only indicates a significant reduction of friction coefficient (Figure 3(a)) for polypropylene impregnated with oils in comparison to this parameter value for nonimpregnated polypropylene. The lowest values were recorded for PP impregnated with RME and engine oil (0.05), but this value for the specimens impregnated with gear oil was not significantly higher (0.07). In comparison to unmodified PP, friction coefficient was reduced by 72% for RME and engine oil and by 61% for gear oil.

In the case of some polymers, twofold growth of linear velocity to $0.26\,m{\cdot}s^{-1}$ resulted in changed shapes of friction curves (see Figure 3(a)). This is especially visible for unmodified PP; the curve first rises (till friction distance ca. 20 m and load 200 N) and next stabilises to descend slightly beyond ca. 70 m at 600 N. For unmodified PP, friction coefficient in the stabilisation range is ca. 0.23, nearly 30% higher than at $0.13\,m{\cdot}s^{-1}$.

For PA 6.6, friction coefficient quickly increases to ca. 0.18 (friction distance ca. 20 m and load nearly 300 N) and then it grows decidedly slower and reaches 0.20 at friction distance 64 m and load 560 N. In further ranges of friction distance

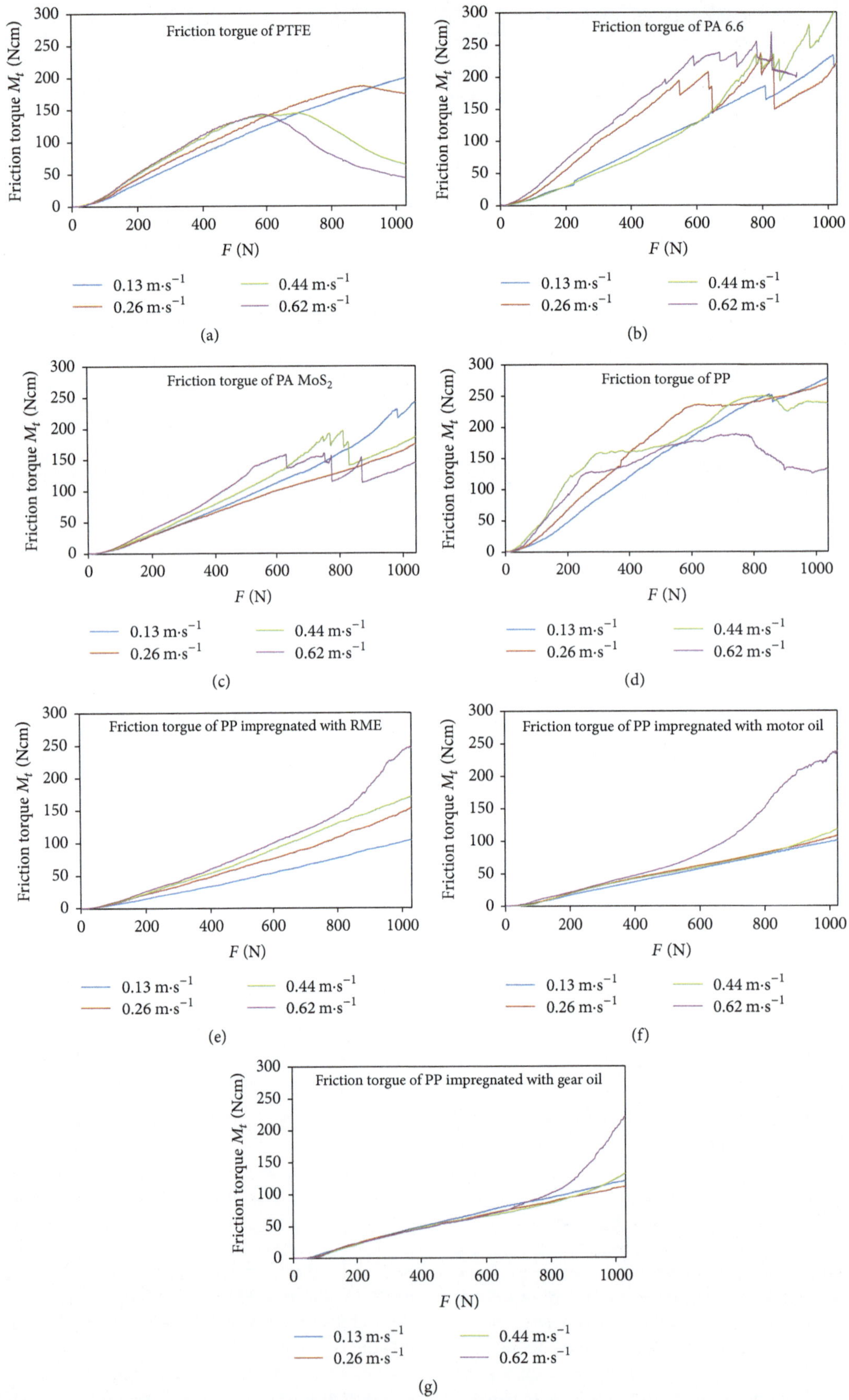

FIGURE 4: Relationship between sliding distance, load, and friction torque of the samples under various sliding velocities, 0.13, 0.26, 0.44, and 0.62 m·s^{-1}, for (a) PTFE, (b) PA 6.6, (c) PA MoS_2, (d) unmodified PP, (e) PP impregnated with RME, (f) PP impregnated with engine oil, and (g) PP impregnated with gear oil.

and load, the curve demonstrating friction coefficient of this material shows the previously described folds that indicate very different values of friction torque in individual runs and thus different values of friction coefficient. At this linear velocity (0.26 $m \cdot s^{-1}$), there are also differences between the curves for two examined polyamides PA 6.6 and PA MoS_2. Friction coefficient of PA MoS_2 is two times smaller than that of PA 6.6. The PA MoS_2 value approaches ca. 0.1 (also at friction distance ca. 20 m and load ca. 300 N) and next stabilises. It should be also noted that the curve for PA MoS_2 at 0.26 $m \cdot s^{-1}$ is in the wide range similar to that obtained at 0.13 $m \cdot s^{-1}$.

In the case of PTFE, no significant differences were found in shapes of the friction coefficient curves. The only differences concerned the values themselves. At 0.26 $m \cdot s^{-1}$, friction coefficient was ca. 20% higher than that at 0.13 $m \cdot s^{-1}$. For this material, at both working velocities, a tendency to decreasing friction coefficient was found.

Impregnation with oils results in reduced friction coefficient of PP also at linear velocity of 0.26 $m \cdot s^{-1}$. Significantly lower values were found for the specimens made of impregnated PP. For lower sliding velocity, no significant changes in friction curves caused by impregnation with RME and engine and gear oils could be observed. Only a small reduction in friction coefficient was found in the case of PP impregnated with gear oil. The lowest value was found for PP impregnated with engine oil, but the value was not much higher than that for the specimens impregnated with RME and gear oil. Reduction of maximum friction coefficient in comparison to unmodified PP was 69% for impregnation with RME and 73% for impregnation with engine and gear oils.

Very different shapes of friction coefficient curves were found for sliding velocity of 0.44 $m \cdot s^{-1}$ in the case of unmodified PP, see Figure 3(c). Maximum value of 0.33 was found after sliding distance of 50 m at 200 N. After that, the coefficient value dropped rapidly, showing two minimum values after sliding distances of 120 m and 180 m, which corresponded to loads of 650 N and 900 N. A characteristic feature of PP is its low thermal conductivity. Due to frictional heat, temperatures can easily reach melting point of the polymer and cause its surface melting. When the polymer melts, its friction coefficient decreases, according to the mechanism of thermally controlled friction [16].

It is also interesting that shapes of friction coefficient curves for PA 6.6 and PA MoS_2 are similar. However, completely different shape of the curve is observed for PTFE. Modified PP still shows lower average maximum friction coefficient in comparison to unmodified PP (reduction after stabilisation is 73% for impregnation with RME and 82% for impregnation with gear and engine oils). In the case of impregnation with gear and engine oils, differences between friction coefficients are still very small. The values are similar to those for lower sliding velocity. Shapes of these curves for PP impregnated with engine and gear oils are nearly identical; small differences occur only for the specimens impregnated with RME, when friction coefficient after stabilisation is the highest. Decrease in friction coefficient can be explained by the Bekhet and Naga theory [17] according to the fact that friction force between surfaces of the thermoplastic and steel decreases during the sliding process due to increasing molecular orientation of the thermoplastic parallel to the sliding direction.

Friction curve for PP at linear velocity of 0.62 $m \cdot s^{-1}$ (Figure 3(d)) is similar to that found at 0.44 $m \cdot s^{-1}$, while maximum friction coefficient is lower at 0.62 $m \cdot s^{-1}$. This maximum value of 0.28 occurs after friction distance ca. 70 m at 250 N. After reaching its maximum, friction coefficient quickly decreases, supposedly due to plasticization (fluidization) of the material, occurring at this working velocity and load higher than 250 N. For PTFE as well, shape and values of friction curve are similar to those found at 0.44 $m \cdot s^{-1}$. In this case, maximum friction coefficient is ca. 0.15.

For two examined polyamide grades, shapes of friction curves are similar, while friction coefficient values for these materials are different (see Figure 3(d)). In the stabilisation range, friction coefficient values for PA 6.6 and PA MoS_2 are, respectively, ca. 0.22 and 0.15. Variability of friction coefficient, demonstrated in the diagrams, shows instable mating of PA 6.6 and PA MoS_2 in the tribological pair at its high load.

Impregnated specimens still show the lowest values of maximum friction coefficient in comparison to unmodified PP, PTFE, and both polyamides. The coefficient values are not much higher than those achieved in the test with the lowest sliding velocity. However, a slight change was recorded in the diagrams of actual friction coefficient in function of friction distance and load. Friction coefficient of PP impregnated with gear oil reached stabilisation latest of all, and its value started increasing after friction distance ca. 250 m at 850 N. The coefficient for the loads over 480 N was the lowest just for PP impregnated with gear oil. In the case of PP impregnated with engine oil, increase of the coefficient after stabilisation occurred earliest of all, as early as at 600 N, which corresponds to friction distance of 150 m. For PP impregnated with RME, stabilisation of friction coefficient lasted till 850 N and, for the whole test time, the value was the highest among the variants of modified PP.

Analysis of friction diagrams showed that impregnation of PP with oils improved friction conditions in contrast to dry friction that occurred in the case of unmodified PP, PTFE, and polyamides. This is clearly visible, especially for the specimens tested at low velocity, by reduced values of friction coefficient and its smaller variability.

4.2. Torque Moment. The quantity directly related to friction coefficient is friction torque. Figures 4(a)–4(g) show friction torque for the examined polymers in function of linear velocity and loading force.

For PTFE (Figure 4(a)) at the lowest linear velocity (0.13 $m \cdot s^{-1}$), it can be assumed that friction torque increases linearly in relation to load of the node. It is similar at the other velocities, although linearity occurs up to some load (at 0.26, 0.44, and 0.62 $m \cdot s^{-1}$ it is, respectively, 850, 600, and 900 N); after that friction torque value decreases.

In the case of PA MoS_2 (Figure 4(b)), friction torque initially increases linearly with increasing load, reaching its lowest value in the test carried out at 0.26 $m \cdot s^{-1}$ and the highest value in the test at 0.62 m·s^{-1}. However, at 0.62 m·s^{-1}, friction torque shows high fluctuations at the load of ca. 600 N and higher. In similar way, at 0.44 m·s^{-1}, high variability of friction torque starts from ca. 700 N (of course, similar fluctuations were recorded for actual values of friction coefficient, which was explained previously).

For PA 6.6 (Figure 4(b)), in each test carried out at any linear velocity, after initial linear increase with increasing load, friction torque shows large fluctuations. It is also interesting that no relation occurs between linear velocity and friction torque. Friction torque reaches its highest value at 0.62 m·s^{-1} and subsequently lower values at 0.26 m·s^{-1}, 0.13 m·s^{-1}, and 0.44 m·s^{-1}. For linear velocity of 0.62 m·s^{-1}, friction torque shows high fluctuations when loading force reaches ca. 550 N. Shape of friction torque curve is similar for linear velocity of 0.26 m·s^{-1}. At 0.13 m·s^{-1}, friction torque changes linearly with load force up to 800 N.

Comparison of the curves for friction coefficient in relation to load for unmodified and impregnated PP (Figures 4(d)–4(g)) shows that impregnation with oils significantly affects friction torque and behaviour of the materials during operation.

Friction torque of unmodified PP (Figure 4(d)) increases rapidly with load. The curves obtained for higher linear velocities show their maximum at ca. 800 N and then friction torque decreases, like it is for PTFE, although the values for PP are higher.

Behaviour of PP impregnated with oil is quite different (see Figures 4(e)–4(g)). Some regularity can be observed after impregnation with RME (Figure 4(e)). Namely, friction torque is the lowest for the lowest linear velocity and next it increases with increasing velocity. It is important to note that increased loading force results in almost directly proportional increase of friction torque and at linear velocity of 0.62 m·s^{-1} this proportionality disappears only from ca. 850 N up. Moreover, high similarity in shapes of torque curves for PP impregnated with engine and gear oils can be observed (Figures 4(f) and 4(g)). The differences appear only between the curves obtained for the highest linear velocity, that is, 0.62 m·s^{-1}. It is important that no significant influence of linear velocity on friction torque was observed. This effect is visible only at linear velocity of 0.62 m·s^{-1} and only above some load values. The most important is that, at 0.13, 0.26, and 0.44 m·s^{-1}, increase of friction torque is approximately directly proportional to loading force; therefore it is possible to foresee value of friction torque and thus behaviour of the so-modified PP on the grounds of load value.

4.3. Linear Wear. Figures 5(a)–5(d) show curves of linear wear of the examined polymers at various working velocities in relation to load force and friction distance that was different at individual velocities.

At 0.13 m·s^{-1}, wear of unmodified PP as well as of PP impregnated with RME and gear and engine oils is in practice directly proportional to friction distance and loading force. However, the other examined materials (PA 6.6, PA MoS_2, and PTFE) do not show such a regularity (Figure 5(a)). Wear is most intensive for PA MoS_2 and PP impregnated with RME but least intensive for unmodified PP. In the cases of PA 6.6 and PA MoS_2, occurrence of short moments of intensive wear was found. Wear of PTFE is proportional to friction distance and loading force up to the moment when load and friction distance reach responsibly ca. 480 N and 29 m, and its value is comparable to those of PA MoS_2 and RME-impregnated PP and thus the materials, which at that sliding velocity were most intensively worn. However, in the further range of load and friction distance, this material was no longer subjected to wear (wear remained at the level of 0.7 mm), which can be related to the phenomenon of strain hardening, described in [18].

Increase of velocity to 0.26 m·s^{-1} results in small changes in linear wear of the polymers (Figure 5(b)). It should be noted that wear curve of PA MoS_2 in Figure 5(b) coincides with that of PP impregnated with engine oil. It is interesting that wear of all the polymers except PTFE is nearly directly proportional to load force and friction distance. Like during the test at lower velocity, RME-impregnated PP and PTFE wear most intensively in the initial test range, while unmodified PP and PP impregnated with gear oil wear least intensively in that range. Initially, PTFE also wears proportionally to load and friction distance, and at ca. 380 N and ca. 45 m its wear is at the level of 0.5 mm and does not change in the further range. It is also interesting that wear values of the materials at 0.13 and 0.26 m/s are similar, for PTFE even higher at lower velocity, though friction distance is then two times shorter. This looks as if load of the node played the key role here.

Important changes in wear intensity appear at linear velocity of 0.44 m·s^{-1} (Figure 5(c)) when, like at 0.26 m·s^{-1}, wear curve of PA MoS_2 coincides with that of PP impregnated with engine oil. PP impregnated with RME and engine and gear oils still wears proportionally to load and friction distance, while PP impregnated with gear oil wears least intensively. However, the most intensively wearing material is PTFE whose wear is the largest from the moment when load force is ca. 100 N (which corresponds to friction distance of ca. 26 m) and lasts to ca. 750 N (distance of ca. 150 m). Linear wear of unmodified PP also proceeds in a different way; it can be assumed that initially PP wears almost proportionally to loading force and thus to friction distance, but the period of its very intensive wear begins after reaching ca. 880 N.

Further changes in wear intensity occur at linear velocity of 0.62 m·s^{-1} (Figure 5(d)). Initially, linear wear of PP impregnated with three oils is nearly directly proportional to loading force. However, at higher load, its wear becomes more intensive (ca. 820 N for RME, ca. 830 N for engine oil, and ca. 950 N for gear oil). It was observed that wear of PA 6.6 and PP impregnated with gear oil proceeds most slowly. Wear curve for PA MoS_2 is very similar to that obtained at 0.13 m·s^{-1}. Shapes of linear wear curves for PTFE and unmodified PP are similar to shapes of the curves obtained at lower velocity, that is, 0.44 m·s^{-1}. However, the value of load at that unmodified PP which wears intensively is lower and amounts to ca. 580 N (distance ca. 200 m). The most intensively wearing material is PTFE whose wear is the largest when loading force reaches ca. 700 N (which corresponds to friction distance of ca. 200 m).

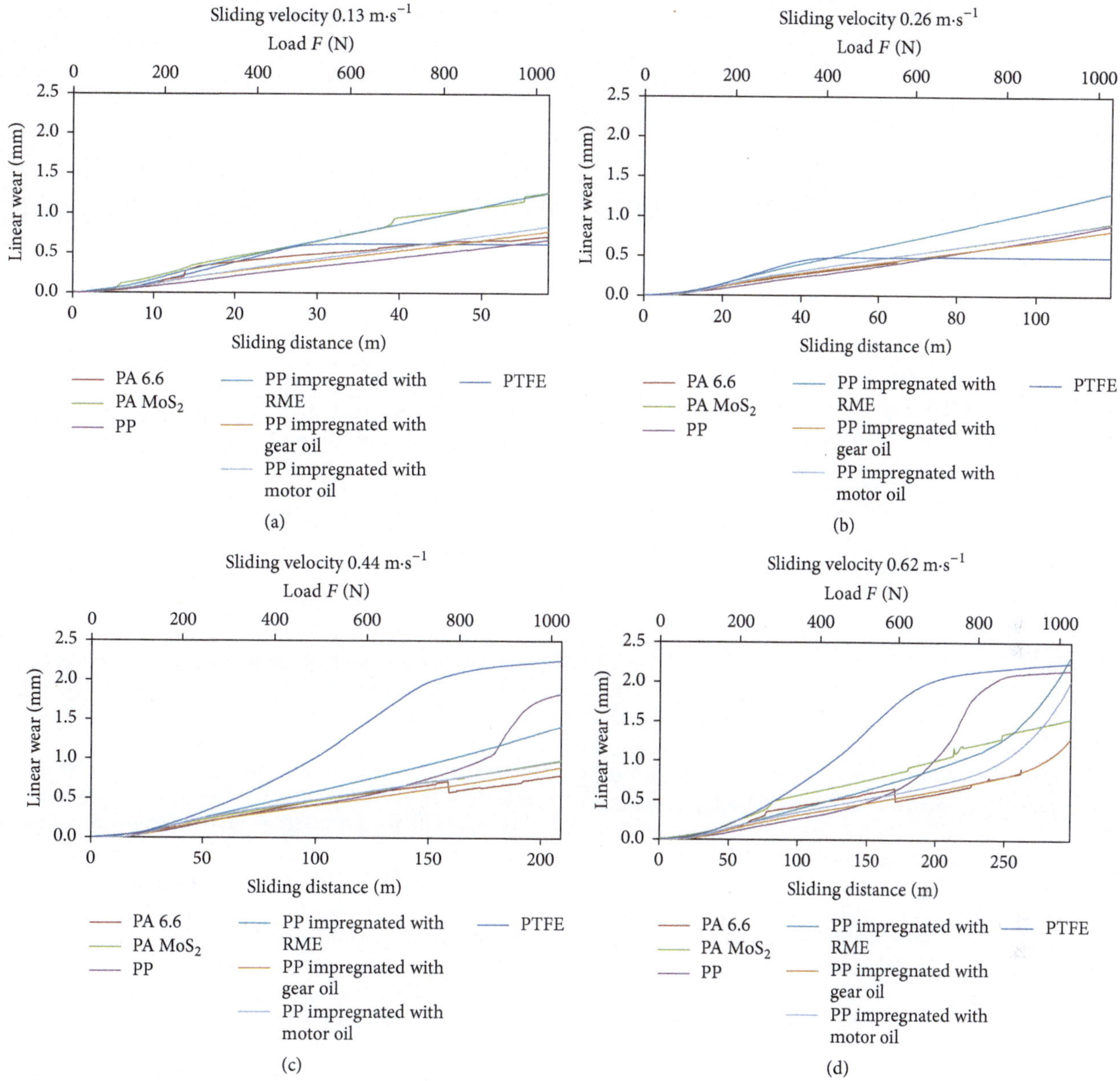

FIGURE 5: Relationship between sliding distance, load, and linear wear of the specimens under various sliding velocities: (a) 0.13, (b) 0.26, (c) 0.44, and (d) 0.62 $m \cdot s^{-1}$.

Therefore, comparison of the results shows that wear of polypropylene impregnated with oils proceeds in more stable way, so it can be foreseen (of course, to some load values) and controlled.

Values of absolute mass consumption of the tested components are shown in Figure 6. At sliding velocities of 0.13, 0.26, and 0.44 $m \cdot s^{-1}$, absolute mass consumption was very high for PTFE. These results confirm also the fact known in tribology that Teflon has low friction coefficient but very poor scratch resistance [19]. In the case of PTFE, intensive wear was observed even at sliding velocity of 0.13 $m \cdot s^{-1}$. It should be also noted that, at 0.13, 0.26, and 0.44 $m \cdot s^{-1}$, PA 6.6, PA MoS_2, and PP impregnated with gear and engine oils showed absolute mass consumption between 0.001 and 0.003 g. Similar values were also recorded for PA 6.6 at 0.62 $m \cdot s^{-1}$. At sliding velocities of 0.44 and 0.62 $m \cdot s^{-1}$, PTFE and unmodified PP reached high values of relative mass consumption, which was caused by significant linear wear of the materials. For unmodified PP, variation of wear is nonlinear and dramatically increases up to 0.44 $m \cdot s^{-1}$. At sliding velocity of 0.62 $m \cdot s^{-1}$, PP impregnated with engine oil and RME showed very high absolute mass consumption. In contrast to unmodified PP, PP impregnated with gear oil was characterized by lower values of linear wear and absolute mass consumption during the tests at 0.44 and 0.62 $m \cdot s^{-1}$. Reduction of wear provided by a film of lubricant is especially visible at linear velocity of 0.62 $m \cdot s^{-1}$.

4.4. Pressures. Comparison of the obtained pressure values occurring on contact surfaces of friction pairs is shown in

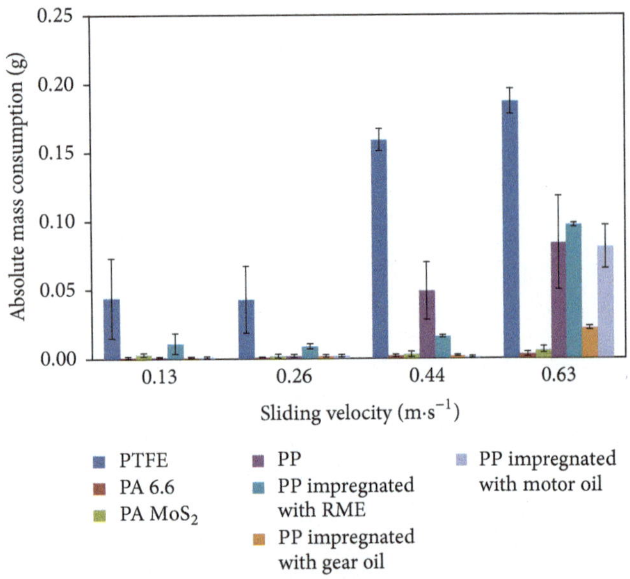

FIGURE 6: Average value of absolute mass consumption of the tested polymers for various sliding velocities: 0.13, 0.26, 0.44, and 0.62 m·s^{-1}.

Figure 7. For PP impregnated with engine and gear oils, at each velocity, the highest pressures were always observed in initial periods of tests. For these two materials, the highest pressure reaching almost 140 MPa occurred during the test carried out at 0.26 m·s^{-1}. Increased pressures for these two modified PPs result from the fact that no linear wear occurs at initial sliding phase and the contact area obtained during grinding-in is kept unchanged. The no-wear period is observed until the load force reaches 50 N to 66 N, depending on linear velocity. This effect could be explained by creation, between the specimen and the counterspecimen, of a boundary layer composed of adsorbed molecules of oil used for impregnation. Such a layer is probably created during grinding-in. The engine and gear oils contain additives improving their lubrication performance and RME contains no such additives. This could explain the fact that no pressure increase was observed for the specimens impregnated with RME. The boundary layer becomes destroyed only after the above-mentioned load is exceeded and then the wear process begins, accompanied by decreasing pressure value.

It is interesting that when loading force exceeds ca. 200 N at each working velocity, pressure value stabilises and remains roughly unchanged in spite of increasing loading force. An exception is the friction pair PTFE-steel during work at the velocity of 0.13 and 0.26 m·s^{-1}, for which, with the increase of load, the pressures slightly increased. In general, stabilisation of pressure during the tests indicates that increase of load is accompanied by adequate increase of contact area of mating materials, resulting in maintaining pressure values on an even level.

For all the examined materials and working velocities, at loading forces higher than ca. 200 N, pressure values did not exceed 30 MPa. During the tests, the lowest pressures during their stabilisation period were recorded at 0.13 and 0.26 m·s^{-1} for PTFE, RME-impregnated PP, and PA MoS$_2$, at 0.44 m·s^{-1} for PTFE, unmodified PP, and RME-impregnated PP, and at 0.62 m·s^{-1} for PTFE and PA MoS$_2$.

4.5. Temperature. During the tests, temperature of each specimen was measured with a thermocouple. For technical reasons, temperature during the tests was measured at some distance from the contact area, which was mentioned in the section concerning methodology. However, it can be assumed that temperature changes at the contact point between a specimen and a counterspecimen are similar to those measured. Of course, temperature values themselves at the contact point are higher. Figures 8(a)–8(d) show temperature changes in relation to friction distance, load, and linear velocity. As can be seen, temperature increased along with elapsed time, which corresponds with increasing load and friction distance. Similarly, at higher working velocities, larger temperature increments were observed. At the lowest linear velocity, temperature increments were roughly comparable, but for the velocity of 0.26 m·s^{-1} up, the largest increments were recorded for unmodified PP and PTFE (Figure 8(b)). The largest temperature increments for unmodified PP and PTFE are maintained also at 0.44 m·s^{-1} (Figure 8(c)). In turn, for the highest linear velocity of 0.62 m·s^{-1}, the largest temperature increments were observed also for PA 6.6 (Figure 8(d)). Moreover, it can be seen that these rapid temperature growths correspond to the above-described changes of friction coefficient and friction torque. Therefore, it can be said that the phenomena occurring in a friction pair are to a large degree controlled by the processes related to temperature increase.

4.6. Wear Surface. Figure 9 shows general condition of surfaces of the tested polymers, and Figures 10–13 show images of worn surfaces, obtained with use of a scanning electron microscope. According to Brostow et al. [4] "sometimes it is mistakenly assumed that interfaces with high friction show a high rate of wear. This is not generally true; both friction and wear have to be determined. There are cases when solid interfaces of polymers show relatively low friction but fairly high wear. Wear of a material can be mechanical and/or chemical; it is generally accelerated by heating during friction. Wear mechanism can be adhesion, abrasion, fatigue and the effect of erosion, chemical reactions including corrosion, or else an induced electric arc. Fairly often there is no single mechanism of wear but a combination of several mechanisms."

4.7. Wear Surface of PTFE. After testing at 0.13 m·s^{-1}, surface of the PTFE specimen shows small marks of wear in form of parallel scratches (see Figure 10(a)). The surface is porous. It was observed at the boundary line between worn and unworn surfaces that the polymer material is plastically deformed by the counterspecimen and next is displaced towards the edge (Figure 10(b)). Twofold increase of linear velocity results in qualitative changes of the wear surface (Figure 10(c)). First of all, increased number of pores was found. Therefore, porosity is not a specific feature of the examined material but results from the existing mechanism of wear. Localised wear areas are present on the surface, indicating that the material is

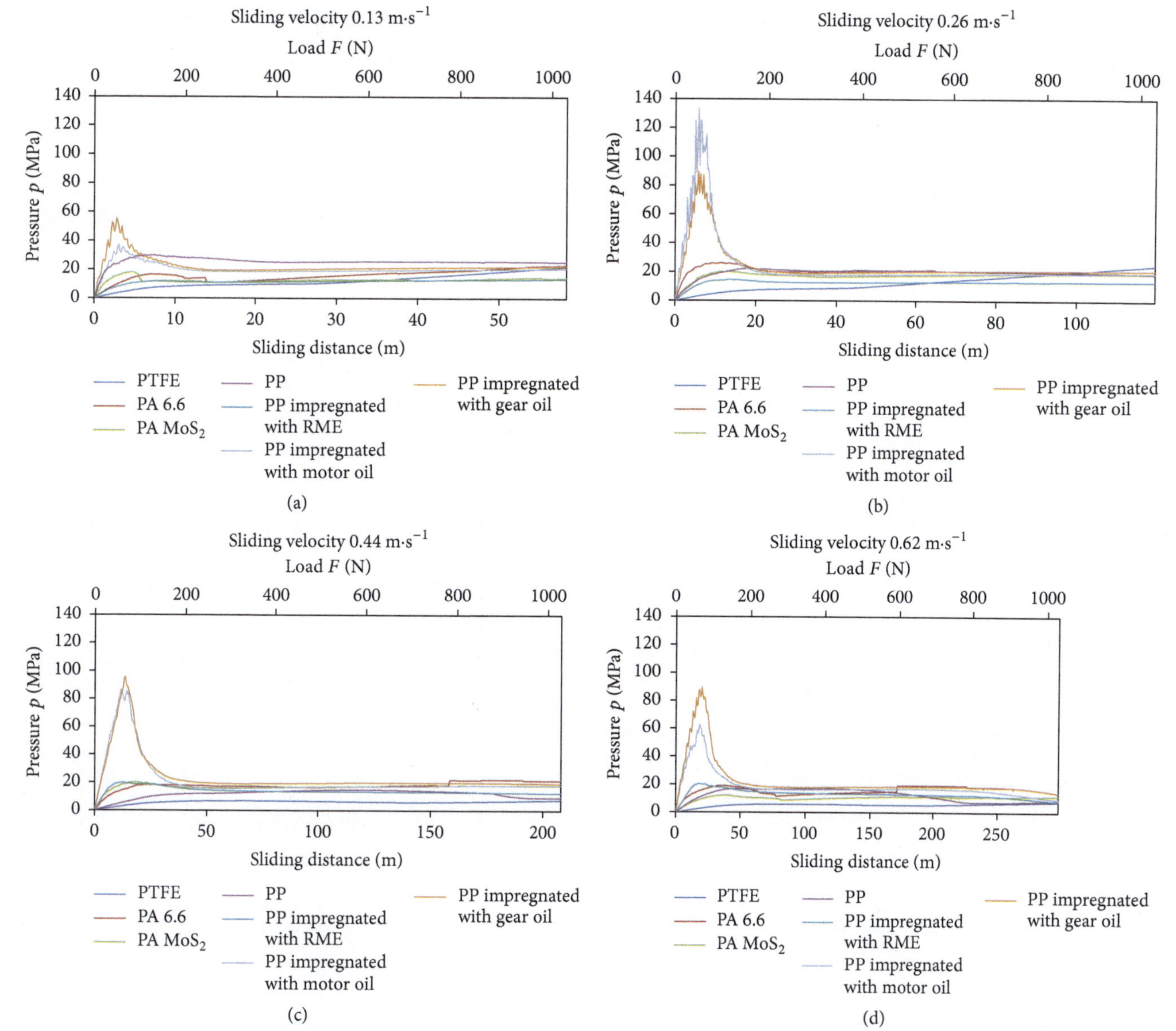

FIGURE 7: Relationships between sliding distance, load, and pressure of the samples under various sliding velocities: (a) 0.13, (b) 0.26, (c) 0.44, and (d) 0.62 $m \cdot s^{-1}$.

plastically deformed, which can result in enlargement of the existing pores. This is especially visible on the exemplary surface after the test at 0.44 $m \cdot s^{-1}$ (Figures 10(d) and 10(e)). It can be seen that the material is plastically deformed and next removed, which results in increased size of discontinuities (Figure 10(e)). Probably, in a further phase, the so-displaced material is torn off as a result of its fatigue. This mechanism of wear results in significant material losses and, consequently, in the largest mass wear among all the tested specimens. Increase of linear velocity to 0.62 $m \cdot s^{-1}$ (Figure 10(f)) did not result in any qualitative change of wear surface in comparison to the surface obtained at 0.44 $m \cdot s^{-1}$.

4.8. Wear Surface of PA 6.6. Cracks, tiny parallel scratches, and wear products pressed in the surface were observed on surfaces of PA 6.6 specimens tested at 0.13 and 0.26 $m \cdot s^{-1}$ (see Figures 11(a) and 11(b)). The cracks could be caused by tearing off the material adhesively joined with the counterspecimen.

Increase of linear velocity to 0.44 $m \cdot s^{-1}$ results in qualitative changes on surface of the examined material (Figure 11(c)). The surface becomes smooth and overlaps of the material appear here and there, named in literature as "roll-shaped particles" [6]. However, their accumulation is small.

On the surface of the specimen tested at the highest velocity (0.62 $m \cdot s^{-1}$), occurrence of a zone with "scaly" structure was found, formed of "roll-shaped particles" that are much more numerous and finer in comparison to those present on the specimen tested at a lower velocity (Figures 11(d) and 11(e)). The authors of [6] state the following: "Frictional work presumably induces the formation of wear particles. Plastic

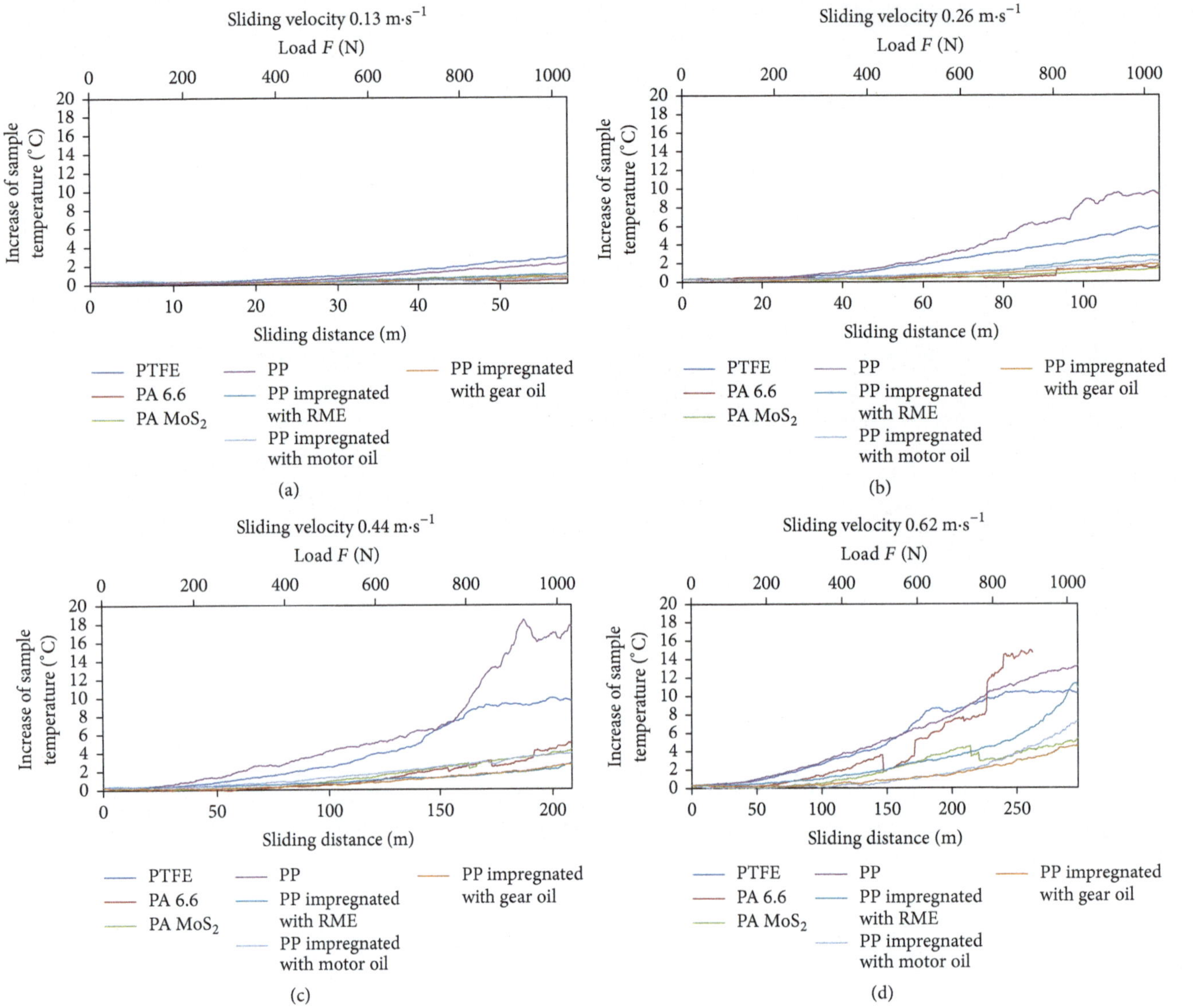

FIGURE 8: Relationships between sliding distance, load, and temperature of the specimens under various sliding velocities: (a) 0.13, (b) 0.26, (c) 0.44, and (d) 0.62 m·s^{-1}.

flow was considered to affect the formation of the 'roll-shaped particles.' However, the main cause of generation of roll-shaped particles was considered to be a local surface breaking because of high friction. These wear particles agglomerated as roll-shaped particles and were removed to the outside of the wear track." This means that, in this case, heat generated during a test at higher linear velocity caused local partial melting and plastic flow of the examined material. So, at 0.62 m·s^{-1}, the material was plastically deformed and then displaced according to the friction direction. In consequence, "roll-shaped particles" appeared on the specimen surface. As a result of fatigue, these particles were next separated from the surface and removed as wear products.

4.9. Wear Surface of PA MoS_2. Analysis of worn surfaces of the tested PA MoS_2 specimens indicates that they are similar to those of the PA 6.6 specimens. So, a similar mechanism of wear can be considered here. A zone with scaly structure is dominating, formed of "roll-shaped particles" much more numerous and larger in comparison to those occurring on surfaces of PA 6.6 specimens (Figures 12(a)–12(d)). It is interesting that the lower the linear velocity is, the larger surface is occupied by the particles but, at the same time, the particles are more flat and adhere to the wear surface (Figures 12(a) and 12(b)). As linear velocity increases, more material is removed from the wear surface, plastically deformed and drawn according to rotation direction of the counterspecimen (Figure 12(d)). As a result, "roll-shaped particles" are formed, as was mentioned before, similar to those created on surfaces of PA 6.6 specimens. It can be supposed that formation of "roll-shaped particles" could contribute to large fluctuations of the previously described friction torque and friction coefficient and thus to unstable behaviour of the examined polyamides.

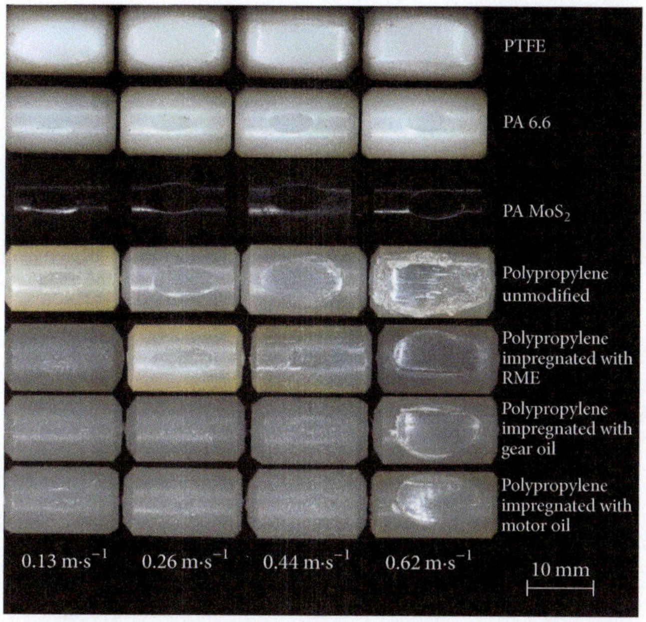

FIGURE 9: Effect of sliding velocity on worn surfaces of the tested specimens.

4.10. Wear Surface of PP. It was found for unmodified PP that, depending on linear velocity, various mechanisms are active, resulting in various intensities of wear. This is especially visible at $0.62\,m{\cdot}s^{-1}$, when the specimen surface showed significant traces of wear.

At $0.13\,m{\cdot}s^{-1}$, wear surface of unmodified PP is smooth and shallow scratches are arranged in parallel (Figure 13(a)). Traces of wear of adhesive nature were also seen, but particles of worn material joined to surface could be only observed, creating characteristic islands protruding over the smooth surface. A similar adhesive mechanism of wear was noted for the specimens tested at higher linear velocities, that is, 0.26 and $0.44\,m{\cdot}s^{-1}$ (Figures 13(b) and 13(c)). At higher velocity of $0.44\,m{\cdot}s^{-1}$, scratches became deeper and local cracks occurred on the wear surface (Figure 13(c)).

Mechanism of wear of unmodified PP changes, when linear velocity increases to $0.62\,m{\cdot}s^{-1}$ (Figures 13(d)–13(f)). At 0.13, 0.26, and $0.44\,m{\cdot}s^{-1}$, it could be seen that the main mechanism of wear was adhesion and microcutting. However, at $0.62\,m{\cdot}s^{-1}$, wear surface of the specimens can be divided into two zones: smooth zone dominated (like before) by adhesion and microcutting (Figure 13(d)) and the zone being a clear fracture with plastic deformation (Figures 13(e) and 13(f)). The zone with a fracture is characterised by highly developed topography with fibrous structure. The fibres create bridges with plastically deformed pores occurring between them. Moreover, it could be seen that a part of the material was cut off from the specimen surface and displaced to the edge of the wear surface.

At low linear velocity, small amount of work is converted to heat. This changes, when sliding velocity increases. Heat generated in this case results in local melting of the material and its plastic flow according to the forcing factors occurring in the friction node. For unmodified PP, thermal wear occurs when linear velocity increases over $0.44\,m{\cdot}s^{-1}$.

Analysis of changes of friction coefficient in function of friction distance at various sliding velocities indicates that thermal wear of PP impregnated with oils did not occur even at $0.62\,m{\cdot}s^{-1}$. So, the conclusion can be drawn that impregnation with oils protects propylene against catastrophic wear that takes place during thermal wear. This phenomenon can be explained by decrease of oil viscosity with increasing operating temperature [20]. For example, Couronné et al. [21] and Cousseau et al. [22] reported significant lubricant starvation of ester-based greases thickened with lithium, since this type of grease has a high flow index.

5. Discussion

Within the presented research, influence of modification of polypropylene by impregnation with oils on friction coefficient and wear was determined during their operation at sliding friction without lubrication. These parameters were compared with those of unmodified polypropylene, Teflon, and polyamides. Impregnation with oils significantly affected friction coefficient and friction torque of the examined polypropylene, decreasing values of these parameters. Figure 14 shows ratios of friction coefficient values found for the examined polymers. The presented values are average values from ratios of the coefficient determined at individual measuring points, that is, occurring at individual loads of the friction pairs. The authors realise the simplification applied at such calculations, but it seems that their results can serve for demonstrative evaluation of influence of impregnating PP with oils on friction coefficient. Clearly visible is reduction of friction coefficient value as a result of such modification of PP: the found general friction coefficient of oil-impregnated polypropylene is 2.2 to 5.3 times lower. At the same time, general friction coefficient of oil-impregnated PP was found to be 1.3 to 3.2 times lower in comparison to PTFE, PA 6.6, and PA MoS_2. It should be added that friction coefficient of unmodified PP is 1.0 to 2.0 times higher in comparison to PTFE, PA 6.6, and PA MoS_2.

However, it was found on the grounds of the recorded linear wear of the examined polymers that, in the range of low and medium loads of the friction pair, unmodified PP was characterised by the highest wear resistance. Wear of oil-impregnated PP was higher than that of unmodified polymer. The smallest differences occurred in the case of impregnation with gear oil and the largest for impregnation with RME, which can be related to absence of additives improving lubricity and to possible influence of RME on properties of the polymer [12]. In general, such regularity occurred at lower sliding velocities (0.13 and $0.26\,m{\cdot}s^{-1}$) and, at low or medium range of loads, at higher sliding velocities (0.44 and $0.62\,m{\cdot}s^{-1}$). The situation was reversed in the range of high loads and higher velocities. In that range, wear of unmodified PP increased drastically, definitely exceeding wear value of impregnated material. It should be added that wear of PP impregnated with engine oil and gear oil was slightly lower or comparable to that of other tested polymers (PTFE, PA MoS_2, and PA 6.6), while PTFE at initial load range was subjected to

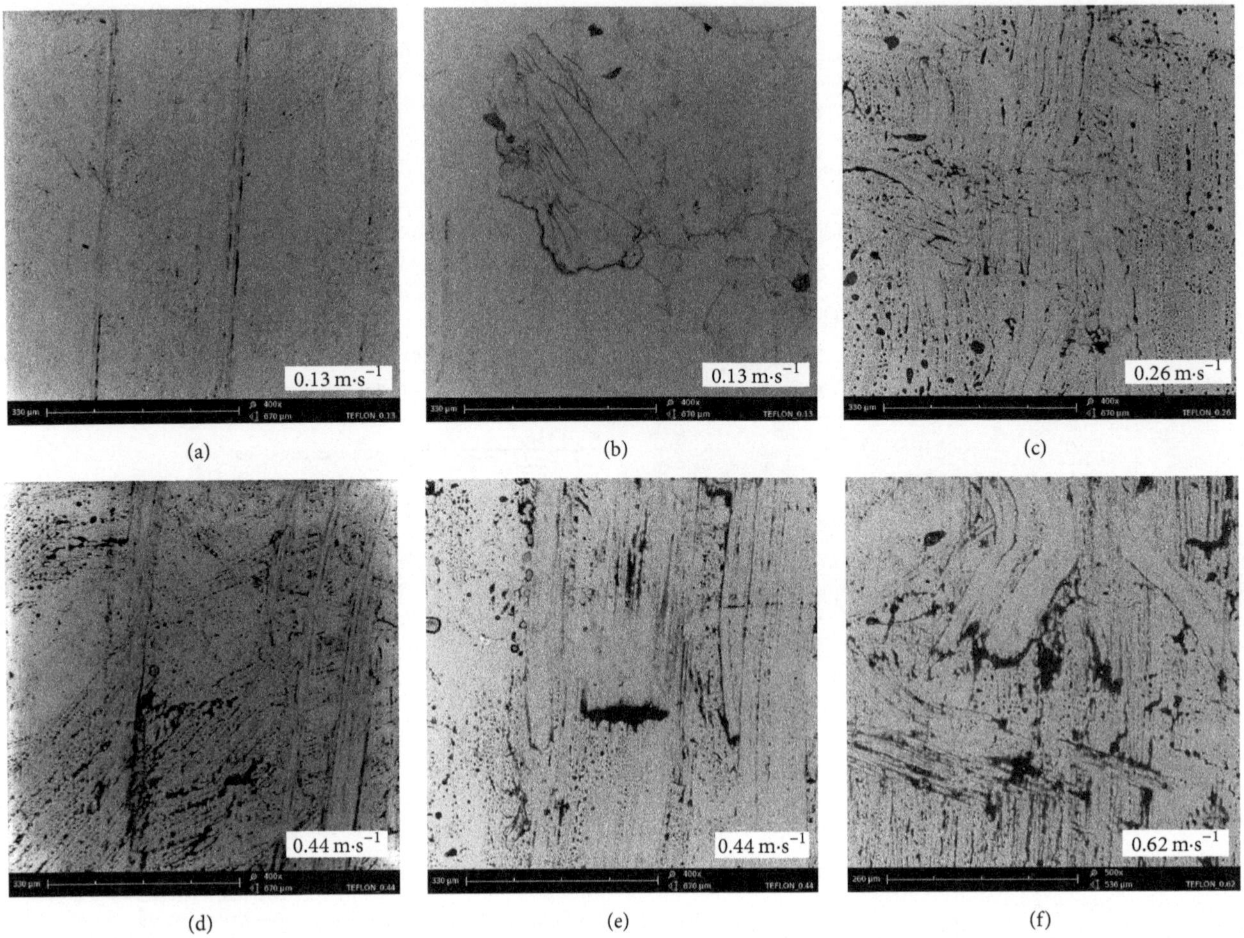

FIGURE 10: Effect of sliding velocity on worn surfaces of the investigated PTFE: (a) scratches and small discontinuities, (b) plastic deformation, (c) localised worn surface areas, (d) localised worn surface areas and numerous pores, (e) marks of fatigue wear, and (f) localised worn surface areas and numerous pores, SEM.

intensive wear (the largest at 0.26, 0.44, and 0.62 m·s^{-1}), but the wear intensity clearly decreased after exceeding a certain load. A similar regularity was observed for unmodified PP at 0.44 and 0.62 m·s^{-1}. In general, measurements of linear wear of the tested polymers indicate complexity of their wear process, multitude of factors playing roles in this process, and their variability depending on working conditions, which constitutes a so-far unrecognised research area.

On the grounds of the obtained results, it is possible to verify the accepted and developed models determining changes of friction coefficient in relation to loading force and linear velocity occurring in the friction pair. Till now, several developed models make it possible to foresee behaviour of polymers depending on their grade and load (Shooter and Tabor [23], Bowers et al. [24], Shooter and Thomas [25], Rees [26], and Myshkin et al. [27]). However, on the grounds of analysis of the above-mentioned ascertainments, it is difficult to state clearly which of these models describes behaviours of the polymers used in the own research. Of course, temperature also rises with increasing linear velocity and load and its effect on behaviour of the polymer and change of friction coefficient cannot be omitted [4, 19, 28, 29], so the interactions occurring in the friction pair are quite complex. It was shown in [27] that friction coefficient value is proportional to the loading force. However, the results obtained in own examinations do not confirm this regularity in full: in most variants applied in own research, after a period of increasing with increasing load, stabilisation of friction coefficient was observed. It is only in final range of the test, at significant loads (different for various polymers and working conditions), that destabilisation of the polymer behaviour was found, resulting probably from additional factors occurring in the friction pair, for example, excessive temperature rise in the contact area of mating materials. It should be emphasised that the examinations were carried out till relatively high forces loading the friction pairs, as appeared, in several cases exceeding practical application. So, comprehensive spectrum of evaluation of the examined polymers was obtained, which was an intention of the authors. In should be also mentioned that load can influence temperatures of viscoelastic transitions in polymers and thus their wear mechanism. Viscoelastic behaviour of polymers is also connected with the relation between linear velocity and friction coefficient and, as shown by Brostow et al., [4, 30, 31]

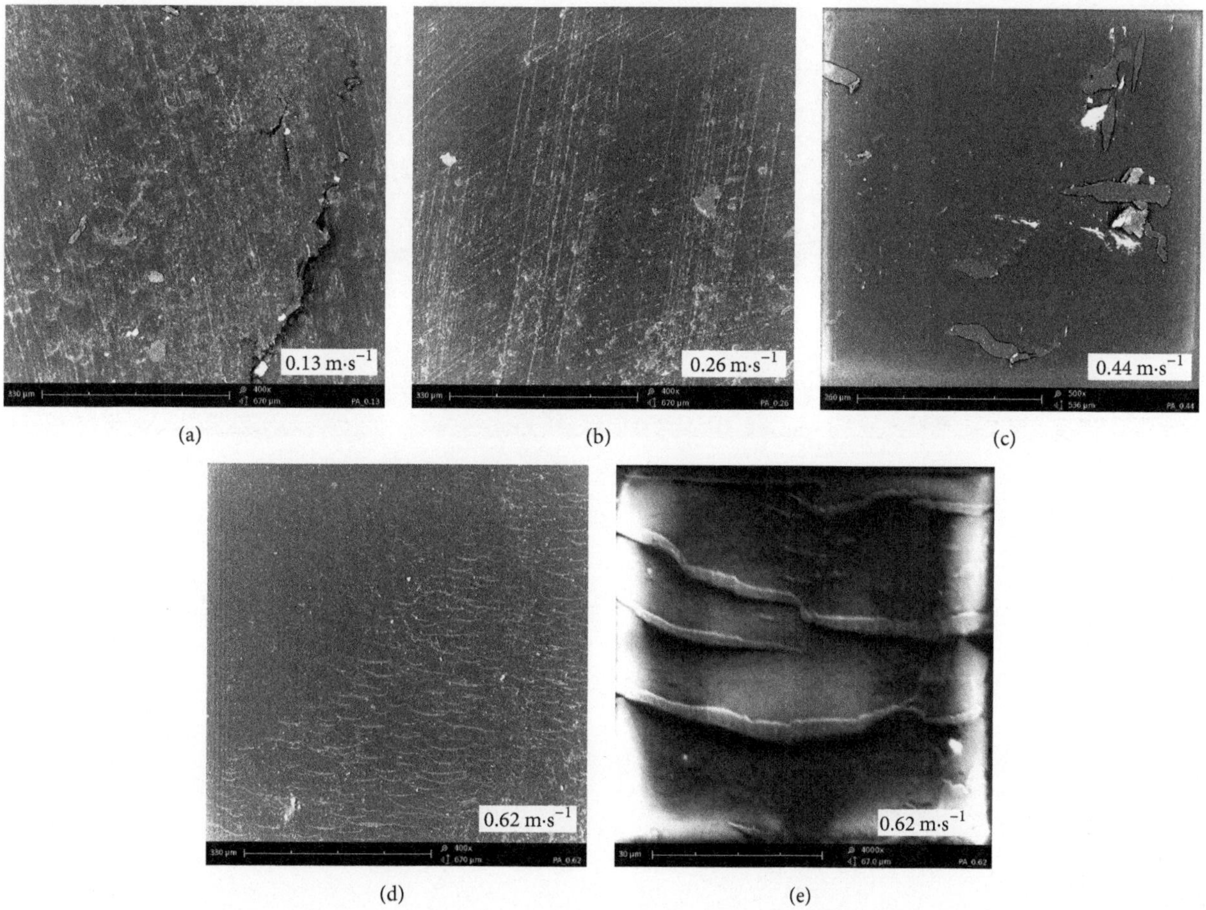

FIGURE 11: Effect of sliding velocity on worn surfaces of PA 6.6 tested at various sliding velocities: ((a) and (b)) cracks, small parallel scratches, and wear products pressed in the specimen surface, (c) sparse overlaps of the material, (d) zone with scaly structure, formed of "roll-shaped particles," and (e) magnified "roll-shaped particles," SEM.

with brittleness of polymers which is one of the most important criteria determining usable properties of polymers. Such models have been developed by Milz and Sargent [32], Tanaka [33], Flom and Porile [34, 35], and White [36]. When tests are carried out at a temperature close to the glass transition temperature, linear velocity clearly affects friction coefficient but, at lower temperatures, this coefficient depends on linear velocity to a small degree only. Of course, increase of linear velocity itself results in increased temperature of the materials composing the friction pair and such a relationship was noticed in the presented own research.

During the examinations, mechanisms of wear of the examined polymers were determined. Surface analysis of unmodified polypropylene confirmed previous observations described in literature, stating that the main mechanisms of wear in a friction pair polymer-metal are adhesion and surface deformation of soft polymer [37, 38]. However, when more heat is generated at higher linear velocity and thus temperature on the contact surface rises, the main wear mechanism becomes thermal wear. This type of wear is connected with deformation and next plastic flow of the material, which leads to its significant loss and, in consequence, to catastrophic wear [39]. As a result of thermal wear, value of friction coefficient decreases, so a change of shape of the friction torque curve can be confirmed by observation. Namely, when the polymer melts, its friction coefficient tends to decrease according to the mechanism of "thermal control of friction" [16]. This phenomenon was additionally supported by Quaglini et al. [5]. These authors noticed that when melting temperature of the polymer is reached, generated heat melts additional volume of polymer rather than causing a temperature rise of the already molten material. In accordance with the model of limiting frictional heat which was shown in [16], friction coefficient rises until a maximum value is achieved. At this point, friction coefficient determined by the "thermal control model" equals that dictated by "solid state friction." "Thermal wear" causes the surfaces to be often spreading or even gluing, which leads to catastrophic wear [5]. In the case of unmodified PP, thermal wear takes place at sliding velocity between 0.44 and 0.62 $m \cdot s^{-1}$, while in the case of impregnated PP it takes place at sliding velocity higher than 0.62 $m \cdot s^{-1}$. Thus, one of the advantages of impregnation of PP is the possibility of using such modified material under higher loads.

Despite some uncertainties in assessment of wear resistance of the tested PP, it can be considered that the most important finding of this research is that impregnation

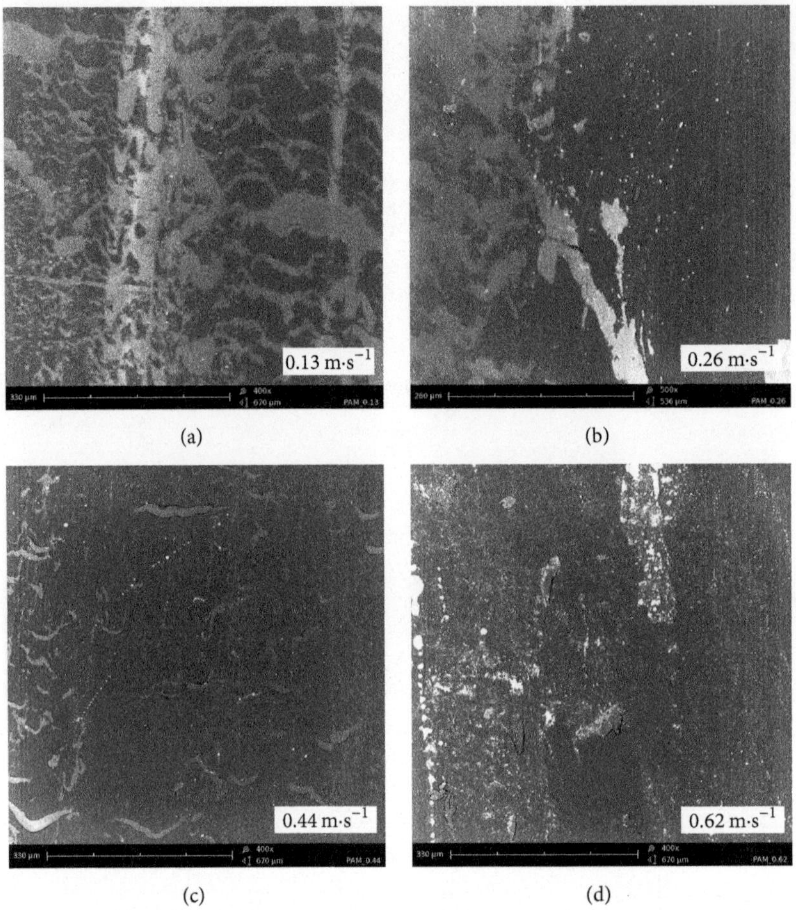

FIGURE 12: Effect of sliding velocity on worn surfaces of the investigated PA MoS_2: ((a), (b), and (c)) surface with scale structure, composed of "roll-shaped particles" at various linear velocities, (d) plastic deformations in scaly zone, SEM.

of PP with oil improves its tribological properties at low sliding velocities. Impregnation also moves thermal wear into higher sliding velocities and causes better resistance of PP to melting and plastic flow. According to these results, it can be concluded that impregnation of polypropylene even with inexpensive RME can give some benefits during its use in low-load frictional pairs under dry-friction conditions.

6. Conclusions

We conclude the following:

(1) Impregnation of polypropylene with engine oil, gear oil, or RME results in significant reduction of friction coefficient and thus of friction torque in relation to not only unmodified polypropylene but also the examined polyamides and Teflon. It is difficult to indicate the most effective impregnating agent, since different results were found for various working velocities of a friction pair (polymeric specimen and steel counterspecimen), while differences in friction coefficient values for individual impregnating oils were small (till ca. 0.07).

(2) Operation of impregnated polypropylene becomes more stable; that is, polypropylene impregnated with oil is able to work correctly under higher loads and, in particular, at increased sliding velocities.

(3) Within low and medium loads of the friction pair, unmodified polypropylene was characterised by the highest resistance to wear from among the examined polymers. Impregnation with oils resulted in its more intensive wear in comparison to unmodified polypropylene. The smallest differences in intensity of wear occurred in the case of polypropylene impregnated with gear oil and the largest differences—in the case of polypropylene impregnated with RME. Such a regularity took place at lower sliding velocities of the friction pair (0.13 and 0.26 $m{\cdot}s^{-1}$) and, at low and medium load range, at higher sliding velocities (0.44 and 0.62 $m{\cdot}s^{-1}$). However, at higher working velocities under high loads, wear of unmodified polypropylene drastically increased, exceeding intensity of wear of impregnated polypropylene.

(4) Intensity of wear of polypropylene impregnated with engine oil and gear oil was slightly lower or comparable to that of other tested polymers (PTFE, PA MoS_2, and PA6.6), while Teflon in initial range of load was subjected to intensive wear (the most intensive at

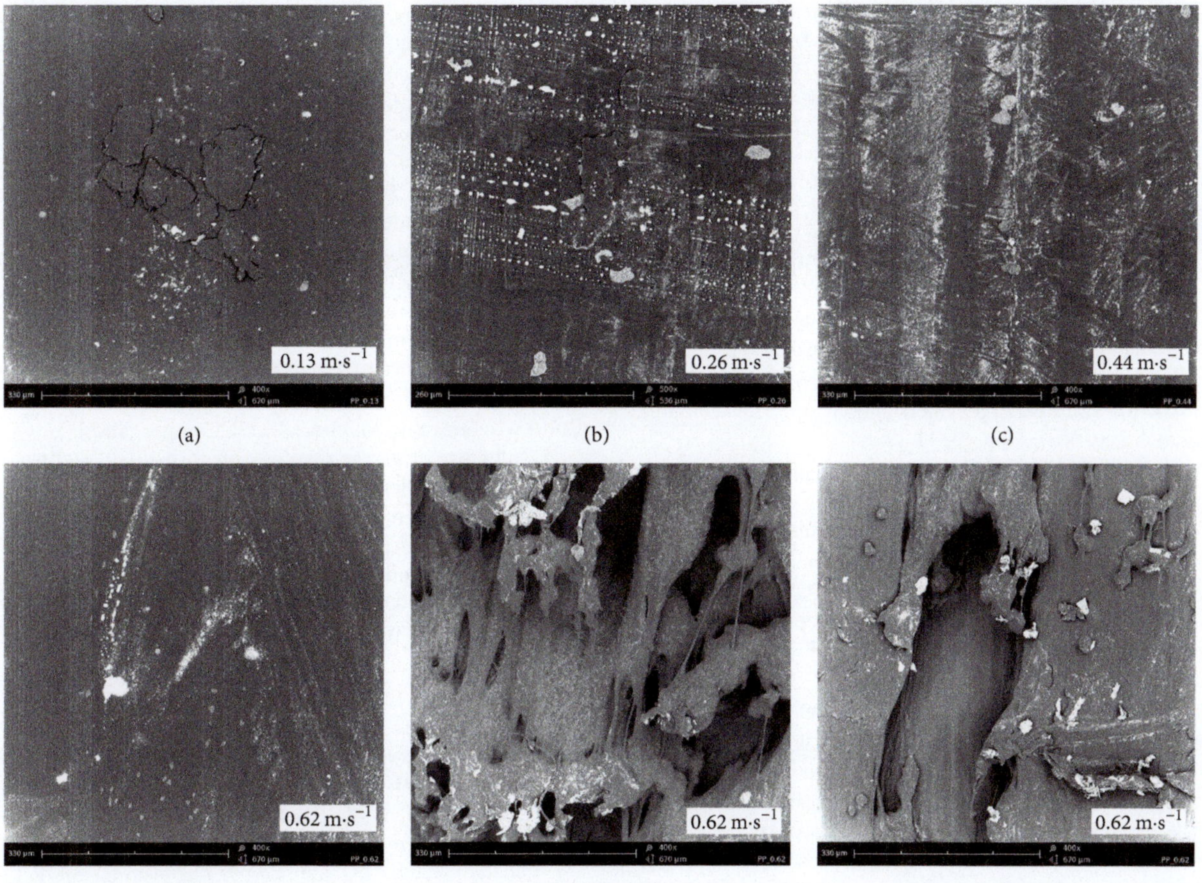

FIGURE 13: Effect of sliding velocity on worn surfaces of the investigated PA 6.6: ((a) and (b)) shallow scratches and wear traces of adhesive nature; (c) scratches and cracks on the surface; (d) smooth zone with dominating adhesion and microcutting; (e) zone with fracture; bridges with plastically deformed pores between them; (f) zone with fracture, SEM.

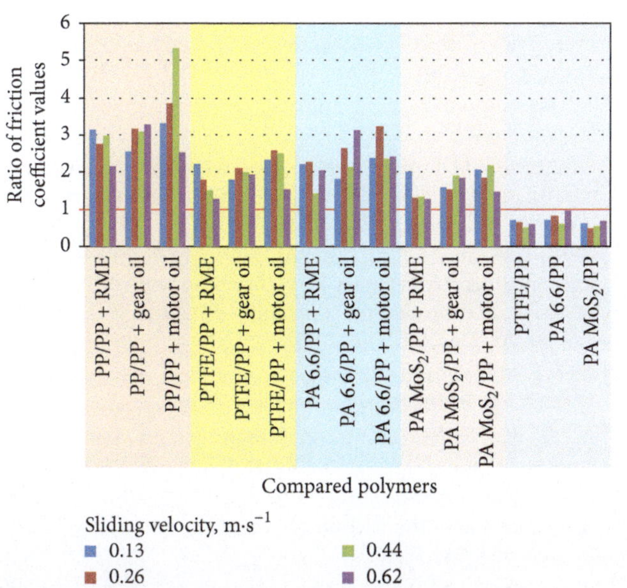

FIGURE 14: Relations between friction coefficient values for the examined polymers depending on sliding velocity.

0.26, 0.44, and 0.62 $m \cdot s^{-1}$), and after exceeding some load value its intensity of wear clearly decreased. A similar course of wear curve was found for unmodified polypropylene at sliding velocities of 0.44 and 0.62 $m \cdot s^{-1}$.

(5) During the research, mechanisms of wear of unmodified polypropylene, Teflon, and polyamide were identified. The mechanism for polypropylene was complex. At low linear velocity, the main mechanism was adhesive wear. Along with increasing velocity, accompanied by generation of larger amounts of heat, thermal wear started playing a role as the main mechanism of wear. Surface observations made it possible to identify two zones: smooth zone with dominating adhesion and the zone with developed topography including a fracture and plastic deformation.

(6) In the case of the remaining tested polymers, the most important mechanism of wear was adhesion (PTFE, PA 6.6, and PA MoS_2), microcutting (PTFE, PA 6.6, and PA MoS_2), fatigue wear (PTFE), and

formation of "roll-shaped particles" combined with plastic deformation (PA 6.6 and PA MoS_2).

References

[1] M. D. Bermudez, W. Brostow, F. J. Carrion-Vilches, and J. Sanes, "Scratch resistance of polycarbonate containing ZnO nanoparticles: effects of sliding direction," *Journal of Nanoscience and Nanotechnology*, vol. 10, no. 10, pp. 6683–6690, 2010.

[2] C. J. Hooke, S. N. Kukureka, P. Liao, M. Rao, and Y. K. Chen, "The friction and wear of polymers in non-conformal contacts," *Wear*, vol. 200, no. 1-2, pp. 83–94, 1996.

[3] D. Rus, L. Capitanu, and L.-L. Badita, "A qualitative correlation between friction coefficient and steel surface wear in linear dry sliding contact to polymers with SGF," *Friction*, vol. 2, no. 1, pp. 47–57, 2014.

[4] W. Brostow, V. Kovačević, D. Vrsaljko, and J. Whitworth, "Tribology of polymers and polymer-based composites," *Journal of Materials Education*, vol. 32, no. 5-6, pp. 273–290, 2010.

[5] V. Quaglini, P. Dubini, D. Ferroni, and C. Poggi, "Influence of counterface roughness on friction properties of engineering plastics for bearing applications," *Materials and Corrosion*, vol. 30, no. 5, pp. 1650–1658, 2009.

[6] K. Shibata, T. Yamaguchi, and K. Hokkirigawa, "Tribological behavior of polyamide 66/rice bran ceramics and polyamide 66/ glass bead composites," *Wear*, vol. 317, no. 1-2, pp. 1–7, 2014.

[7] H. Unal, A. Mimaroglu, U. Kadioglu, and H. Ekiz, "Sliding friction and wear behaviour of polytetrafluoroethylene and its composites under dry conditions," *Materials and Corrosion*, vol. 25, no. 3, pp. 239–245, 2004.

[8] H. Unal and F. Findik, "Friction and wear behaviours of some industrial polyamides against different polymer counterparts under dry conditions," *Industrial Lubrication and Tribology*, vol. 60, no. 4, pp. 195–200, 2008.

[9] B. S. Ünlü, E. Atik, and S. S. Yilmaz, "Tribological behaviors of polymer based journal bearings manufactured from particle reinforced bakelite composites," *Materials and Corrosion*, vol. 30, no. 9, pp. 3896–3899, 2009.

[10] V. K. Thakur and A. S. Singha, "Physico-chemical and mechanical characterization of natural fibre reinforced polymer composites," *Iranian Polymer Journal*, vol. 19, no. 1, pp. 3–16, 2010.

[11] B.-B. Jia, T.-S. Li, X.-J. Liu, and P.-H. Cong, "Tribological behaviors of several polymer-polymer sliding combinations under dry friction and oil-lubricated conditions," *Wear*, vol. 262, no. 11-12, pp. 1353–1359, 2007.

[12] P. Sędłak, B. Białobrzeska, and T. Stawicki, "Friction coefficient and wear resistance of a modified polypropylene impregnated with different oils," *Iranian Polymer Journal*, vol. 25, no. 3, pp. 263–275, 2016.

[13] Z. Lawrowski, "Polymers in the construction of serviceless sliding bearings," *Archives of Civil and Mechanical Engineering*, vol. 7, no. 4, pp. 139–150, 2007.

[14] Z. Rymuza, "Tribology of polymers," *Archives of Civil and Mechanical Engineering*, vol. 7, no. 4, pp. 177–184, 2007.

[15] K. R. Makinson and D. Tabor, "Friction and transfer of polytetrafluoroethylene," *Nature*, vol. 201, no. 4918, pp. 464–466, 1964.

[16] C. M. M. Ettles, "Polymer and elastomer friction in the thermal control regime," *Tribology Transactions*, vol. 30, no. 2, pp. 149–159, 1987.

[17] N. E. Bekhet and S. A. Naga, "An assessment of the frictional and wear behaviour of polyethylener steel couple," in *Proceedings of the EGTRIB-92, the 3rd Conference of the Egyptian Society of Tribology*, pp. 41–53, 1992.

[18] M. D. Bermudez, W. Brostow, F. J. Carrion-Vilches, J. J. Cervantes, G. Damarla, and J. M. Perez, "Scratch velocity and wear resistance," *e-Polymers*, vol. 3, pp. 1–10, 2005.

[19] W. Brostow, J.-L. Deborde, M. Jaklewicz, and P. Olszynski, "Tribology with emphasis on polymers: friction, scratch resistance and wear," *Journal of Materials Education*, vol. 25, no. 4–6, pp. 119–132, 2003.

[20] T. Cousseau, B. M. Graça, A. V. Campos, and J. H. O. Seabra, "Influence of grease rheology on thrust ball bearings friction torque," *Tribology International*, vol. 46, pp. 106–113, 2012.

[21] I. Couronné, P. Vergne, D. Mazuyer, N. Truong-Dinh, and D. Girodin, "Effects of grease composition and structure on film thickness in rolling contact," *Tribology Transactions*, vol. 46, no. 1, pp. 31–36, 2003.

[22] T. Cousseau, B. Graça, A. Campos, and J. Seabra, "Friction torque in grease lubricated thrust ball bearings," *Tribology International*, vol. 44, no. 5, pp. 523–531, 2011.

[23] K. V. Shooter and D. Tabor, "The frictional properties of plastics," *Proceedings of the Physical Society, Section B*, vol. 65, no. 9, pp. 661–671, 1952.

[24] R. C. Bowers, W. C. Clinton, and W. A. Zisman, "Frictional behavior of polyethylene, polytetrafluorethylene and halogenated derivatives," *Lubrication Engineering*, vol. 9, pp. 204–209, 1953.

[25] K. Shooter and R. H. Thomas, "Frictional properties of some plastics," *Research*, vol. 2, pp. 533–539, 1952.

[26] B. L. Rees, "Static friction of bulk polymers over a temperature range," *Research*, vol. 10, pp. 331–338, 1957.

[27] N. K. Myshkin, M. I. Petrokovets, and A. V. Kovalev, "Tribology of polymers: adhesion, friction, wear, and mass-transfer," *Tribology International*, vol. 38, no. 11-12, pp. 910–921, 2005.

[28] G. V. Vinogradov, G. M. Bartenev, A. I. El'kin, and V. K. Mikhaylov, "Effect of temperature on friction and adhesion of crystalline polymers," *Wear*, vol. 16, no. 3, pp. 213–219, 1970.

[29] W. Brostow and H. E. H. Lobland, *Materials: Introduction and Applications*, John Wiley & Sons, New York, NY, USA, 2017.

[30] W. Brostow, H. E. Hagg Lobland, and M. Narkis, "Sliding wear, viscoelasticity, and brittleness of polymers," *Journal of Materials Research*, vol. 21, no. 9, pp. 2422–2428, 2006.

[31] W. Brostow and H. E. H. Lobland, "Survey of relations of chemical constituents in polymer-based materials with brittleness and its associated properties," *Chemistry & Chemical Technology*, vol. 10, no. 4, pp. 595-560, 2016.

[32] W. C. Milz and L. E. Sargent, "Frictional characteristic of plastics," *Lubrication Engineering*, vol. 11, pp. 313–317, 1955.

[33] K. Tanaka, "Kinetic friction and dynamic elastic contact behaviour of polymers," *Wear*, vol. 100, no. 1-3, pp. 243–262, 1984.

[34] D. G. Flom and N. T. Porile, "Effects of temperature and high-speed sliding on the friction of 'teflon' on 'teflon'," *Nature*, vol. 175, pp. 682–685, 1955.

[35] D. G. Flom and N. T. Porile, "Friction of Teflon sliding on Teflon," *Journal of Applied Physics*, vol. 26, no. 9, pp. 1088–1092, 1955.

[36] H. S. White, "Small oil-free bearings," *Journal of Research of the National Bureau of Standards*, vol. 57, no. 4, pp. 185–189, 1956.

[37] N. E. Bekhet, "Tribological behaviour of drawn polypropylene," *Wear*, vol. 236, no. 1-2, pp. 55–61, 1999.

[38] S. Bahadur and K. C. Ludema, "The viscoelastic nature of the sliding friction of polyethylene, polypropylene and copolymers," *Wear*, vol. 18, no. 2, pp. 109–128, 1971.

[39] W. Wieleba, "The mechanism of tribological wear of thermoplastic materials," *Archives of Civil and Mechanical Engineering*, vol. 7, no. 4, pp. 185–199, 2007.

Effect of Surface Modification of Palygorskite on the Properties of Polypropylene/ Polypropylene-g-Maleic Anhydride/ Palygorskite Nanocomposites

David Cisneros-Rosado and Jorge Alonso Uribe-Calderon

Centro de Investigación Científica de Yucatán, A.C., Unidad de Materiales, Calle 43 No. 130 x 32 y 34, Colonia Chuburná de Hidalgo, 97205 Mérida, YUC, Mexico

Correspondence should be addressed to Jorge Alonso Uribe-Calderon; jorge.uribe@cicy.mx

Academic Editor: Domenico Acierno

The effect of surface modification of palygorskite (Pal) on filler dispersion and on the mechanical and thermal properties of polypropylene (PP)/polypropylene grafted maleic anhydride (PP-g-MAH)/palygorskite (Pal) nanocomposites was evaluated. A natural Pal mineral was purified and individually surface modified with hexadecyl tributyl phosphonium bromide and (3-Aminopropyl)trimethoxysilane; the pristine and modified Pals were melt-compounded with PP to produce nanocomposites using PP-g-MAH as compatibilizer. The grafting of Pal surface was verified by FT-IR and the change in surface hydrophilicity was estimated by the contact angle of sessile drops of ethylene glycol on Pal tablets. The extent of Pal dispersion and the degree of improvement in both the mechanical and thermal properties were related to the surface treatment of Pal. Modified Pals were better dispersed during melt processing and improved Young's modulus and strength; however, maximum deformation tended to decrease. The thermal stability of PP/PP-g-MAH/Pal nanocomposites was considerably improved with the content of modified Pals. The degree of crystallinity increased with Pal content, regardless of the surface modification. Surfactant modified Pal exhibited better results in comparison with silane Pal; it is possible that longer alkyl chains from surfactant molecules promoted interactions with polymer chains, thereby improving nanofiller dispersion and enhancing the properties.

1. Introduction

Palygorskite (Pal) is a type of clay with a unique three-dimensional structure and fibrous morphology, exhibiting chemical and thermal stability, high aspect ratio, large specific surface area (150–200 m^2/g), and mechanical strength [1, 2].

The chemical formula for palygorskite is reported as $[(Mg, Al, Fe)_5\ Si_8O_{20}(OH)_2(OH_2)_4]\bullet 4H_2O$ [3]; the clay structure is composed of octahedral and tetrahedral units of silicates that form interconnected channels with dimensions of 3.7×10.6 Å. Substitutions of Si for Al atoms in the tetrahedral units provide to clay a cation exchange capacity (CEC) ranging from 20 to 30 meq/100 g [2]. These interesting properties suggest that Pal can be used as polymer nanofiller [4], although it is mainly employed as a catalyst carrier [5–7] or as adsorbent material [8, 9].

The hydrophilic nature of Pal could represent a challenge for the production of polymer/Pal nanocomposites. A recurrent procedure to promote Pal dispersion within the polymer is to modify its surface with organic modifiers in order to reduce the interactions among silicate fibers and to promote compatibility with the polymer matrix. Pal has been traditionally modified with a silane coupling agent [1, 10–13] to produce polymer/Pal nanocomposites based on epoxy resin [3], Nylon 6 [14–17], natural rubber [18], ethylene-vinyl acetate copolymers [19], or poly-aniline [20]. However, the use of cationic surfactants to modify Pal has been less frequently reported [21–25].

On the other hand, PP is one of the most widely used polyolefins due to its low cost, low density, and extraordinary versatility in terms of applications and recycling [26]. Several studies have been devoted to improving the mechanical

[27–30] and thermal properties of PP [31] by incorporating mineral fillers (i.e., clay) in an effort to diversify the use of PP. The intrinsic nonpolar nature of PP makes it incompatible with clay fillers; the use of modified clays and compatibilizers (such as polypropylene grafted maleic anhydride [32–34]) is a common approach to produce PP/clay nanocomposites. A number of studies have focused on the preparation of PP/PP-g-MAH/Pal nanocomposites [35–38], but despite these efforts, the factors promoting the dispersion of Pal and improvement of the properties are not yet completely understood.

In this study, the effect of surface modification of Pal on nanofiller dispersion and the enhancement of properties was evaluated on polypropylene (PP) nanocomposites prepared by melt compounding using polypropylene grafted maleic anhydride (PP-g-MAH) as compatibilizer. Thus, a Pal mineral was purified and characterized and separately modified with a conventional silane coupling agent (3-Aminopropyl)trimethoxysilane (APTMS) and a cationic surfactant (hexadecyl tributyl phosphonium bromide, HDTBP) to produce Pals with different surface energies. To the best of our knowledge, there are no previous reports concerning Pal modification with HDTBP to produce polymer nanofiller. It was expected that surface modification promotes the deagglomeration of silicate fibers, thereby improving dispersion during melt compounding and enhancing some properties. Pal characterization involved Scanning Electron Microscopy (SEM) and Fourier Transform Infrared Spectroscopy (FTIR); Pal dispersion within PP was determined by SEM and TEM; the mechanical and thermal properties (Thermogravimetric Analysis, TGA and Differential Scanning Calorimetry (DSC)) of the resulting nanocomposites were also determined.

2. Experimental

2.1. Materials. Palygorskite mineral was directly collected from a seam in Chapab, Mexico. Pal was purified with hydrochloric acid (J. T. Baker, 37% w/v). Ethylene glycol from Sigma-Aldrich was used to determine the surface energy of modified Pal. Isotactic PP (Valtec HP423M, melt flow index of 3.8 g/10 min, Mw = 232,400 g/mole, and density of 0.9 g/cm^3) was used as a matrix, and PP-g-MAH of Eastman G Polymers™ (Eastman G-3015, maleic anhydride 1.32 wt.%, acid number of 15 mg KOH/g, and Mw = 47,000 g/mole) was used as compatibilizer. (3-Aminopropyl)trimethoxysilane (APTMS 99% purity) from Sigma-Aldrich and hexadecyl tributyl phosphonium bromide $HDTBP^+Br^-$, 98% purity) from Alpha Aesar were used to modify the Pal surface.

2.2. Clay Purification. Pal mineral (100 g), as received, was dispersed mechanically with 3 L of distilled water for 1 h. Subsequently, the slurry was filtered through a 70 mesh sieve (0.224 μm, standard ASTM E 11-70) and 100 mL of hydrochloric acid solution (1 N) was added to the Pal slurry and stirred mechanically for 1 h. Several cycles of washing/centrifuge were carried out until a pH ~7 was achieved [39]. Finally, the purified Pal (Pal-1) was dried in a convection oven at 100°C for 24 h. A cation exchange capacity of 26.2 meq/100 g was determined for Pal, following the procedure reported in the literature [40].

2.3. Pal Modification with Silane Coupling Agent. Conventional silane modification was performed on Pal-1 [8]: 10 g of Pal-1 was dispersed with 170 mL of toluene using an ultrasonic bath (10 min) and then it was filtered through a 70 mesh sieve. APTMS was added dropwise to the mixture (under vigorous mechanical stirring) at the ratio Pal : silane of 1 : 1 g : ml. The reaction was conducted under reflux for 2 h at 45°C under mechanical stirring. The resulting modified Pal was washed with toluene and anhydrous ethanol to remove unreacted silane coupling agent. Finally, the silane modified Pal (Pal-2) was dried in a convection oven at 100°C for 24 h, ground, and sieved (70 mesh).

2.4. Pal Modification with the Phosphonium Surfactant. To produce a new organo-Pal, 25 g of Pal was mechanically dispersed in 2.5 L of distilled water for 1 h and then was filtered through a 70 mesh sieve and sonicated for 30 min (ultrasonic tip Vibra Cell, model 750 VC, at 20 kHz). Subsequently, HDTBP solution was slowly added to the Pal suspension; the amount of surfactant corresponded to the 1.1 CEC of Pal. The reaction was then maintained under mechanical stirring at 80°C for 5 h [9]. The modified Pal was washed several times with distilled water to remove the bromide ions. Finally, the modified Pal (Pal-3) was dried in a convection oven at 100°C for 24 h, ground, and sieved (70 mesh).

2.5. Preparation of PP/PP-g-MAH/Pal Nanocomposites. The dried Pals were melt-compounded separately with PP/PP-g-MAH (containing 5 wt.% of PP-g-MAH) in a batch mixer Plasticorder, Brabender at 190°C for 5 min at 100 rpm to obtain PP/PP-g-MAH/Pal nanocomposites. Pal contents were 0.5, 1, and 2 wt.%. Other studies have reported that the tensile mechanical properties of PP did not increase considerably at concentrations higher than 2 wt.% of Pal [35, 37]. The composites were compression molded at 190°C in a lab press for 3 min at 45 kN clamping load; specimens for further characterization were obtained from the above plates (120 × 120 mm, 1 mm thick).

3. Characterization

3.1. Scanning Electron Microscopy (SEM) and Transmission Electron Microscopy (TEM) Analysis. Pals and the fracture surfaces of PP/PP-g-MAH/Pal nanocomposites (under cryogenic conditions) were observed using a variable-pressure Scanning Electron Microscope Jeol model JSM-6360 LV (20 kV); samples were coated with a thin layer of gold. TEM analysis was carried out with a Titán 80–300 k microscope, on ultrathin samples obtained by a microtome under cryogenic conditions.

3.2. Fourier Transform Infrared Spectroscopy (FTIR). FTIR analysis was carried out on Pals with a FTIR Nicolet 8700. KBr/Pal tablets (2% wt. clay) were obtained using a KBr pellet die applying a 20 kN clamping load for 5 min. The scanning

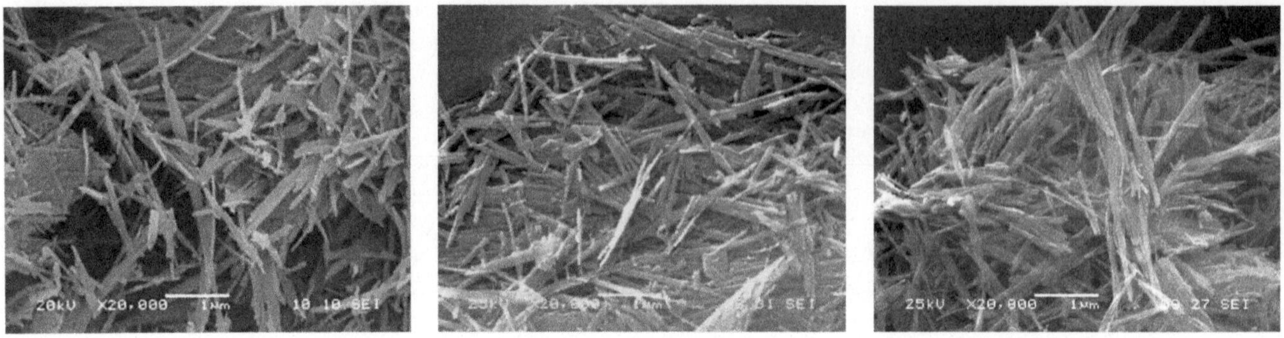

FIGURE 1: SEM micrographs for palygorskite samples: (a) Pal-1, (b) Pal-2, and (c) Pal-3.

range was from 4000 to 400 cm^{-1} with a resolution of 4 cm^{-1}; the results are the average of 100 scans.

3.3. Thermogravimetric Analysis (TGA). The thermal stability of PP/PP-g-MAH/Pal nanocomposites was determined with a Thermogravimetric Analyzer TGA-7 Perkin Elmer, from 50°C to 700°C with a heating rate of 10°C/min under a Nitrogen atmosphere; the sample weight was 5 mg.

3.4. Contact Angle Measurements. To assess the change in the hydrophilicity of Pal, the contact angles of sessile drops of ethylene glycol (Sigma-Aldrich) on the surface of Pal tablets were measured at room temperature. Pal tablets were prepared by compressing 300 mg of Pal using a KBr pellet die applying 50 kN clamping force for 5 min (Lab Carver Press). The values of contact angles were determined using the dynamic sessile drop method with a goniometer-Ramé-Hart Instrument Co model 250. The average contact angles were calculated from twenty measurements of four different drops.

3.5. Mechanical Tensile Properties. Mechanical tensile testing was carried out using a universal tensile machine (Mini SHIMADZU model AGS-1KN); specimens were deformed at 5 mm/min at room temperature according to the standard test method ASTM-638D. Elastic modulus, tensile strength, and elongation at break were determined. The total length for the dog-bone specimen was 115 mm, gauge length was 25 mm, and the narrow section width was 6 mm.

3.6. Differential Scanning Calorimetry (DSC). DSC analysis was carried out using a Perkin Elmer DSC-7 calorimeter under a Nitrogen atmosphere. The samples (6 mg) were sealed in aluminum pans and heated from 50°C to 250°C at 10°C/min and cooled at 50°C/min; two heating runs were conducted.

4. Results and Discussion

4.1. Morphology of Pals. The morphology of Pals was assessed by SEM micrographs (Figure 1). The morphology of Pal-1 is presented as a fibrillar structure (Figure 1(a)); the diameter ranges from 20 to 30 nm and the length shows a wide range from hundreds of nanometers to few microns. In addition, Pal was observed as agglomerates due to the strong interactions among individual silicate fibers [15]. SEM micrographs of Pal-2 (Figure 1(b)) revealed no substantial changes in the clay fiber morphology after silane treatment. Clay aggregates seem to be more compact; it may be possible therefore that silane molecules can contribute to the aggregation among clay particles. Finally, Figure 1(c) shows Pal-3 fibers; some clay particles appear to be separated from the agglomerates, possibly due to the weakened clay-clay interactions after the surface modification. The SEM observations suggest that the experimental conditions used during surface modification of Pal did not significantly modify the aspect ratio morphology.

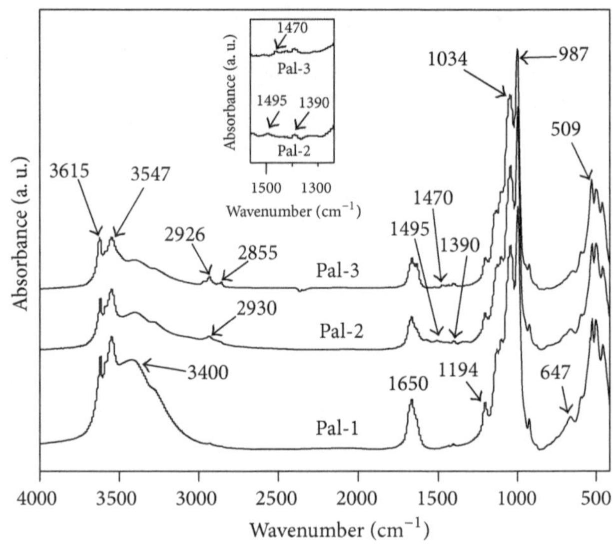

FIGURE 2: FT-IR spectra of Pal-1, Pal-2, and Pal-3.

4.2. FT-IR Characterization. FT-IR analysis was conducted to verify the surface modification of Pal; the FT-IR spectra of Pal-1, Pal-2, and Pal-3 are presented in Figure 2. The band in 3615 cm^{-1} for the Pal-1 spectrum can be attributed to the stretching vibrations of Mg-OH from the dioctahedral

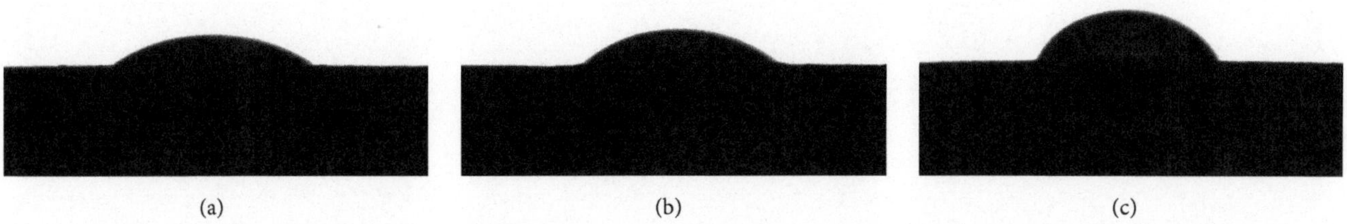

FIGURE 3: Profiles of sessile drops of ethylene glycol on the surfaces of: (a) Pal-1, (b) Pal-2, and (c) Pal-3, respectively.

units. A band associated with the stretching vibrations of the coordinated water is observed at 3547 cm^{-1} [41]. The band at 3400 cm^{-1} corresponds to the stretching vibrations of absorbed and zeolitic water. The band at 1650 cm^{-1} was attributed to OH bending vibrations of coordinated, absorbed, and zeolite water [3]; it is probable that the above band could overlap the signal associated with vibrations of the amino groups from silane molecules at 1608 cm^{-1}. The bands at 1194 cm^{-1} and 1034 cm^{-1} were attributed to asymmetric stretching vibrations from Si-O. The band in 987 cm^{-1} is due to vibrations stretching perpendicularly to the plane of Si-O, and the band at 647 cm^{-1} could be attributed to stretching vibrations from Mg bonded water (zeolitic and surface absorbed) [42]. The signal in 509 cm^{-1} corresponds to the deformation of Si-O bonds from tetrahedral sheets [41]. The IR signals are consistent with the results reported in other studies.

The FT-IR spectrum for Pal-2 shows additional bands, for example, a band at 2930 cm^{-1} corresponding to the stretching vibrations of C-H bonds from the silane molecule [43, 44]. Bands associated with the symmetric bending vibrations for N-H due to the amino groups of the silane grafted on the surface of Pal were observed at 1495 cm^{-1} and 1390 cm^{-1} [44]. On the other hand, the bands at 2926 and 2855 cm^{-1} in the FT-IR spectrum for Pal-3 correspond to the symmetric and antisymmetric stretching vibrations from -CH_2- groups of surfactant, respectively [9, 20]. The band in 1470 cm^{-1} corresponds to the deformation of C-H bonds [45] from the alkyl chains of surfactant. The experimental evidence suggests that the silane and cation surfactant molecules were successfully grafted to the Pal surface.

4.3. Contact Angle Measurements. Contact angle measurements have been reported for other mineral clays such as montmorillonite (MMT) [46–48], kaolinite [49], and smectite [50] for the purpose of verifying the surface modification of clays. Figure 3 shows profiles of the sessile droplets of ethylene glycol on the surfaces of Pal tablets. Ethylene glycol was chosen as test liquid due to its viscosity and polarity. The average values of the contact angle for the Pals are shown in Table 1, which tend to increase with the surface treatment. Pal-1 belongs to a type of natural hydrophilic silicate showing hydrophilic nature; as a consequence, the contact angle of a sessile droplet of a partial hydrophilic was small. The increment in the contact angle for other Pals can be interpreted as an increase in the hydrophobicity of the Pal surface due to the hydrophobic silane coupling agent or the cationic surfactant grafted to the surface. The enhancement of hydrophobicity in the organically modified clays could lead to an improvement in the dispersion/exfoliation of clay within the polymer [51].

TABLE 1: Contact angles for ethylene glycol on three types of Pal.

Sample	θ (°)
Purified-Pal	14.01 ± 0.99
APTMS-Pal	20.09 ± 2.27
HDTBP-Pal	23.04 ± 2.87

4.4. Pal Dispersion within the PP/PP-g-MAH/Pal Nanocomposites. The dispersion of Pal within PP matrix was determined by SEM observations (Figure 4) of the fracture surfaces of PP/PP-g-MAH/Pal nanocomposites with three different types of Pal (2 wt.% Pal content). Figures 4(a)–4(c) correspond to Pal-1 nanocomposites at different magnifications, in which Pal-1 was distributed as aggregates in micrometer size; the light dots are the fractured ends of fibrous crystals of Pal. Figures 4(d)–4(f) correspond to PP/PP-g-MA/Pal-2 nanocomposites where Pal-2 was distributed as smaller agglomerates; the results suggest that Pal dispersion was improved with the surface treatment due to the weakening of interactions among palygorskite single crystals. Weaker interactions among fibers allow the separation of Pal crystals from the bundles during the melt compounding [16]. Finally, Figures 4(g)–4(i) correspond to PP/PP-g-MA/Pal-3 nanocomposites. Interestingly, small aggregates were found on the fractured surface indicating good distribution and dispersion of Pal. The experimental evidence suggests that the surfactant modification of Pal effectively improved the compatibility among components, resulting in a uniform distribution and dispersion of Pal and inhibition of Pal reagglomeration.

TEM images of the PP/PP-g-MA/Pal nanocomposites at 2% w/w Pal content are shown in Figure 5. In Figures 5(a)–5(c), aggregates of Pal-1 with diameters around 193 nm are observed; the measurements were carried out using ImageJ software. On the other hand, in Figures 5(d)–5(f), the silicate fibers of Pal-2 are dispersed within the polymer matrix with aggregates with diameter around 144 nm. The Figure 5(g) shows the transversal view of Pal-3 fibers well dispersed within the PP/PP-g-MAH blend. Figures 5(h)-5(i) show details of the distribution and dispersion of Pal-3; the

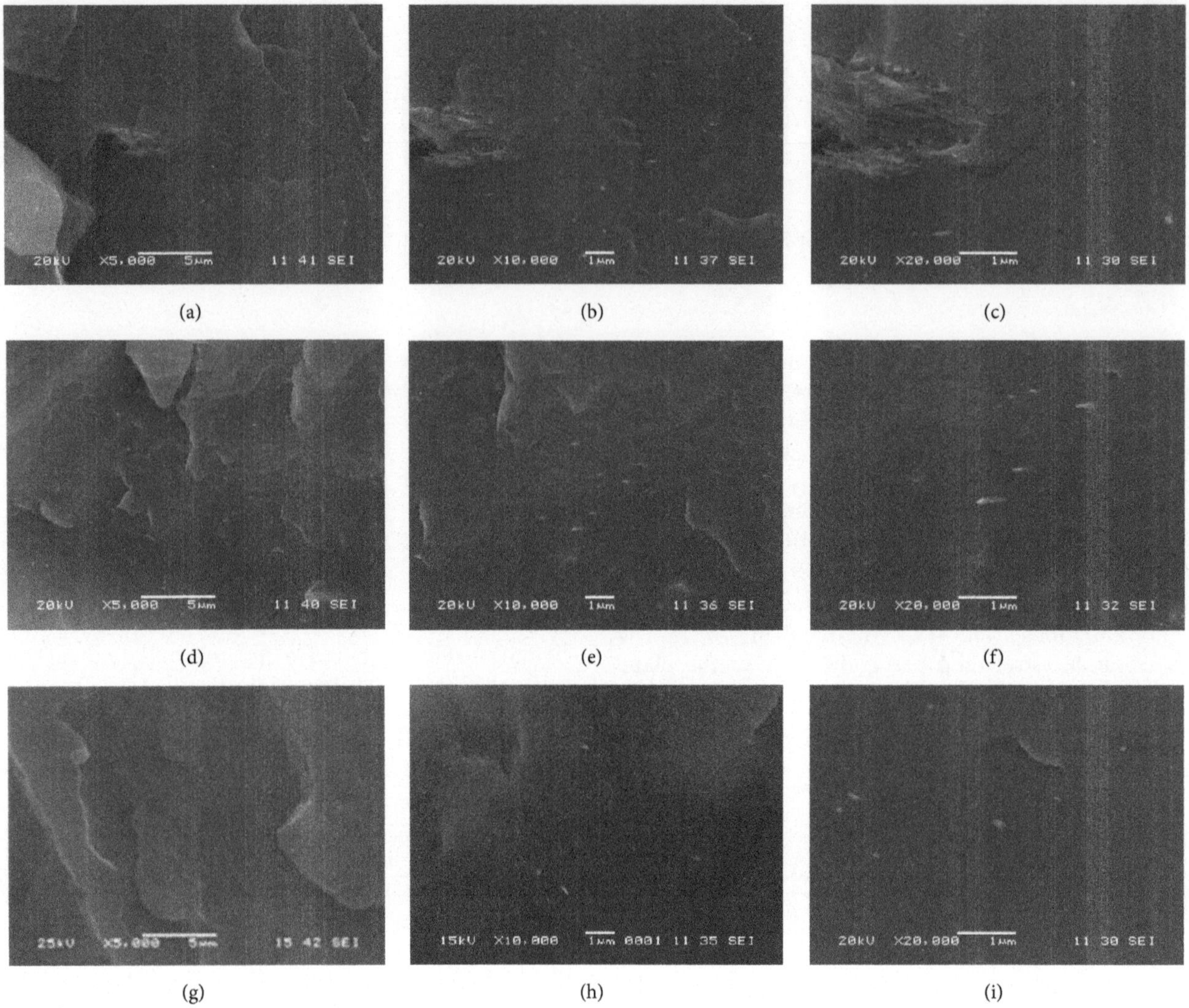

FIGURE 4: SEM micrographs of the fractured surface at different magnifications of PP/PP-g-MAH/Pal nanocomposites (2 wt.%): Pal-1 (a), (b), and (c); Pal-2 (d), (e), and (f); Pal-3 (g), (h), and (i).

diameter of Pal-3 aggregate ranges from 20 to 40 nm. The results indicate that the distribution and dispersion of silicate fibers depend on the reduction of surface energy of Pal: Pal-3 modified with longer alkane molecules shows better dispersion and distribution.

4.5. Mechanical Properties. The tensile mechanical properties of PP/PP-g-MAH/Pal nanocomposites are shown in Figure 6. In general, Young's modulus (Figure 6(a)) increased with Pal concentration due to the stiffness of silicate fibers, depending on the surface modification of Pal. For instance, the modulus of PP/PP-g-MAH/Pal-1 nanocomposites increase rapidly with a Pal load at low concentration (0.5 and 1 wt.%), while, at higher Pal content, the modulus reaches a *plateau*. It may be that, at this relatively high concentration of Pal, the agglomeration of silicate fibers acts as a stress concentrator, as reported in other studies [35, 37].

However, the nanocomposites containing Pal-2 and Pal-3 exhibited higher values of Young's modulus compared with Pal-1 nanocomposites; the results suggest that the reduction of surface energy of silicate fibers promoted its dispersion and compatibility with the polymeric matrix, increasing the stiffness of nanocomposites at low concentration of Pals. There are no significant differences in the modulus with the experimental error for Pal-2 and Pal-3 nanocomposites.

On the other hand, the tensile strength of PP (Figure 6(b)) increased slightly with the addition of Pals; however, no significant differences were observed with the experimental error. It is important to point out that the strength of the polymer matrix was not only preserved but actually slightly improved; it has also been reported that the inclusion of mineral filler within PP tends to reduce the matrix strength [52]. Further studies are required in order to promote the polymer-Pal interactions by increasing the content of PP-g-MAH to improve the mechanical properties.

It is widely accepted that the incorporation of nanoreinforcers in polymeric materials tends to reduce the elongation at break of nanocomposites. The experimental results

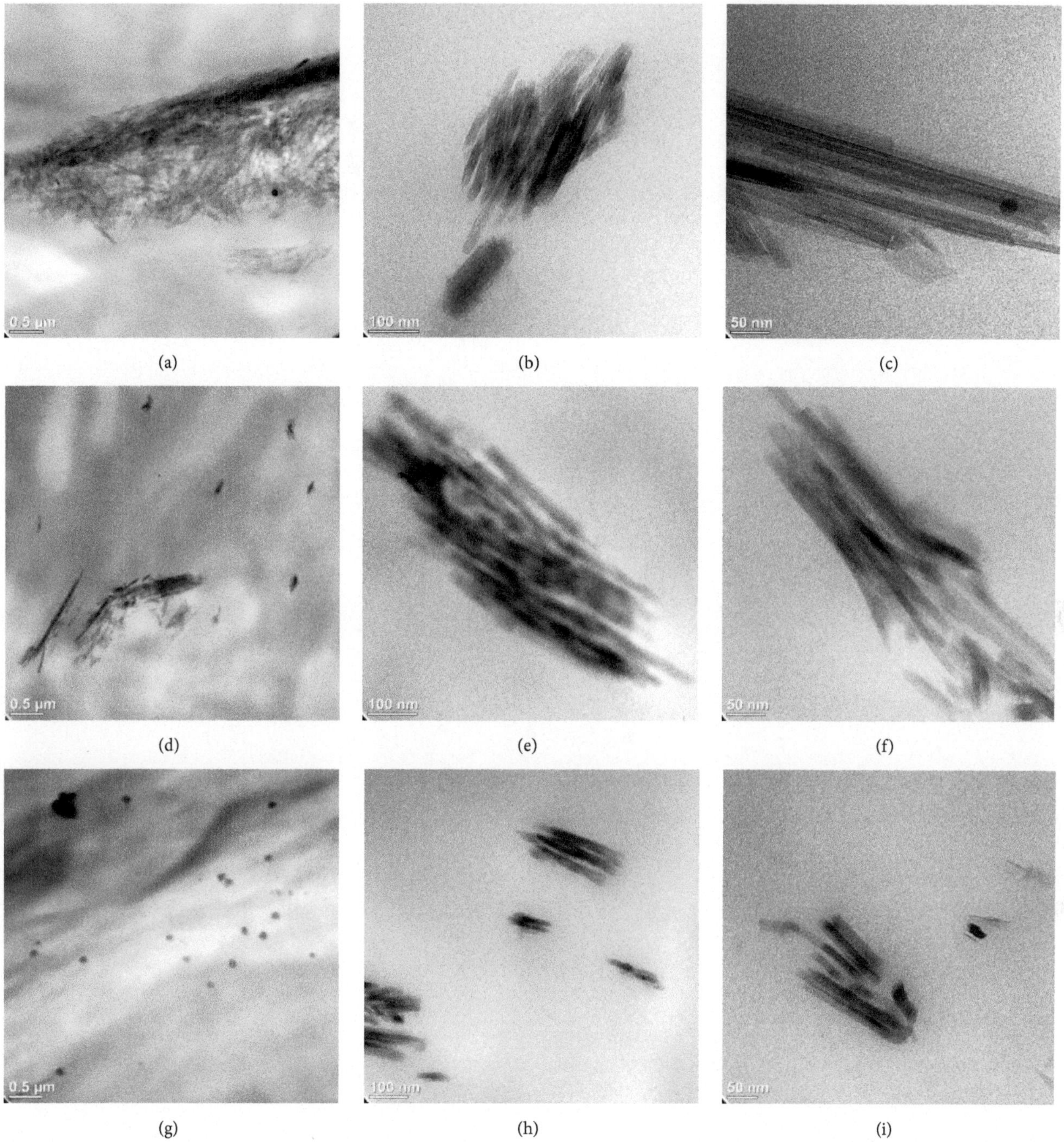

FIGURE 5: TEM micrographs of PP/PP-g-MAH/Pal nanocomposites: Pal-1 ((a) to (c)), Pal-2 ((d) to (f)), and Pal-3 ((g) to (i)).

concerning elongation at break (Figure 6(c)) indicate that the property decreases with the Pal content (regardless of the type of Pal used), possibly due to the restriction in molecular movement of the polymer chains between silicate fibers and/or Pal agglomeration [31, 53].

4.6. Thermal Properties. The TGA analysis provides relevant information concerning the thermal stability of the nanocomposites. The TGA traces in Figure 7 depict the variations of the residual mass as a function of temperature for pure PP, PP/PP-g-MAH, and PP/PP-g-MAH/Pal nanocomposites. TGA traces for PP and PP PP-g-MAH (Figure 7(a)) show that the thermal decomposition of the materials occurs in one step; PP/PP-g-MAH seems to be slightly less thermally stable than PP due to the lower molecular weight of the compatibilizer. TGA traces of Pal-1 nanocomposites were shifted at higher temperatures especially at 1 and 2%wt. of Pal-1. On the other hand, the TGA traces of nanocomposites containing Pal-2 showed a drastic shift toward higher temperatures with the inclusion of Pal-2 even at the lowest

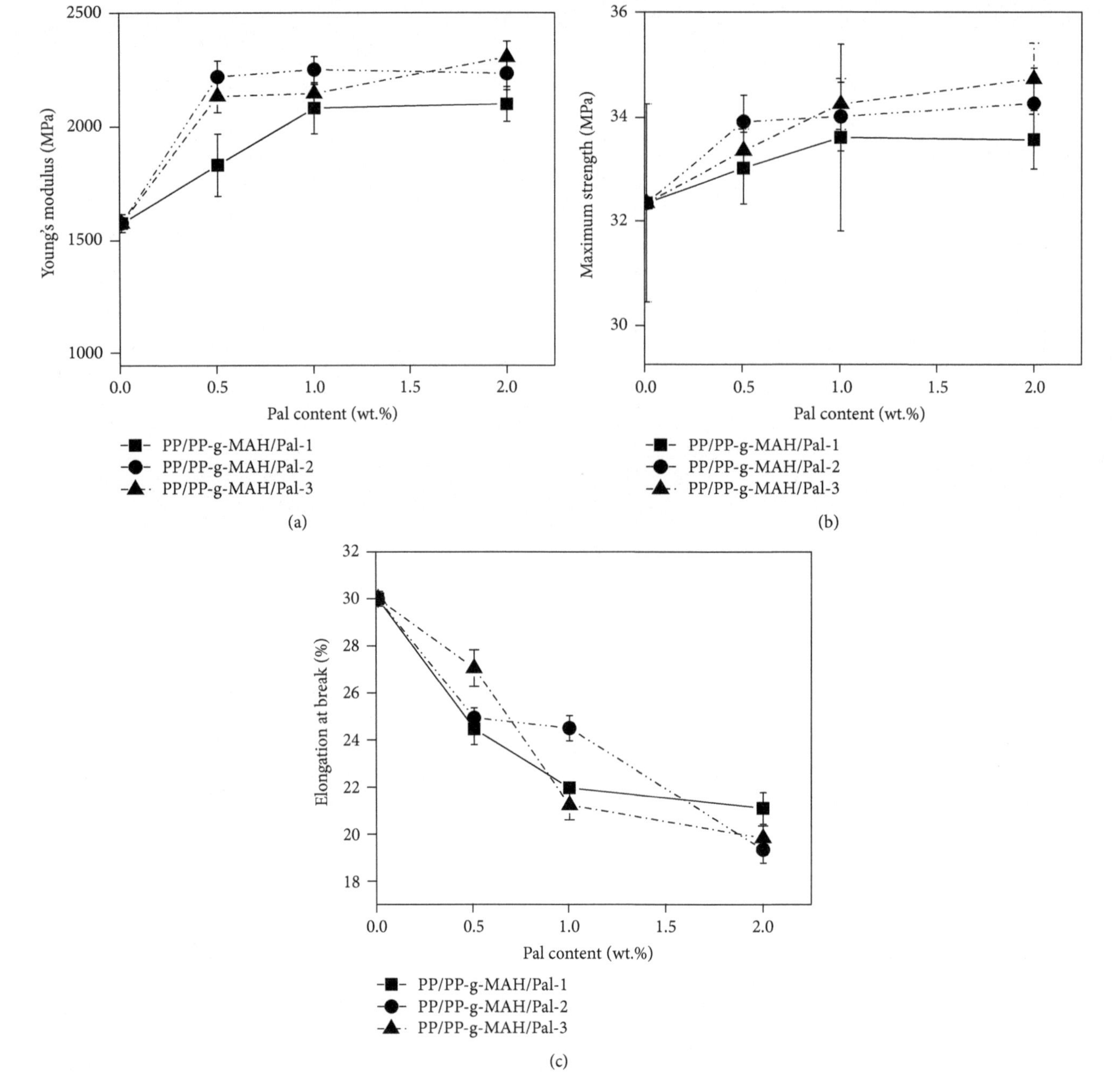

FIGURE 6: Mechanical properties of PP/PP-g-MAH/Pal nanocomposites: (a) Young's modulus, (b) maximum strength, and (c) elongation at the break.

concentration of Pal-2 and thermal stability increased with filler concentration (Figure 7(b)). The thermal stability of PP/PP-g-MAH/Pal-3 nanocomposites (Figure 7(c)) presents an interesting behavior: the TGA traces were shifted to higher temperatures at low content of Pal-3 and the thermal stability remained, regardless of the Pal-3 content. The results suggest that the thermal stability of PP nanocomposites depended on the degree of Pal dispersion promoted by the surface modification, with the Pal modified with the phosphonium surfactant producing the best results. The improvement in the thermal stability of PP/clay nanocomposites is attributed to a barrier effect of the dispersed particles of the clay, which can minimize the permeability of the volatile degradation products from the nanocomposites [31, 53], as well as the higher thermal stability of clays modified with phosphonium surfactants [54, 55].

The effect of Pal dispersion on the crystallinity of PP was evaluated by DSC. Figure 8 shows the DSC traces (corresponding to the second heating run) for PP, PP/PP-g-MAH, and PP/PP-g-MAH/Pal nanocomposites as a function of type and concentration of Pal. PP exhibited a melting peak at 165.1°C and PP/PP-g-MAH at 165.8°C. In general,

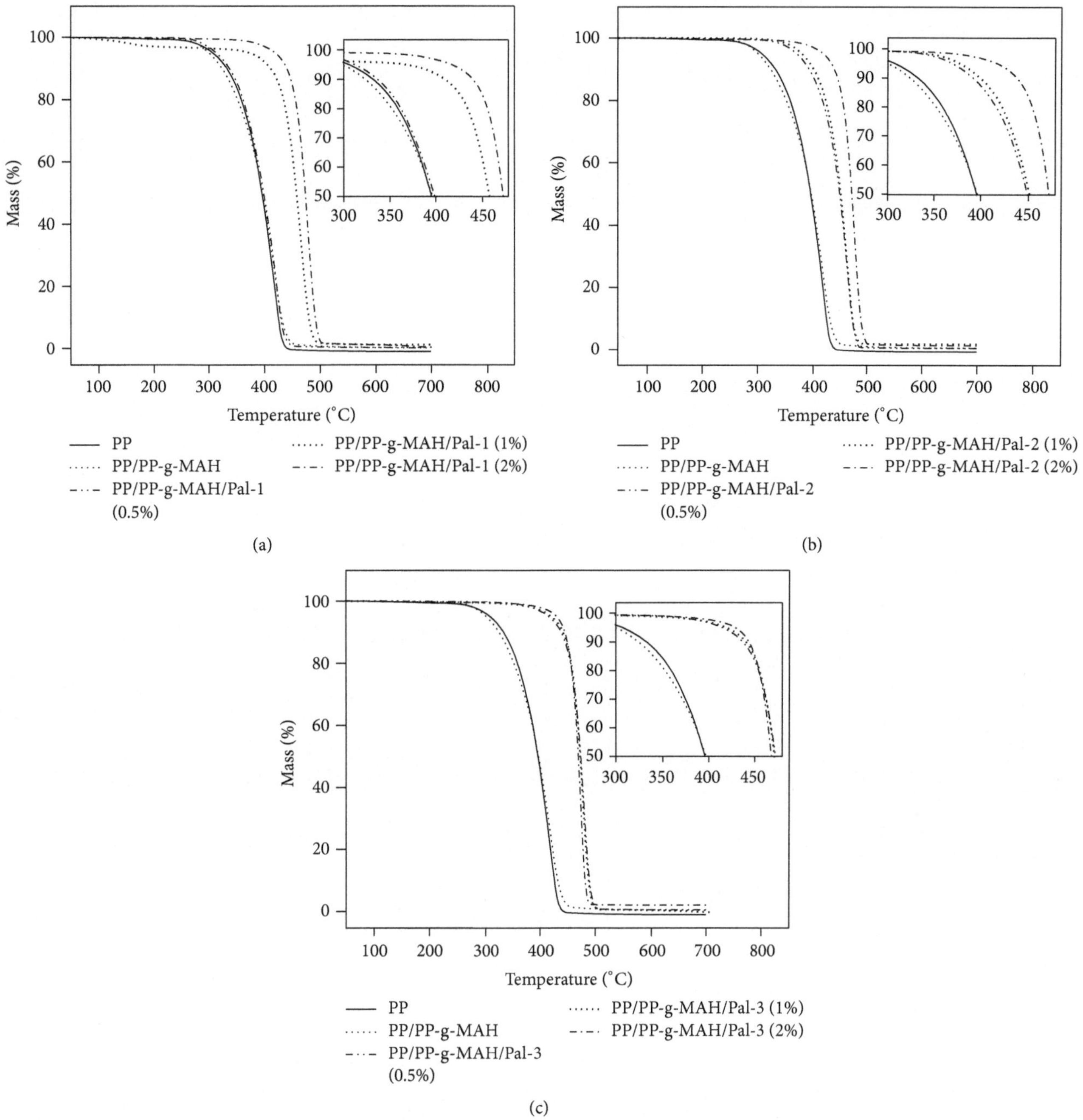

FIGURE 7: TGA traces for PP and PP/PP-g-MAH/Pal nanocomposites (0.5, 1, and 2 wt.% Pal content): (a) Pal-1, (b) Pal-2, and (c) Pal-3.

Pal nanocomposites exhibited the melting peak in the same temperature range. Table 2 depicts the melting point (T_m), melting enthalpy (ΔH_c), and the degree of crystallinity (X_c). The degree of crystallinity was calculated considering the melting enthalpy for 100% crystalline PP to be 190.6 J/g [56]. A slight shift of the melting temperatures of nanocomposites to lower temperatures was observed, which might be related to the reduction in the crystallite size due to the presence of palygorskite [57].

The degree of crystallinity of PP matrix from nanocomposites tends to increase with Pal content, regardless of the surface treatment. PP/PP-g-MAH/Pal-1 shows lower values in the degree of crystallinity given that the Pal agglomerates reduce the concentration of active sites for nucleation [58]. Thus, the better dispersed Pal-2 and Pal-3 in nanocomposites act as effective nucleation agents, increasing slightly the degree of crystallinity.

5. Conclusions

The effect of surface modification of Pal on filler dispersion and the mechanical and thermal properties was evaluated.

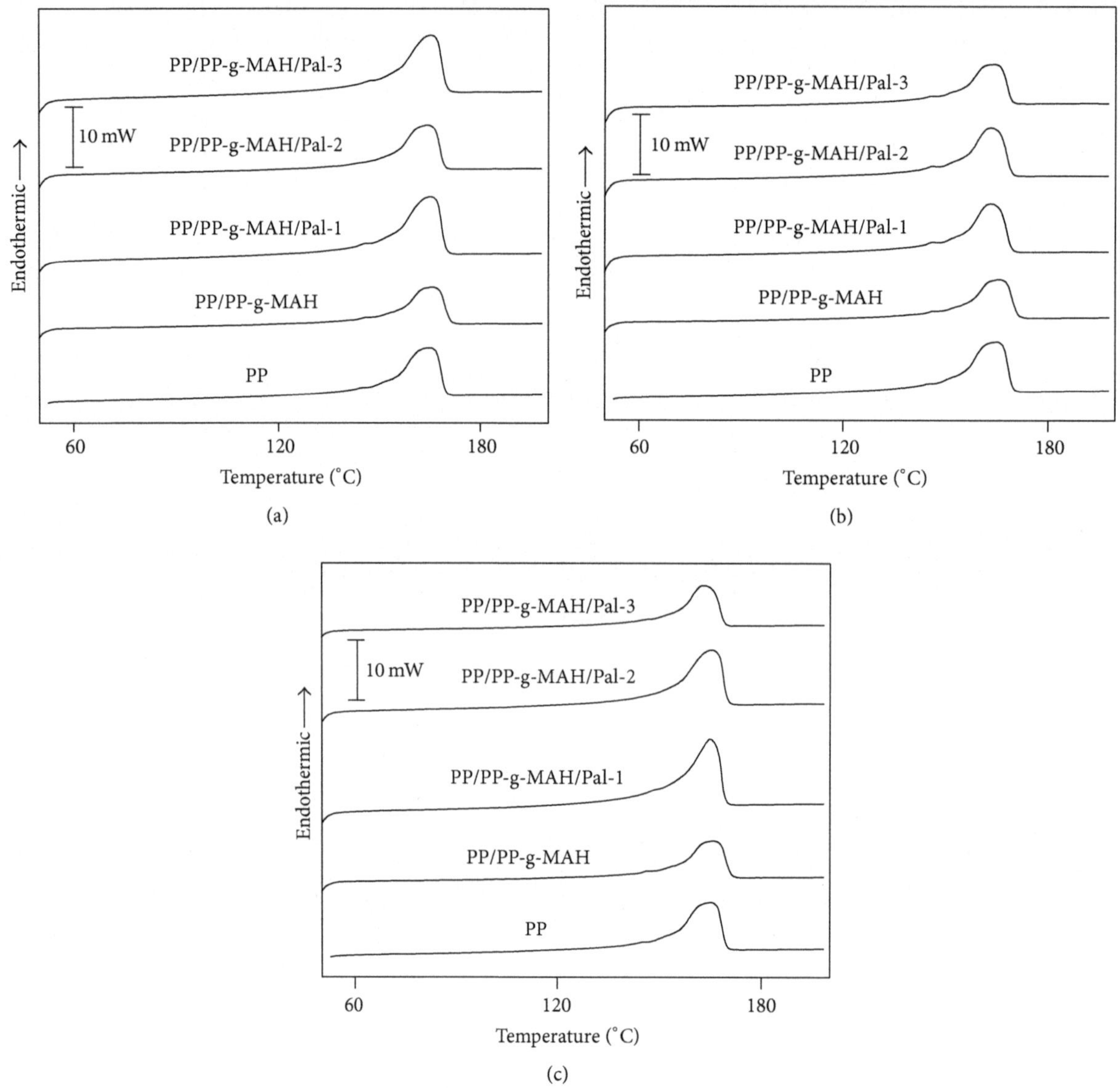

FIGURE 8: DSC thermograms for PP, PP/PP-g-MAH, and PP/PP-g-MAH/Pal: (a) 0.5, (b) 1, and (c) 2 wt.% Pal content, respectively.

TABLE 2: DSC parameters for PP and PP/PP-g-MAH/Pal nanocomposites.

Sample	Tm (°C)	ΔH_c (J/g)	X_c (%)
PP	165.1	79.7	41.8
PP/PP-g-MAH	165.8	79.9	42.7
PP/PP-g-MAH/Pal-1 (0.5% w/w)	165.2	79.3	41.8
PP/PP-g-MAH/Pal-1 (1% w/w)	164.2	80.4	42.4
PP/PP-g-MAH/Pal-1 (2% w/w)	162.9	79.8	41.9
PP/PP-g-MAH/Pal-2 (0.5% w/w)	165.3	80.2	42.5
PP/PP-g-MAH/Pal-2 (1% w/w)	164.3	80.6	42.7
PP/PP-g-MAH/Pal-2 (2% w/w)	163.5	81.4	43.6
PP/PP-g-MAH/Pal-3 (0.5% w/w)	164.3	82.6	44.2
PP/PP-g-MAH/Pal-3 (1% w/w)	163.1	80.9	42.7
PP/PP-g-MAH/Pal-3 (2% w/w)	163.2	83.6	44.8

Natural palygorskite was purified and surface modified separately to produce a silane modified Pal and a new phosphonium surfactant modified Pal. PP/PP-g-MAH/Pal nanocomposites were melt-compounded with different concentration of Pals. FT-IR analysis confirmed the surface modification of Pal and the contact angle measurements suggested that hydrophilicity of Pal was slightly reduced with the surface modification.

Pal particles were dispersed in different manners depending on the surface treatment. Pristine Pal was poorly distributed/dispersed within PP; however, Pal dispersion was improved when the surface of Pal was modified with a silane coupling agent or cationic surfactant, and the latter was better dispersed. It is possible that the longer alkyl chains of the cationic surfactant promoted interactions among polymer chains and modified Pal, thereby improving the dispersion. Young's modulus increased and maximum strength slightly improved with Pal content, and elongation at the break tended to decrease. As expected, samples with improved dispersion of Pal exhibited relatively improved mechanical properties. On the other hand, the thermal stability of nanocomposites depended on the degree of Pal dispersion; contrarily, the crystallinity of the nanocomposites is affected slightly by Pal, at least under the conditions described in this study.

Acknowledgments

The present study was financially supported by FOMIX-Yucatan Grant 6030200001 and the Conacyt-Mexico Grant no. 222908. The technical support from Rossana Vargas-Coronado and Santiago Duarte-Aranda on FTIR analysis and SEM observations is highly appreciated.

References

[1] L. Wang and J. Sheng, "Preparation and properties of polypropylene/org-attapulgite nanocomposites," *Polymer*, vol. 46, no. 16, pp. 6243–6249, 2005.

[2] A. K. Helmy, S. G. de Bussetti, and E. A. Ferreiro, "The surface energy of palygorskite," *Powder Technology*, vol. 171, no. 2, pp. 126–131, 2007.

[3] R. Wang, Z. Li, Y. Wang et al., "Effects of modified attapulgite on the properties of attapulgite/epoxy nanocomposites," *Polymer Composites*, vol. 34, no. 1, pp. 22–31, 2013.

[4] E. Ruiz-Hitzky, M. Darder, A. C. S. Alcântara, B. Wicklein, and P. Aranda, "Functional nanocomposites based on fibrous clays," *RSC Smart Materials*, pp. 1–53, 2016.

[5] J.-L. Cao, G.-S. Shao, Y. Wang, Y. Liu, and Z.-Y. Yuan, "CuO catalysts supported on attapulgite clay for low-temperature CO oxidation," *Catalysis Communications*, vol. 9, no. 15, pp. 2555–2559, 2008.

[6] S. Guo and L. Shi, "Synthesis of succinic anhydride from maleic anhydride on Ni/diatomite catalysts," *Catalysis Today*, vol. 212, pp. 137–141, 2013.

[7] P. Pushpaletha and M. Lalithambika, "Modified attapulgite: an efficient solid acid catalyst for acetylation of alcohols using acetic acid," *Applied Clay Science*, vol. 51, no. 4, pp. 424–430, 2011.

[8] A. Xue, S. Zhou, Y. Zhao, X. Lu, and P. Han, "Effective NH2-grafting on attapulgite surfaces for adsorption of reactive dyes," *Journal of Hazardous Materials*, vol. 194, pp. 7–14, 2011.

[9] Y. Xi, M. Mallavarapu, and R. Naidu, "Adsorption of the herbicide 2,4-D on organo-palygorskite," *Applied Clay Science*, vol. 49, no. 3, pp. 255–261, 2010.

[10] G. Tian, Y. Kang, B. Mu, and A. Wang, "Attapulgite modified with silane coupling agent for phosphorus adsorption and deep bleaching of refined palm oil," *Adsorption Science & Technology*, vol. 32, no. 1, pp. 37–48, 2014.

[11] X. Yuan, C. Li, G. Guan, X. Liu, Y. Xiao, and D. Zhang, "Synthesis and characterization of poly(ethylene terephthalate)/attapulgite nanocomposites," *Journal of Applied Polymer Science*, vol. 103, no. 2, pp. 1279–1286, 2007.

[12] L. Wang and J. Sheng, "Graft polymerization and characterization of butyl acrylate onto silane-modified attapulgite," *Journal of Macromolecular Science - Pure and Applied Chemistry*, vol. 40, no. 11, pp. 1135–1146, 2003.

[13] L. Zhang, Q. Jin, J. Huang, Y. Liu, L. Shan, and X. Wang, "Modification of palygorskite surface by organofunctionalization for application in immobilization of H 3 PW 12 O 40," *Applied Surface Science*, vol. 256, no. 20, pp. 5911–5917, 2010.

[14] B. Pan, Q. Yue, J. Ren et al., "A study on attapulgite reinforced PA6 composites," *Polymer Testing*, vol. 25, no. 3, pp. 384–391, 2006.

[15] C. Subramani, V. S. Jamnik, and S. T. Mhaske, "Effect of attapulgite filler on the properties of nylon-6," *Polymer Composites*, vol. 29, no. 8, pp. 890–893, 2008.

[16] J. Shi, X. Yang, Q. Han, X. Wang, and L. Lu, "Polyurethane grafted attapulgite as novel fillers for nylon 6 nanocomposites," *Journal Wuhan University of Technology, Materials Science Edition*, vol. 26, no. 4, pp. 615–619, 2011.

[17] L. Xia, B. Shentu, and Z. Weng, "Effect of dendrimer-like PAMAM grafted attapulgite on the microstructure and morphology of Nylon-6," *Polymer Composites*, vol. 35, no. 4, pp. 627–635, 2014.

[18] N. Othman, S. N. A. Muttalib, and N. I. Ismail, "The effect of surface modification on the properties of palygorskite filled natural rubber nanocomposite," *Macromolecular Symposia*, vol. 371, no. 1, pp. 35–43, 2017.

[19] L. Shao, Z. Ji, J. Ma, C. Xue, and F. Deng, "Morphology and interaction of nanocomposite foams formed with organo-palygorskite and ethylene-vinyl acetate copolymers," *Polymer Bulletin*, vol. 74, no. 2, pp. 413–429, 2016.

[20] X. Lei, Y. Liu, and Z. Su, "Synthesis and characterization of organo-attapulgite/ polyaniline- dodecylbenzenesulfonic acid based on emulsion polymerization method," *Polymer Composites*, vol. 29, no. 3, pp. 239–244, 2008.

[21] J. Huang, X. Wang, Q. Jin, Y. Liu, and Y. Wang, "Removal of phenol from aqueous solution by adsorption onto OTMAC-modified attapulgite," *Journal of Environmental Management*, vol. 84, no. 2, pp. 229–236, 2007.

[22] Y. Chang, X. Lv, F. Zha, Y. Wang, and Z. Lei, "Sorption of p-nitrophenol by anion-cation modified palygorskite," *Journal of Hazardous Materials*, vol. 168, no. 2-3, pp. 826–831, 2009.

[23] B. Sarkar, M. Megharaj, Y. Xi, and R. Naidu, "Surface charge characteristics of organo-palygorskites and adsorption of p-nitrophenol in flow-through reactor system," *Chemical Engineering Journal*, vol. 185-186, pp. 35–43, 2012.

[24] B. Sarkar, Y. Xi, M. Megharaj, and R. Naidu, "Orange II adsorption on palygorskites modified with alkyl trimethylammonium and dialkyl dimethylammonium bromide—an isothermal and kinetic study," *Applied Clay Science*, vol. 51, no. 3, pp. 370–374, 2011.

[25] R. Dong, Y. Liu, X. Wang, and J. Huang, "Adsorption of sulfate ions from aqueous solution by surfactant-modified palygorskite," *Journal of Chemical & Engineering Data*, vol. 56, no. 10, pp. 3890–3896, 2011.

[26] H. A. Maddah, "Polypropylene as a promising plastic: a review," *American Journal of Polymer Science*, vol. 6, no. 1, pp. 1–11, 2016.

[27] M. Kawasumi, N. Hasegawa, M. Kato, A. Usuki, and A. Okada, "Preparation and mechanical properties of polypropylene–clay hybrids," *Macromolecules*, vol. 30, no. 20, pp. 6333–6338, 1997.

[28] J. A. Tarapow, C. R. Bernai, and V. A. Alvarez, "Mechanical properties of polypropylene/clay nanocomposites: effect of clay content, polymer/clay compatibility, and processing conditions," *Journal of Applied Polymer Science*, vol. 111, no. 2, pp. 768–778, 2009.

[29] J. J. Chen, J. Y. Chen, S. P. Zhu, Y. Cao, and H. L. Li, "Mechanical properties, morphology, and crystal structure of polypropylene/chemically modified attapulgite nanocomposites," *Journal of Applied Polymer Science*, vol. 121, no. 2, pp. 899–908, 2011.

[30] L. G. Furlan, C. I. Ferreira, C. Dal Castel et al., "Effect of processing conditions on the mechanical and thermal properties of high-impact polypropylene nanocomposites," *Materials Science and Engineering: A*, vol. 528, no. 22, pp. 6715–6718, 2011.

[31] W. C. Liaw, P. C. Huang, C. S. Chen, C. L. Lo, and J. L. Chang, "PPgMA/APTS compound coupling compatabilizer in PP/Clay hybrid nanocomposite," *Journal of Applied Polymer Science*, vol. 109, no. 3, pp. 1871–1880, 2008.

[32] N. G. Karsli and A. Aytac, "Effects of maleated polypropylene on the morphology, thermal and mechanical properties of short carbon fiber reinforced polypropylene composites," *Materials and Design*, vol. 32, no. 7, pp. 4069–4073, 2011.

[33] S. S. Ray and M. Okamoto, "Polymer/layered silicate nanocomposites: a review from preparation to processing," *Progress in Polymer Science*, vol. 28, no. 11, pp. 1539–1641, 2003.

[34] J. Hari, Z. Dominkovics, E. Fekete, and B. Pukánszky, "Kinetics of structure formation in PP/layered silicate nanocomposites," *Express Polymer Letters*, vol. 3, no. 11, pp. 692–702, 2009.

[35] J. Chen, J. Chen, S. Zhu, Y. Cao, and H. Li, "Mechanical properties, morphology, and crystal structure of polypropylene/chemically modified attapulgite nanocomposites," *Journal of Applied Polymer Science*, vol. 121, no. 2, pp. 899–908, 2011.

[36] Q. G. Tang, F. Wang, M. R. Tang, and F. Q. Zhang, "Effect of attapulgite modification on properties of polypropylene nanocomposites," *Applied Mechanics and Materials*, vol. 320, pp. 407–412, 2013.

[37] J. Chen, Y. Yu, J. Chen, H. Li, J. Ji, and D. Liu, "Chemical modification of palygorskite with maleic anhydride modified polypropylene: Mechanical properties, morphology, and crystal structure of palygorskite/polypropylene nanocomposites," *Applied Clay Science*, vol. 115, pp. 230–237, 2015.

[38] S. M. G. Antonio, G. C. P. Iván, and G. R. J. Luis, "Influence of chemically treated palygorskite over the rheological behavior of polypropylene nanocomposites," *Ingeniería, Investigación y Tecnología*, vol. 16, no. 4, pp. 491–501, 2015.

[39] A. Li, A. Wang, and J. Chen, "Studies on poly(acrylic acid)/attapulgite superabsorbent composite. I. Synthesis and characterization," *Journal of Applied Polymer Science*, vol. 92, no. 3, pp. 1596–1603, 2004.

[40] S. D. Ross and Q. Ketterings, "Recommended methods for determining soil cation exchange capacity," in *Cooperative Bulletin No. 493*.

[41] Y. Cai, J. Xue, and D. A. Polya, "A Fourier transform infrared spectroscopic study of Mg-rich, Mg-poor and acid leached palygorskites," *Spectrochimica Acta - Part A: Molecular and Biomolecular Spectroscopy*, vol. 66, no. 2, pp. 282–288, 2007.

[42] W. Yan, P. Yuan, M. Chen, L. Wang, and D. Liu, "Infrared spectroscopic evidence of a direct addition reaction between palygorskite and pyromellitic dianhydride," *Applied Surface Science*, vol. 265, pp. 585–590, 2013.

[43] S. Xue, M. Reinholdt, and T. J. Pinnavaia, "Palygorskite as an epoxy polymer reinforcement agent," *Polymer*, vol. 47, no. 10, pp. 3344–3350, 2006.

[44] A. Xue, S. Zhou, Y. Zhao, X. Lu, and P. Han, "Adsorption of reactive dyes from aqueous solution by silylated palygorskite," *Applied Clay Science*, vol. 48, no. 4, pp. 638–640, 2010.

[45] J. Madejová and P. Komadel, "Baseline studies of the clay minerals society source clays: infrared methods," *Clays and Clay Minerals*, vol. 49, no. 5, pp. 410–432, 2001.

[46] F. C. Liu, E. H. Han, W. Ke et al., "Polar influence of the organic modifiers on the structure of montmorillonite in epoxy nanocomposites," *Journal of Materials Science and Technology*, vol. 29, no. 11, pp. 1040–1046, 2013.

[47] L. A. Savas and S. Savas, "Interrelation between contact angle and interlayer spacing of montmorillonite clay used in polymer nanocomposites," *Polymers and Polymer Composites*, vol. 21, no. 3, pp. 129–132, 2013.

[48] Z. Zhang, J. Zhang, L. Liao, and Z. Xia, "Synergistic effect of cationic and anionic surfactants for the modification of Ca-montmorillonite," *Materials Research Bulletin*, vol. 48, no. 5, pp. 1811–1816, 2013.

[49] Y. Hu, X. Liu, and Z. Xu, "Role of crystal structure in flotation separation of diaspore from kaolinite, pyrophyllite and illite," *Minerals Engineering*, vol. 16, no. 3, pp. 219–227, 2003.

[50] J. Ballah, M. Chamerois, S. Durand-Vidal, N. Malikova, P. Levitz, and L. J. Michot, "Effect of chemical and geometrical parameters influencing the wettability of smectite clay films," *Colloids and Surfaces A: Physicochemical and Engineering Aspects*, vol. 511, pp. 255–263, 2016.

[51] M. Hetzer and D. De Kee, "Wood/polymer/nanoclay composites, environmentally friendly sustainable technology: A review," *Chemical Engineering Research and Design*, vol. 86, no. 10, pp. 1083–1093, 2008.

[52] E. Manias, A. Touny, L. Wu, K. Strawhecker, B. Lu, and T. C. Chung, "Polypropylene/montmorillonite nanocomposites. Review of the synthetic routes and materials properties," *Chemistry of Materials*, vol. 13, no. 10, pp. 3516–3523, 2001.

[53] H. Baniasadi, A. Ramazani S.A., and S. Javan Nikkhah, "Investigation of in situ prepared polypropylene/clay nanocomposites properties and comparing to melt blending method," *Materials and Design*, vol. 31, no. 1, pp. 76–84, 2010.

[54] C. B. Hedley, G. Yuan, and B. K. G. Theng, "Thermal analysis of montmorillonites modified with quaternary phosphonium and ammonium surfactants," *Applied Clay Science*, vol. 35, no. 3-4, pp. 180–188, 2007.

[55] J. U. Calderon, B. Lennox, and M. R. Kamal, "Thermally stable phosphonium-montmorillonite organoclays," *Applied Clay Science*, vol. 40, no. 1–4, pp. 90–98, 2008.

[56] L. Zhang, C. Wan, and Y. Zhang, "Investigation on the multiwalled carbon nanotubes reinforced polyamide 6/polypropylene composites," *Polymer Engineering and Science*, vol. 49, no. 10, pp. 1909–1917, 2009.

[57] L. N. Carli, J. S. Crespo, and R. S. Mauler, "PHBV nanocomposites based on organomodified montmorillonite and halloysite: the effect of clay type on the morphology and thermal and mechanical properties," *Composites A: Applied Science and Manufacturing*, vol. 42, no. 11, pp. 1601–1608, 2011.

[58] E. Bilotti, R. Zhang, H. Deng, F. Quero, H. R. Fischer, and T. Peijs, "Sepiolite needle-like clay for PA6 nanocomposites: an alternative to layered silicates?" *Composites Science and Technology*, vol. 69, no. 15-16, pp. 2587–2595, 2009.

15

Fabrication of Core-Shell PLGA-Chitosan Microparticles Using Electrospinning: Effects of Polymer Concentration

Nguyen Thi Hiep,[1] Nguyen Dai Hai,[2,3] and Vo Van Toi[1]

[1] *Tissue Engineering and Regenerative Medicine Laboratory, Department of Biomedical Engineering, International University, Vietnam National University, Ho Chi Minh City 700000, Vietnam*
[2] *Institute of Applied Materials Science, Vietnam Academy of Science and Technology, 01 Mac Dinh Chi, District 1, Ho Chi Minh City, Vietnam*
[3] *Graduate University of Science and Technology, Vietnam Academy of Science and Technology, Hanoi, Vietnam*

Correspondence should be addressed to Nguyen Thi Hiep; nthiep1981@gmail.com

Academic Editor: Zhe Qiang

This investigation aims to fabricate the core-shell microparticles composed of poly(lactic-co-glycolic acid) and chitosan (PLGA-CS MPs) using electrospinning. The challenge of using electrospinning is that it has many parameters which change product outcome if any single parameter is changed. However, the advantage of this system is that we can fabricate either micro/nanofibers or micro/nanoparticles. To learn about the effect of liquid concentration, the electrospinning parameters (voltage, needle sizes, distance from needle to collector, and ejection speed) were fixed while the concentration of PLGA or chitosan was varied. The results showed that PLGA microparticles can be fabricated successfully when the concentration of PLGA is smaller than 10 wt%. Presence of the chitosan shell was confirmed by zeta potential measurements, FT-IR, optical observation, and fluorescence observation. Thickness of the chitosan shell can be controlled by changing the concentration of chitosan and measured by fluorescamine labeling method. Moreover, SEM observation showed that concentration of chitosan affected the size of PLGA-CS microparticles. The MTT (3-(4,5-dimethylthiazol-2-yl)-2,5-diphenyl tetrazolium bromide) assay test showed that PLGA-CS microparticles possess excellent biocompatibility.

1. Introduction

Poly(lactic-co-glycolic acid) microparticles (PLGA MPs) attract considerable therapeutic interest in drug and antigen delivery systems [1–3], mainly because of their sustained release, localized release, and vivo stabilization [3]. However, PLGA is a hydrophobic polymer. Thus, it is difficult to be submerged into water. So, a hydrophilic surface modification is needed to improve the effectiveness of microparticulate delivery systems. To modify the PLGA MPs surface, Pluronics® [4] or chitosan was suggested as a suitable materials to coat conventional PLGA particles [5–8]. Chitosan is normally selected to coat PLGA MPs for drug/genes delivery system because of its hydrophilic property, cationic charge, biodegradability, and mucoadhesive properties [6–10]. There are many ways to coat PLGA surface with chitosan such as solvent double or triple emulsion method [11], static micromixer of O/W-dispersion of PLGA and chitosan [5, 12], droplet microfluidic solvent evaporation and extraction [13], and supercritical fluid technology [14]. Among these methods, the double or triple emulsion method is a conventional approach to coat chitosan onto PLGA microparticles [15]. However, this method has some disadvantages such as (1) organic solvents and shear stress can denature or deactivate the PLGA partly during the preparation process; (2) PLGA particles also generate an acidic internal environment after they gradually hydrolyzed into lactic acid and glycolic acid, which is detrimental to biopolymer stability; and (3) the encapsulation efficiency of PLGA in the particles is essentially low, thus resulting in the consumption of a large quantity of PLGA in the preparation.

Table 1: Parameters of electrospinning.

	PLGA 85/15 (wt%)	Chitosan (CS, wt%)	Needle (gauge)	Distance (cm)	Voltage (kV)	Ejection speed (ml/h) PLGA	Ejection speed (ml/h) CS
PLGA1	5	—	25	15	27	0.5	—
PLGA2	7	—	25	15	27	0.5	—
PLGA3	10	—	25	15	27	0.5	—
PLGA-CS1	5	0.01	25	15	25	0.5	0.5
PLGA-CS2	5	0.15	25	15	25	0.5	0.5
PLGA-CS3	5	0.3	25	15	25	0.5	0.5

Electrosprayer is a well-controlled method to generate PLGA microparticles [4, 16]. Surface hydrophilization of hydrophobic PLGA MPs using electrosprayer has been reported by blending with hydrophilic polymer, Pluronic [4]. Moreover, electrosprayer has been known as a high technology to fabricate core-shell electrospun fibers using the core-shell spinneret needle [16–18]. Thus, the purpose of this study is to employ electrospinning machine in fabricating the core-shell poly(lactic-co-glycolic acid)-chitosan microparticles (PLGA-CS MPs) with the spinneret needle's assistance. The advantage of this approach in comparison with conventional methods is the easiness to modify the hydrophilic chitosan onto the hydrophobic PLGA MPs and the thickness of chitosan by changing the concentration of the polymer. In addition, this process, unlike conventional emulsion, prevents polymer waste in washing steps. Solvents are removed during operation time.

The main goal of this investigation is to study about the effect of PLGA and chitosan concentration on fabricating PLGA-CS MPs. Morphology of PLGA-CS MPs was observed by SEM. FT-IR spectrum, zeta potential, and fluorescent observation were employed to evaluate the coating achievements. The biocompatibility of PLGA-CS microparticles was evaluated by cytotoxicity test and cell viability test. This approach is not only fast and safe but also has a high potential for biomedical applications.

2. Materials and Methods

2.1. Materials. Chitosan (C3646, ≥75% deacetylated), poly(lactic-co-glycolic) acid (PLGA) (85 : 15), hexamethyldisilazane (HDMS), tetrahydrofuran (THF, minimum 99%), and dimethylformamide (DMF, 99%) were purchased from Sigma-Aldrich, USA. Acetic acid (CH_3COOH, 99.0%) was purchased from Duksan Pure Chemical Co., Korea. Glutaraldehyde was purchased from DaeJung Co., Korea. Dimethylsulfoxide (DMSO, 99.0%, Samchun Pure Chemical Co., Ltd., Korea), ethanol (Merck, Germany), fetal bovine serum (FBS), PS (penicillin/streptomycin (antibiotics)), Dulbecco's phosphate buffered saline (D-PBS) without calcium or magnesium, 3-(4,5-dimethylthiazol-2-yl)-2,5-diphenyl tetrazolium bromide (MTT), and trypsin-EDTA were purchased from GIBCO (Carlsbad, CA). The L-929 cells line was obtained from ATCC, USA.

2.2. Methods

2.2.1. Fabrication of PLGA-CS MPs. Before applying polymer solutions in the electrospinning system, chitosan solutions were prepared by dissolving chitosan in 1.5 wt% acetic acid where PLGA solutions were prepared in THF:DMF with ratio 1 : 1. Then, the PLGA-CS MPs were formed by ejecting chitosan solution into the shell of the core-shell spinneret needle while PLGA solution was ejected to the core of spinneret as shown in Figure 1. To investigate the effect of the concentration of polymer, other parameters of electrospinning such as the voltage, the distance from needle to collector, and the speed of ejecting solution were fixed (see Table 1).

Table 2: Particles size and zeta potential of PLGA-CS MPs.

System	Particles size (μm)	Zeta potential (mV)
PLGA1	0.2–1	−4.8 ± 0.1
PLGA2	0.2–2	−4.6 ± 0.2
PLGA3	—	−4.6 ± 0.2
PLGA-CS1	0.2–3	+14.9 ± 0.5
PLGA-CS2	2–10	+19.1 ± 0.4
PLGA-CS3	3–20	+37.0 ± 0.4

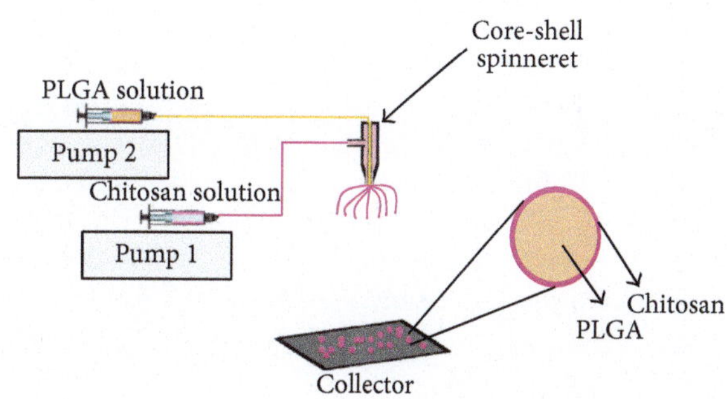

Figure 1: Scheme of electrospinning process for fabrication of microparticles.

2.2.2. Characterization of PLGA-CS MPs. The morphology of the core-shell PLGA-CS MPs was examined using scanning electron microscope (SEM, SM-65F, JEOL, Japan). The particle size distribution was determined by counting on SEM images (see Table 2).

Presence of chitosan on PLGA surface was detected by the zeta potential measurement, the FT-IR analysis, the fluorescent observation, the optical observation, and fluorescamine method. The zeta potential measurements of PLGA MPs and PLGA-CS MPs were taken using the zeta potential analyzer. The infrared spectra of PLGA-CS MPs were analyzed by attenuated reflectance Fourier transform spectroscopy (FT-IR) (Spectrum GX, PerkinElmer, USA) over a wavelength range between 4,000 and 500 cm^{-1}, and the spectra were collected from 64 scans with a resolution of 4 cm^{-1}. Optical images were observed by immersing MPs in water for 24 hours prior to taking photos using a microscope. To obtain fluorescent chitosan, the PLGA-CS MPs were immersed in a 0.1 mg/ml solution of tetramethylrhodamine (TRITC) in ethanol/water/sodium bicarbonate (0.1 M) and allowed to react for 24 h. Optical images and confocal images of PLGA-CS MPs were observed with an inverted microscope (Leica, Germany) and a confocal microscope (Olympus, FV 1000). To determine the amount of chitosan coated on the surface of PLGA-CS MPs, fluorescamine (Acros Organics; Morris Plains, NJ) was applied [19]. Briefly, the PLGA-CS suspension was obtained by adding 50 mg PLGA-CS MPs to 100 μl distilled water. Then, the PLGA-CS suspensions were centrifuged at 16,000 g in a microcentrifuge (Microcentrifuge, Thermo Scientific). 100 μL fluorescamine in DMSO solution (2.0 g/L) was added to 20 μL of the supernatant in a black 96-well fluorescent detection microplate. After a three-hour reaction in a dark room, fluorescence was measured at an excitation wavelength of 390 nm and emission wavelength of 515 nm using a microplate spectrofluorometer reader. A chitosan calibration curve was run simultaneously with each experiment. The amount of chitosan attached to the PLGA-CS MPs was calculated by dissolving PLGA-CS MPs in acetic acid. The fluorescence intensity was converted to the concentration of chitosan by comparison with a standard curve of pure chitosan. Then, the mass of chitosan coated PLGA-CS MPs was determined as shown in Figure 5.

2.2.3. *In Vitro Experiments*

Sample Sterilization. In order to prepare samples for in vitro experimental processes, the microparticles were placed on an aluminum foil and sterilized with UV irradiation for 15 minutes.

Cell Maintenance. The fibroblast cells (L-929) were subcultured in RPMI-1640 medium and supplemented with 10% fetal bovine serum (FBS), 1% penicillin-streptomycin (P/S), and 5% FBS. Cells were incubated at 37°C in a humidified atmosphere of 5% CO_2 (incubator, ASTEC, Japan). Cells were dissociated with trypsin-DETA (GIBCO), centrifuged, and then resuspended in medium. The culture medium was changed every 2 days, supplying enough nutrients to keep cells healthy.

Cytotoxicity Test. To confirm the biocompatibility of microparticles, the cytotoxicity of PLGA, PLGA-CS1, PLGA-CS2, and PLGA-CS3 MPs was evaluated using MTT assay method described in our previous reports [20–24]. Briefly, the fibroblast cells were seeded with a density of 1×10^4 cells/ml in a 96-well tissue culture plate and incubated for 24 h at 37°C and 5% CO_2. Diluted extract solutions (200 μl) of every well at various concentrations (100, 50, 25, 12.5, and 0%) were then added. The fibroblast cells were treated for 7 days and then 20 μl of 5 mg/ml 3-(4,5-dimethylthiazol-2-yl)-2, 5-diphenyltetrazolium bromide (MTT) solution was added to each well. The plate was then incubated for 4 h. After the medium containing MTT solution was removed, DMSO was added to each well to dissolve any insoluble formazan crystals. The optical density (OD) was measured at the wavelength of 595 nm with an enzyme-linked immunosorbent assay (ELISA) reader (Infinite F50, Tecan Austria GmbH, Austria).

Cell Proliferation Test. In order to evaluate cell proliferation, the fibroblast cells L929 (at a cell density of 10^4 cells/1 ml in RPMI-1640) were seeded on PLGA and PLGA-CS MPs and cultured for various times at 37°C in humidified atmosphere with 5% CO_2. Then, cell proliferation was examined using MTT assay [23, 24].

3. Results and Discussion

PLGA-CS microparticles have been applied for biomacromolecule delivery because of their advantages such as the sustainable and controllable release of PLGA synthesized biopolymer, their cationic charge, biodegradability and mucoadhesive properties of natural chitosan [6, 25]. So far, there are several methods used for fabricating PLGA-CS MPs; W/O emulsion is kind of a famous method [1]. However, W/O method is a consumable method where both biopolymer and drug are easy to be lost in the fabricating process. Hence, we are trying to fabricate PLGA-CS MPs by using electrospinning. Electrospinning is a simple and easy method for fabricating nano/microfibers but it has many parameters to be controlled. Thus, in this investigation, the electrospinning parameters were fixed as mentioned in Table 1 while the concentration of polymers was changed. The purpose of this investigation is to obtain microparticles from electrospinning machine. So the applied voltage of electrospinning machine was changed due to the changing polymer solutions. For fabricating of PLGA MPs, the voltage was 27 kV; this high voltage was suitable to obtain the particles; otherwise, the fibers formed. For fabricating the core-shell PLGA-CS MPs, the suitable voltage was 25 kV.

To investigate the effect of PLGA concentration on fabricating PLGA MPs, the following electrospinning parameters were fixed as follows: voltage: 27 kV, distance from needle to collector: 15 cm, needle size: 25 gauge, and speed of ejecting solution: 0.5 ml/h (see Table 1) while the concentration of PLGA solutions was selected as 5 wt%, 7 wt%, and 10 wt%, respectively. Results show that the size and shape of the particles depend on PLGA concentration. Specifically, with concentration of 5 wt% and 7 wt%, the spheres are fabricated (named as PLGA1 MPs and PLGA3 MPs, resp.); however, with concentration of 10 wt%, the fibers were formed (named PLGA3 MPs). Overall the sizes of particles were changed according to changing PLGA concentration. Indeed, when concentration of PLGA increased from 5 wt% to 7 wt%, the particle sizes increased from approximately 0.2–1 μm (Figures 2(a1) and 2(a2)) to 0.2–2 μm (Figures 2(b1) and 2(b2)). However, at 10 wt% of PLGA concentration, the mixing of nanofibers and microparticles formed (Figures 2(c1) and 2(c2)). Therefore, the concentration of 5 wt% is going to be used for the fabricating core-shell PLGA-CS MPs.

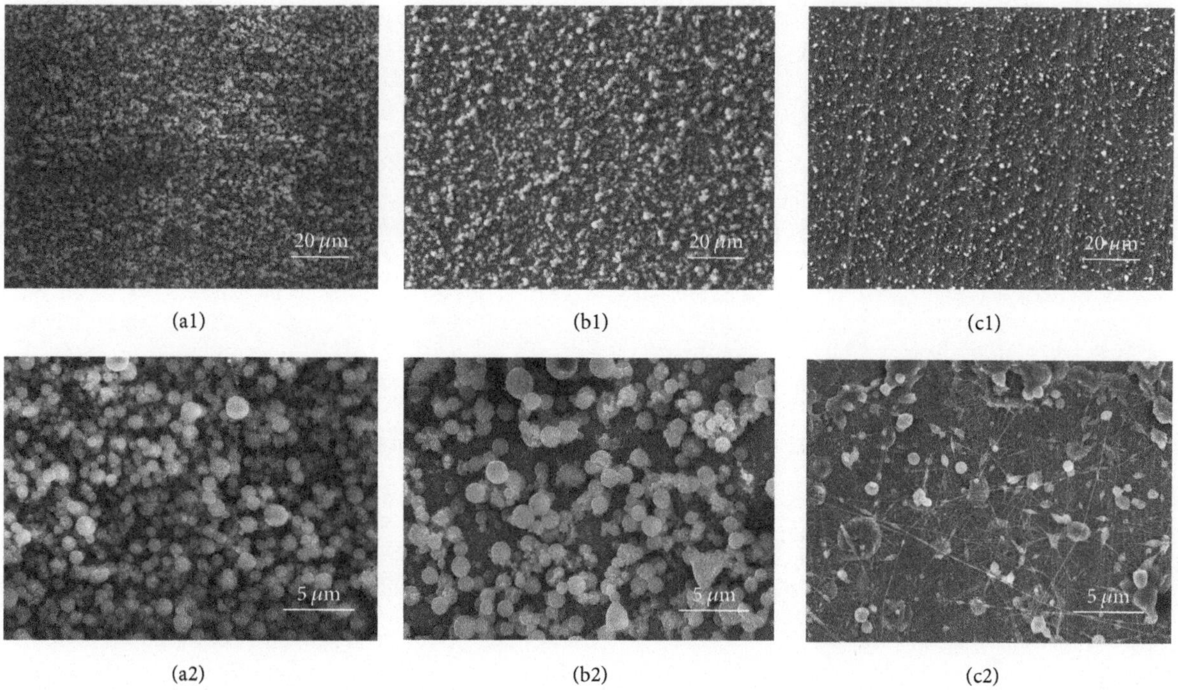

Figure 2: SEM observation of PLGA MPs in different concentration of PLGA: PLGA1 (a1, a2), PLGA2 (b1, b2), and PLGA3 (c1, c2).

To fabricate the core-shell PLGA-CS MPs, 5 wt% of PLGA concentration was chosen. The initial electrospinning parameters were fixed as before (distance from needle to collector: 15 cm, needle size: 25 gauge, and speed of ejecting solution: 0.5 ml/h). However, the voltage was decreased from 27 kV to 25 kV to optimize the formation of particles. Similar to the investigation of PLGA MPs, here, the PLGA-CS MPs sizes and shapes were controlled solely by the concentration of chitosan. Figure 3 shows SEM morphology of PLGA-CS MPs with 5 wt% PLGA for different chitosan concentration as follows: 0.01 wt% (Figures 3(a1) and 3(a2)), 0.15 wt% (Figures 3(b1) and 3(b2)), and 0.3 wt% (Figures 3(c1) and 3(c2)), dissolved in 1.5 wt% acetic acid. For 5 wt% of PLGA and 0.01 wt% of chitosan (PLGA-CS1), particle sizes of core-shell PLGA-CS1 MPs were in the range of 0.2–3 μm. In addition, Figures 3(a1) and 3(a2) show that the chitosan-shells of PLGA-CS1 MPs were melted and deposited around the PLGA-core. However, when the concentration of chitosan increased fifteenfold, that is, for 5 wt% of PLGA and 0.15 wt% of chitosan (PLGA-CS2), the particle sizes increased to around 2–10 μm, and the surface of PLGA-CS2 MPs became porous. For 5 wt% of PLGA and 0.3 wt% of chitosan (PLGA-CS3), the morphology of PLGA-CS3 MPs was similar to that of PLGA-CS2 MPs with porous structure on the surface and the size of PLGA-CS3 MPs increased dramatically to around 3–20 μm.

In addition, the zeta potential of the microparticles increased with higher chitosan concentrations in the external aqueous. It increased from approximately -4.8 ± 0.1 mV (chitosan-free PLGA particles) to $+14.9 \pm 0.5$ mV when the chitosan concentration in the shell of the core-shell spinneret was 0.01 wt%. The zeta potential increased significantly toward positive charge of $+19.1 \pm 0.4$ mV and $+37.0 \pm 0.4$ mV when the chitosan concentration was 0.15 wt% and 0.3 wt%, respectively (Table 2). The higher the degree of zeta potential, the more the stability of particles. This result shows that the increasing of chitosan concentration creates a stability of PLGA-CS MPs in solution [26].

Investigations were conducted to determine if PLGA-CS MPs were indeed composed of PLGA and CS. The presence of PLGA and chitosan in the core-shell PLGA-CS MPs was confirmed by analyzing the presence of specific peaks of single pure materials in the PLGA-CS spectrum such as the strong characteristic absorption bands at about 1746 cm^{-1} attributed to the stretching vibration of C–O bond, and the bands at 1130 cm^{-1} and 1412 cm^{-1} arose from C–O bond and methyl group C–H bond of PLGA, respectively [27]. Also in Figure 4 the absorptions bands of chitosan are shown: 1094 cm^{-1} (O–C stretch), 1159 cm^{-1} (bridge–O stretch), 1313–1481 cm^{-1} (–CH bend), 1627 cm^{-1} (C=O stretch), 2923–2867 cm^{-1} (C–H stretch), 1598–1600 cm^{-1} (N–H bend), and 3431 cm^{-1} (O–H stretch) [28, 29]. Hence, the FT-IR spectrum (Figure 4) confirmed that the presence of chitosan in PLGA-CS MPs and thus PLGA-CS MPs might be fabricated successfully.

To determine if PLGA-CS MPs were indeed coated by chitosan, the fluorescent image of chitosan, optical images of PLGA MPs and PLGA-CS MPs, and weight of chitosan in PLGA-CS MPs were observed and showed in Figure 5. The series fluorescent images (Figures 5(a)–5(d)) proved that the PLGA MPs were coated by a thin layer of chitosan. Optical images show that the surface of PLGA-CS MP was rough (Figure 5(f)) compared to the smooth surface of PLGA MP (Figure 5(e)). Figure 5(g) is the graph of weight change of chitosan, which shows that chitosan weight increases if the

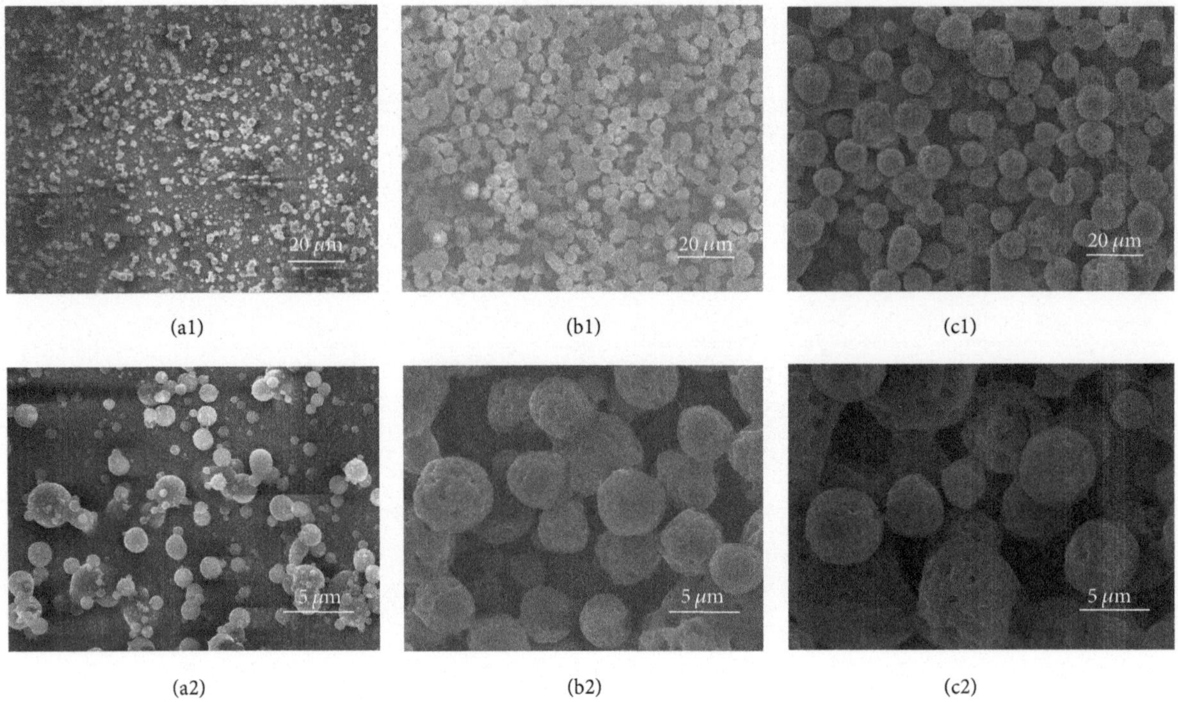

FIGURE 3: SEM observation of PLGA-CS MPs in different concentration of chitosan: PLGA-CS1 (a1, a2), PLGA-CS2 (b1, b2), and PLGA-CS3 (c1, c2).

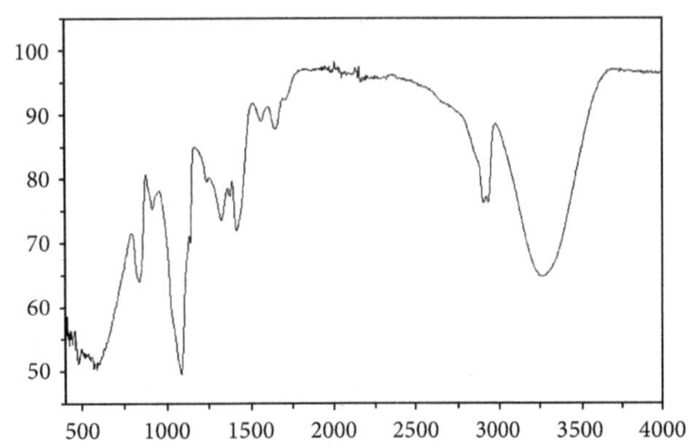

FIGURE 4: FT-IR spectrum of PLGA-CS MPs.

concentration of chitosan increases (PLGA-CS3 > PLGA-CS2 > PLGA-CS1).

The cytotoxicity test is important for new biomaterials to ensure their biocompatibility before conducting in vitro tests. Although both PLGA and CS are known for excellent biocompatibility, in this study, the cytotoxicity of PLGA, PLGA-CS1, PLGA-CS2, and PLGA-CS3 microparticles was examined using MTT assay to ensure safe method. MTT assay is a colorimetric assay that measures the optical density of purple color after the crystallized formazan dyes are dissolved using DMSO. The crystallized formazan was created by yellow MTT solution. The principle of MTT can be briefly described as follows: the MTT enters the cells and passes into the mitochondria where it is reduced to an insoluble and colored (dark purple) formazan product. Since the reduction of MTT can only occur in metabolically active cells, the level of activity is a measure of the viability of the cells. According to the ISO-10993-5, the MTT assay refers to MTT cytotoxicity employed to evaluate new composite materials as previously reported [20]. In this investigation, MTT assay was employed to examine the cytotoxicity and cell proliferation on PLGA and PLGA/CS MPs. The results showed that all of samples had excellent biocompatibility. Indeed, the cell viability of the 100% extracted solution for all samples (PLGA, PLGA-CS1, PLGA-CS2, and PLGA-CS3) was over 80% and not significantly different from each other as shown in Figure 6. Thus, the cytotoxicity test revealed that all samples have nontoxic. Washing steps to remove residual solvent are unnecessary in this method. The products are therefore safe for use in biomedical applications.

From the cell proliferation test, Figure 7 shows the optical density of purple color that is produced by DMSO after dissolving the deposited crystallized formazan dyes on samples [23, 24]. Because MTT detects the activity power of the cells, the results of MTT assay in this examination can be used to evaluate the cell health. Therefore, the viability of cells on the PLGA, PLGA-CS1, PLGA-CS2, and PLGA-CS3 MPs could be compared using the MTT assay methods [20]. In this examination, microparticles were placed on the aluminum foil and were cultured for 1, 3, 7, and 14 days in static environments in order to observe the growth behavior of the cells as shown in Figure 7. Overall, in comparison with the PLGA MPs group, the optical density of the PLGA-CS MPs groups decreased when the chitosan concentration increased. Thus results are similar with several reports which indicated polysaccharides such as chitosan derivatives were able to inhibit cell proliferation. However, the investigator can

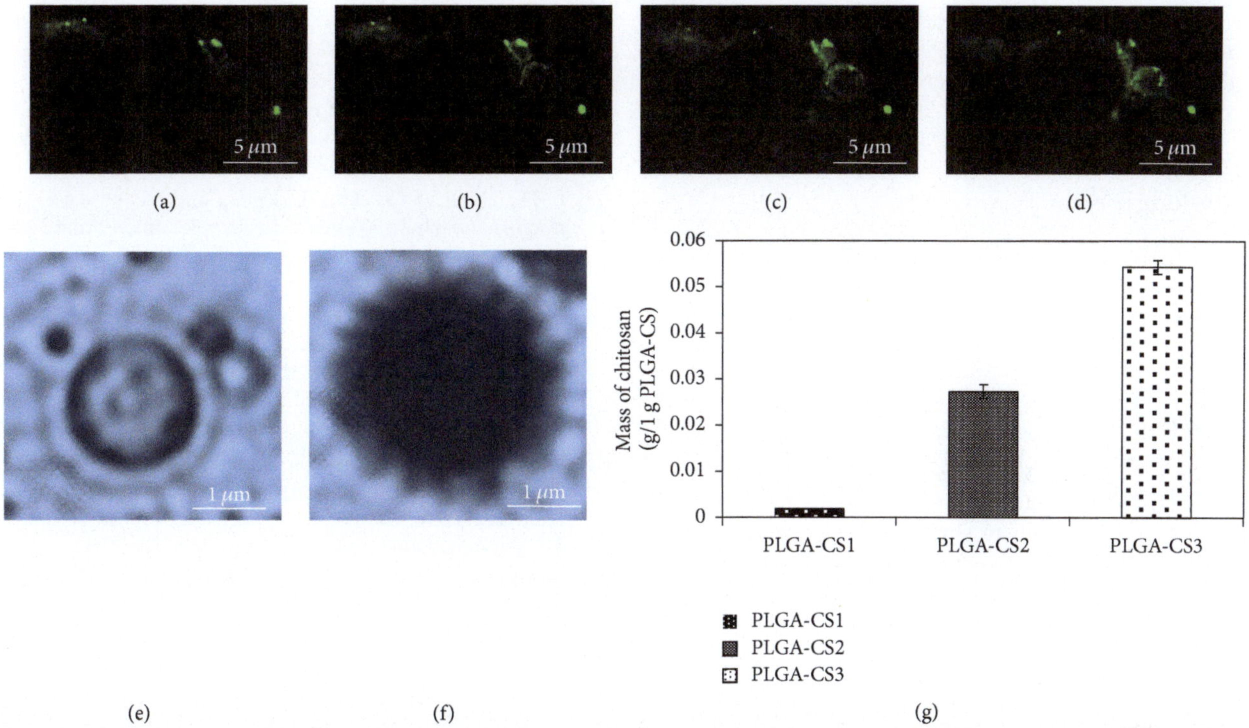

Figure 5: Series of images of fluorescein-labeled chitosan coating PLGA was visualized by confocal laser scanning microscopy (a–d). Microscope observation of PLGA MP (e) and PLGA-CS2 MP (f) immersed in distilled water for 24 hours. Mass of chitosan coated PLGA MPs in 1 gram of PLGA-CS MPs.

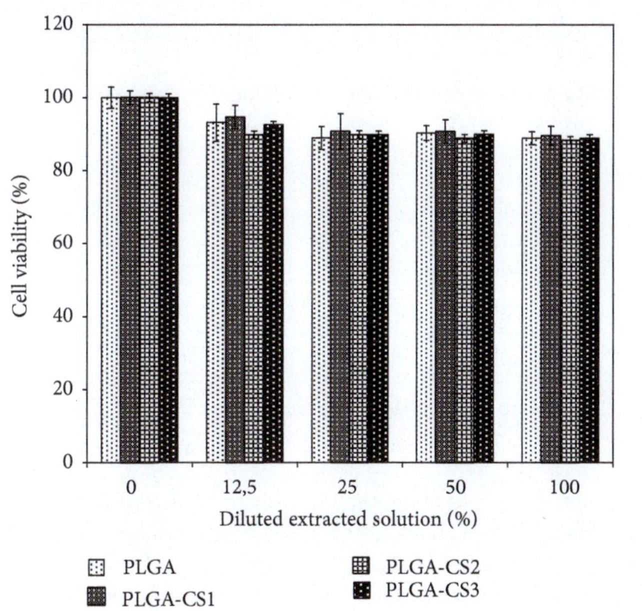

Figure 6: Percentage of cell viability on the series of the diluted extracted solution of PLGA MPs and PLGA-CS MPs.

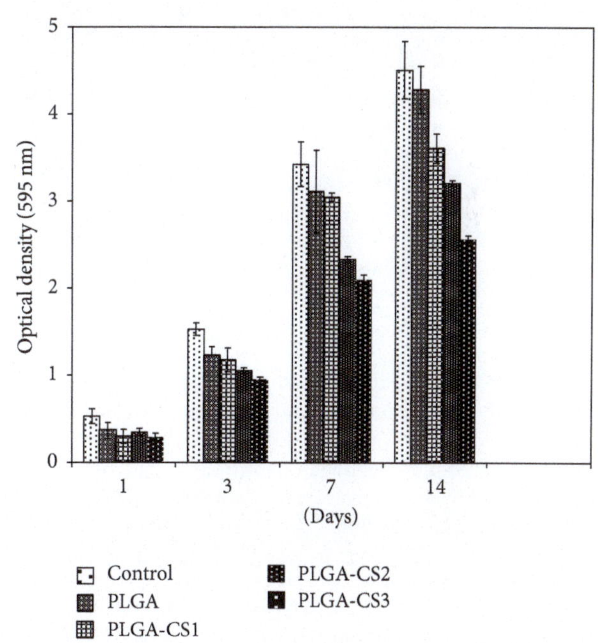

Figure 7: Cellular proliferation on the control, PLGA MPs, and PLGA-CS MPs 1, 3, 7, and 14 days cultured.

manipulate and take advantage of this for making a stable carrier to carry an anticancer drug [30].

4. Conclusion

In this study, the core-shell PLGA-CS MPs have been fabricated successfully using electrospinning machine. The results revealed that the success of PLGA MPs fabrication was determined by concentration of PLGA solution (below 10 wt%); the thickness of the chitosan shell can be controlled by the concentration of chitosan. However, PLGA-CS MPs show excellent biocompatibility and cell proliferation. Hence, the

PLGA MPs or PLGA-CS MPs produced by electrospinning machine promise wide and high potential applications in biomedical fields due to its convenience, cost effectiveness, and excellent biocompatibility.

Acknowledgments

The research facility was supported by Vietnam National University, Ho Chi Minh City, under Grant no. 1161/QĐ-ĐHQG-KHCN.

References

[1] T.-Q. Bao, N.-T. Hiep, Y.-H. Kim, H.-M. Yang, and B.-T. Lee, "Fabrication and characterization of porous poly(lactic-co-glycolic acid) (PLGA) microspheres for use as a drug delivery system," *Journal of Materials Science*, vol. 46, no. 8, pp. 2510–2517, 2011.

[2] C. Vilos and L. A. Velasquez, "Therapeutic strategies based on polymeric microparticles," *Journal of Biomedicine and Biotechnology*, vol. 2012, Article ID 672760, 2012.

[3] M. Figueiredo and R. Esenaliev, "PLGA Nanoparticles for Ultrasound-Mediated Gene Delivery to Solid Tumors," *Journal of Drug Delivery*, vol. 2012, pp. 1–20, 2012.

[4] A. Seth and D. S. Katti, "A one-step electrospray-based technique for modulating morphology and surface properties of poly(lactide-co-glycolide) microparticles using Pluronics®," *International Journal of Nanomedicine*, vol. 7, pp. 5129–5136, 2012.

[5] S. Fischer, C. Foerg, S. Ellenberger, H. P. Merkle, and B. Gander, "One-step preparation of polyelectrolyte-coated PLGA microparticles and their functionalization with model ligands," *Journal of Controlled Release*, vol. 111, no. 1-2, pp. 135–144, 2006.

[6] N. Nafee, S. Taetz, M. Schneider, U. F. Schaefer, and C.-M. Lehr, "Chitosan-coated PLGA nanoparticles for DNA/RNA delivery: effect of the formulation parameters on complexation and transfection of antisense oligonucleotides," *Nanomedicine: Nanotechnology, Biology, and Medicine*, vol. 3, no. 3, pp. 173–183, 2007.

[7] L. Chronopoulou, M. Massimi, M. F. Giardi et al., "Chitosan-coated PLGA nanoparticles: a sustained drug release strategy for cell cultures," *Colloids and Surfaces B: Biointerfaces*, vol. 103, pp. 310–317, 2013.

[8] F. Ganji and M. J. Abdekhodaie, "Chitosan-g-PLGA copolymer as a thermosensitive membrane," *Carbohydrate Polymers*, vol. 80, no. 3, pp. 740–746, 2010.

[9] S.-E. Jin, C.-K. Kim, and Y.-B. Kim, "Cellular delivery of cationic lipid nanoparticle-based SMAD3 antisense oligonucleotides for the inhibition of collagen production in keloid fibroblasts," *European Journal of Pharmaceutics and Biopharmaceutics*, vol. 82, no. 1, pp. 19–26, 2012.

[10] S. Jafarinejad, K. Gilani, E. Moazeni, M. Ghazi-Khansari, A. R. Najafabadi, and N. Mohajel, "Development of chitosan-based nanoparticles for pulmonary delivery of itraconazole as dry powder formulation," *Powder Technology*, vol. 222, pp. 65–70, 2012.

[11] Y.-Y. Yang, H.-H. Chia, and T.-S. Chung, "Effect of preparation temperature on the characteristics and release profiles of PLGA microspheres containing protein fabricated by double-emulsion solvent extraction/evaporation method," *Journal of Controlled Release*, vol. 69, no. 1, pp. 81–96, 2000.

[12] M. L. Manca, G. Loy, M. Zaru, A. M. Fadda, and S. G. Antimisiaris, "Release of rifampicin from chitosan, PLGA and chitosan-coated PLGA microparticles," *Colloids and Surfaces B: Biointerfaces*, vol. 67, no. 2, pp. 166–170, 2008.

[13] L.-H. Hung, S.-Y. Teh, J. Jester, and A. P. Lee, "PLGA micro/nanosphere synthesis by droplet microfluidic solvent evaporation and extraction approaches," *Lab on a Chip*, vol. 10, no. 14, pp. 1820–1825, 2010.

[14] L. Casettari, E. Castagnino, S. Stolnik, A. Lewis, S. M. Howdle, and L. Illum, "Surface characterisation of bioadhesive PLGA/chitosan microparticles produced by supercritical fluid technology," *Pharmaceutical Research*, vol. 28, no. 7, pp. 1668–1682, 2011.

[15] R. Jalil and J. R. Nixon, "Biodegradable poly(lactic acid) and poly(lactide-co-glycolide) microcapsules: problems associated with preparative techniques and release properties," *Journal of Microencapsulation*, vol. 7, no. 3, pp. 297–325, 1990.

[16] B. Almería, W. Deng, T. M. Fahmy, and A. Gomez, "Controlling the morphology of electrospray-generated PLGA microparticles for drug delivery," *Journal of Colloid and Interface Science*, vol. 343, no. 1, pp. 125–133, 2010.

[17] L. T. Son and K. Takaomi, "Hollow-fiber membrane absorbents embedded molecularly imprinted polymeric spheres for bisphenol A target," *Journal of Membrane Science*, vol. 384, no. 1-2, pp. 117–125, 2011.

[18] C. Feng, K. C. Khulbe, T. Matsuura, S. Tabe, and A. F. Ismail, "Preparation and characterization of electro-spun nanofiber membranes and their possible applications in water treatment," *Separation and Purification Technology*, vol. 102, pp. 118–135, 2013.

[19] R. Håkanson, L.-I. Larsson, and F. Sundler, "Fluorescamine: A novel reagent for the histochemical detection of amino groups," *Histochemistry*, vol. 39, no. 1, pp. 15–23, 1974.

[20] N. T. Hiep and B.-T. Lee, "Electro-spinning of PLGA/PCL blends for tissue engineering and their biocompatibility," *Journal of Materials Science: Materials in Medicine*, vol. 21, no. 6, pp. 1969–1978, 2010.

[21] T. H. Nguyen and B. T. Lee, "Fabrication and characterization of cross-linked gelatin electro-spun nano-fibers," *Journal of Biomedical Science and Engineering*, vol. 3, no. 12, pp. 1117–1124, 2010.

[22] T.-H. Nguyen and B.-T. Lee, "The effect of cross-linking on the microstructure, mechanical properties and biocompatibility of electrospun polycaprolactone-gelatin/PLGA-gelatin/PLGA-chitosan hybrid composite," *Science and Technology of Advanced Materials*, vol. 13, no. 3, Article ID 035002, 2012.

[23] T.-H. Nguyen, T. Q. Bao, I. Park, and B.-T. Lee, "A novel fibrous scaffold composed of electrospun porous poly(ε-caprolactone) fibers for bone tissue engineering," *Journal of Biomaterials Applications*, vol. 28, no. 4, pp. 514–528, 2013.

[24] T.-H. Nguyen and B.-T. Lee, "In vitro and in vivo studies of rhBMP2-coated PS/PCL fibrous scaffolds for bone regeneration," *Journal of Biomedical Materials Research—Part A*, vol. 101, no. 3, pp. 797–808, 2013.

[25] A. Grenha, C. Remuñán-López, E. L. S. Carvalho, and B. Seijo, "Microspheres containing lipid/chitosan nanoparticles complexes for pulmonary delivery of therapeutic proteins," *European Journal of Pharmaceutics and Biopharmaceutics*, vol. 69, no. 1, pp. 83–93, 2008.

[26] T. A. Ahmed and B. M. Aljaeid, "Preparation, characterization, and potential application of chitosan, chitosan derivatives, and chitosan metal nanoparticles in pharmaceutical drug delivery," *Drug Design, Development and Therapy*, vol. 10, pp. 483–507, 2016.

[27] M. V. Jose, V. Thomas, D. R. Dean, and E. Nyairo, "Fabrication and characterization of aligned nanofibrous PLGA/Collagen blends as bone tissue scaffolds," *Polymer*, vol. 50, no. 15, pp. 3778–3785, 2009.

[28] M. Monier, Y. Wei, A. A. Sarhan, and D. M. Ayad, "Synthesis and characterization of photo-crosslinkable hydrogel membranes based on modified chitosan," *Polymer*, vol. 51, no. 5, pp. 1002–1009, 2010.

[29] J. Brugnerotto, J. Lizardi, F. M. Goycoolea, W. Argüelles-Monal, J. Desbrières, and M. Rinaudo, "An infrared investigation in relation with chitin and chitosan characterization," *Polymer*, vol. 42, no. 8, pp. 3569–3580, 2001.

[30] M. Jiang, H. Ouyang, P. Ruan et al., "Chitosan derivatives inhibit cell proliferation and induce apoptosis in breast cancer cells," *Anticancer Research*, vol. 31, no. 4, pp. 1321–1328, 2011.

Method Validation for Progesterone Determination in Poly(methyl methacrylate) Nanoparticles Synthesized via Miniemulsion Polymerization

Odinei Fogolari,[1] Arlindo Cristiano Felippe,[2] Fernanda Vitória Leimann,[3] Odinei Hess Gonçalves,[3] Claudia Sayer,[1] and Pedro Henrique Hermes De Araújo[1]

[1]*Department of Chemical Engineering and Food Engineering, Federal University of Santa Catarina, 88010 970 Florianópolis, SC, Brazil*
[2]*Federal University of South Frontier (UFFS), Chapecó, SC, Brazil*
[3]*Post-Graduation Program of Food Technology (PPGTA), Federal University of Technology–Paraná (UTFPR), Campo Mourão, PR, Brazil*

Correspondence should be addressed to Pedro Henrique Hermes De Araújo; pedro.h.araujo@ufsc.br

Academic Editor: Ulrich Maschke

Exogenous progesterone has several applications in human health and in veterinary medicine, especially in fixed-time artificial insemination protocol. Progesterone nanoencapsulation in biocompatible polymers, such as poly(methyl methacrylate) (PMMA), is an alternative to substitute silicone-based release device traditionally used for estrus control. Progesterone concentration inside the nanoparticles must be precisely known; for that reason, a validation methodology must be applied to ensure reliable results, suitable for nanoparticles application. In this work, an UV-Vis spectrophotometric method was validated for the determination of progesterone in PMMA nanoparticles synthesized by miniemulsion polymerization. Chloroform was used as solvent, showing selectivity to the encapsulated drug and the components of the polymeric matrix did not influence progesterone recovery. Detection and quantitation limits (DL and QL) obtained were 0.32 and 0.96 $mg{\cdot}L^{-1}$, respectively, and precision tests (between different analysts and equipment) indicated acceptable Relative Standard Deviations (RSD $<$ 5%). Miniemulsion polymerization reactions were carried out producing two different morphologies: nanospheres (NS) and nanocapsules (NC), with average intensity diameters (Dz) of 150–200 nm and 240–300 nm, respectively. Polymerization gravimetric conversions obtained for both cases were higher than 95% and encapsulation efficiencies greater than 69% and 90% for the nanospheres and nanocapsules, respectively.

1. Introduction

Progesterone is a steroid hormone with essential functions to reproduction. Drugs with progestogens are used in humans for endometrial protection, dysfunctional bleeding, treatments in pre- or postmenopause, pregnancy maintenance in assisted reproduction treatment, and prevention of premature birth [1]. In veterinary medicine, exogenous progesterone is used especially for cattle in fixed-time artificial insemination protocol, aimed at the synchronization of estrus in females and improvements in fertilization rates. The use of estrus cycle control methods, besides facilitating the management of livestock, allows expanding the use of artificial insemination, accelerating genetic improvement and bringing improvements to the production of meat and milk [2, 3].

The incorporation of progesterone in a nanometric matrix can promote prolonged release and be beneficial in various applications [4]. Biopolymer micro- and nanoparticles have been proposed as an alternative to encapsulate progesterone by different techniques, such as glutaraldehyde crosslinking of chitosan or casein dispersed in a nonaqueous phase [5, 6] inclusion complex with β-cyclodextrin or 2-hydroxypropyl-β-cyclodextrin [7], miniemulsification/solvent evaporation method [8, 9], solvent precipitation [10, 11], supercritical CO_2 antisolvent expansion of emulsions [12], and water in oil in water double emulsion [13].

A technique that stands out in the synthesis of polymeric nanoparticles is miniemulsion polymerization, since it allows several applications for synthesized nanoparticles, including production of high solids and low viscosity latexes, hybrid polymer particles, polymerization in non-aqueous system, step polymerization in aqueous dispersed media, production of low-molecular weight polymers in dispersed media, incorporation of hydrophobic monomers, encapsulation of inorganic solids, and encapsulation of drugs [14]. The encapsulation of hydrophobic compounds by miniemulsion polymerization has been applied for drugs such as indomethacin [15], paclitaxel [16], and tamoxifen [17].

In direct miniemulsion polymerization, the dispersed phase contains the monomer, the costabilizer, and, depending on the application, other compounds of hydrophobic characteristics (e.g., drug to be encapsulated). In order to produce nanodroplets of monomer, a high-shear device is used (sonifier, rotor–stator system as Ultra-Turrax or high-pressure homogenizer) producing a miniemulsion with average droplet sizes between 50 and 500 nm, which is then polymerized [18]. This technique allows the synthesis of solid polymeric particles (nanospheres, NS) and core-shell particles (nanocapsules, NC), with oily core. Drug incorporation and release profile may differ in NS and NC, because they depend on the drug physicochemical properties as well as on the interaction with the encapsulant and the release medium. Hydrophobic drugs such as progesterone tend to be solubilized in the oily nucleus of the NC, increasing the release time.

In a study with amphiphilic β-cyclodextrin particles dispersed in aqueous medium, Memisoglu-Bilensoy et al. [19] identified that in NS the physicochemical properties of hydrophobic steroid drugs (progesterone, hydrocortisone, and testosterone) play a crucial role in drug loading and release. Authors confirm that the higher the hydrophobicity of the compound, the slower the release process. The release of hydrophobic IR-780 iodide dyes encapsulated in NS and NC prepared with biodegradable polymer no water soluble poly(D,L-lactide) (PLA) or polycaprolactone (PCL) was evaluated by Bazylińska et al. [20]. They observed slower release rates in NC and confirmed that hydrophobic dye can be more effective when enclosed in the oleic core of a NC than when enclosed only in the polymeric matrix of a NS. Similar results were obtained for indomethacin ethyl ester release in NS and NC of PCL. The presence of the polymer prolonged the ester burst release, while the presence of the oil prolonged the ester sustained release [21].

Particle size is given by the initial droplet dispersion, and both surfactant coverage and surface tension do not significantly change during the process, which makes the nucleation of the droplets the main polymerization mechanism [22]. Degradation phenomena as coalescence and monomeric diffusion (Ostwald ripening) can make the miniemulsion unstable, leading to an increase of droplets size. Nevertheless, this phenomena can be controlled with the use of an emulsifier and costabilizer at appropriate concentrations [14].

The development of polymeric nanoparticles for use in drug delivery should consider the biocompatibility and biodegradability of components. Therefore, the use of biocompatible polymers and solvent-free, natural, and nonionic emulsifiers such as lecithin, in addition of costabilizers based in oils and fatty acids, is preferred for biomedical applications [9, 23–26].

PMMA is considered safe for applications in various biomedical products as intraocular lenses, bone cement, and dental material, and it is registered in several cosmetic products [27]. Toxicological studies with the use of PMMA nanoparticles with satisfactory results were obtained by Lekshmi et al. [28] when evaluating in vivo toxicity in albino rats. Authors did not detect any changes in anatomopathological, haematological, and biochemical parameters. PMMA nanoparticles have also demonstrated lower toxicity in vitro assays with human cell cultures (K562 [29], TPH1 e A549 [26]).

Techniques such as Liquid Chromatography [30–32] and Gas Chromatography [33–35] have been used to quantify encapsulated and released progesterone. Few studies have performed progesterone quantification using UV spectrophotometry [4, 36–38]. However, none of them presented the method validation according to international regulations, which is essential to ensure the precision and accuracy of results.

A frequent problem that restricts the use of the spectrophotometric method to compounds quantification is related to low selectivity [30]. However, in matrices in which the formulation components do not interfere significantly in drug absorption spectrum, spectrophotometric methods can provide precision and accuracy, indexes similar to the chromatographic methods, up to a certain concentration range. Maliwal et al. [38] have compared the HPLC method with the UV spectrophotometric method for progesterone determination in commercial formulations (tablets). The methods showed no significant difference and they were considered suitable for routine analyses in tablets.

Since progesterone needs to be extracted from nanoparticles for drug recovery and encapsulation efficiency determinations, a good solvent must be applied to guarantee an accurate result. This implies that polymeric nanoparticle also need to be solubilized. Alcoholic solutions of methanol [36, 38, 39] and propanol [30] often used in progesterone determination protocols are not suitable to the present study, which proposes progesterone encapsulation in PMMA nanoparticles, since these solvents do not favor the dissolution of PMMA polymer chains. Chloroform dissolves PMMA easily and can be an appropriate solvent to progesterone extraction from nanoparticles.

Considering the above, this study aimed at validating a UV spectrophotometric method for progesterone determination in PMMA nanoparticles. Method was validated according to criteria established by the International Council for Harmonisation (specificity, linearity, precision, accuracy, detection limit, quantitation limit, and robustness) and applied to determine the progesterone concentration in PMMA nanoparticles. Nanoparticles parameters such as average diameter and monomer gravimetric conversion were also determined.

2. Materials and Methods

2.1. Materials. Methyl methacrylate (monomer, MMA), *2,2*-azobisisobutyronitrile (initiator, AIBN), progesterone (P4, 99%), chloroform, and tetrahydrofuran (THF) were purchased from Sigma-Aldrich. Lecithin (Alpha Aesar) was used as surfactant. Crodamol GTCC® (caprylic/capric triglyceride, costabilizer) was purchased from Alpha Química. Deionized water was used in all experiments. All reagents and solvents were analytical grade. AIBN was previously recrystallized in methanol (Sigma-Aldrich), filtered, and vacuum dried.

2.2. Method Validation for Progesterone Determination in PMMA Nanoparticles. The validation of spectrophotometric method was performed considering the following parameters: specificity, linearity, precision, accuracy, detection limit (DL), quantitation limit (QL), and robustness, according to the criteria established by the International Council for Harmonisation of Technical Requirements for Pharmaceuticals for Human Use [40]. Two spectrophotometers were used for the method validation: (1) Spectrophotometer Hitachi model U-1900 (spectral bandwidth 4 nm; wavelength accuracy ±0.5 nm; wavelength repeatability ±0.3 nm; photometric accuracy ±0.002 Abs; photometric repeatability 0.001 Abs) and (2) Spectrophotometer Rayleigh, model UV-2601 (spectral bandwidth 2 nm; wavelength accuracy ±0.3 nm; wavelength repeatability ±0.15 nm; photometric accuracy ±0.002 Abs; photometric repeatability 0.001 Abs).

2.2.1. Specificity. The evaluation of the specificity was performed by comparison of the absorption spectra obtained for progesterone with that obtained for nanoparticles (with or without progesterone) solubilized in chloroform. In order to obtain the absorption spectra of the polymer nanoparticles, latex samples were subjected to drying in an oven at 60°C until they reached constant weight. Dried samples were then solubilized in chloroform, diluted, and scanned from 200 to 300 nm.

2.2.2. Linearity. For linearity determination, three calibration curves were prepared at different days. For each calibration curve, a standard progesterone solution was prepared in chloroform and subsequently diluted in eight different concentrations, ranging from 2 to 50 mg·L^{-1} ($n = 24$) and the absorbance measurements were conducted in triplicate for each concentration at 253 nm. Linearity was assessed using linear regression and the quality of model fit was verified by analysis of variance (ANOVA).

2.2.3. Detection Limit (DL) and Quantitation Limit (QL). Detection and quantification limits were obtained based on the slope (S) and standard deviation (σ) of the intercept with the axis Y (absorbance) of linearity curves, according to [40]

$$\mathrm{LD} = \frac{3.3\sigma}{S}, \qquad \mathrm{LQ} = \frac{10\sigma}{S}. \tag{1}$$

2.2.4. Precision and Accuracy. Precision was evaluated based on the Relative Standard Deviation (RSD) of a measurement series. This analysis took into consideration repeatability, intermediate precision, and reproducibility [40]. Repeatability was performed by repeating the analytical procedure for 3 concentration levels covering the minimum and maximum concentrations of linearity test. Each concentration was analyzed 7 times, on the same day, by the same analyst and equipment. For intermediate precision, the same procedure was followed, but on different days and by different analysts. Reproducibility was evaluated with the reproduction of the analysis at the intermediate level of concentration in a second laboratory on two different days. Similarly, the accuracy was evaluated based on the Recovery Index (Rec, (2)), in the above described analysis, where $[\mathrm{P4}]_{\mathrm{experimental}}$ (mg·L^{-1}) represents the concentration detected during the analysis at 253 nm and $[\mathrm{P4}]_{\mathrm{standard}}$ represents the theoretical concentration of the evaluated dilution.

$$\mathrm{Rec} = \frac{[\mathrm{P4}]_{\mathrm{experimental}}}{[\mathrm{P4}]_{\mathrm{standard}}} \times 100. \tag{2}$$

2.2.5. Robustness. Robustness evaluates the ability of the analytical procedure to remain unchanged when subjected to small variations in process parameters. Thus, two brands of the solvent (chloroform, analytical grade) were evaluated to verify a possible interference on P4 quantification. A standard progesterone solution of 520 mg·L^{-1} was prepared. Then, aliquots of 250 μL of this solution were diluted to 5 mL of each brand of chloroform ("A" and "B"), resulting in a concentration of 26 mg·L^{-1}. Finally, samples were analyzed in 5 replicates at 253 nm.

2.3. PMMA Nanoparticles Synthesis for Progesterone Encapsulation. PMMA nanoparticles were produced by the miniemulsion polymerization technique and the formulations were defined according to previous studies conducted in the research group in the synthesis of biocompatible nanoparticles for drug delivery [23–26]. Formulations used in particles synthesis are presented in Table 1 for nanospheres (NS) and nanocapsules (NC). Considering that MMA monomer presents certain solubility in water (150 mmol·L^{-1}), a hydrophobic initiator (AIBN) was used to prevent secondary nucleation and the formation of pure PMMA particles in the aqueous phase [41].

The organic phase of the miniemulsion was prepared with MMA, lecithin, Crodamol GTCC, AIBN, and drug (P4) when it was specified. These components were homogenized and then mixed during 40 minutes with water, in order to form a macroemulsion. Then, the macroemulsion was subjected to the process of droplet size reduction by sonication (Fisher Scientific, Sonic Dismembrator Model 500, 400 W) for 4 minutes at 60% amplitude (10 s pulse on and 5 s pulse off). The miniemulsion prepared was equally divided into glass test tubes, which were filled with gaseous nitrogen to remove oxygen from the head space. Finally, the test tubes were closed and immersed in a thermostatic bath at 70°C to start the polymerization reaction. Each test tube was removed from

TABLE 1: Standard miniemulsion formulation for progesterone encapsulation in PMMA nanospheres (NS) or in PMMA nanocapsules (NC).

Formulation	Water (%)	AIBN (%)	Lecithin (%)	Crodamol GTCC (%)	MMA (%)	P4($mg_{P4}\cdot g_{latex}^{-1}$)
NS	79.5	0.200	0.300	2.00	18.0	—
NS P4 1 mg	79.5	0.200	0.300	2.00	18.0	1
NS P4 2 mg	79.5	0.200	0.300	2.00	18.0	2
NS P4 20 mg	79.5	0.200	0.300	2.00	18.0	20
NC	79.5	0.200	0.300	10.0	10.0	—
NC P4 20 mg	79.5	0.200	0.300	10.0	10.0	20

the bath at a specified time interval and cooled to obtain conversion kinetics and size diameter results.

2.4. Nanoparticles Characterization. Throughout the polymerization reactions, samples were withdrawn to evaluate nanoparticles average diameter (Dz) and diameter polydispersity index (PDI), which were determined by Dinamic Light Scattering using a Zetasizer equipment (Nano Series, Malvern Instruments). Prior to reading, samples were diluted approximately at 1 : 20 (v/v) with previously prepared MMA saturated water.

The morphological characterization of polymeric nanoparticles was performed by Transmission Electron Microscopy (TEM, JEOL model JEM 1011 at 100 kV). For this analysis, drops of the diluted samples (0.05%, v/v) were placed on a 300-mesh Formvar/carbon copper grid (Electron Microscopy Science). After drying, samples were sputter-coated with a thin carbon film to avoid degradation of the PMMA under the electron beam.

Methyl methacrylate gravimetric conversion (X_g) was determined as described by Bernardy et al. (2010) [25]. Thus, latex samples (2 g) were taken at different time intervals from the reaction media and transferred to previously weighted aluminum capsules containing 0.2 g of 1 wt.% hydroquinone aqueous. After that, capsules were dried at 60°C until reaching constant weight. Conversion was determined as the ratio between experimental and theoretical polymer contents. The fraction of nonvolatile components (emulsifier, costabilizers, hydroquinone, and progesterone) was deducted from the polymer fraction.

The residual monomer content was determined in latex samples at the end of the polymerization reaction (280 min). Analyses were performed by Headspace Gas Chromatography (GC 2010AF Shimadzu) equipped with a Flame Ionization Detector (FID) using the total evaporation technique of the volatile fraction from the sample. A calibration curve was prepared with standard samples containing the monomer to be measured, analyzed in triplicate.

PMMA molecular weight distributions were determined by Gel Permeation Chromatography (GPC) using High-Performance Liquid Chromatography (HPLC, Shimadzu, model LC-20A), with three columns Shimadzu Shim Pack GPC 800 Series 300 × 8 mm (GPC 801, GPC 804 e GPC 807), refractive index detector (model RID-10A) and autosampler (model SIL-20A). Latex samples (0.02 g) were initially solubilized in tetrahydrofuran (THF, 4 mL), filtered (0.45 μm), and finally analyzed. THF was used as mobile phase at a flow rate of 1 $mL\cdot min^{-1}$ and 35°C. The polymer molecular weights were determined from a calibration curve produced with polystyrene standards with molecular weights between 580 $g\cdot mol^{-1}$ and 9.835×10^6 $g\cdot mol^{-1}$.

2.5. Progesterone Recovery Yield (RY) and Encapsulation Efficiency (EE) in PMMA Nanoparticles. In order to evaluate possible interferences of the components of the polymeric nanoparticles in drug recovery yield (RY), PMMA nanoparticles were prepared, according to Table 1, with progesterone concentrations of 1, 2, and 20 $mg_{P4}\cdot g_{latex}^{-1}$. With the purpose of determining RY, latex samples (1.5 g) were dried at 60°C and solubilized in 10 mL of chloroform. Aliquots were diluted to obtain absorbance values between 0.3 and 1.0 at 253 nm. Each formulation was analyzed in triplicate and the RY was calculated with (3), where C_{rec} ($mg\cdot mL^{-1}$) is the concentration of recovered progesterone after nanoparticles synthesis and $C_{theoretical}$ ($mg\cdot mL^{-1}$) is the progesterone concentration added to the nanoparticles synthesis formulation.

For determining the encapsulation efficiency (EE), polymer particles were separated from the aqueous phase by centrifugation at 13.528 ×g for 45 min. The precipitate was dried at 60°C, dissolved in 10 mL of chloroform, diluted, and filtered (PTFE, 0.2 μm) and the absorbance was determined at 253 nm. Encapsulation efficiency (EE) was then obtained according to (4), where M_{rec} is the mass of recovered progesterone in precipitate, M_{lt} is latex mass, and F_{P4} is the progesterone fraction in the latex. Each formulation was measured in triplicate and latex samples without progesterone were subjected to the same treatment and served as reference (blank sample) in spectrophotometric readings.

$$\text{RY} = \frac{C_{rec}}{C_{theoretical}} \times 100 \quad (3)$$

$$\text{EE}\,(\%) = \frac{M_{rec}}{F_{P4} \times M_{lt}} \times 100. \quad (4)$$

3. Results and Discussion

3.1. Method Validation for Progesterone Determination in PMMA Nanoparticles

3.1.1. Specificity. In quantification tests, specificity results should ensure that the procedure is not affected by the presence of impurities or excipients [42]. Figure 1 presents the absorption spectra of pure progesterone (P4) and of the

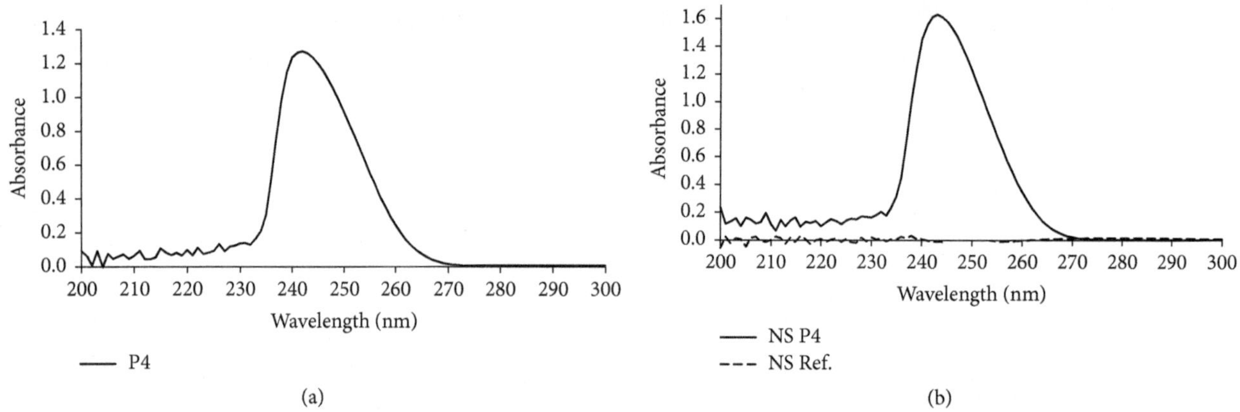

FIGURE 1: UV absorption spectra of (a) pure P4 and (b) P4-loaded NS (20 mg·g^{-1}) and NS reference (no P4 added).

PMMA nanoparticles. In the ultraviolet region, progesterone presents only one absorption peak located between 230 and 270 nm. The maximum absorption is between 241 and 243 nm (Figure 1(a)); however, the cut off of the solvent is 245 nm, making it infeasible to quantify the analyte at wavelengths below this value. Therefore, the methodology was validated using absorbance at 253 nm, where drug absorption is high and the interference of the solvent is considerably reduced.

Progestogen free nanospheres (blank sample) when solubilized in chloroform have absorbance near to zero in the measured spectrum region (Figure 1(b)) and nanocapsules presented the same behavior. Thus, it is demonstrated that the method is selective to the progesterone among the components of PMMA nanoparticles (polymer, emulsifier, and costabilizer) at 253 nm.

3.1.2. Linearity, Detection Limit (DL), and Quantitation Limit (QL). The resulting standard curve "Abs = 0.0277∗C + 0.0035" showed adjusted coefficient of determination (R^2_{adj}) of 0.997, indicating high proportionality index between concentration and absorbance in the range of 2 and 50 mg·L^{-1}. The analysis of variance (ANOVA, unilateral test) of the results confirmed that the linear regression model was significant (for $p < 0.05$), presenting no lack of fit.

DL and QL estimated by (1) were 0.32 e 0.96 mg·L^{-1}, respectively, showing that the method is able to detect and quantify small concentrations of the P4 in the sample. Considering an optimal working range between 10 and 40 mg·L^{-1}, QL corresponds to 10% of the less value of this range being therefore acceptable for the method.

3.1.3. Precision and Accuracy. Results of the repeatability and intermediate precision are shown in Table 2, which contains analyses at 3 concentration levels, performed by different analysts in different days. In Table 3 calculated and critical t values obtained through Student's t-test are presented.

The results of RSD were less than 5% in the evaluation of intermediate precision and repeatability, including the maximum and minimum concentrations evaluated.

In all possible comparisons, the means found for the maximum and intermediate concentration showed no difference to the confidence level of 95%. For the minimum concentration the comparison between means of "Day 1/Analyst A versus Day 2/Analyst A" and "Day 1/Analyst A versus Day 2/Analyst B" showed statistical differences for the aforementioned confidence level. However, variations are expected and acceptable for concentrations near to the limit of quantification.

The results obtained for reproducibility in a second laboratory (Lab. B) are shown in Table 4. The comparison of interlaboratory means, performed by Student's t-test is shown in Table 5.

Means have RSD less than 5%, which is an acceptable value. According to the data, there is no statistical difference at a confidence level of 95% for all possible comparisons. With respect to the accuracy, one may observe that the difference between theoretical and recovered values, for intermediate and maximum concentrations, was less than 7% (Tables 2 and 4). Therefore, in general, the recovery data were satisfactory and compatible with the proposed method.

3.1.4. Robustness. The robustness was carried out considering the possibility of acquiring solvents of different trademarks. Table 6 shows the comparison results of the analyses conducted with two different solvents suppliers.

The results showed that the use of analytical grade solvents from different suppliers did not interfere in the analysis. The means showed no statistical difference at a significance level of 0.05 for Student's t-test. Also, no statistical difference regarding the standard concentration of 26.0 mg·L^{-1} was detected.

3.2. PMMA Nanoparticles Characterization and Progesterone Determination. The proportion of 1 : 9 and 1 : 1 between MMA : Crodamol favored, respectively, the formation of particles with nanospheres (NS) and nanocapsules (NC) morphology, as it was verified by other authors [23, 25, 43]. The images presented in Figure 2 confirm these morphologies. In Figure 2(a) spherical solid particles with lower size dispersion can be observed. On the other hand, in Figure 2(b) one may observe that particles formed a core composed by Crodamol and an outer layer composed by PMMA, characteristics of

TABLE 2: Repeatability evaluation and intermediate precision results.

$C_{Standard}$ (mg·L^{-1})	Calculated concentration (mg·L^{-1})[1]	RSD (%)[2]	Rec (%)[3]	Calculated concentration (mg·L^{-1})	RSD (%)	Rec (%)
	Day 1/Analyst A			Day 2/Analyst A		
52.0	53.4 ± 0.6	1.2	102.6	52.6 ± 0.7	1.4	101.2
26.0	27.1 ± 0.2	0.8	104.3	27.2 ± 0.6	2.1	104.7
2.1	2.3 ± 0.1	4.6	109.4	2.4 ± 0.1	4.8	116.1
	Day 1/Analyst B			Day 2/Analyst B		
52.0	53.3 ± 1.3	2.4	102.5	52.8 ± 1.0	1.8	101.5
26.0	27.7 ± 0.8	2.9	106.6	27.5 ± 0.5	1.9	105.9
2.1	2.3 ± 0.1	3.7	111.4	2.4 ± 0.1	2.7	114.8

[1]Mean ± standard deviation ($n = 7$); [2]Relative Standard Deviation; [3]recovery.

TABLE 3: Student's t-test to the mean data in the evaluation of the repeatability and intermediate precision.

$C_{Standard}$ (mg·L^{-1})	Conditions	$t_{calculated}$ Day 2/Analyst A	Day 1/Analyst B	Day 2/Analyst B	$t_{critical}$*
52.0	Day 1/Analyst A	2.060	0.151	1.328	2.178
	Day 2/Analyst A	—	1.187	0.376	2.178
	Day 1/Analyst B	—	—	0.812	2.178
26.0	Day 1/Analyst A	0.461	1.951	2.006	2.178
	Day 2/Analyst A	—	1.345	1.068	2.178
	Day 1/Analyst B	—	—	0.528	2.178
2.1	Day 1/Analyst A	*2.368*	0.812	*2.449*	2.178
	Day 2/Analyst A	—	1.803	0.514	2.178
	Day 1/Analyst B	—	—	1.783	2.178

*t value for bilateral test ($p < 0.05$; degrees of freedom = 12).

TABLE 4: Results of the reproducibility of the method.

$C_{Standard}$ (mg·L^{-1})	Conditions	Mean (mg/L)*	RSD (%)	Rec (%)
26.0	Day 1/Lab. B	26.9 ± 0.8	3.1	103.3
	Day 2/Lab. B	27.2 ± 1.2	4.5	104.6

*Mean ± standard deviation ($n = 7$).

TABLE 5: Student's t-test to the mean data in the evaluation of the reproducibility.

	$t_{calculated}$ Day 2/Lab. A	Day 1/Lab. B	Day 2/Lab. B	$t_{critical}$*
Day 1/Lab. A	0.461	0.758	0.335	2.178
Day 2/Lab. A	—	0.922	0.119	2.178
Day 1/Lab. B	—	—	0.697	2.178

*t value for bilateral test ($p < 0.05$; degrees of freedom = 12).

nanocapsules. NCs (Figure 2(b)) presented higher particles sizes dispersion, confirming data obtained by DLS.

Polymerization reactions were carried out with the formulations shown in Table 1, varying only the concentration of progesterone. Table 7 shows the concentration of progesterone based on the mass added to the formulation, the concentration obtained by spectrophotometric analysis in recovery yield (RY), and the encapsulation efficiency (EE).

Recovery yield (RY) determined by (3) was between 94 and 106%, showing that the method is robust with respect to formulation components, which did not influence the quantification of the drug. Encapsulation efficiency (EE) was

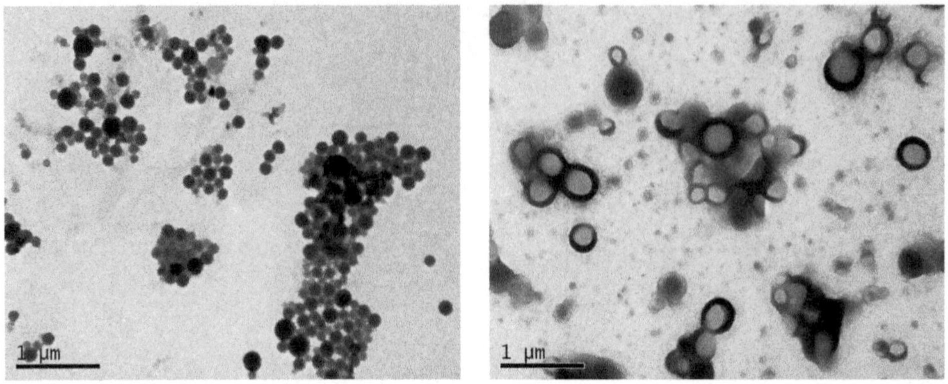

(a) (b)

FIGURE 2: TEM micrographs of PMMA (a) nanospheres and (b) nanocapsules, both using $1\ mg_{P4}{\cdot}g_{latex}^{-1}$.

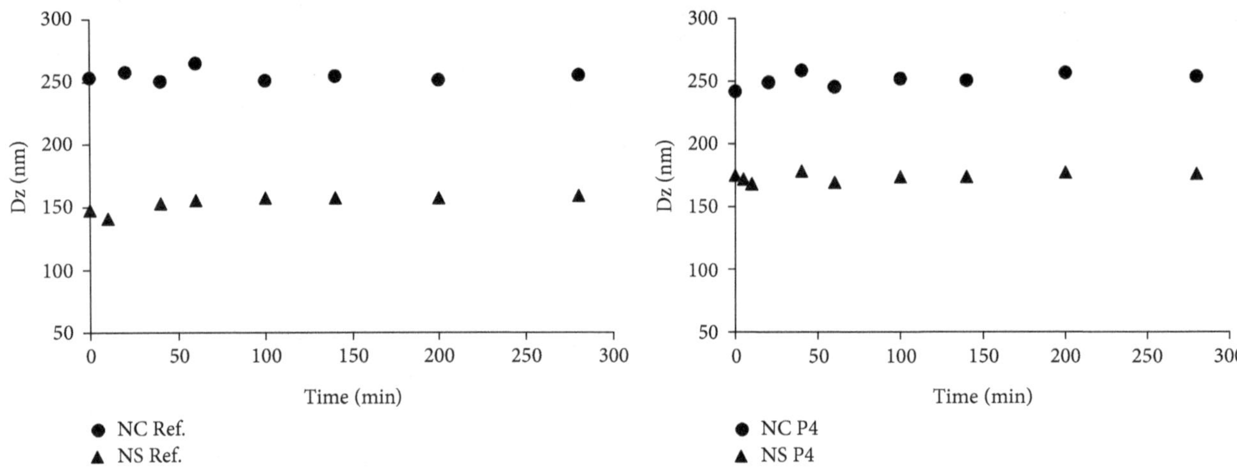

(a) (b)

FIGURE 3: Average intensity diameter (Dz) of nanoparticles during miniemulsion polymerization for the formation of PMMA blank nanoparticles (Ref.) (a) and with progesterone (P4; $1\ mg_{P4}{\cdot}g_{latex}^{-1}$) (b). NC (nanocapsules) and NS (nanospheres).

TABLE 6: Robustness results for analyses with different brands of chloroform.

Sample	Chloroform A	Chloroform B
Mean ($mg{\cdot}L^{-1}$)*	26.6 ± 0.8[a]	25.7 ± 0.6[a]
RSD (%)	3.2	2.5
Rec (%)	102.3	98.8

*Mean ± standard deviation (n = 5) for standard 26.0 mg/L. [a]There is no significant difference between the means by Student's t-test (p < 0,05).

high when low P4 concentrations were used, but it was decreased for the highest P4 concentration used meaning that there is a limit in the amount of P4 that could be incorporated in the nanoparticles. In fact, the reduction of the EE for the concentration of $20\ mg{\cdot}g^{-1}$ was expected, because it is a relatively high charge for the formulation (about 10% solids fraction). It was found that nanocapsules have higher encapsulation efficiency than nanospheres, probably due to the higher solubility of progesterone in Crodamol than in PMMA.

Polymerization reactions with the proposed formulations (Table 1) were evaluated with $1\ mg_{P4}{\cdot}g_{latex}^{-1}$. The average intensity diameter results (Dz) of the nanodroplets/nanoparticles throughout the polymerization reactions are shown in Figure 3.

PMMA nanocapsules (NC) presented average diameter between 240 and 290 nm, and polydispersity indexes (PDI) between 0.20 and 0.25. On the other hand, PMMA nanosphere (NS) presented lower Dz results, between 150 and 200 nm, as well as a monodisperse size distribution, with PDI below 0.13. Figures 3(a) and 3(b) show that the presence of progesterone did not considerably change particles diameter, reflecting the stability of the miniemulsion even in the presence of progesterone. The ratio between the final particle diameter (250 min of reaction) and the nanodroplets diameter (beginning of the reaction) was 1.05 for NC and 1.01 for NS. Therefore, the increase in the average size was

TABLE 7: Added and calculated P4 concentration and encapsulation efficiency (EE) in PMMA nanoparticles.

Formulation[1]	P4 added ($mg{\cdot}g^{-1}$)	P4 calculated ($mg{\cdot}g^{-1}$)[2]	EE (%)
NS 1 mg P4	1.09	1.16 ± 0.27	94 ± 1
NS 2 mg P4	1.99	2.05 ± 0.14	100 ± 3
NS 20 mg P4	20.36	19.21 ± 0.36	69 ± 3
NC 20 mg P4	19.99	20.20 ± 1.71	90 ± 1

[1]Nanospheres (NS) and nanocapsules (NC) formulations prepared as shown in Table 1, except for the concentration of progesterone (P4); [2]P4 concentration calculated using the validated method.

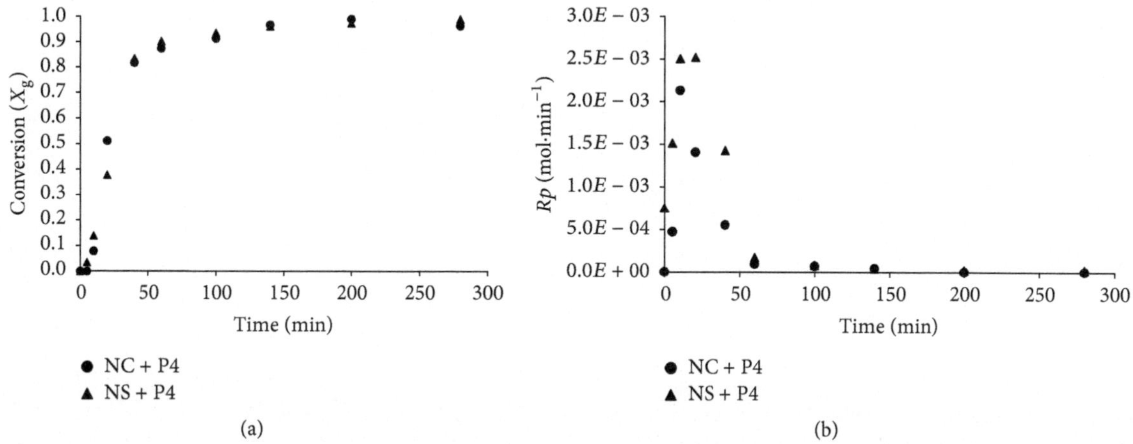

FIGURE 4: (a) Conversion and (b) reaction rate (Rp) during miniemulsion polymerization for the formation of PMMA NC (nanocapsules) and NS (nanospheres) using 1 $mg_{P4}{\cdot}g_{latex}^{-1}$.

TABLE 8: Residual monomer content obtained by gas chromatography at the end of the reactions (t = 280 min).

Formulation	P4 ($mg{\cdot}g^{-1}$)	Residual monomer (%)
NS	—	0.60
NS	20	0.20
NC	—	0.10
NC	20	0.20

negligible, indicating the absence of degradation phenomena such as coalescence or diffusional degradation. Particle size stability also suggests that nucleation of the droplets was the main reaction mechanism of polymerization [22].

In Figure 4 the evolution of the gravimetric conversion (a) and the reaction rate (b) are presented. Residual monomer results are shown in Table 8 and in Figure 5 molar masses distributions are presented.

The behavior of the polymerization kinetics for the formation of NC and NS with 1 $mg_{P4}{\cdot}g_{latex}^{-1}$ was similar and, after 280 min, the reaction reached a conversion up to 97%. The reduction in the reaction rate (Rp) for NC observed in Figure 4(b) did not damage the polymer conversion. The reduction in the Rp is a consequence of the lower number of particles in the system for NC (higher Dz; see Figure 3) causing an increase in the number of radicals/particle proportion and, consequently, in the rate of radical termination. The presence of residual monomer in the latex samples is undesirable, because it may increase toxicity. Data indicated that the monomer content in the samples is low, less than 0.6%. These results agree with those obtained by Feuser et al. [44], which did not detect residual monomer in PMMA samples synthesized by miniemulsion polymerization, with a similar formulation.

In addition to changing the reaction rate, the small particle size can increase the molar mass of the polymer. Systems with higher number of particles and smaller Dp have a lower ratio of number of radical/particles, allowing the growth of the polymer chain and the increase of the molar mass. This can be verified on the weight average molar mass *(Mw)* of the nanospheres, whose value was 510 kDa, while for nanocapsules it was close to 440 kDa. These results are in accordance with Tiarks et al. [41] who observed increase of Dp and molar mass reduction with increasing stabilizer concentration in PMMA nanocapsules. Samples with progesterone had average weight of 445 kDa and 380 kDa, respectively, for nanospheres and nanocapsules. Despite this reduction, the analysis of the molar mass distribution, shown in Figure 5, indicated that the drug effect is insignificant in this parameter.

4. Conclusions

The validated UV spectrophotometric method was selective, precise, accurate, and robust enough for the determination of progesterone in polymeric nanoparticles of PMMA. Relative Standard Deviation was lower than 5% in all experimental

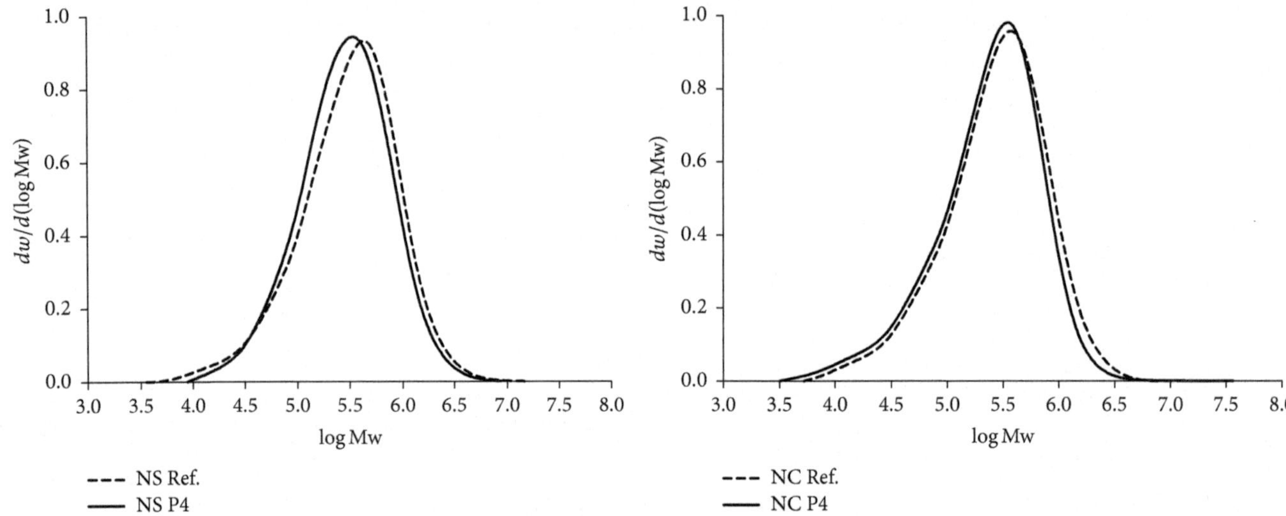

FIGURE 5: Distribution of the molar masses in PMMA nanoparticles reference (Ref.; drug free) and with 20 $mg_{P4} \cdot g_{latex}^{-1}$ (NS: nanospheres; NC: nanocapsules).

conditions evaluated. With respect to the accuracy, the maximum percentage difference obtained between progesterone added and recovered was 7% for intermediate working concentration (26 $mg \cdot L^{-1}$). The methodology was applied to determine encapsulation efficiency of progesterone in PMMA nanocapsules and nanospheres. The tests performed with nanoparticles showed satisfactory results, indicating that the polymer matrix and their components, when solubilized in chloroform, did not interfere in the quantification of the progesterone. The biocompatible formulations used in the synthesis of PMMA nanoparticles demonstrated stability and potential for the incorporation of progesterone. Moreover, the presence of progesterone did not cause significant changes in parameters such as particle diameter, monomer conversion, and molar mass.

References

[1] X. Ruan and A. O. Mueck, "Systemic progesterone therapy - Oral, vaginal, injections and even transdermal?" *Maturitas*, vol. 79, no. 3, pp. 248–255, 2014.

[2] P. S. Baruselli, E. L. Reis, M. O. Marques, L. F. Nasser, and G. A. Bó, "The use of hormonal treatments to improve reproductive performance of anestrous beef cattle in tropical climates," *Animal Reproduction Science*, vol. 82-83, pp. 479–486, 2004.

[3] R. Islam, "Synchronization of Estrus in Cattle: a review," *Veterinary World*, vol. 4, no. 3, pp. 136–141, 2009.

[4] J. E. Oliveira, E. S. Medeiros, L. Cardozo et al., "Development of poly(lactic acid) nanostructured membranes for the controlled delivery of progesterone to livestock animals," *Materials Science and Engineering C*, vol. 33, no. 2, pp. 844–849, 2013.

[5] S. R. Jameela, T. V. Kumary, A. V. Lal, and A. Jayakrishnan, "Progesterone-loaded chitosan microspheres: A long acting biodegradable controlled delivery system," *Journal of Controlled Release*, vol. 52, no. 1-2, pp. 17–24, 1998.

[6] M. S. Latha, A. V. Lal, T. V. Kumary, R. Sreekumar, and A. Jayakrishnan, "Progesterone release from glutaraldehyde cross-linked casein microspheres: in vitro studies and in vivo response in rabbits," *Contraception*, vol. 61, no. 5, pp. 329–334, 2000.

[7] E. Lemos-Senna, D. Wouessidjewe, S. Lesieur, and D. Duchêne, "Preparation of amphiphilic cyclodextrin nanospheres using the emulsification solvent evaporation method. Influence of the surfactant on preparation and hydrophobic drug loading," *International Journal of Pharmaceutics*, vol. 170, no. 1, pp. 119–128, 1998.

[8] J. Matsumoto, Y. Nakada, K. Sakurai, T. Nakamura, and Y. Takahashi, "Preparation of nanoparticles consisted of poly(L-lactide)-poly(ethylene glycol)-poly(L-lactide) and their evaluation in vitro," *International Journal of Pharmaceutics*, vol. 185, no. 1, pp. 93–101, 1999.

[9] F. V. Leimann, M. H. Biz, K. C. Kaufmann et al., "Characterization of progesterone loaded biodegradable blend polymeric nanoparticles," *Ciencia Rural*, vol. 45, no. 11, pp. 2082–2088, 2015.

[10] S. Ueda, S. Matsumoto, A. Takagi, and T. Yamane, "Synthesis of poly(3-hydroxybutyrate-co-3-hydroxyvalerate) from methanol and n-amyl alcohol by the methylotrophic bacteria Paracoccus denitrificans and Methylobacterium extorquens," *Applied and Environmental Microbiology*, vol. 58, no. 11, pp. 3574–3579, 1992.

[11] H. F. Salem, "Sustained-release progesterone nanosuspension following intramuscular injection in ovariectomized rats," *International Journal of Nanomedicine*, vol. 5, no. 1, pp. 943–954, 2010.

[12] M. R. Guilherme, M. R. Mauricio, E. T. Tenório-Neto et al., "Polycaprolactone nanoparticles containing encapsulated progesterone prepared using a scCO2 emulsion drying technique," *Materials Letters*, vol. 124, pp. 197–200, 2014.

[13] J. F. S. F. Silva, P. Cnop, R. J. R. Sßnchez et al., "Avaliação da dinâmica útero-ovárica da égua sob o efeito de um implante

subcutâneo de micro-cápsulas de polihidroxibutirato contendo progesterona," *Rev. Port. Ciências Veterinárias*, vol. 101, pp. 225–230, 2006.

[14] J. M. Asua, "Miniemulsion polymerization," *Progress in Polymer Science*, vol. 27, no. 7, pp. 1283–1346, 2002.

[15] I. Rajot, S. Bône, C. Graillat, and T. Hamaide, "Nonionic nanoparticles by miniemulsion polymerization of vinyl acetate with oligocaprolactone macromonomer or miglyol as hydrophobe. application to the encapsulation of indomethacin," *Macromolecules*, vol. 36, no. 20, pp. 7484–7490, 2003.

[16] C.-Y. Huang, C.-M. Chen, and Y.-D. Lee, "Synthesis of high loading and encapsulation efficient paclitaxel-loaded poly(n-butyl cyanoacrylate) nanoparticles via miniemulsion," *International Journal of Pharmaceutics*, vol. 338, no. 1-2, pp. 267–275, 2007.

[17] T. S. Moreira, M. A. de Oliveira, L. M. da Rocha e Lima, M. N. de Souza, and J. C. da Silva Pinto, "Synthesis of nanoparticles loaded with tamoxifen by in situ miniemulsion RAFT polymerization," *Macromolecular Symposia*, vol. 344, no. 1, pp. 101–107, 2014.

[18] A. Guyot, K. Landfester, F. Joseph Schork, and C. Wang, "Hybrid polymer latexes," *Progress in Polymer Science (Oxford)*, vol. 32, no. 12, pp. 1439–1461, 2007.

[19] E. Memisoglu-Bilensoy, M. Şen, and A. A. Hincal, "Effect of drug physicochemical properties on in vitro characteristics of amphiphilic cyclodextrin nanospheres and nanocapsules," *Journal of Microencapsulation*, vol. 23, no. 1, pp. 59–68, 2006.

[20] U. Bazylińska, A. Lewińska, Ł. Lamch, and K. A. Wilk, "Polymeric nanocapsules and nanospheres for encapsulation and long sustained release of hydrophobic cyanine-type photosensitizer," *Colloids and Surfaces A: Physicochemical and Engineering Aspects*, vol. 442, pp. 42–49, 2014.

[21] L. Cruz, L. U. Soares, T. D. Costa et al., "Diffusion and mathematical modeling of release profiles from nanocarriers," *International Journal of Pharmaceutics*, vol. 313, no. 1-2, pp. 198–205, 2006.

[22] F. J. Schork, Y. Luo, W. Smulders, J. P. Russum, A. Butté, and K. Fontenot, "Miniemulsion polymerization," *Advances in Polymer Science*, vol. 175, pp. 129–255, 2005.

[23] A. P. Romio, C. Sayer, P. H. H. Araújo, M. Al-Haydari, L. Wu, and S. R. P. Da Rocha, "Nanocapsules by miniemulsion polymerization with biodegradable surfactant and hydrophobe," *Macromolecular Chemistry and Physics*, vol. 210, no. 9, pp. 747–751, 2009.

[24] F. R. Steinmacher, N. Bernardy, J. B. Moretto, E. I. Barcelos, P. H. H. Araújo, and C. Sayer, "Kinetics of MMA and VAc miniemulsion polymerizations using miglyol and castor oil as hydrophobe and liquid core," *Chemical Engineering and Technology*, vol. 33, no. 11, pp. 1877–1887, 2010.

[25] N. Bernardy, A. P. Romio, E. I. Barcelos et al., "Nanoencapsulation of quercetin via miniemulsion polymerization," *Journal of Biomedical Nanotechnology*, vol. 6, no. 2, pp. 181–186, 2010.

[26] P. E. Feuser, P. C. Gaspar, E. Ricci-Júnior et al., "Synthesis and characterization of poly(methyl methacrylate) pmma and evaluation of cytotoxicity for biomedical application," *Macromolecular Symposia*, vol. 343, no. 1, pp. 65–69, 2014.

[27] L. C. Becker, W. F. Bergfeld, D. V. Belsito et al., "Final Report of the cosmetic ingredient review expert panel safety assessment of polymethyl methacrylate (PMMA), methyl methacrylate crosspolymer, and methyl methacrylate/glycol dimethacrylate crosspolymer," *International Journal of Toxicology*, vol. 30, no. 3, 2011.

[28] U. M. Dhana lekshmi, G. Poovi, N. Kishore, and P. N. Reddy, "In vitro characterization and in vivo toxicity study of repaglinide loaded poly (methyl methacrylate) nanoparticles," *International Journal of Pharmaceutics*, vol. 396, no. 1-2, pp. 194–203, 2010.

[29] A. N. Mendes, I. Hubber, M. Siqueira et al., "Preparation and cytotoxicity of poly(methyl methacrylate) nanoparticles for drug encapsulation," *Macromolecular Symposia*, vol. 319, no. 1, pp. 34–40, 2012.

[30] V. Pucci, F. Bugamelli, R. Mandrioli, B. Luppi, and M. A. Raggi, "Determination of progesterone in commercial formulations and in non conventional micellar systems," *Journal of Pharmaceutical and Biomedical Analysis*, vol. 30, no. 5, pp. 1549–1559, 2003.

[31] R. M. T. Fernandes, G. C. Gomes, A. M. Porcari et al., "LC-MS/MS quantitation of plasma progesterone in cattle," *Theriogenology*, vol. 76, no. 7, pp. 1266–e2, 2011.

[32] G. R. Pereira, J. M. Marchetti, and M. V. L. B. Bentley, "A rapid method for determination of progesterone by reversed-phase liquid chromatography from aqueous media," *Analytical Letters*, vol. 33, no. 5, pp. 881–889, 2000.

[33] H. M. Leith, P. L. Truran, and S. J. Gaskell, "Quantification of progesterone in human saliva," *Biological Mass Spectrometry*, vol. 13, no. 5, pp. 257–261, 1986.

[34] P. Liere, Y. Akwa, S. Weill-Engerer et al., "Validation of an analytical procedure to measure trace amounts of neurosteroids in brain tissue by gas chromatography-mass spectrometry," *Journal of Chromatography B: Biomedical Sciences and Applications*, vol. 739, no. 2, pp. 301–312, 2000.

[35] T. Lukkainen, W. Vandenheuvel, and E. Horning, "An estrogen determination method using gas chromatography," *Biochimica et Biophysica Acta*, vol. 62, no. 1, pp. 153–159, 1962.

[36] H.-I. Chang, M. R. Williamson, Y. Perrie, and A. G. A. Coombes, "Precipitation casting of drug-loaded microporous PCL matrices: Incorporation of progesterone by co-dissolution," *Journal of Controlled Release*, vol. 106, no. 3, pp. 263–272, 2005.

[37] G. V. Salmoria, P. Klauss, K. M. Zepon, and L. A. Kanis, "The effects of laser energy density and particle size in the selective laser sintering of polycaprolactone/progesterone specimens: morphology and drug release," *International Journal of Advanced Manufacturing Technology*, vol. 66, no. 5–8, pp. 1113–1118, 2013.

[38] D. Maliwal, P. Jain, A. Jain, and V. Patidar, "Determination of progesterone in capsules by high-performance liquid chromatography and UV- spectrophotometry," *Journal of Young Pharmacists*, vol. 1, no. 4, p. 371, 2009.

[39] Z. Wu, C. Zhang, C. Yang, X. Zhang, and E. Wu, "Simultaneous quantitative determination of norgestrel and progesterone in human serum by high-performance liquid chromatography-tandem mass spectrometry with atmospheric pressure chemical ionization," *Analyst*, vol. 125, no. 12, pp. 2201–2206, 2000.

[40] ICH, "ICH Topic Q2 (R1) validation of analytical procedures: text and methodology," in *Proceedings of the international conference on harmonisation of technical requirements for registration of pharmaceuticals for human use*, 2005, http://www.ich.org/fileadmin/Public_Web_Site/ICH_Products/Guidelines/Quality/Q2_R1/Step4/Q2_R1__Guideline.pdf.doi.

[41] F. Tiarks, K. Landfester, and M. Antonietti, "Preparation of polymeric nanocapsules by miniemulsion polymerization," *Langmuir*, vol. 17, no. 3, pp. 908–918, 2001.

[42] United State Pharmacopoeia, Validation of Compendial Procedures, United States Pharmacopoeia XXXVII Natl. Formul. XXXII. (2014) 1–5.

[43] A. P. Romio, N. Bernardy, E. Lemos Senna, P. H. H. Araújo, and C. Sayer, "Polymeric nanocapsules via miniemulsion polymerization using redox initiation," *Materials Science and Engineering C*, vol. 29, no. 2, pp. 514–518, 2009.

[44] P. E. Feuser, P. C. Gaspar, A. V. Jacques et al., "Synthesis of ZnPc loaded poly(methyl methacrylate) nanoparticles via miniemulsion polymerization for photodynamic therapy in leukemic cells," *Materials Science and Engineering C*, vol. 60, pp. 458–466, 2016.

Facile TEMPO Immobilization onto Poly(acrylic acid)-Modified Magnetic Nanoparticles: Preparation and Property

Guohui Huang,[1] Xiaoxuan Liu,[1] Yiling Bei,[2] and Huiqing Ma[2]

[1]*School of Materials and Energy, Guangdong University of Technology, Guangzhou, Guangdong 510006, China*
[2]*School of Chemistry and Chemical Engineering, Shandong University, Jinan, Shandong 250100, China*

Correspondence should be addressed to Xiaoxuan Liu; p-xxliu@gdut.edu.cn and Yiling Bei; beiyiling@sdu.edu.cn

Academic Editor: Mehdi Salami-Kalajahi

Adding catalysts to magnetic polyvalent supports facilitating catalyst recycling and recovery seems feasible. Polymer-modified magnetic nanocomposites for organocatalyst immobilization are a plausible approach to this technology. Here, we present facile and efficient method for 2,2′,6,6′-tetramethylpiperidinyl-1-oxy (TEMPO) immobilization onto polymer-modified magnetic nanoparticles under mild reaction conditions. Poly(acrylic acid) was chosen to graft from magnetic nanoparticle through a simple inverse emulsion polymerization technique. The resulting poly(acrylic acid) magnetic nanocomposite is an ideal material to immobilize the organocatalyst 4-hydroxy-2,2′,6,6′-tetramethylpiperidinyl-1-oxy (H-TEMPO) via an esterification reaction with pendant carboxyl group on the polymer chain. Instrumental analysis confirmed that poly(acrylic acid) chain was grafted on the silica-coated magnetic particles by this simple method while maintaining their magnetic properties; elemental analysis indicated that TEMPO was efficiently immobilized onto the polymer chain. The catalysis tests under both Anelli and Minisci system showed that the nanocomposite catalyst exhibits proper selectivity and activity for the alcohol/aldehyde transformation. Recycling experiments showed that stability and reusability of the nanocomposite catalyst were satisfying.

1. Introduction

Green catalysis for synthetic protocols and reactions is a fascinating and a major challenge in sustainable chemistry. There has been an explosion of interest in development of various methods to improve catalyst efficiency [1]. One popular route is to immobilize catalyst to enable the reuse of catalyst and simplify the purification process [2]. Immobilizing catalyst onto different support materials not only improves catalyst productivity but also prevents hazardous catalysts from leaking into the environment [3].

Recently, different types of catalyst support materials such as polymer supports [4–10], ionic liquid supports [10–12], inorganic supports [13–18], metal organic frame supports [19], and nanocomposite supports [20–22] have been invented for immobilizing catalysts on beads, surfaces, and membranes, and so forth [23]. Of these, magnetic iron oxide nanocomposites have been the most widely used support material [2] because they offer enormous surface area with more binding sites; this reduces the disadvantages of heterogeneous catalysis [24].

Typically, organocatalyst TEMPO can utilize harmless and friendly oxidants [25, 26] such as molecular oxygen, hydrogen peroxide, or air for alcohols/aldehydes or ketones stoichiometric transformation. Immobilizing TEMPO on magnetic nanocomposites heightens catalyst-recycling efficiency; however magnetic nanoparticles are inorganic, and it is challenging to directly immobilize organic catalysts. This can lead to low catalyst loading efficiency. Currently, different methods have been successfully used to immobilize TEMPO on magnetic nanoparticles, such as "click" chemistry [27, 28], amine reduction [29–31], and encapsulation and polymer/nanoparticle blends [32]. The design and synthesis of magnetic nanoparticles with polymer chains on its surface [33] for specific catalyst immobilization not only improves immobilization efficiency but also avoids the use of potential harmful substances such azides or cyanoborohydrides that are employed in previous works [27–32]. The polymer chains on the magnetic nanoparticles surface offer pendant functional groups for organocatalyst anchoring reaction.

Here, we synthesized magnetic iron oxide nanoparticles with surface vinyl groups and employed reverse emulsion polymerization for surface polymerization of acrylic acid on the nanoparticle. This approach combines a poly(acrylic acid) chain grafted from magnetic iron oxide nanoparticles and forms an ideal support material for 4-hydroxy-2,2′,6,6′-tetramethylpiperidinyl-1-oxy (H-TEMPO). This novel method uses simple approaches and avoids potential harmful substances [27–31]; hence it is a facile and green approach to prepare iron oxide-based magnetic nanocomposite materials with efficient immobilization of TEMPO. The product was investigated as catalyst for alcohols/aldehydes or ketone transformation under both the Anelli [34] and Minisci [35] protocols.

2. Experimental

2.1. Raw Materials. All reagents and solvents were of analytical grade and used as received without further purification unless otherwise noted. The organic catalyst 4-hydroxy-2,2′,6,6′-tetramethylpiperidinyl-1-oxy (H-TEMPO) and trichlorovinylsilane were purchased from Energy Chemical (China). Dicyclohexylcarbodiimide was purchased from J&K Scientific. Monomer acrylic acid was carefully distilled to remove inhibitor and preserved in refrigerator before use.

2.2. Characterization. Elemental analysis was performed on a CE instruments EA 1110 elemental analyzer. The infrared spectroscopy of all products was recorded in the range 4000–400 cm^{-1} with Bruker Tensor 27 FT-IR spectrometer. Transmission electron microscopy (JEM 2100) was operated at 200 kV for morphological analysis. Magnetic properties of synthesis materials were studied with vibrating sample magnetometer (9600 VSM, BOJ Electronics, Troy, MI) at RT. Gas chromatography analysis was measured on Agilent GC-MS chromatograph (model type: 7890A-5975C).

2.2.1. Preparation of Silica-Coated Magnetite Nanoparticles (Si-Fe_3O_4). Bare iron oxide (Fe_3O_4) nanoparticles were prepared by coprecipitation method with minor modifications [36]. In a typical process, ferric chloride hexahydrate (11.70 g) and ferrous chloride tetrahydrate (4.28 g) were dissolved in 120 mL deionized water at room temperature under nitrogen inert atmosphere with vigorous stirring until the salts were completely dissolved; the temperature was then raised to 60°C. Next, aqueous ammonia (25%, 80 mL) was quickly poured into the solution and the temperature was increased to 90°C with continual stirring for 30 minutes with a mechanical stirrer. Bare Fe_3O_4 nanoparticles were carefully collected with a neodymium magnet and washed with deionized water until they were neutralized.

Bare Fe_3O_4 nanoparticles from the previous step were directly suspended in a solution of ethanol and deionized water (V/V = 4/1, 200 mL) and sonicated for 15 min. Next, 5 mL of tetraethoxysilane (TEOS) was added slowly to the mixture and sonicated for 10 min. The reaction mixture was transferred to a single-neck flask equipped with a mechanical stirring device. Aqueous ammonia (25%, 18 mL) was added within 10 min. The mixture reacted at room temperature for 12 h. Iron oxide nanoparticles coated with silica (Fe_3O_4@SiO_2) were collected and washed three times with water and ethanol using a neodymium magnet and then dried for 24 hours.

2.2.2. Preparation of Magnetic Nanoparticles Coated with Vinyl Group (Vinyl-MNPs). Dry Si-Fe_3O_4 powder (3.0 g) and 6 mL of trichlorovinylsilane were suspended in 150 mL dry toluene under inert atmosphere followed by sonication of the mixture for 30 minutes. The products reacted on a 105°C oil bath and for 10 hours. The product powder was separated using a neodymium magnet, washed with ethanol three times and dried under vacuum.

2.2.3. Surface Polymerization Acrylic Acid on Vinyl-MNPs to Prepare Poly(acrylic acid)@ SiO_2@ Fe_3O_4 (PAA-MNPs). We prepared a mixture of vinyl-MNPs (0.2 g) and acrylic acid (3.0 g) in a 50 mL beaker followed by 10.0 mL of deionized water. This was sonicated for 20 minutes. The suspension was slowly added dropwise to Span-80 cyclohexane solution (20 g Span-80 in 250 mL cyclohexane) with stirring. This was stirred for another hour to preemulsify under an inert atmosphere. The temperature was adjusted to 60°C for 15 min, and then the initiator 2,2′-azobisisobutyronitrile (AIBN, 0.1 g) was added. To ensure complete conversion of the monomers, the reaction proceeded for another 4 hours. Methanol was added for demulsification after the reaction was over. The nanocomposite product was refluxed in ethanol for 2 hours before using a neodymium magnet to collect and wash with methanol; the product was then dried under vacuum.

2.2.4. Immobilization of TEMPO on PAA-MNPs Magnetic Nanoparticles (TEMPO-MNPs). The PAA-MNPs (1 g) and H-TEMPO (2 g) and N,N′-dicyclohexylcarbodiimide (DCC, 1.2 equivalent to H-TEMPO) were suspended in dry THF (100 mL). The reaction mixture was refluxed for 8 hours. The target product was collected using a neodymium magnet and washed three times with ethanol and dried under vacuum.

2.2.5. Determination of Catalysis Activity and Recycling Test

General Procedure for Catalysis Activity Test under Minisci Protocol. Alcohol (20 mmol), TEMPO-MNPs powder (200 mg), $Mn(NO_3)_2$(120 mg, 50% aqueous solution), $Co(NO_3)_2$(120 mg, 50% aqueous solution), and 208.3 mg internal standard (acetophenone) in 15 mL of CH_3COOH were stirred in a two-necked flask at 40°C. The atmosphere of flask was then replaced with pure oxygen (balloon filled). The resulting suspension was stirred at 50°C under this normal atmosphere pressure. Samples were regularly removed for GC-MS analysis.

General Procedure for Catalysis Activity Test under Anelli Protocol. A 25 mL reaction flask was charged with alcohol (20 mmol), 1.0 mL KBr solution (1.0 M) 10 mL dichloromethane, 208.3 mg internal standard (acetophenone), and 150 mg supported TEMPO-MNPs material. The resulting suspension was stirred in an ice-water bath for 10 min before gradual addition of NaClO (0.35 M, pH 8.7, 1.2 eq to alcohols).

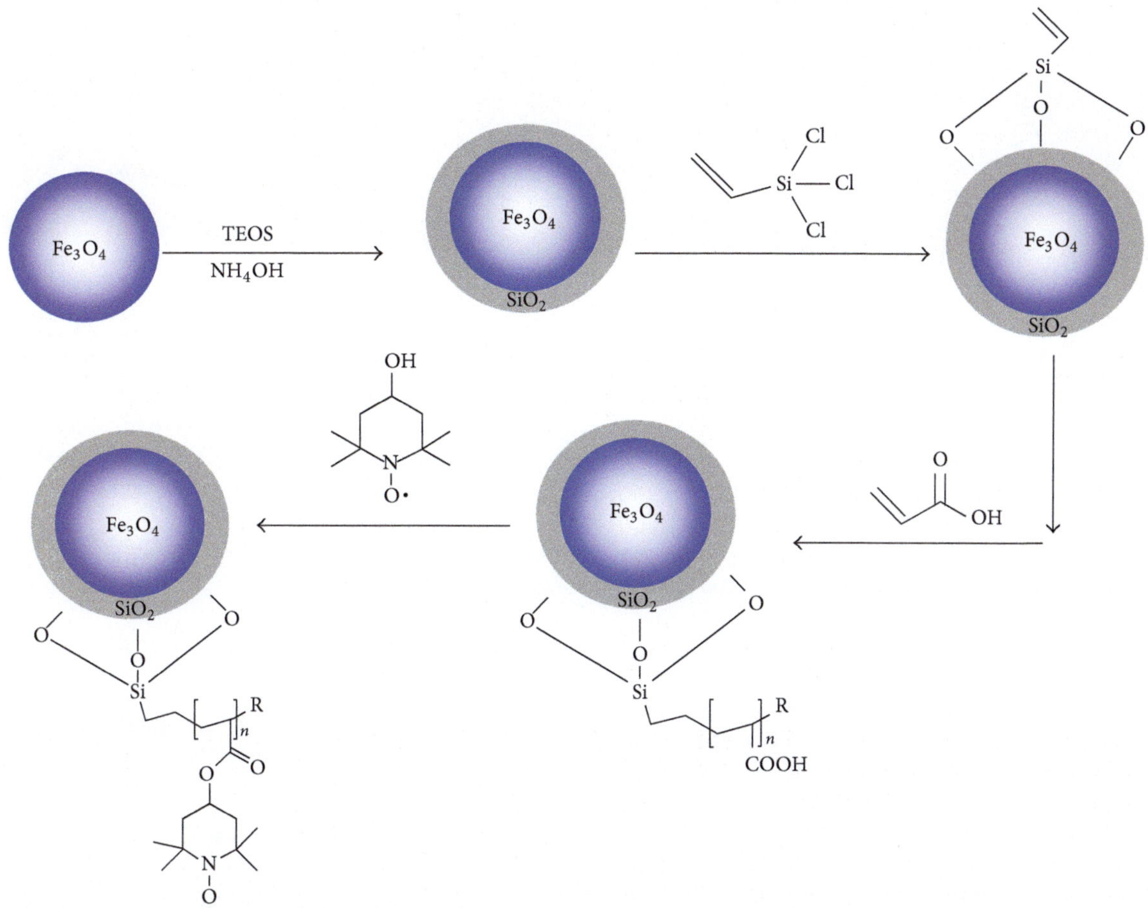

SCHEME 1: Synthetic procedure.

Samples were regularly removed with an external magnet followed by Et_2O (3×5 mL) extraction and direct GC-MS analysis.

Gas chromatography (GC) results for all catalysis experiments were included in the supporting information (see Supplementary Material pages 1 and 2 available online at https://doi.org/10.1155/2017/9621635).

2.2.6. Procedure for Catalysis Recycling Property of Supported TEMPO-MNPs Material. Recycling of supported TEMPO-MNPs materials used a Minisci oxidation method without an internal standard. After the oxidation reaction was complete, the TEMPO-MNPs were magnetically separated and washed subsequently with ethyl dichloromethane, acetone, and H_2O and directly used for subsequent catalysis without further purification. A new reaction was then performed with fresh reactants under the same conditions.

Gas chromatography-mass spectrometry (GC-MS) results of recycling test were included in supporting information (see Supplementary Material page 3).

3. Results and Discussion

The synthetic route for the nanocomposite materials and TEMPO anchoring procedure are illustrated in Scheme 1. The key step was polymerizing acrylic acid onto the magnetic particle surface. In the first step, relatively monodisperse bare (Fe_3O_4) nanoparticles were synthesized by coprecipitation of iron (II) and iron (III) salts under basic solution at 90°C using Huang et al.'s method [36]. There are diverse routes for obtaining magnetic iron oxide particles. The simplest and most efficient chemical pathway is probably coprecipitation. Fortunately, particles prepared by this method contain extra surface hydroxyl (-OH) groups for chemical modification. Tetraethyl orthosilicate (TEOS) was used to coat the bare magnetic particles. The silica shell prevents Fe_3O_4 nanoparticle self-oxidation and leaching of ferric ions from the material core under severe conditions. The outer layer of the shell also provides many more Si-OH units for derivatization with silane coupling agent [37]. In the third step, the silica-coated Fe_3O_4 nanoparticles were dispersed in dry toluene and a trichlorovinylsilane mixture solution and vigorously stirred. This resulted in a surface containing vinyl groups. The vinyl-coated nanocomposite particles were further modified with acrylic acid via an inverse emulsion polymerization resulting in polymer-magnetic support material (Scheme 2).

Reverse emulsion polymerization takes place in latex particles, in which the surface containing vinyl group nanocomposite particles was confined. Vinyl silane compounds are traditionally used in the coating industry to improve adhesiveness on glass product; hence, surface vinyl groups on

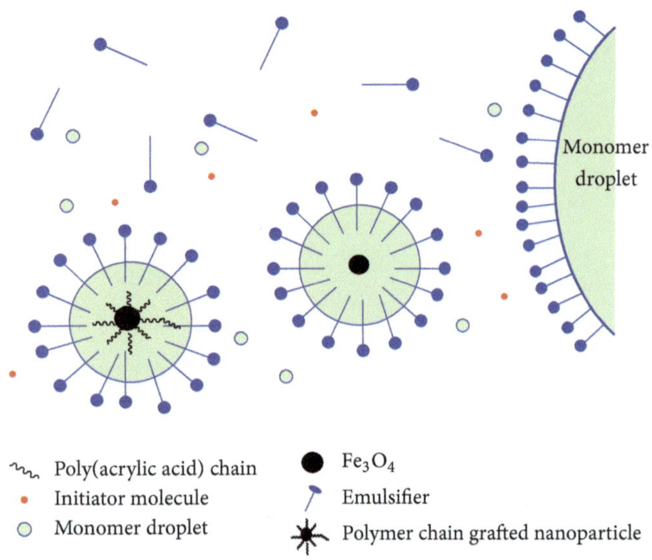

SCHEME 2: Schematic illustration for the inverse emulsion polymerization.

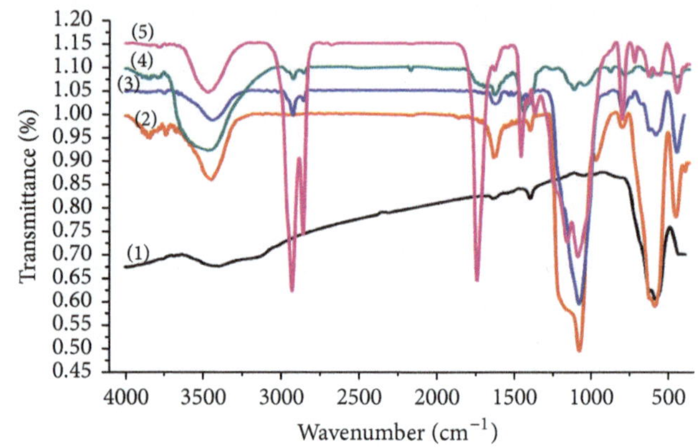

FIGURE 1: FT-IR spectra of prepared Fe_3O_4 materials.

TABLE 1: Elemental analysis results and catalyst loading calculation.

Sample	C (wt%)	H (wt%)	N (wt%)	TEMPO loading
Bare Fe_3O_4	0	0	0	0
Si-Fe_3O_4	0	0	0	0
Vinyl-Fe_3O_4	0.2	0.01	0	0
PAA-Fe_3O_4	9.83	1.10	0	0
TEMPO-MNPs	32.2	3.55	3.51	2.5 mmol/g

nanoparticle surrounded by the monomer could react with radical reactive center to form an acrylic acid polymer chain. Versus bulk polymerization and suspension polymerization, this synthesis procedure can efficiently prevent acrylic acid from crosslinking to form an insoluble magnetized gel. This novel support material has many carboxylic acid groups on the surface because of every side chain of poly(acrylic acid) (PAA) and is well-suited for TEMPO immobilization. The anchoring of TEMPO onto magnetic supports was easily and mildly conducted using N,N′-dicyclohexylcarbodiimide (DCC)/4-dimethyl-amino-pyridine (DMAP) and esterification catalysis in tetrahydrofuran solution.

3.1. FT-IR Analysis and Elemental Analysis. FT-IR spectroscopy was used to verify the conversion of organic functional groups on the particle surface. The inorganic nature of the iron oxide nanoparticle changed the polymer's composite nature. FT-IR data at each modification step clearly revealed this shift in the inorganic to organic composite. The FT-IR spectra of PAA/SiO_2/Fe_3O_4 (Figure 1(4)) showed characteristic saturated hydrocarbons stretching vibration absorption (3000~2900 cm^{-1}, C-H stretching) and a band at 1710 cm^{-1} (carbonyl). These data confirmed PAA segment bonding to the surface of magnetic nanoparticles.

After esterification, the anchoring of TEMPO onto the polymer chains of magnetic supports enhanced the organic absorption peaks. Comparison before and after the anchoring of TEMPO (Figure 1(4, 5)) shows a red shift in the carboxyl group peak (1710 cm^{-1}–1718 cm^{-1}) further indicating that the TEMPO was successfully immobilized onto the magnetic carrier.

Elemental analysis confirmed TEMPO loading was approximately 2.5 mmol/g (elemental results were provided in Table 1 and catalyst loading was calculated according to nitrogen content). This result also indicated that TEMPO was simply and conveniently attached to the particle surface polymer chain with respect to the carboxyl group. Previous studies have immobilized TEMPO on solid surfaces via "click" reactions, amide reduction reactions, and so forth. Recently, Karimi and coworkers showed that TEMPO was linked to the surface of the aminopropyltriethoxysilane-modified magnetic nanoparticles through amide reduction using sodium cyanoborohydrides [29–31]. Another previously known example reported by Schätz et al. [27] and Tucker-Schwartz and Garrell [28] is through "click" reaction using azide modified nanoparticles with TEMPO derivatives. However, those synthesis procedures require harsh reaction conditions or expensive catalysts. In contrast to the reports, our method requires no unstable azide compounds or toxic cyanoborohydrides substances but also increases catalyst loading to 2.5 mmol/g (0.1 mmol/g in the literature [28]). Hence, this synthetic route is simple and efficient.

3.2. Magnetic Properties and TEM Analysis. Transmission electron microscopy (TEM) (Figure 2) characterization of the magnetic materials indicated that bare nanoparticles were relatively uniform with an average size range of 10~30 nm. The image also revealed that the Fe_3O_4 nanoparticles were coated with mesoporous organic compound shells. Room-temperature magnetization curves (Figure 2(c)) showed that, after several modifications, the nanoparticle remained superparamagnetic.

3.3. Catalyst Activity and Recycling Test. With the TEMPO-MNPs in hand, catalytic application and selective oxidation

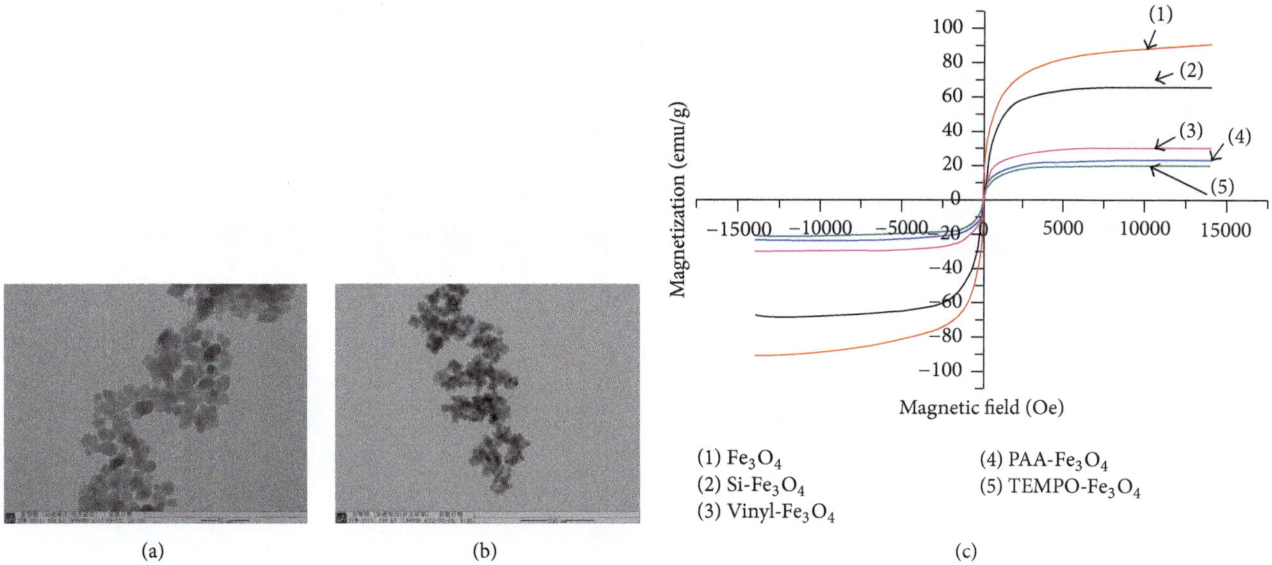

FIGURE 2: TEM images of bare Fe_3O_4 materials (a); TEMPO-MNPs material (b); magnetization curves of the resulting Fe_3O_4 materials (c).

TABLE 2: TEMPO-MNPs mediated oxidation of alcohols to aldehydes/ketones.

ENTRY	Alcohol	Conversion[c] [%]	Yield[c] [%]	Selectivity[c] [%]
1[a]	Benzyl alcohol	99	99	99.0
2[a]	4-Chlorobenzyl alcohol	98	98	98.0
3[a]	4-Methoxylbenzyl alcohol	99	84.8	85.6
4[a]	4-Phenyl-2-butanol	99	95.1	96.1
5[b]	Benzyl alcohol	99	96.9	97.8
6[b]	4-Chlorobenzyl alcohol	99	83.1	84.0
7[b]	4-Methoxylbenzyl alcohol	99	99	99.0

[a]Alcohol (1 mmol) in DCM, potassium bromide solution (1.0 M, 1 mL), 4-methoxyacetophenone (208.0 mg, internal standard), and TEMPO-MNPs (150 mg) were added to a round-bottom flask. The suspension was stirred in ice-water bath before addition of NaClO solution (0.35 M, pH 8.7, 1.05 eq). Sample aliquots were regularly taken and analyzed by GC. [b]A 25 mL Schlenk-flask was charged with a solution of 1 mmol alcohol, TEMPO-MNPs (200 mg), $Mn(NO_3)_2$ (50%, 120 mg), and $Co(NO_3)_2$·6H2O (0.4 mmol, 100 mg) and in 15 mL of acetic acid was stirred at 40°C under O_2 at atmospheric pressure. Sample aliquots were regularly taken and analyzed by GC. [c]Conversion, yield, and selectivity were calculated according to GC or GS-MS results in the supporting information.

of alcohol were further investigated along with its recycling efficiency. There are many catalytic systems employing nitroxides as an active catalyst in cooperation with metal or metal-free cocatalysts for alcohol oxidation. They can catalyze alcohols to the corresponding aldehydes/ketones in high yield and with chemoselectivity [34, 35, 38–40]. For example, Anelli [34] first used TEMPO as a catalyst in cooperation with hypochlorite and bromide for benzyl alcohols selective oxidation. The Anelli protocol proposed by Montanari and Anelli used a two-phase (CH_2Cl_2/H_2O) system and sodium hypochlorite as the oxidant. This convenient approach then uses bromide ion as a cocatalyst for selective oxidation of alcohols. Minisci and coworkers (Minisci protocol) used an environmentally friendly oxidant such as oxygen or air and reported aerobic oxidation catalyzed by TEMPO in cooperation with trace amounts of $Mn(NO_3)_2$ and $Co(NO_3)_2$ in acetic acid solution with excellent conversion and chemoselectivity [34]. Thus, the catalysis procedures based on TEMPO present a practical method for the selective transformation of alcohols to aldehydes/ketones. To determine if the TEMPO-MNPs retained their catalytic activity of small molecular catalysts, we investigated catalysis test under Minisci and Anelli protocols (Table 2).

This approach catalytically transformed alcohols to the corresponding carbonyl compounds (Table 2) using the insoluble magnetically supported TEMPO. Under the Anelli oxidation protocol, primary benzyl alcohols were quickly and efficiently converted into the aromatic aldehydes (5 min) (Table 2, entries 1–3). Secondary alcohols can also be transformed to ketones almost as quickly as the benzyl alcohol (entry 4, Table 2) with TEMPO catalysis.

Analogously, aldehydes were prepared by the selective transformation of primary alcohols under mild conditions proposed by Minisci; the yield was good. The catalytic activity of supported TEMPO materials combining cocatalysts Mn(II)-Co(II) nitrates was also applied to the aerobic oxidation of alcohols/aldehydes with excellent yields (entries 5–7, Table 2). The catalytic oxidation experiments under two well-known procedures showed that the organic catalyst TEMPO was successfully immobilized on the nanocomposite material

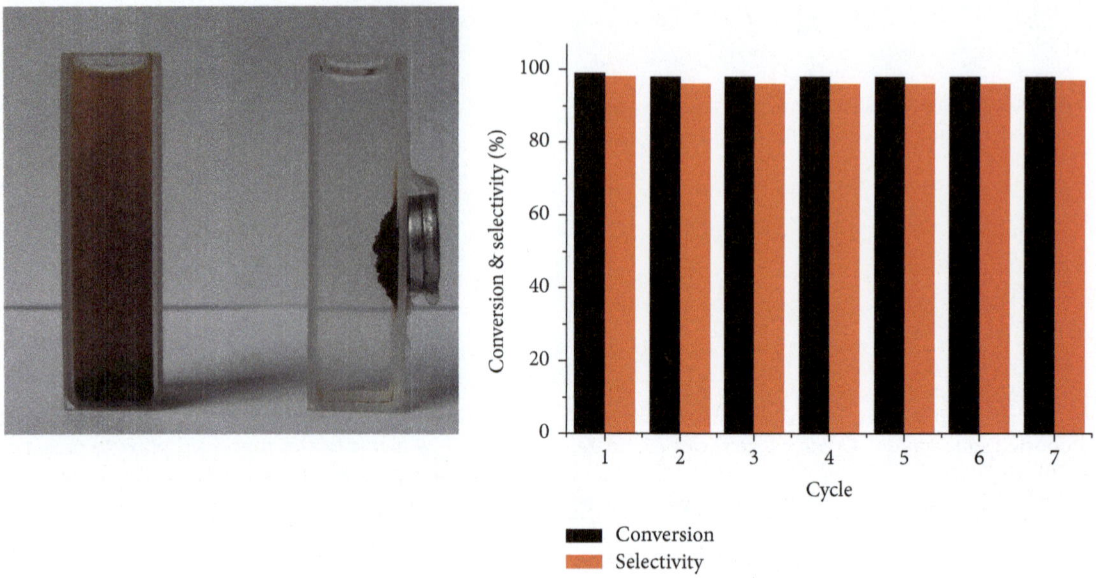

(a) (b)

FIGURE 3: (a) Image for the magnetic separation of the nanoparticles in the Montanari oxidation of benzyl alcohol. (b) Recyclability of TEMPO supported on nanocomposite particles in the Montanari oxidation of benzyl alcohol.

while still maintaining good catalytic ability throughout the synthesis process.

Encouraged by these results, we next investigated TEMPO-MNPs recyclability and reusability via oxidation of 4-methoxylbenzyl alcohol under the Montanari oxidation procedure. TEMPO-MNPs (150 mg scale) recyclability was achieved via separation through successive magnetic decantation as shown in Figure 3(a). After washing thoroughly with CH_2Cl_2, acetone, and H_2O, successively, TEMPO-MNPs continued in the next run. The recycling and reuse demonstrated the efficiency of magnetic separation of the prepared material. The reusability of TEMPO-supported material was illustrated by GC conforming 4-methoxylbenzyl alcohol conversion in the 7th cycle. Oxidation conversion of 7th run was 97% after a reaction time of 5 min and with a selectivity of >96% (Figure 3(b), cycle 7). GC-MS analysis of the separated supernatant (as shown in Figure 3(a)) indicated no trace of TEMPO residue during oxidation. TEMPO remained stable under variety of catalytic environments despite being attached to a poly(acrylic acid) chain through an ester bond TEMPO-MNPs.

4. Conclusion

In conclusion, we developed a novel method to prepare magnetic polymer nanoparticles as an ideal support material for immobilizing organic catalysts 2,2′,6,6′-tetramethylpiperidinyl-1-oxy (TEMPO). According to characterization results, acrylic acid is efficiently polymerized onto the silica-coated magnetic particles by applying inverse emulsion polymerization, and TEMPO is immobilized onto the particles. The synthesis procedures are simple and facile. They efficiently immobilized TEMPO. TEMPO-MNPs still retained the advantages of magnetic materials (simple separation with no mechanical loss) and showed excellent catalytic capabilities in alcohols for selective oxidation under the Anelli and Minisci procedures. The recyclability study on TEMPO-MNPs shows neither significant loss of activity nor selectivity dropping after 7 consecutive cycles.

Additional Points

Supporting Information. Gas chromatography (GC) and gas chromatography-mass spectrometry (GC-MS) results for catalysis experiments are included in the supporting information.

Acknowledgments

This work was financially supported by the National Nature Science Foundation of China (Grant nos. 51403041 and 51641302) and China Postdoctoral Science Foundation (2014M562183). The authors thank LetPub (www.letpub.com) for its linguistic assistance during the preparation of this manuscript.

References

[1] L. L. Chng, N. Erathodiyil, and J. Y. Ying, "Nanostructured catalysts for organic transformations," *Accounts of Chemical Research*, vol. 46, no. 8, pp. 1825–1837, 2013.

[2] V. Polshettiwar, R. Luque, A. Fihri, H. B. Zhu, M. Bouhrara, and J.-M. Basset, "Magnetically recoverable nanocatalysts," *Chemical Reviews*, vol. 111, no. 5, pp. 3036–3075, 2011.

[3] R. B. N. Baig and R. S. Varma, "Magnetically retrievable catalysts for organic synthesis," *Chemical Communications*, vol. 49, no. 8, pp. 752–770, 2013.

[4] R. B. N. Baig and R. S. Varma, "Copper on chitosan: a recyclable heterogeneous catalyst for azide-alkyne cycloaddition reactions in water," *Green Chemistry*, vol. 15, no. 7, pp. 1839–1843, 2013.

[5] M. Bradshaw, J. Zou, L. Byrne, K. Swaminathan Iyer, S. G. Stewart, and C. L. Raston, "Pd(ii) conjugated chitosan nanofibre mats for application in Heck cross-coupling reactions," *Chemical Communications*, vol. 47, no. 45, pp. 12292–12294, 2011.

[6] P. Kasaplar, C. Rodríguez-Escrich, and M. A. Pericàs, "Continuous flow, highly enantioselective Michael additions catalyzed by a PS-supported squaramide," *Organic Letters*, vol. 15, no. 14, pp. 3498–3501, 2013.

[7] Y. Li, Y. Wu, Q. Xu et al., "Facile and controllable synthesis of polystyrene/palladium nanoparticle@polypyrrole nanocomposite particles," *Polymer Chemistry*, vol. 4, no. 17, pp. 4655–4662, 2013.

[8] G. Strappaveccia, D. Lanari, D. Gelman et al., "Efficient synthesis of cyanohydrin trimethylsilyl ethers via 1,2-chemoselective cyanosilylation of carbonyls," *Green Chemistry*, vol. 15, no. 1, pp. 199–204, 2013.

[9] C. Yang, C.-H. Choi, C.-S. Lee, and H. Yi, "A facile synthesis-fabrication strategy for integration of catalytically active viral-palladium nanostructures into polymeric hydrogel microparticles via replica molding," *ACS Nano*, vol. 7, no. 6, pp. 5032–5044, 2013.

[10] D. P. Hickey, R. D. Milton, D. Chen, M. S. Sigman, and S. D. Minteer, "TEMPO-modified linear poly(ethylenimine) for immobilization-enhanced electrocatalytic oxidation of alcohols," *ACS Catalysis*, vol. 5, no. 9, pp. 5519–5524, 2015.

[11] M. V. Khedkar, T. Sasaki, and B. M. Bhanage, "Immobilized palladium metal-containing ionic liquid-catalyzed alkoxycarbonylation, phenoxycarbonylation, and aminocarbonylation reactions," *ACS Catalysis*, vol. 3, no. 3, pp. 287–293, 2013.

[12] Z.-L. Shen, H.-L. Cheong, Y.-C. Lai, W.-Y. Loo, and T.-P. Loh, "Application of recyclable ionic liquid-supported imidazolidinone catalyst in enantioselective Diels-Alder reactions," *Green Chemistry*, vol. 14, no. 9, pp. 2626–2630, 2012.

[13] H. Zhao, W. Hao, Z. Xi, and M. Cai, "Palladium-catalyzed cross-coupling of PhSeSnBu3 with aryl and alkyl halides in ionic liquids: a practical synthetic method of diorganyl selenides," *New Journal of Chemistry*, vol. 35, no. 11, pp. 2661–2665, 2011.

[14] K. Arya, U. C. Rajesh, and D. S. Rawat, "Proline confined FAU zeolite: heterogeneous hybrid catalyst for the synthesis of spiroheterocycles via a mannich type reaction," *Green Chemistry*, vol. 14, no. 12, pp. 3344–3351, 2012.

[15] M. Genelot, N. Villandier, A. Bendjeriou, P. Jaithong, L. Djakovitch, and V. Dufaud, "Palladium complexes grafted onto mesoporous silica catalysed the double carbonylation of aryl iodides with amines to give α-ketoamides," *Catalysis Science and Technology*, vol. 2, no. 9, pp. 1886–1893, 2012.

[16] S. Wang, Q. Zhao, H. Wei et al., "Aggregation-free gold nanoparticles in ordered mesoporous carbons: Toward highly active and stable heterogeneous catalysts," *Journal of the American Chemical Society*, vol. 135, no. 32, pp. 11849–11860, 2013.

[17] A. Machado, M. H. Casimiro, L. M. Ferreira et al., "New method for the immobilization of nitroxyl radical on mesoporous silica," *Microporous and Mesoporous Materials*, vol. 203, pp. 63–72, 2015.

[18] N. Mittal, G. M. Nisola, J. G. Seo, S.-P. Lee, and W.-J. Chung, "Organic radical functionalized SBA-15 as a heterogeneous catalyst for facile oxidation of 5-hydroxymethylfurfural to 2,5-diformylfuran," *Journal of Molecular Catalysis A: Chemical*, vol. 404-405, pp. 106–114, 2015.

[19] D. T. Genna, A. G. Wong-Foy, A. J. Matzger, and M. S. Sanford, "Heterogenization of homogeneous catalysts in metal-organic frameworks via cation exchange," *Journal of the American Chemical Society*, vol. 135, no. 29, pp. 10586–10589, 2013.

[20] A. Pourjavadi, S. H. Hosseini, M. Doulabi, S. M. Fakoorpoor, and F. Seidi, "Multi-Layer functionalized poly(ionic liquid) coated magnetic nanoparticles: highly recoverable and magnetically separable brønsted acid catalyst," *ACS Catalysis*, vol. 2, no. 6, pp. 1259–1266, 2012.

[21] S. Wang, Z. Zhang, B. Liu, and J. Li, "Silica coated magnetic Fe3O4 nanoparticles supported phosphotungstic acid: a novel environmentally friendly catalyst for the synthesis of 5-ethoxymethylfurfural from 5-hydroxymethylfurfural and fructose," *Catalysis Science & Technology*, vol. 3, no. 8, pp. 2104–2112, 2013.

[22] J. Zhu, P.-C. Wang, and M. Lu, "Synthesis of novel magnetic silica supported hybrid ionic liquid combining TEMPO and polyoxometalate and its application for selective oxidation of alcohols," *RSC Advances*, vol. 2, no. 22, pp. 8265–8268, 2012.

[23] J. Lu and P. H. Toy, "Organic polymer supports for synthesis and for reagent and catalyst immobilization," *Chemical Reviews*, vol. 109, no. 2, pp. 815–838, 2009.

[24] Q. Zhang, J. Kang, B. Yang, L. Zhao, Z. Hou, and B. Tang, "Immobilized cellulase on Fe_3O_4 nanoparticles as a magnetically recoverable biocatalyst for the decomposition of corncob," *Chinese Journal of Catalysis*, vol. 37, no. 3, pp. 389–397, 2016.

[25] Q. Cao, L. M. Dornan, L. Rogan, N. L. Hughes, and M. J. Muldoon, "Aerobic oxidation catalysis with stable radicals," *Chemical Communications*, vol. 50, no. 35, pp. 4524–4543, 2014.

[26] S. Wertz and A. Studer, "Nitroxide-catalyzed transition-metal-free aerobic oxidation processes," *Green Chemistry*, vol. 15, no. 11, pp. 3116–3134, 2013.

[27] A. Schätz, R. N. Grass, W. J. Stark, and O. Reiser, "TEMPO supported on magnetic C/Co-nanoparticles: a highly active and recyclable organocatalyst," *Chemistry-A European Journal*, vol. 14, no. 27, pp. 8262–8266, 2008.

[28] A. K. Tucker-Schwartz and R. L. Garrell, "Simple preparation and application of TEMPO-coated Fe3O4 superparamagnetic nanoparticles for selective oxidation of alcohols," *Chemistry-A European Journal*, vol. 16, no. 42, pp. 12718–12726, 2010.

[29] B. Karimi and E. Farhangi, "A highly recyclable magnetic core-shell nanoparticle-supported TEMPO catalyst for efficient metal- and halogen-free aerobic oxidation of alcohols in water," *Chemistry - A European Journal*, vol. 17, no. 22, pp. 6056–6060, 2011.

[30] B. Karimi and E. Farhangi, "One-pot oxidative passerini reaction of alcohols using a magnetically recyclable TEMPO under metal- and halogen-free conditions," *Advanced Synthesis and Catalysis*, vol. 355, no. 2-3, pp. 508–516, 2013.

[31] B. Karimi, H. M. Mirzaei, and E. Farhangi, "Fe_3O_4@SiO_2-TEMPO as a magnetically recyclable catalyst for highly selective aerobic oxidation of 5-hydroxymethylfurfural into 2,5-diformylfuran under metal-and halogen-free conditions," *ChemCatChem*, vol. 6, no. 3, pp. 758–762, 2014.

[32] Z. Zheng, J. L. Wang, M. Zhang, L. Xu, and J. B. Ji, "Magnetic polystyrene nanosphere immobilized TEMPO: a readily prepared, highly reactive and recyclable polymer catalyst in the selective oxidation of alcohols," *ChemCatChem*, vol. 5, no. 1, pp. 307–312, 2013.

[33] M. Benaglia, A. Puglisi, and F. Cozzi, "Polymer-supported organic catalysts," *Chemical Reviews*, vol. 103, no. 9, pp. 3401–3429, 2003.

[34] P. L. Anelli, C. Biffi, F. Montanari, and S. Quici, "Fast and selective oxidation of primary alcohols to aldehydes or to carboxylic-acids and of secondary alcohols to ketones mediated by oxoammonium salts under two-phase conditions," *Journal of Organic Chemistry*, vol. 52, no. 12, pp. 2559–2562, 1987.

[35] A. Cecchetto, F. Fontana, F. Minisci, and F. Recupero, "Efficient Mn-Cu and Mn-Co-TEMPO-catalysed oxidation of alcohols into aldehydes and ketones by oxygen under mild conditions," *Tetrahedron Letters*, vol. 42, no. 38, pp. 6651–6653, 2001.

[36] Y.-F. Huang, Y.-F. Wang, and X.-P. Yan, "Amine-functionalized magnetic nanoparticles for rapid capture and removal of bacterial pathogens," *Environmental Science and Technology*, vol. 44, no. 20, pp. 7908–7913, 2010.

[37] D. Wang, L. Salmon, J. Ruiz, and D. Astruc, "A recyclable ruthenium(ii) complex supported on magnetic nanoparticles: a regioselective catalyst for alkyne-azide cycloaddition," *Chemical Communications*, vol. 49, no. 62, pp. 6956–6958, 2013.

[38] P. Gamez, I. W. C. E. Arends, J. Reedijk, and R. A. Sheldon, "Copper(II)-catalysed aerobic oxidation of primary alcohols to aldehydes," *Chemical Communications*, vol. 9, no. 19, pp. 2414-2415, 2003.

[39] A. M. Khenkin and R. Neumann, "Oxygen transfer from sulfoxides: Selective oxidation of alcohols catalyzed by polyoxomolybdates," *Journal of Organic Chemistry*, vol. 67, no. 20, pp. 7075–7079, 2002.

[40] M. F. Semmelhack, C. R. Schmid, D. A. Cortes, and C. S. Chou, "Oxidation of alcohols to aldehydes with oxygen and cupric ion, mediated by nitrosonium ion," *Journal of the American Chemical Society*, vol. 106, no. 11, pp. 3374–3376, 1984.

18

Effect of Polymer Binder on the Transparent Conducting Electrodes on Stretchable Film Fabricated by Screen Printing of Silver Paste

Chan Kyu Lim,[1] Yo Seb Lee,[2] Sung Hoon Choa,[2] Deuk Young Lee,[3] Lee Soon Park,[4] and Su Yong Nam[1]

[1] *Department of Graphic Arts Information Engineering, Pukyong National University, Busan 48547, Republic of Korea*
[2] *Graduate School of NID Fusion Technology, Seoul National University of Science and Technology, Seoul 01811, Republic of Korea*
[3] *ELK Corporation, Daejeon 34013, Republic of Korea*
[4] *School of Material Science and Engineering, Ulsan National Institute of Science and Technology (UNIST), Ulsan 44919, Republic of Korea*

Correspondence should be addressed to Su Yong Nam; suynam@pknu.ac.kr

Academic Editor: Zhe Qiang

Smart wearable devices and sensors have been fabricated by screen printing of metal paste as functional circuits since the metal interconnects exhibited much less electrical resistance than other conducting materials such as carbon nanotube or conducting polymers (PEDOT:PSS). In this study, we chose silver particle as conductive material in the form of silver paste and used screen printing to fabricate a stretchable touch screen panel utilizing metal mesh method for the transparent electrode patterning. The rheological study of Ag pastes showed that the binder polymer with high molecular weight and low glass transition temperature (T_g) can stabilize the Ag paste with Ag particle content over 80% by weight. The stretching and bending tests of Ag electrode films obtained by screen printing indicated that good conductivity of Ag electrodes is related to the stability of Ag paste obtained with the high molecular weight binder polymer.

1. Introduction

Transparent conducting electrode is widely used in touch screen panels (TSPs), flexible displays, and wearable devices [1–3]. The printed electronics enable a fast and continuous process in a large scale at ambient temperature [4–6]. Screen printing is a low-cost process to fabricate a transparent conducting film for TSPs compared to the photolithographic process which involves metal thin film deposition, UV exposure, and development conducted in the clean room. Various materials including indium-tin oxide (ITO), carbon nanotube (CNT), silver nanoparticle/nanowire, and conducting polymer (PEDOT:PSS) have been used as interconnects for the fabrication of smart displays and wearable devices [4, 5]. In contrast to the conventional electronics based on rigid circuit boards or flexible films such as polyester or polyimide film, the stretchable electronic devices need to be fabricated on elastomeric polymer films such as polyurethane and polydimethylsiloxane.

As for the conductive materials, the CNT compounds and conducting polymer (PEDOT:PSS) can be stretched to high elongations but suffer from high electrical resistance. However, metal interconnects based on metal nanopowder or metal nanowire show much less electrical resistance although they have limitation in the high elongation [7, 8]. In this study, we chose silver particle as conductive material in the form of silver paste and used screen printing to fabricate a stretchable touch screen panel utilizing metal mesh method for the transparent electrode patterning.

2. Materials and Method

2.1. Materials and Preparation of Silver Paste. The substrates tested include both rigid and stretchable films with high

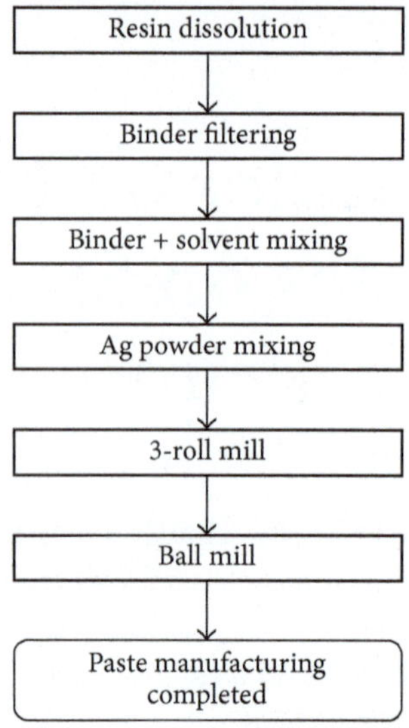

FIGURE 1: Preparation process of Ag paste.

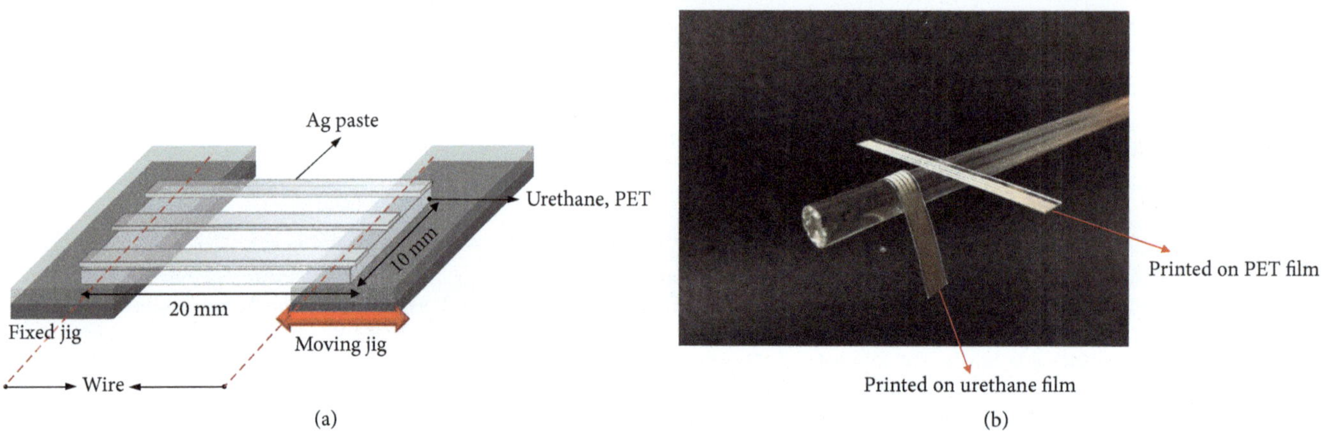

FIGURE 2: Screen printing of Ag paste: (a) electrode pattern and (b) Ag electrodes on polyester and polyurethane films.

visible light transmittance in the range of 92.2%–92.5% such as polyester and thermoplastic polyurethane films. The unsaturated polyesters with different molecular weight and glass transition temperatures (T_g) were used as binder polymer to make silver paste. The flake type silver powder had average diameter of 1.93 μm and top density of 3.52 g/cc. 2-Ethoxyethyl acetate (ECA) and benzoyl peroxide were used as solvent and thermal initiator. 3-Aminopropyltriethoxysilane was used as coupling agent for improved adhesion of the silver paste. The preparation of silver paste is shown in Figure 1. First, the unsaturated polyester resin and ECA solvent were added to three-neck flask at 50 : 50 wt% ratio and stirred at 65°C for 24 hour followed by filtration with 300 mesh screen. The concentrated binder polymer solution was diluted with ECA solvent and mixed with silver powder followed by addition of dispersant (BYK-180), coupling agent, and thermal initiator. The resulting Ag pastes were milled 5 times with 3-roll mill (Gyeongyong Machinery Co., Korea) and further stabilized by roll miller at 10 rpm for 24 hr before subjecting to the rheological tests and screen printing.

2.2. Electrode Patterning by Screen Printing and Measurements. Silver electrode patterns (line/space = 1 mm/1 mm) were obtained by screen printing method on the stretchable polyurethane (thickness 200 μm) and relatively rigid polyester (175 μm) films as shown in Figure 2. Silver paste printed films were cured at 130°C for 30 min in the box heater. The rheological property of the silver paste was measured at 23°C by rotating disk rheometer (Thermohaake, Kheowin Pro. 2.92, Germany).

The variations of electrical resistance with stretching and bending modes were measured by using the polyurethane and polyester films, respectively. For the bending test, the bending radius of curvature was calculated by the following equations as shown in Figure 3:

$$\text{Bending radius } (r_1) = \frac{L}{2\pi\sqrt{dL/L - \pi^2 hs^2/12L^2}}, \quad (1)$$

$$\text{Bending radius } (r_2) = \frac{L - dL}{L}. \quad (2)$$

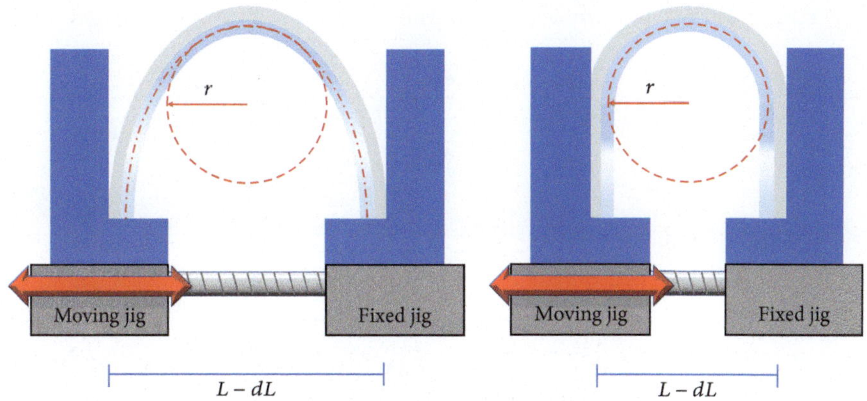

FIGURE 3: Schematic diagram of bending test.

Here, L is the length of the substrate film, dL is moving distance, and hs is the thickness of the test films. Equation (1) is used in the bending test where the test film is not contacted with the wall of the bending machine, while (2) is used when the film is contacted with the wall.

3. Results and Discussion

3.1. Rheological Properties of the Silver Paste. In the screen printing process, the composition of silver paste, especially the binder polymer, plays important role on both the patterning of silver electrodes and the stretching and bending property of the resulting transparent Ag electrode films. In this work, unsaturated polyester resin was used as binder polymer which has good dispersing property of silver particles over 80% by weight. The unsaturated polyester samples were obtained from SK Chemical Co., Korea, of which molecular weights and glass transition temperatures were measured by GPC and DSC, respectively. Table 1 shows the three different unsaturated polyester resins in terms of molecular weight and glass transition temperature and the formulation of three silver pastes made with the polyester resins.

Here ethoxyethyl acetate (ECA) was used as solvent and the three additives include dispersant (BYK-180, 0.7 wt%), silane coupling agent (Shin-Etsu, KBE-903:3, 0.6 wt%), and thermal initiator (benzoyl peroxide, 0.6 wt%). The dispersant BYK-180 was obtained from BKY Co., Germany. It is an acrylate copolymer with carboxylic groups (acid value of 94 mgKOH/g) neutralized with alkyl ammonium salt which help dispersion of Ag particles in the ECA solvent.

The viscosity versus shear rate plot of the Ag pastes obtained at room temperature (23°C) is shown in Figure 4. The viscosity of the Ag paste was measured first with increasing shear rate and then with decreasing shear rate to get hysteresis loops. The hysteresis loop obtained by the plot of Ag paste viscosity versus shear rate is closely related to the stability of the well-dispersed Ag paste; the smaller the hysteresis loop the more stable the paste. Although Ag paste (1) and paste (2) had similar hysteresis loop, Ag paste (1) with higher molecular weight (MW) binder polymer than the other two exhibited well-dispersed Ag particles as shown in the optical microscope image (Figure 5) of screen printed electrode. This suggests that the binder polymer with higher MW can stabilize the Ag paste better than the ones with lower MW under the condition of excessive Ag particles (over 80% by weight) in the paste, which is a requirement setup for the good conductivity of the silver electrodes under high deformation through the percolation mechanism of electrical conduction with silver particles.

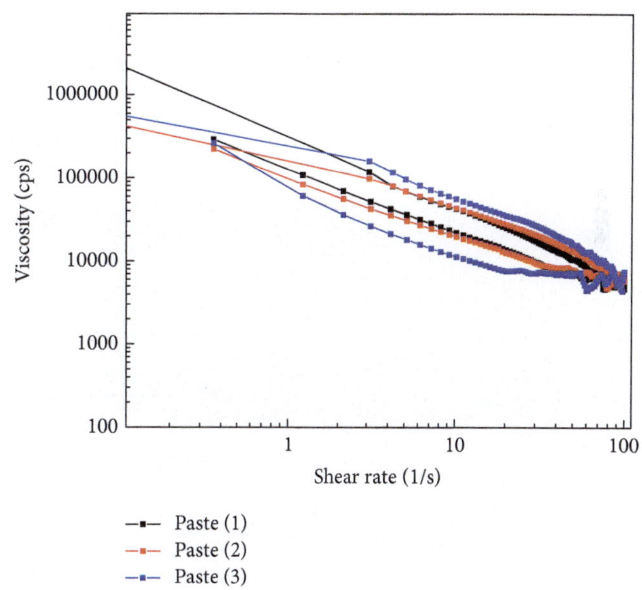

FIGURE 4: Viscosities versus shear rate plot of Ag paste.

The thixotropic index (TI) values defined as TI = (viscosity of paste at 5 s^{-1})/(viscosity of paste at 50 s^{-1}) also support the viscosity versus shear rate plot as shown in Table 2 in which Ag paste (1) made with high MW binder polymer had higher TI values than the other two Ag paste samples.

The storage (G′) and loss (G″) modulus of silver pastes are shown in Figure 6. Ag paste (1) with high MW binder polymer maintained higher storage modulus up to high shear stress than the other two Ag pastes. The high storage modulus of the Ag paste results in increased sharpness of Ag electrodes in the screen printing, which can be maintained after heat curing. This is an important point in the electrical

Table 1: Physical property of binder polymers and formulation of silver pastes.

(a)

Binder polymers	Glass transition temp. (T_g)	Mole weight (M_n, g/mol)	Softening temp (°C)
Polyester (1)	16°C	40,000	120°C
Polyester (2)	70°C	28,000	145°C
Polyester (3)	10°C	26,000	90°C

(b)

Ag pastes (wt%)	Ag powder	Polyester (1)	Polyester (2)	Polyester (3)	Solvent	Additives
Ag paste (1)	83.4	4.1			10.6	1.9
Ag paste (2)	83.4		4.1		10.6	1.9
Ag paste (3)	83.4			4.1	10.6	1.9

Table 2: Viscosity and thixotropic index (TI) values of silver pastes.

Ag paste	Viscosity (50 s^{-1}, cps)	TI (5/50 s^{-1})
Ag paste (1)	10,495	6.67
Ag paste (2)	13,323	5.16
Ag paste (3)	16,128	5.81

Figure 5: Optical microscope image of screen printed Ag electrode (×600) by using Ag paste (1).

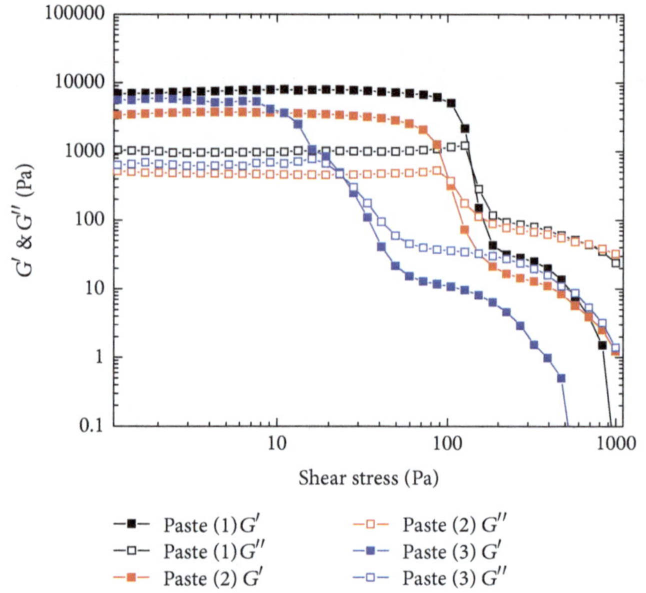

Figure 6: Storage moduli and loss moduli versus shear stress plot of Ag paste.

conductivity of the Ag electrodes under stretching or bending deformations, since the uniform shape of multiple layered Ag particles obtained as a result of increased sharpness during screen printing can meet the requirement of electrical conduction by percolation mechanism. Ag paste (2) showed higher storage modulus than Ag paste (3) while the MW of binder polymers were nearly same. This could be explained by much higher T_g value of binder polymer (2) than binder polymer (3), corresponding to more extended polymer conformation or effective molecular weight, thus resulting in higher storage modulus.

3.2. Stretching Test of Silver Electrodes Films. Flexible polyurethane films with patterned silver electrodes and without pattern (Ag paste printed all surface) were subjected to the electrical resistance versus elongation tests as shown in Figures 7 and 8, respectively. The higher the molecular weight of binder polymer in Ag pastes was, the lower the electrical resistance was observed in both the patterned and unpatterned Ag electrode films. It was noted that Ag electrode films printed on all surface exhibited much lower resistance at higher elongation compared to the patterned Ag electrode films. The order of low electrical resistance upon stretching was the same as the MW of binder polymer in the Ag pastes. This could be explained as follows although the electrical resistance versus elongation of Ag electrode film tests was performed after curing of the screen printed Ag pastes at 130°C for 30 min. As the degree of unsaturation in the diacid monomer used to make binder polymer is the same, it can be assumed that the increase of MW of binder polymer after curing at 130°C for 30 min will show same trend in the order of MW as the uncured binder polymer. This suggests that the good dispersion and stability of the silver paste are very important in the mechanical properties of the resulting transparent conducting electrode films.

3.3. Bending Test of Silver Electrode Films. Bending tests were conducted with the relatively rigid polyester (PET) films on

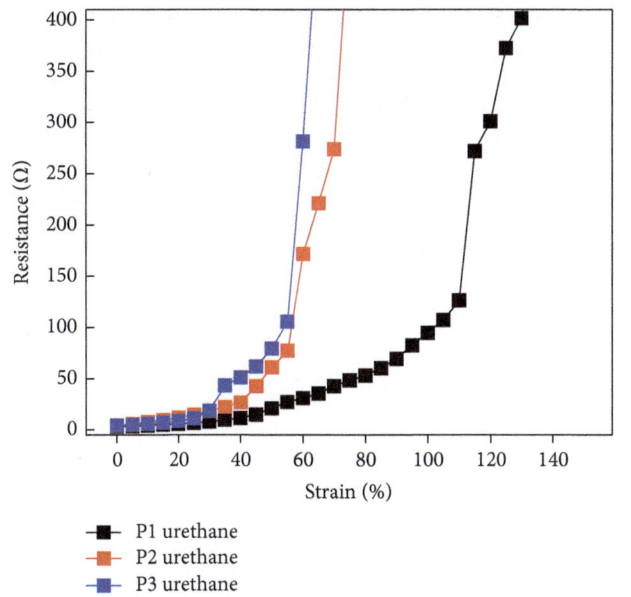

FIGURE 7: Electrical resistance versus tensile strain plot of patterned Ag electrode films.

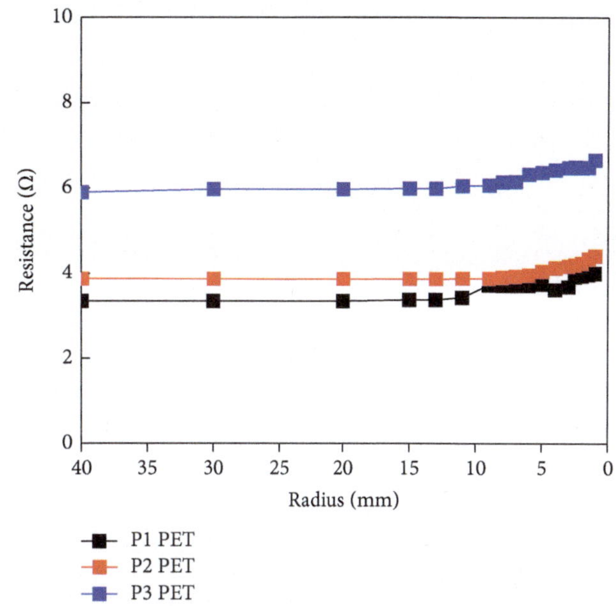

FIGURE 9: Electrical resistance versus radius of curvature plot for outer bending test.

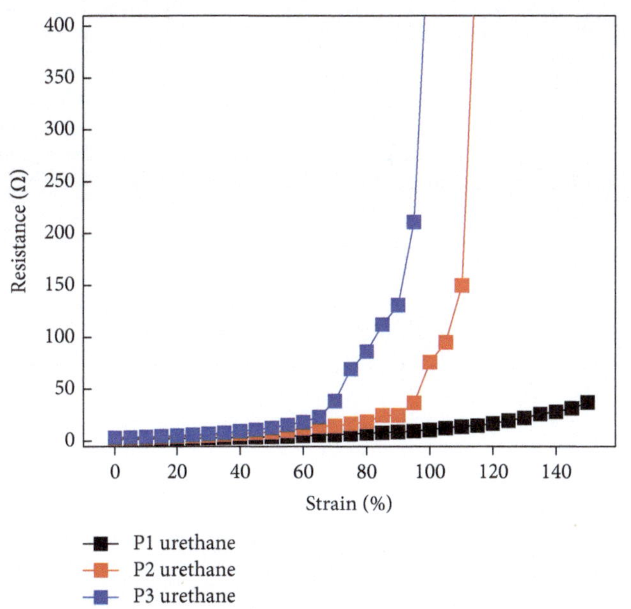

FIGURE 8: Electrical resistance versus tensile strain plot of unpatterned Ag electrode films.

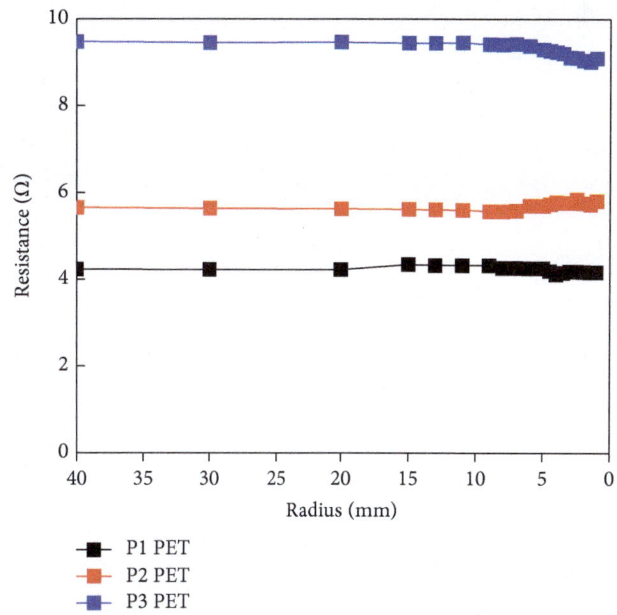

FIGURE 10: Electrical resistance versus radius of curvature plot for inner bending test.

which silver electrodes were patterned by screen printing. The bending radius versus electrical resistance plots were obtained in two different ways, that is, bending the Ag electrode films outside and inside as shown in Figures 9 and 10, respectively.

The electrical resistance increased more in the outer bending than inner bending which indicated that the microcrack occurred in the outer bending. It was however noted that in the very small bending radius close to 1 mm the electrical resistance was increased in the outer bending while being decreased in the inner bending, suggesting that crack damage is larger in the outside bending in case of extreme bending.

The bending fatigue tests were also conducted both by outer and inner bending modes as shown in Figures 11 and 12, respectively. As expected, the resistance was higher with increasing bending cycles in case of outer bending than inner bending, the same trend as the bending radius experiments. It is also noted that the bending fatigue tests conducted under 85°C/85% relative humidity was found to be “pass” status.

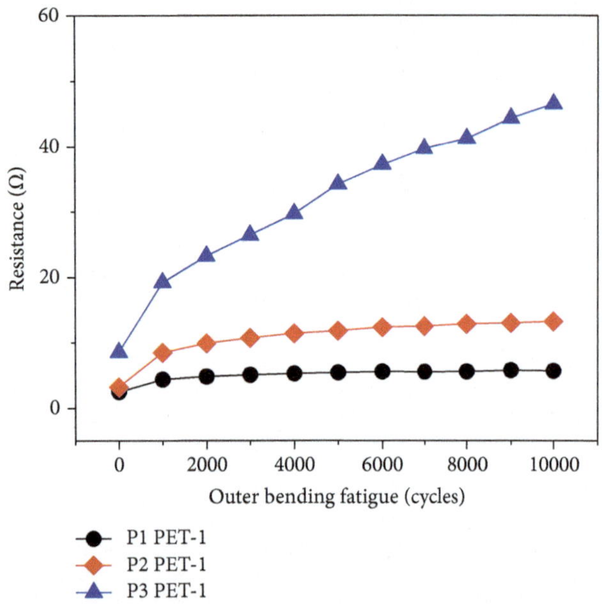

FIGURE 11: Electrical resistance versus bending cycle plot for outer bending fatigue test.

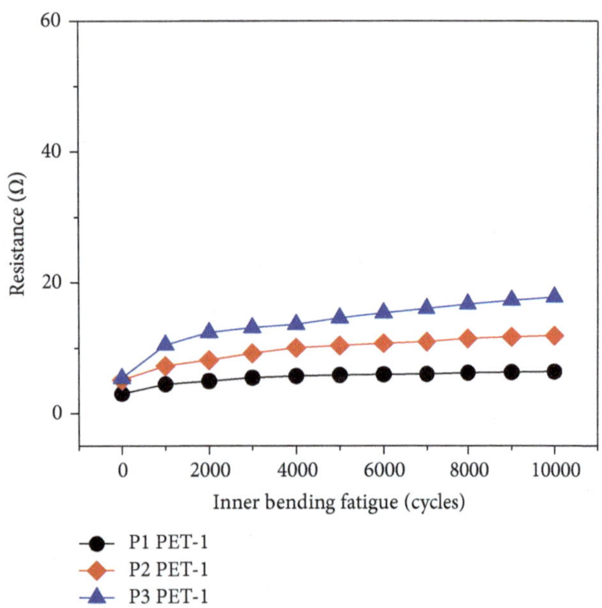

FIGURE 12: Electrical resistance versus bending cycle plot for inner bending fatigue test.

4. Conclusions

Silver paste was prepared by using unsaturated polyester resin as binder polymers with different molecular weights and glass transition temperatures. The transparent silver electrode patterns were obtained by screen printing on the flexible polyurethane and relatively rigid polyester films as substrates. The rheological study of Ag pastes showed that the binder polymer with high molecular weight and low glass transition temperature (T_g) can stabilize Ag paste more effectively with the Ag particle content over 80% by weight. The stretching and bending tests of Ag electrode films obtained by screen printing indicated that good conductivity of Ag electrodes is also related to the stability of Ag paste obtained with the high molecular weight binder polymer.

Acknowledgments

This material is based upon work supported by the Ministry of Trade, Industry & Energy (MOTIE, Korea) under Industrial Technology Innovation Program "Development of Stretchable Transparent Electrode Film with 3 μm Line Space Based on Cu/Ag Composite Nanoparticles" (no. 10062220).

References

[1] T. Q. Trung and N.-E. Lee, "Flexible and stretchable physical sensor integrated platforms for wearable human-activity monitoringand personal healthcare," *Advanced Materials*, vol. 28, no. 22, pp. 4338–4372, 2016.

[2] Y.-M. Choi, K.-Y. Kim, E. Lee, J. Jo, and T.-M. Lee, "Fabrication of a single-layer metal-mesh touchscreen sensor using reverse-offset printing," *Journal of Information Display*, vol. 16, no. 1, pp. 37–41, 2015.

[3] K. Kim, B. G. Hyun, J. Jang, E. Cho, Y.-G. Park, and J.-U. Park, "Nanomaterial-based stretchable and transparent electrodes," *Journal of Information Display*, vol. 17, no. 4, pp. 131–141, 2016.

[4] W. Honda, S. Harada, T. Arie, S. Akita, and K. Takei, "Wearable, human-interactive, health-monitoring, wireless devices fabricated by macroscale printing techniques," *Advanced Functional Materials*, vol. 24, no. 22, pp. 3299–3304, 2014.

[5] N. Matsuhisa, M. Kaltenbrunner, T. Yokota et al., "Printable elastic conductors with a high conductivity for electronic textile applications," *Nature Communications*, vol. 6, 2015.

[6] S.-Y. Nam, C.-G. Lim, H.-M. Nam, and B.-S. Gyun, "The Properties of Printed Electrode Pattern on the Stretchable Films," *Journal of Korean Society for Imaging Science and Technology*, vol. 22, pp. 1–6, 2016.

[7] J. Suikkola, T. Björninen, M. Mosallaei et al., "Screen-printing fabrication and characterization of stretchable electronics," *Scientific Reports*, vol. 6, Article ID 25784, 2016.

[8] I. Lee, J. Lee, S. H. Ko, and T.-S. Kim, "Reinforcing Ag nanoparticle thin films with very long Ag nanowires," *Nanotechnology*, vol. 24, no. 41, Article ID 415704, 2013.

19

Use of TBzTD as Noncarcinogenic Accelerator for ENR/SiO_2 Nanocomposites: Cured Characteristics, Mechanical Properties, Thermal Behaviors, and Oil Resistance

Laksamon Raksaksri,[1] Saowaroj Chuayjuljit,[1]
Phasawat Chaiwutthinan,[2] and Anyaporn Boonmahitthisud[1]

[1] *Department of Materials Science, Faculty of Science, Chulalongkorn University, Bangkok 10330, Thailand*
[2] *MTEC, National Science and Technology Development Agency (NSTDA), Thailand Science Park, Khlong Luang, Pathum Thani 12120, Thailand*

Correspondence should be addressed to Anyaporn Boonmahitthisud; anyaporn.b@chula.ac.th

Academic Editor: Domenico Acierno

This study reported the use of tetrabenzylthiuram disulphide (TBzTD) as a noncarcinogenic accelerator in a traditional sulfur curing system of epoxidized natural rubber (ENR)/nanosilica ($nSiO_2$) composites. ENR used in this work was synthesized via in situ epoxidation of natural rubber (NR) in the presence of performic acid generated from the reaction of formic acid and hydrogen peroxide at 50°C for 8 h to acquire the epoxide content of about 40 mol%. Accordingly, the resulting ENR was referred to as ENR 40. The curing characteristics, mechanical properties, thermal behaviors, dynamic mechanical properties, and oil resistance of ENR 40/$nSiO_2$ nanocomposites filled with three loadings of $nSiO_2$ (1, 2, and 3 parts per hundred parts of rubber) were investigated and compared with NR and neat ENR 40. The results revealed that the scorch and cure times of ENR 40/$nSiO_2$ nanocomposites were slightly longer than those of NR but slightly shorter than those of ENR 40. The tensile properties and tear strength for both before and after aging of all ENR 40/$nSiO_2$ nanocomposites were higher than those of ENR 40, while the glass transition temperature, storage modulus at −65°C, thermal stability, and oil resistance of ENR 40/$nSiO_2$ nanocomposites were higher than those of NR and ENR 40.

1. Introduction

Natural rubber (NR) is one of the most important natural biosynthesis polymers and renewable resources that possesses excellent mechanical properties owing to its ability to crystallize upon stretching (strain-induced crystallization), but its thermal, oxidation, and oil resistance cannot compete with some special purpose synthetic rubbers [1–12]. This is a normal consequence of its highly unsaturated hydrocarbon chain structure and nonpolar nature [2–4]. However, the certain properties of NR can be improved via chemical modifications at the carbon-carbon double bonds (C=C) in its isoprene unit to produce functionalized specialty rubbers with an expanded range of physical and chemical properties. The controlled partial epoxidation of NR to form epoxidized NR (ENR) is one of the important methods that introduce a new reactive group into the NR backbone, leading to the new and useful properties. This involves the random introduction of polar epoxide (oxirane) groups onto the C=C in the repeat unit of NR by the two consecutive reactions. The first one included the "in situ" formed peracid generated from the endothermic reaction of hydrogen peroxide (H_2O_2) and the acid, followed by the exothermic epoxidation reaction, leading to an increased polarity and a reduction in the C=C in the rubber chains [5, 8, 9, 12–18]. As a consequence, ENR would have oil and oxidation resistance comparable to some of the synthetic rubbers, depending on the extent of epoxidation (mol% epoxidation) [5, 9, 10, 12]. It has been reported that ENR with 50 mol% epoxidation exhibits oil resistance similar to that of medium-acrylonitrile-content nitrile rubber and

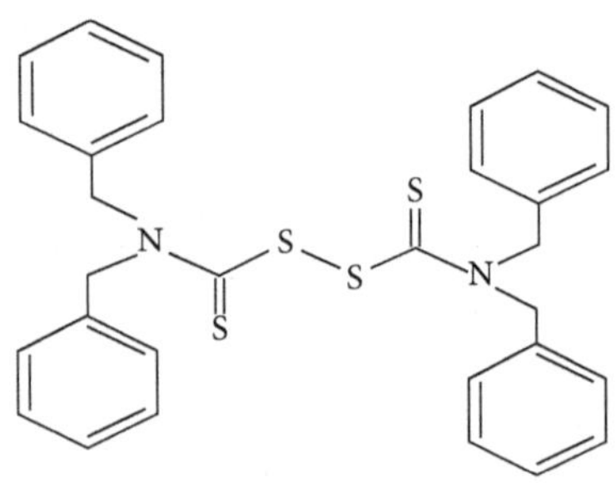

SCHEME 1: Molecular structure of tetrabenzylthiuram disulphide (TBzTD).

superior to that of chloroprene rubber [9, 12, 19]. As the level of epoxidation increases more than 50 mol%, the ENR products become harder and of less resilience and elasticity at room temperature (RT), whereas the products of lower epoxide content are typical elastomers with less oil resistance [8, 10, 17]. In this study, ENR with medium epoxide content (approximately 40 mol%) was synthesized from concentrated NR latex to impart higher oil resistance compared to the NR while maintaining both elasticity and strain crystallization of the products. The resulting ENR was then vulcanized by a traditional sulfur curing system using an organic accelerator in a usual manner applicable for NR. Tetrabenzylthiuram disulphide (TBzTD, $C_{30}H_{28}N_2S_4$, Scheme 1) was selected for using as a safe (noncarcinogenic) accelerator to reduce the generation of hazardous nitrosamines in rubber compounds. Nitrosamines are generally generated from the rubber products during mixing, curing, and storage when using amine containing accelerators such as tetramethylthiuram disulfide (TMTD) [7, 20]. With this respect, the TBzTD, a fast curing accelerator with no secondary amine, has been developed to replace the TMTD, resulting in the nitrosamine free cure.

Recently, reinforcement of rubbers with inorganic nanoparticles, even at a very low loading, can significantly improve their mechanical and thermal properties [9, 10, 19, 21–25]. To achieve the desired properties, the nanofillers must sufficiently disperse in the rubber matrix, as poor dispersion leads to the obvious decrease in the performance properties of the nanocomposites. Nanosilica ($nSiO_2$) is an inorganic filler with a large number of hydroxyl (-OH) groups on its surface and thus provides strong filler-filler interactions to generate $nSiO_2$ aggregates and agglomerates. However, the epoxide groups of ENR can effectively cause a specific interaction between $nSiO_2$ and ENR via hydrogen bonding, leading to a decrease in filler-filler interaction and an increase in compatibilization [9, 22–24]. With this respect, ENR nanocomposites can be directly prepared using $nSiO_2$ as a nonblack reinforcement without the aid of any coupling agents or pretreatments [9, 12]. Therefore, their unique properties can be combined in order to improve the application properties of the resulting nanocomposites. To the best of our knowledge, use of the TBzTD as the nitrosamine free accelerator in the sulfur curing of the ENR/$nSiO_2$ nanocomposites has never been reported. Therefore, the mechanical properties and thermal and dynamic mechanical properties, as well as oil resistance of the prepared ENR/$nSiO_2$ nanocomposites, were investigated and compared with those of the neat NR and ENR.

2. Experimental Section

2.1. Materials. High ammonia concentrated NR latex with a 60% dry rubber content (DRC), polyalcohol ethylene oxide condensate (Terric 16A-16), sulfur, zinc oxide (ZnO), stearic acid, Voltamol, and bentonite were supplied from the Rubber Research Institute of Thailand (Thailand). TBzTD manufactured by Performance Additive Co. (Italy) was purchased from Behn Meyer Chemicals Co. (Thailand). Formic acid and HCOOH (98%) manufactured by Fisher Chemical Co. (UK) and H_2O_2 (35%) manufactured by Chem-Supply Pty Co. (Australia) were purchased from Myscience Co. (Thailand) and RCI Lab Scan Co. (Thailand), respectively. Nano-SiO_2 having particle size of about 12 nm and specific surface area of 175–225 $m^2\,g^{-1}$ manufactured by Sigma Aldrich (Singapore) was purchased from A.C.S. Xenon Co. (Thailand). Sodium carbonate and methanol were purchased from Merck Thailand Co. (Thailand) and RCI Lab Scan Co. (Thailand), respectively. ASTM oil number 1 (IRM 901) and ASTM oil number 3 (IRM 903) were purchased from Chemical Innovation Co. (Thailand). All materials were used as received without further purification.

2.2. Preparation and Characterization of ENR. The ENR was synthesized via in situ latex epoxidation of NR by performic acid generated from the reaction between HCOOH and H_2O_2. Prior to epoxidation, NR latex was diluted with distilled water to reduce the DRC from 60% to 20% and then stabilized with the nonionic surfactant Terric 16A-16 (3 parts by weight per hundred of rubber, phr) at RT under stirring for 1 h. The assigned amounts of HCOOH (1 M of isoprene unit) and H_2O_2 (0.75 M of isoprene unit) were then added to the stabilized NR latex. The HCOOH was fed dropwise over 10–15 min at 40°C, followed by raising the temperature to 50°C within 10 min prior to adding the H_2O_2 over 5–10 min with continuous stirring. The reaction was allowed to proceed at 50°C for 8 h. The obtained products were sampled and coagulated by an excess 95% (v/v) methanol, filtered, washed well with distilled water, soaked in 10% (w/w) sodium carbonate solution for 30 min, washed again until being neutral, pressed into a thin sheet, and dried at 60°C for 48 h until at a constant weight before analysis.

The mol% epoxidation of the resulting ENR was examined by Fourier transform infrared (FT-IR) spectroscopy on a Nicolet 6700-FT-IR (USA) over a frequency range of 400–4000 cm^{-1} and by proton nuclear magnetic resonance spectroscopy (^{1}H-NMR) on an Avance III HD 500 MHz NMR spectrometer (Bruker, Germany). The mol% epoxidation was calculated from (1) using data from the FT-IR peaks at 870 and 835 cm^{-1} [14, 26].

$$\text{Mol\% epoxidation} = \left[\frac{A_{870}}{(A_{870} + A_{835})}\right] \times 100, \quad (1)$$

where A_{870} and A_{835} are the absorbance at wavenumbers 870 and 835 cm^{-1}, respectively.

To estimate the mol% epoxidation from the ^{1}H-NMR analysis, two characteristic peaks were considered in relation with (2) [8, 9, 13, 24]:

$$\text{Mol\% epoxidation} = \left[\frac{A_{2.7}}{(A_{2.7} + A_{5.14})}\right] \times 100, \qquad (2)$$

where $A_{2.7}$ and $A_{5.14}$ are the integrate areas under the ^{1}H-NMR peaks at about 2.7 ppm, representing the epoxide proton, and at about 5.14 ppm, representing an olefinic proton, respectively.

The glass transition temperature (T_g) of the sample was evaluated by differential scanning calorimetry (DSC) on a Mettler Toledo DSC 1 STARe System (Switzerland) at a heating/cooling rate of 5°C min^{-1} under a N_2 flow of 20 mL min^{-1} from −80 to 20°C.

2.3. Preparation and Characterization of ENR/nSiO$_2$ Nanocomposites. The as-synthesized ENR was compounded with nSiO$_2$ (0, 1, 2, and 3 phr) by latex compounding. The obtained nanocomposites were sampled and coagulated by an excess 95% (v/v) methanol, filtered, washed well with distilled water, soaked in 10% (w/w) sodium carbonate solution for 30 min, and washed again until being neutral. The ENR/nSiO$_2$ nanocomposites were mixed with ZnO (5 phr), stearic acid (2 phr), TBzTD (6 phr), and sulfur (0.5 phr) in an internal mixer (MX500-D75L90, Thailand) at 70°C for 9 min and then sheeted on an open two-roll mill at RT for 1 min. The cure characteristics of the different stocks were determined using a moving die rheometer (MDR, A0225-rheo Tech Techpro, USA) at 150°C according to ASTM D5289. The respective scorch time (t_{s2}), cure time (t_{90}), minimum torque (M_L), maximum torque (M_H), and torque difference ($M_H - M_L$) were obtained from the rheograph. The cure rate index (CRI) was then calculated from (3) [1, 7, 18]:

$$\text{CRI} = \frac{100}{(t_{90} - t_{s2})}. \qquad (3)$$

The ENR well-mixed compound was then press-cured into 3 mm thick sheet using a compression molding machine (LP-S-20, LabTech Engineering, Thailand) at 150°C according to the t_{90} obtained from the MDR. For comparison, NR vulcanizate was also prepared using the same sulfur curing process as above.

2.4. Testing and Characterization of the Vulcanizates. The tensile properties (tensile strength, modulus at 300% strain (M300), and elongation at break) of the vulcanizates were measured on the standard dumbbell-shape specimen according to ISO 37 Type II using an Instron universal testing machine (5840 series single column systems, USA) with a 1 kN load cell and at a crosshead speed of 500 mm min^{-1}. The tear strength was measured according to ASTM D624 on an angle-shaped sample with the same universal testing machine at a strain rate of 500 mm min^{-1}. At least five specimens were examined for each test.

The thermal aging of the sample was performed in a universal oven (Model UF55, Germany) with air circulating system at 70°C for 72 h according to ASTM D573. After aging, the specimen was conditioned at RT for at least 16 h prior to determining the tensile properties and tear strength in the same manner as the unaged specimen. The percentage retention in properties was calculated from the following equation [4]:

$$\text{Retention (\%)} = \left(\frac{\text{Value after aging}}{\text{Value before aging}}\right) \times 100. \qquad (4)$$

The thermal stability of the sample was measured by means of thermogravimetric analysis (TGA) on a Mettler Toledo, TGA/SDTA 851^{e} instrument (Switzerland), operating under a N_2 atmosphere at a heating rate of 20°C min^{-1} within the temperature range of 50–1,000°C. The dynamic mechanical analysis (DMA) of the sample was performed on a Mettler Toledo DMA/SDTA 861^{e} instrument (Switzerland) under a shear mode at a constant frequency of 1 Hz and a heating rate of 3°C min^{-1} over a temperature range of −50 to 50°C and under a N_2 flow.

The oil resistance was determined by measuring the percentage volume change of the sample after a 72 h immersion period in two types of oil, ASTM oil number 1 and number 3, at 100°C. Prior to immersion, the test specimen was weighted in both air and distilled water at RT. After immersion, the test specimen was removed from the respective oils, cooled to RT by transferring to a cool and clean test oil for about 30 min, and then dipped quickly in acetone at RT and wiped lightly with filter paper to remove the excess oil from the surface prior to weighing again in both air and distilled water. The percent change in volume was then evaluated using (5) according to ASTM D471-79:

$$\%\,\Delta V = \frac{[(M_3 - M_4) - (M_1 - M_2)]}{(M_1 - M_2)} \times 100, \qquad (5)$$

where M_1 and M_2 are the initial mass (g) of the specimen in air and distilled water, respectively, and M_3 and M_4 are the mass (g) of the specimen in air and distilled water after immersion, respectively.

3. Results and Discussion

3.1. Characterization of Unvulcanized Samples. Figure 1 shows the representative FT-IR spectra of the samples over a frequency range of 4000–400 cm^{-1}. The characteristic peaks of NR (Figure 1(a)) that appeared at 3037, 2850, 1660, 1450, 1375, and 835 cm^{-1} are assigned to the =C–H stretching, C–H stretching, C=C stretching, C–H bending of CH_2, C–H bending of CH_3, and =C–H bending, respectively [3, 8, 15, 24]. The new characteristic peaks that appeared at 1240 and 870 cm^{-1} (Figure 1(b)) are assigned to the symmetric stretching and asymmetric stretching of epoxide ring, respectively [5, 8, 15, 16, 18, 24–27]. This confirmed the formation of epoxide or oxirane rings from the reaction of performic (generated in situ by the reaction of HCOOH and H_2O_2) with the C=C on the NR molecules. The absorbance at 870 cm^{-1} was then compared with that of the residue olefinic peak (C=C *cis*) at 835 cm^{-1} to evaluate the mol% epoxidation. In

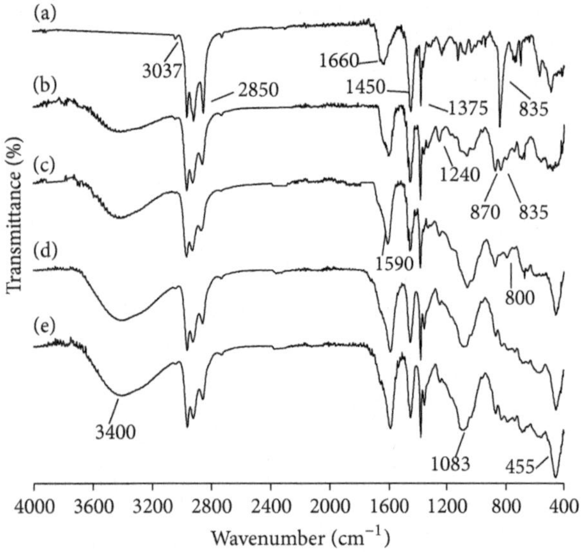

FIGURE 1: FTTR spectra of (a) NR, (b) ENR 40, and nanocomposites of (c) 100/1 ENR 40/nSiO_2, (d) 100/2 ENR 40/nSiO_2, and (e) 100/3 ENR 40/nSiO_2.

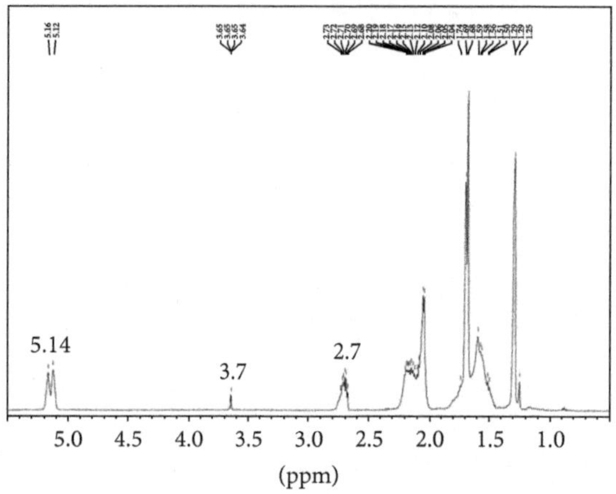

FIGURE 2: ^{1}H-NMR spectrum of ENR 40.

addition, the broad peak at about 3400 cm^{-1} (Figure 1(b)) indicates the presence of –OH groups in the ENR molecules, resulting from a partial opening of epoxide ring during ENR preparation [5, 15, 16, 24]. Figures 1(c) and 1(d) also show broad peaks at about 3400 cm^{-1}, which are mainly owing to the O–H stretching vibration of –OH groups on nSiO_2 surfaces and from the opening of epoxide ring, and so the peak size increased with increasing nSiO_2 loading. The peak at 1590 cm^{-1} is attributed to the bending vibration of the O–H bond in the adsorbed water molecules on nSiO_2 surfaces. The intensive peaks at about 1083 cm^{-1}, together with the less intensive peaks at 800 and 455 cm^{-1}, are attributed to the asymmetric stretching vibration, symmetric stretching vibration, and asymmetric bending vibration of Si-O-Si, respectively [24–26].

The presence of epoxide rings in ENR molecules was also confirmed by the ^{1}H-NMR analysis. The ^{1}H-NMR spectrum of the synthesized ENR sample revealed two peaks at 2.7 and 1.3 ppm (Figure 2), which represented the proton attached to the epoxide ring and the proton from a methyl group attached to the epoxide ring, respectively, while the peak at 5.14 ppm represented the olefinic proton of the unreacted NR. In addition, the peak at 3.7 ppm was attributed to some functional groups such as –OH and hydrofuran obtained from the opening of epoxide rings [8, 9, 14, 24]. This is in agreement with the result of the FT-IR analysis that was previously mentioned.

DSC analysis was also used as a quantitative measurement for the mol% epoxidation from the increased T_g of the synthesized ENR compared with that of the pure NR [5, 12–14]. It has been reported that, for the ENR with side ring-opening reaction, each mol% epoxidation gives a 1.15°C increase in T_g [12]. The increased T_g of ENR was owing to the polar epoxide group that provided strong intermolecular interaction and so restricted the mobility of the ENR chains [5, 16, 24]. The DSC thermograms of the unvulcanized samples based on the second heating scan performed between −80 and 20°C are shown on the left side of Figure 3 and the obtained T_g values are summarized in Table 1. As can be seen, T_g of the synthesized ENR (−26.2°C) was higher than that of the NR (−70°C) about 43.8°C. Although T_g of ENR was much higher than that of NR, it was not significantly changed after adding different nSiO_2 loadings, suggesting that the presence of nSiO_2 had a small influence on T_g of the nanocomposites.

The calculated mol% epoxidation of the synthesized ENR from the FT-IR (using (1)), ^{1}H-NMR (using (2)), and DSC analyses was about 43, 39, and 38%, respectively. Based on the mol% epoxidation, the as-synthesized ENR was then denoted as ENR 40.

The MDR rheographs of the samples conducted at 150°C are shown in Figure 4, and the main cure characteristics (t_{s2}, t_{90}, CRI, M_L, M_H, and $M_H - M_L$) are all summarized in Table 2. The sulfur vulcanization with unsaturated rubbers generally occurred through a free radical reaction initiated by accelerator. The resulting sulfur radicals further attack reactive allylic hydrogens in rubber chains to form sulfur crosslinks, and so the vulcanization depends upon the number of the allylic hydrogens [28]. t_{s2} is the time required for the torque value to increase two units from M_L, which corresponds to the premature vulcanization of rubber compound, while t_{90} is the vulcanization time for optimum cure torque [1, 3]. As can be seen, NR exhibited shorter t_{s2} and t_{90} than ENR 40 after being cured under the same condition. This is because NR possessed a higher level of unsaturation and reactive sites than ENR 40, and so the crosslinks started and reached the optimum cure torque in a shorter time [1, 18, 28]. Moreover, the high content of benzylthiuram substituents in the TBzTD accelerator may provide a steric hindrance to the limited crosslink sites of the ENR 40. However, the ENR 40 had higher CRI than the NR because the isolated C=C in ENR reacted more rapidly than the contiguous C=C in NR, and also owing to the activation of the adjacent C=C by the epoxide group in the ENR [9, 10, 13].

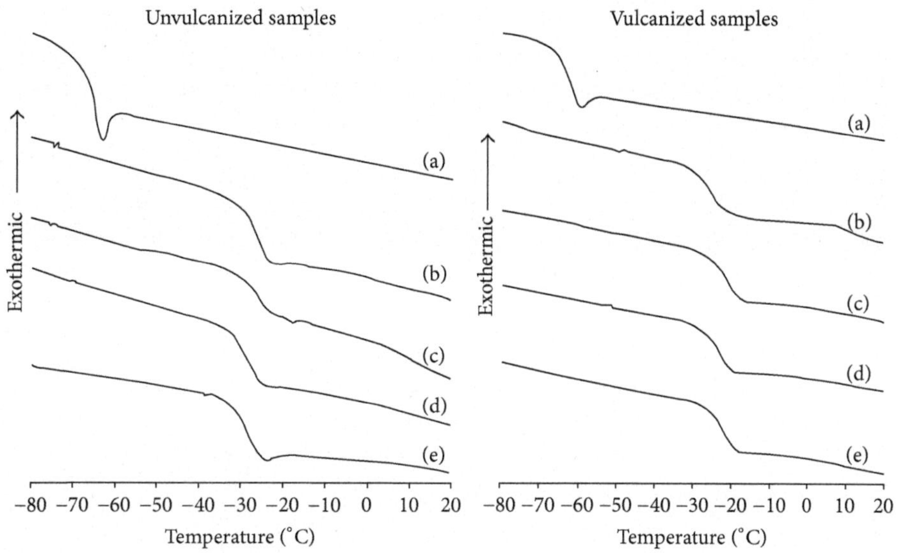

FIGURE 3: DSC thermograms derived from the second heating scan of (a) NR, (b) ENR 40, and nanocomposites of (c) 100/1 ENR 40/nSiO_2, (d) 100/2 ENR 40/nSiO_2, and (e) 100/3 ENR 40/nSiO_2.

TABLE 1: Thermal properties of NR, ENR 40, and ENR 40/nSiO_2 nanocomposites obtained from DSC, DMA, and TGA analyses.

Thermal properties	NR	ENR 40	ENR 40/nSiO_2		
			100/1	100/2	100/3
DSC					
T_g (°C): unvulcanized	−68.2	−26.2	−24.3	−27.2	−25.7
T_g (°C): vulcanized	−64.2	−25.0	−21.6	−23.5	−22.5
DMA					
E' at −65°C (MPa)	404.2	670.1	905.5	1408.8	808.7
T_g (°C) at E'' peak	−53.8	−15.9	−14.8	−19.3	−15.5
T_g (°C) at tan δ peak	−48.0	−7.6	−7.0	−8.3	−7.5
tan δ_{max}	2.5	2.2	2.1	2.4	2.0
TGA and DTG					
$T_{onset\ 1}$ (°C)	226	251	252	252	248.7
$T_{end\ set\ 1}$ (°C)	287	290	291.5	296	279.3
$T_{max\ 1}$ (°C)	268.7	281.7	275	284.8	268.5
$T_{onset\ 2}$ (°C)	362.7	369.8	378.4	379.7	377
$T_{end\ set\ 2}$ (°C)	419.4	434	434.9	435	434.6
$T_{max\ 2}$ (°C)	384.7	391.7	402.7	399.2	401.3
$T_{50\%}$ (°C)	388.4	403	407	408	410.6
char (%)	6	6	13.4	15.9	16.6

t_{s2} and t_{90} of the NR were also shorter than those of the ENR 40/nSiO_2 nanocomposites. This again attributed to the higher level of the reactive sites in the NR molecules. Compared with the neat ENR 40, the nanocomposite containing 1 phr nSiO_2 had lower t_{s2} and t_{90} because of the self-crosslinks or additional crosslinks from the specific interaction between the epoxide groups in the ENR 40 and –OH groups on the nSiO_2 surface through hydrogen bonding [9, 22–24, 29]. However, the CRI of the ENR 40 nanocomposite with 1 phr nSiO_2 was lower than that of the neat ENR 40, indicating a slower rate of vulcanization. This is attributed to the adsorption of the basic accelerator on the –OH groups of the nSiO_2 surface and so lowered the vulcanization rate of the nanocomposites [9, 25]. Further increase in the nSiO_2 loading to 2 phr resulted in longer t_{s2} and t_{90}, and a lower CRI because of the higher level of the adsorbed accelerator on the nSiO_2 surface, which subsequently reduced the accelerator efficiency. However, at 3 phr nSiO_2, t_{s2} and t_{90} were again shorter and the CRI was again increased owing to the agglomeration of nSiO_2 particles in the nanocomposite that decreased the free Si–OH groups and so reduced the accelerator being adsorbed on the nSiO_2 surface.

M_L is commonly related to the viscosity of the uncured compound, while M_H is regarded to the elastic modulus

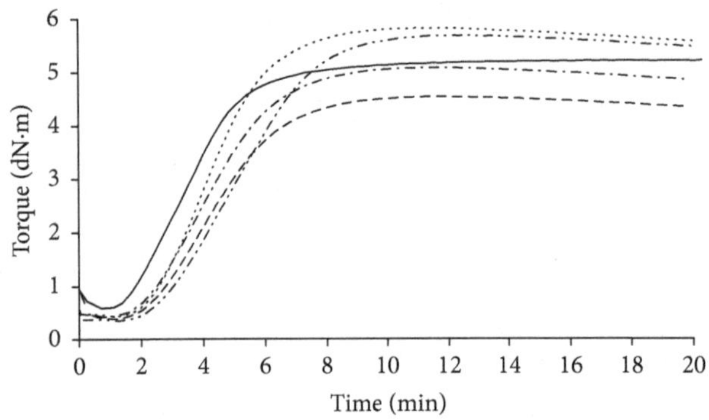

FIGURE 4: MDR rheographs of NR, ENR 40, and ENR 40/$nSiO_2$ nanocomposites.

TABLE 2: Cure characteristics of NR, ENR 40, and ENR 40/$nSiO_2$ nanocomposites.

Cure characteristic	NR	ENR 40	ENR 40/$nSiO_2$		
			100/1	100/2	100/3
t_{s2} (min)	3.6	5.1	4.3	4.8	3.9
t_{90} (min)	6.0	7.2	6.9	7.9	6.8
CRI (min^{-1})	41.7	47.6	38.5	31.3	35.7
M_L (dN·m)	0.7	0.4	0.5	0.3	0.4
M_H (dN·m)	5.2	4.4	5.1	5.7	5.8
$M_H - M_L$ (dN·m)	4.5	4.0	4.6	5.4	5.4

of the vulcanizate [3, 9, 10]. $M_H - M_L$ is a measure of the dynamic shear modulus, which indirectly indicates the extent of crosslinking (crosslink density) of the vulcanizate [9, 10, 18, 25]. The M_L, M_H, and $M_H - M_L$ values of the ENR 40 were found to be lower than those of the NR (Table 2), indicating the lower viscosity of the uncured ENR 40 and also the lower elastic modulus and crosslink density of the ENR 40 vulcanizate. The lower viscosity of the uncured compound may be owing to the possible opening of epoxide rings and the degradation of ENR 40 molecules either during epoxidation or during compounding at high temperature, while the lower elastic modulus and crosslink density of the ENR 40 vulcanizate may be owing to the lower number of the C=C in the ENR 40 compared to that in the NR [9]. However, the M_L values of ENR 40 and its nanocomposites were nearly the same, indicating that the $nSiO_2$ at a very low loading had no effect on the viscosity of the uncured compounds, while the M_H and $M_H - M_L$ values of the nanocomposites noticeably increased compared to those of the ENR 40 as a consequence of the strong interaction between ENR 40 and $nSiO_2$ as previously mentioned.

3.2. Characterization of Vulcanized Samples. The mechanical properties of the samples both before and after aging at 100°C for 72 h in air are summarized in Table 2. It is seen that the tensile properties (tensile strength, M300, and elongation at break) and tear strength of the NR were all higher than those of the ENR 40. This is owing to the higher ability of NR to undergo strain crystallization as a result of its high molecular stereoregularity and also owing to the higher elasticity of NR that provided superior mechanical properties. Moreover, the epoxidation resulted in a reduced tendency to crystallize at high elongation of the synthesized ENR 40. Table 2 also shows a reduction in all the studied mechanical properties of NR and ENR 40 after aging as a result of the molecular chain scission and/or crosslink breakage at high temperature and oxygen content [3, 4]. It is seen that the percentage retentions of tensile strength, M300, elongation at break, and tear strength of the NR were all lower than those of the ENR 40. This is because of a higher susceptibility to thermal oxidative degradation of NR, which was mainly owing to the high level of the C=C in the main chain that gave rise to the lower retention values [4]. Therefore, changes in the mechanical properties of rubber products depended on the extent of such reactions that limited their useful lifetime at elevated temperature in the presence of atmospheric oxygen. However, the high rubber-rubber interaction could improve the aging resistance of the ENR 40.

The effects of $nSiO_2$ on the mechanical properties of ENR 40 nanocomposites are also shown in Table 2 and Figure 5. As can be seen, the tensile properties and tear strength of the ENR 40 nanocomposites with three loadings of $nSiO_2$ (1, 2 and 3 phr) were all higher than those of the neat ENR 40, owing to the reinforcing effect of $nSiO_2$, but at 2 phr $nSiO_2$ was the highest for all the mechanical properties. This may be attributed to a better dispersion of $nSiO_2$ in the ENR 40 matrix, providing a greater filler-rubber interaction between the silanol and epoxide groups. This is an essential role for achieving the good mechanical properties. However, the decrease in the mechanical properties of the nanocomposite at 3 phr $nSiO_2$ may be owing to the agglomeration of an excess $nSiO_2$, leading to a poor dispersion of $nSiO_2$ in the ENR 40 matrix, a reduction in contact area between $nSiO_2$ and ENR 40, and also a filler-rubber interaction. However, the percentage retention in mechanical properties of all the ENR 40 nanocomposites was higher than that of the neat ENR 40, owing to a very high thermal resistance of the $nSiO_2$ [25, 26].

The DSC curves of the vulcanized NR, ENR 40, and ENR 40/$nSiO_2$ nanocomposites are shown on the right side of Figure 3, and the obtained T_g values are also presented in Table 1. The T_g values of all the vulcanizates were higher than those of the corresponding unvulcanized samples according to the formation of crosslinks in the rubber matrix that restricted the flexibility of the rubber chains; hence T_g increased. In addition, all the T_g values of the vulcanized samples followed the same trend as those of the unvulcanized samples (Table 1). Among the nanocomposites, the ENR 40 filled with 2 phr $nSiO_2$ exhibited the lowest T_g owing to a better dispersion of $nSiO_2$ that caused a decrease of the ENR 40 being trapped within the $nSiO_2$ network and so resulted in a higher mobility of the rubber chain segments; hence T_g decreased [30, 31].

Figure 6 shows the storage modulus (E'), loss modulus (E''), and loss tangent (tan δ) curves obtained from the DMA analysis of the samples, and their E' (at −65°C), T_g, and

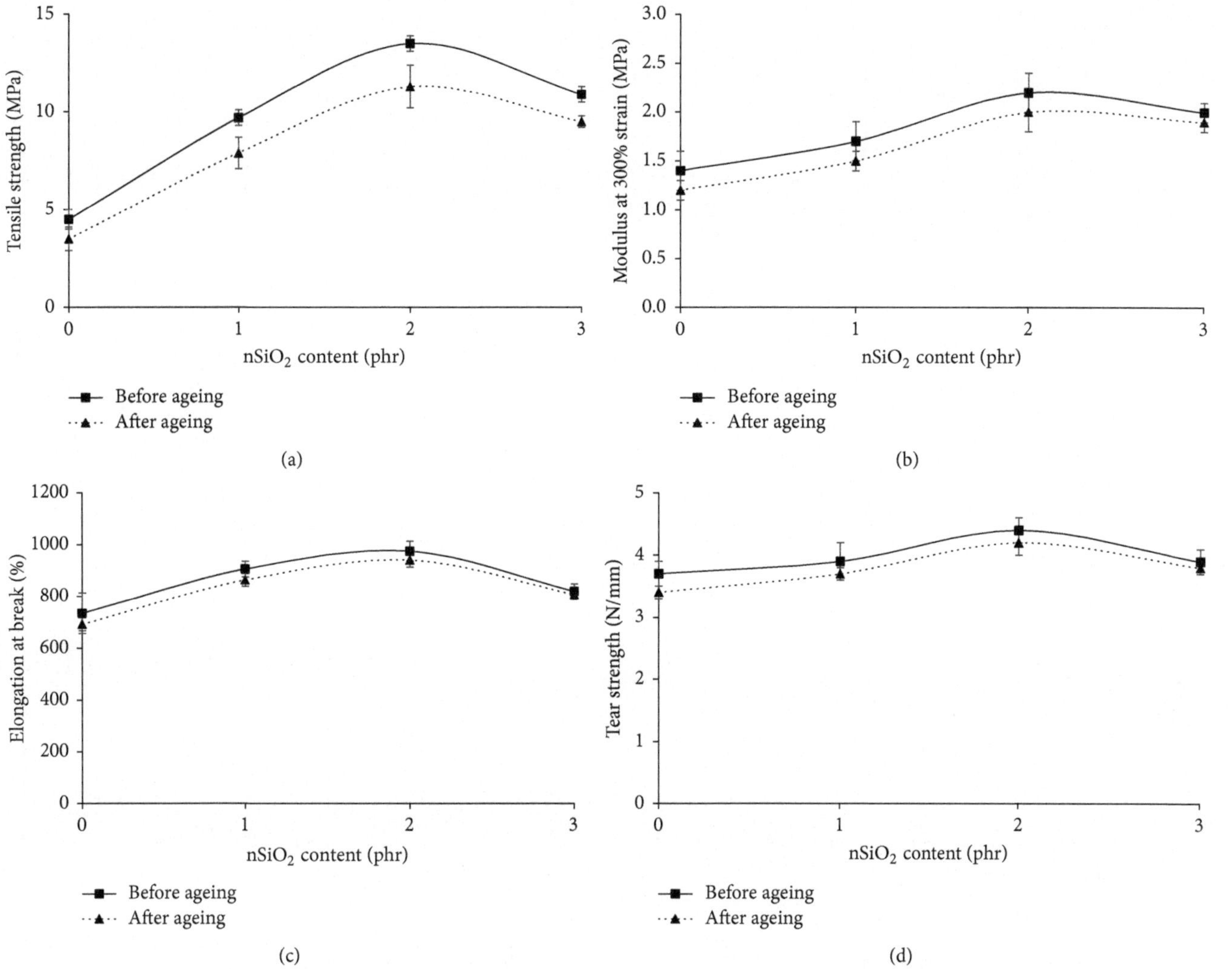

FIGURE 5: Mechanical properties of ENR 40/nSiO_2 nanocomposites in terms of (a) tensile strength, (b) M300, (c) elongation at break, and (d) tear strength.

tan δ_{max} (height of tan δ peak) values are also listed in Table 1. According to Figure 6(a), the ENR 40 and ENR 40/nSiO_2 nanocomposites had much higher E' values than the NR. The higher E' value of ENR 40 is again owing to the strong intermolecular force between epoxide groups that restricted the mobility of the ENR 40 chain segments. Additionally, all the ENR 40/nSiO_2 nanocomposites exhibited higher E' values than the neat ENR 40. This may be owing to the good interaction between the nSiO_2 and ENR 40 and also to the high stiffness of nSiO_2 that restricted the mobility of the rubber chains as previously discussed. According to Figures 6(b) and 6(c), the T_g values derived from the temperatures at the E'' and tan δ peaks followed the same behavior as that obtained from the DSC curves (Figure 3). It is seen that the T_g values of the ENR 40 and ENR 40/nSiO_2 nanocomposites were also much higher than those of the NR according to the same reason as previously discussed. However, the tan δ_{max} values of the ENR 40 and ENR 40/nSiO_2 nanocomposites were all lower than those of the NR. This indicated that the ENR 40 and ENR 40/nSiO_2 nanocomposites were stiffer than the NR as a consequence of the strong intermolecular attraction or rubber-filler interaction as previously mentioned.

The thermal stability of the samples was evaluated by TGA, and their representative TG and derivative thermogravimetry (DTG) curves are shown in Figure 7. The derived T_{onset}, $T_{50\%}$, $T_{end\ set}$, T_{max}, and % char values are also listed in Table 1. All the samples had similar TG and DTG curves and each one presented a two-step degradation. The degradation of the first step for all samples observed at 226–252°C ($T_{onset\ 1}$), 287–296°C ($T_{end\ set\ 1}$), and 268.5–284.8°C ($T_{max\ 1}$) was attributed to the low molecular weight compounds such as nonrubber components and impurities as well as additives [15, 16] and has a weight loss of 6.7–8.1%, whereas the second step observed at 362.7–379.7°C ($T_{onset\ 2}$), 419.4–435°C ($T_{end\ set\ 2}$), and 385–401°C ($T_{max\ 2}$) was attributed to the main degradation of rubber hydrocarbon, corresponding to the weight loss of 76.2–87.1%. Additionally, $T_{50\%}$ for all samples was in the range of 388.4–410.6°C. As can be seen, $T_{onset\ 2}$, $T_{50\%}$, $T_{end\ set\ 2}$, and $T_{max\ 2}$ of the ENR 40 were all higher than those of the NR, indicating that the thermal

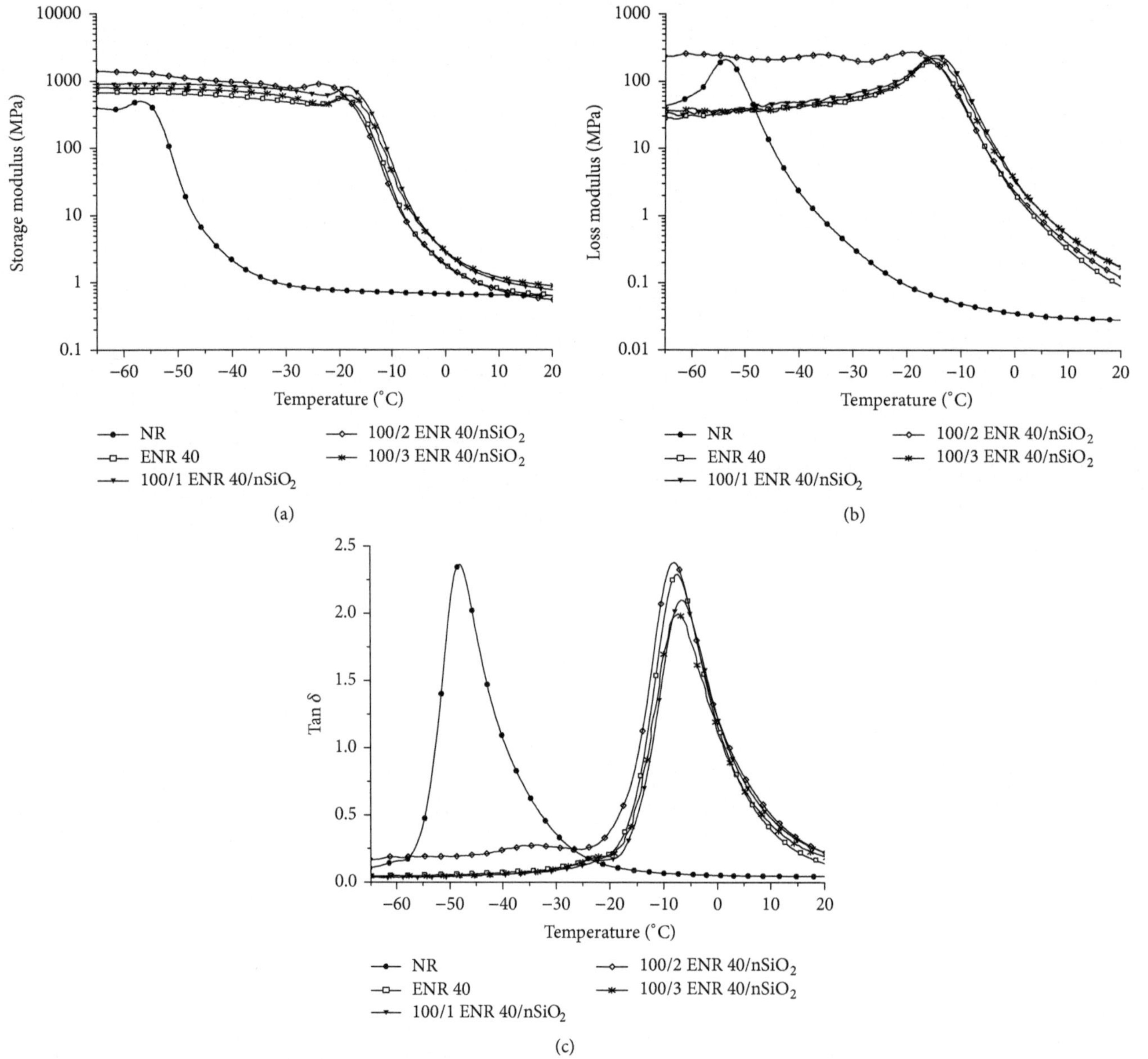

Figure 6: Dynamic mechanical properties of NR, ENR 40, and ENR 40/nSiO_2 nanocomposites in terms of (a) storage modulus, (b) loss modulus, and (c) tan δ.

stability of ENR 40 was improved. This is owing to the strong intermolecular attraction in ENR 40 caused by the insertion of oxygen atoms in the NR molecules and also to the high susceptibility to degradation of NR as previously mentioned. Moreover, the degradation temperatures from Table 3 also revealed that the addition of nSiO_2 to the ENR 40 exhibited a dose-dependent increase in the thermal stability of the nanocomposites by shifting $T_{\text{onset } 2}$, $T_{50\%}$, $T_{\text{end set } 2}$, and $T_{\max 2}$ towards higher temperatures compared to those of the neat ENR 40. This is owing to the high thermal stability of nSiO_2 accompanied with a strong filler-rubber interaction through the hydrogen bonding that caused the degradation to occur at a higher temperature and that also hindered the mass transport during the thermal degradation process [22–25]. It was also found that the char formation of the nanocomposites (13.4–16.6 wt%) remarkably increased with increasing nSiO_2 loading compared to that of the neat NR and ENR 40 (both at 6 wt%) as shown in Figure 7 and Table 1. This is because nSiO_2 is an inorganic nanofiller that can withstand high temperature and remain as residue even at 1,000°C and so delays and reduces the thermal degradation of the vulcanizates by restricting the permeability of volatile decomposition products out of the samples [4, 25].

The percentage volume change of the NR, ENR 40, and ENR 40/nSiO_2 nanocomposites after being immersed in ASTM oil number 1 (aniline point ~124°C) and ASTM oil number 3 (aniline point ~90°C) at 100°C for 72 h is presented in Table 4. It is seen that the percentage volume

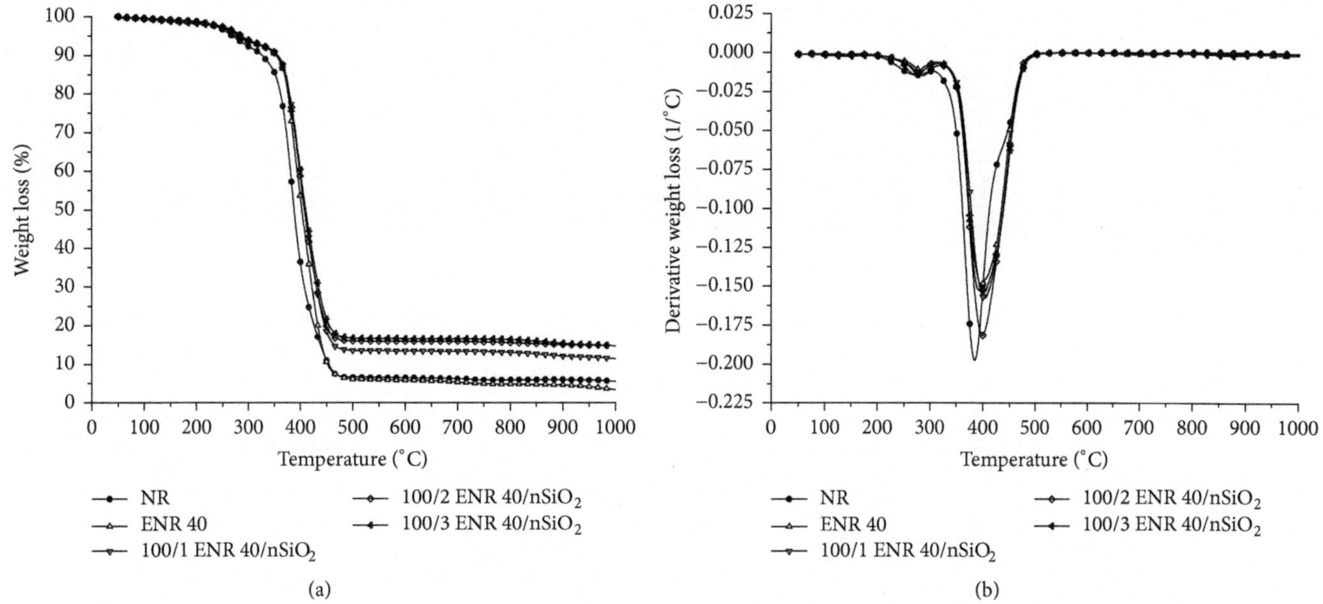

FIGURE 7: (a) TGA and (b) DTG thermograms of NR, ENR 40, and ENR 40/$nSiO_2$ nanocomposites.

TABLE 3: Mechanical properties and % retention of NR, ENR 40, and ENR 40/$nSiO_2$ nanocomposites.

Mechanical properties	NR	ENR 40	ENR 40/$nSiO_2$		
			100/1	100/2	100/3
Tensile strength					
Before aging (MPa)	13.4 ± 0.7	4.5 ± 0.5	9.7 ± 0.4	13.5 ± 0.4	10.9 ± 0.4
After aging (MPa)	8.8 ± 0.5	3.5 ± 0.6	7.9 ± 0.8	11.3 ± 1.1	9.5 ± 0.3
Retention (%)	65.7	77.8	81.4	83.7	87.2
M300					
Before aging (MPa)	2.4 ± 0.2	1.4 ± 0.2	1.7 ± 0.2	2.2 ± 0.2	2.0 ± 0.1
After aging (MPa)	1.8 ± 0.2	1.2 ± 0.1	1.5 ± 0.1	2.0 ± 0.2	1.9 ± 0.1
Retention (%)	75	85.7	88.2	90.9	95
Elongation at break					
Before aging (%)	949.0 ± 26.3	735.5 ± 78.5	906.5 ± 30.1	975.8 ± 38.1	820.3 ± 29.2
After aging (%)	739.6 ± 42.5	692.7 ± 25.1	864.4 ± 24.5	941.5 ± 26.8	808.2 ± 15.5
Retention (%)	77.9	94.2	95.4	96.5	98.5
Tear strength					
Before aging (N/mm)	4.6 ± 0.1	3.7 ± 0.2	3.9 ± 0.3	4.4 ± 0.2	3.9 ± 0.2
After aging (N/mm)	4.0 ± 0.2	3.4 ± 0.1	3.7 ± 0.1	4.2 ± 0.2	3.8 ± 0.1
Retention (%)	87	91.9	94.9	95.4	97.4

change of NR in both oils was much higher than that of the ENR 40 and ENR 40/$nSiO_2$ nanocomposites, indicating that the epoxidation significantly improved their oil resistance. Moreover, the ENR 40/$nSiO_2$ nanocomposites exhibited a $nSiO_2$ load-dependent higher oil resistance than the neat ENR 40 according to the interaction between the epoxide groups of the ENR 40 and the Si–OH groups on the $nSiO_2$ and also to their higher crosslink density that limited the penetration of oils into the rubber matrix. However, the volume change of the samples in ASTM oil number 3 was much higher than that in ASTM oil number 1. This is owing to the fact that the lower the aniline point the more severe the swelling action.

4. Conclusions

In this work, the ENR 40 was successfully synthesized via in situ performic epoxidation of NR in the latex state using a fixed molar ratio of $HCOOH/H_2O_2$ to isoprene unit of 1/0.75 in the presence of a surfactant at 50°C for 8 h. Beside the epoxide groups, the ring-opened structures were also observed in the as-synthesized ENR 40. The resulting ENR 40 was melt mixed with three loadings of $nSiO_2$ (1, 2, and 3 phr) and then vulcanized as usual in a sulfur curing system using TBzTD as a noncarcinogenic accelerator, which resulted into nitrosamine free cure. The curing characteristics, mechanical properties, thermal behaviors, dynamic mechanical

TABLE 4: Oil resistance of NR, ENR 40, and ENR 40/nSiO_2 nanocomposites.

Oil resistance	NR	ENR 40	ENR 40/nSiO_2		
			100/1	100/2	100/3
Volume change					
In IRM 901 oil (%)	111.3	7.6	4.0	6.5	3.7
In IRM 903 oil (%)	244.3	96.9	74.1	87.5	75.4

properties, and oil resistance of the prepared ENR 40/nSiO_2 nanocomposites were investigated and compared with NR and neat ENR 40. The results revealed that t_{s2} and t_{90} of ENR 40/nSiO_2 nanocomposites were slightly longer than those of NR but slightly shorter than those of ENR 40. However, the CRI of all ENR 40/nSiO_2 nanocomposites was lower than that of NR and ENR 40, while the crosslink density ($M_H - M_L$) was slightly higher. The mechanical properties (tensile properties and tear strength) of ENR 40 for both before and after aging at 100°C for 72 h were found to be improved as a result of the incorporation of nSiO_2 (all, but 2 phr optimal), while the % retention in all mechanical properties of ENR 40/nSiO_2 nanocomposites was higher than that of NR and ENR 40 in a dose-dependent manner. The T_g, E' at −65°C, thermal stability, and oil resistance (in ASTM Oil number 1 and number 3) of ENR 40 were also enhanced with the addition of nSiO_2 (all, but 1 phr optimal).

Acknowledgments

The authors gratefully acknowledge the 100th Anniversary Chulalongkorn University Fund for Doctoral Scholarship and also the 90th Anniversary Chulalongkorn University Fund (Ratchadaphiseksomphot Endowment Fund) for financial support.

References

[1] M. M. Hassan, A. A. Abd El-Megeed, and N. A. Maziad, "Evaluation of curing and physical properties of NR/SBR blends using radiation-grafting copolymer," *Polymer Composites*, vol. 30, no. 6, pp. 743–750, 2009.

[2] D. Derouet, P. Intharapat, Q. N. Tran, F. Gohier, and C. Nakason, "Graft copolymers of natural rubber and poly(dimethyl(acryloyloxymethyl)phosphonate) (NR-g-PDMAMP) or poly(dimethyl(methacryloyloxyethyl)phosphonate) (NR-g-PDMMEP) from photopolymerization in latex medium," *European Polymer Journal*, vol. 45, no. 3, pp. 820–836, 2009.

[3] Y. S. R. Yahya, A. R. Azura, and Z. Ahmad, "Effect of curing systems on thermal degradation behaviour of natural rubber (SMR CV 60)," *Journal of Physical Science*, vol. 22, no. 2, pp. 1–14, 2011.

[4] Z. X. Ooi, H. Ismail, and A. A. Bakar, "A comparative study of aging characteristics and thermal stability of oil palm ash, silica, and carbon black filled natural rubber vulcanizates," *Journal of Applied Polymer Science*, vol. 130, no. 6, pp. 4474–4481, 2013.

[5] B. Thongnuanchan, K. Nokkaew, A. Kaesaman, and C. Nakason, "Epoxidized natural rubber-bonded para rubber wood particleboard," *Polymer Engineering and Science*, vol. 47, no. 4, pp. 421–428, 2007.

[6] K. Jai-Eau, E. Wimolmala, and N. Sombatsompop, "Cure behavior and antimicrobial performance of sulfur-cured natural rubber vulcanizates containing 2-hydroxypropyl-3-piperazinylquinolinecarboxylic acid methacrylate or silver-substituted zeolite," *Journal of Vinyl and Additive Technology*, vol. 19, no. 2, pp. 123–131, 2013.

[7] K. Abhitha and K. Thomas, "Safe vulcanization system for heat resistant natural rubber products for engineering applications," *American Journal of Engineering Research*, vol. 3, p. 13, 2013.

[8] R. Yoksan, "Epoxidized natural rubber for adhesive applications," *Kasetsart Journal (Natural Science)*, vol. 42, pp. 325–332, 2008.

[9] C. M. Vu, H. T. Vu, and H. J. Choi, "Fabrication of natural rubber/epoxidized natural rubber/nanosilica nanocomposites and their physical characteristics," *Macromolecular Research*, vol. 23, no. 3, pp. 284–290, 2015.

[10] P. L. Teh, Z. A. M. Ishak, A. S. Hashim, J. Karger-Kocsis, and U. S. Ishiaku, "Effects of epoxidized natural rubber as a compatibilizer in melt compounded natural rubber-organoclay nanocomposites," *European Polymer Journal*, vol. 40, no. 11, pp. 2513–2521, 2004.

[11] A. S. M. Bashir, Y. Manusamy, T. L. Chew, H. Ismail, and S. Ramasamy, "Mechanical, thermal, and morphological properties of (eggshell powder)-filled natural rubber latex foam," *Journal of Vinyl and Additive Technology*, vol. 23, no. 1, pp. 3–12, 2015.

[12] V. Tanrattanakul, B. Wattanathai, A. Tiangjunya, and P. Muhamud, "In situ epoxidized natural rubber: improved oil resistance of natural rubber," *Journal of Applied Polymer Science*, vol. 90, no. 1, pp. 261–269, 2003.

[13] S. Gnecco, A. Pooley, and M. Krause, "Epoxidation of low-molecular-weight Euphorbia lactiflua natural rubber with "in situ" formed performic acid," *Polymer Bulletin*, vol. 37, no. 5, pp. 609–615, 1996.

[14] J. E. Davey and M. J. R. Loadman, "Chemical demonstration of the randomness of epoxidation of natural rubber," *British Polymer Journal*, vol. 16, no. 3, pp. 134–138, 1984.

[15] Z.-Q. Zeng, H.-P. Yu, Q.-F. Wang, and G. Lu, "Effects of coagulation processes on properties of epoxidized natural rubber," *Journal of Applied Polymer Science*, vol. 109, no. 3, pp. 1944–1949, 2008.

[16] W. A. K. Mahmood and M. H. Azarian, "Thermal, surface, nanomechanical and electrical properties of epoxidized natural rubber (ENR-50)/polyaniline composite films," *Current Applied Physics*, vol. 15, no. 5, pp. 599–607, 2015.

[17] N. V. Bac, L. Terlemezyan, and M. Mihailov, "On the stability and in situ epoxidation of natural rubber in latex by performic acid," *Journal of Applied Polymer Science*, vol. 42, no. 11, pp. 2965–2973, 1991.

[18] Y. Wang, X. Zhang, J. Oh, and K. Chung, "Effect of carbon black on self-crosslinking network structure of polychloroprene rubber and epoxidized natural rubber blends," *Polymer Composites*, 2015.

[19] P. K. Chattopadhyay, U. Basuli, and S. Chattopadhyay, "Studies on novel dual filler based epoxidized natural rubber nanocomposite," *Polymer Composites*, vol. 31, no. 5, pp. 835–846, 2010.

[20] P. Sheth and R. N. Desai, "Replacing TMTD with nitrosamine free TBzTD accelerator in curing of rubber," *International Journal for Scientific Research & Development*, vol. 1, no. 3, pp. 532–535, 2013.

[21] T. E. Motaung, A. S. Luyt, and S. Thomas, "Morphology and properties of NR/EPDM rubber blends filled with small amounts of titania nanoparticles," *Polymer Composites*, vol. 32, no. 8, pp. 1289–1296, 2011.

[22] M. Jarnthong, C. Nakason, N. Lopattananon, and Z. Peng, "Influence of incorporation sequence of silica nanoparticles on morphology, crystallization behavior, mechanical properties, and thermal resistance of melt blended thermoplastic natural rubber," *Polymer Composites*, vol. 33, no. 11, pp. 1911–1920, 2012.

[23] V. A. Doan, S. Nobukawa, S. Ohtsubo, T. Tada, and M. Yamaguchi, "Selective migration of silica particles between rubbers," *Journal of Polymer Research*, vol. 20, no. 5, article 145, 2013.

[24] W. Kaewsakul, K. Sahakaro, W. K. Dierkes, and J. W. M. Noordermeer, "Verification of interactions between silica and epoxidised squalene as a model for epoxidised natural rubber," *Journal of Rubber Research*, vol. 17, no. 3, pp. 129–142, 2014.

[25] M. R. Pourhossaini and M. Razzaghi-Kashani, "Grafting hydroxy-terminated polybutadiene onto nanosilica surface for styrene butadiene rubber compounds," *Journal of Applied Polymer Science*, vol. 124, no. 6, pp. 4721–4728, 2012.

[26] N. Jon, N. A. Samad, N. A. Abdullah, I. Abdullah, and R. Othaman, "Influence of silica addition on the properties of epoxidised natural rubber/polyvinyl chloride composite membrane," *Journal of Applied Polymer Science*, vol. 129, no. 5, pp. 2789–2795, 2013.

[27] A. Thitithammawong, N. Ruttanasupa, and C. Nakason, "Preparation and properties of chlorinated epoxidised natural rubber latex and its latex-based adhesive," *Journal of Rubber Research*, vol. 15, no. 1, pp. 19–34, 2012.

[28] S.-H. Chough and D.-H. Chang, "Kinetics of sulfur vulcanization of NR, BR, SBR, and their blends using a rheometer and DSC," *Journal of Applied Polymer Science*, vol. 61, no. 3, pp. 449–454, 1996.

[29] H. Ismail, R. B. Abdul Majid, and R. Mat Taib, "Effects of dynamic vulcanization on tensile, morphological, and swelling properties of poly(vinyl chloride) (PVC)/epoxidized natural rubber (ENR)/(Kenaf core powder) composites," *Journal of Vinyl and Additive Technology*, vol. 22, no. 3, pp. 206–212, 2016.

[30] P. Saramolee, K. Sahakaro, N. Lopattananon, W. K. Dierkes, and J. W. M. Noordermeer, "Comparative properties of silica- and carbon blackreinforced natural rubber in the presence of epoxidized lowmolecular weight polymer," *Rubber Chemistry and Technology*, vol. 87, no. 2, pp. 320–339, 2014.

[31] L. Qu, G. Yu, L. Wang, C. Li, Q. Zhao, and J. Li, "Effect of filler-elastomer interactions on the mechanical and nonlinear viscoelastic behaviors of chemically modified silica-reinforced solution-polymerized styrene butadiene rubber," *Journal of Applied Polymer Science*, vol. 126, no. 1, pp. 116–126, 2012.

Production of Pectin-Cellulose Biofilms: A New Approach for Citrus Waste Recycling

Veronika Bátori, Mostafa Jabbari, Dan Åkesson, Patrik R. Lennartsson, Mohammad J. Taherzadeh, and Akram Zamani

Swedish Centre for Resource Recovery, University of Borås, 50190 Borås, Sweden

Correspondence should be addressed to Veronika Bátori; veronika.batori@hb.se

Academic Editor: Arthur J. Ragauskas

While citrus waste is abundantly generated, the disposal methods used today remain unsatisfactory: they can be deleterious for ruminants, can cause soil salinity, or are not economically feasible; yet citrus waste consists of various valuable polymers. This paper introduces a novel environmentally safe approach that utilizes citrus waste polymers as a biobased and biodegradable film, for example, for food packaging. Orange waste has been investigated for biofilm production, using the gelling ability of pectin and the strength of cellulosic fibres. A casting method was used to form a film from the previously washed, dried, and milled orange waste. Two film-drying methods, a laboratory oven and an incubator shaker, were compared. FE-SEM images confirmed a smoother film morphology when the incubator shaker was used for drying. The tensile strength of the films was 31.67 ± 4.21 and 34.76 ± 2.64 MPa, respectively, for the oven-dried and incubator-dried films, which is within the range of different commodity plastics. Additionally, biodegradability of the films was confirmed under anaerobic conditions. Films showed an opaque appearance with yellowish colour.

1. Introduction

Plastics production has increased enormously in the past 100 years, and a global production of 322 million tons was reported by Statista [1] for 2015. This vast number of plastic products caused severe plastic pollution by now and they are typically made from nonrenewable sources. On the contrary, bioplastics are made from renewable sources or they are biodegradable; in the best-case scenario, they are both. Today, biopolymers are produced from cultivated crops; however, the land used for bioplastic is still negligible [2].

An example for a biobased and biodegradable material that is built up of different biopolymers, with no land use, is citrus waste. Citrus waste is a globally abundant and environmentally challenging waste that is underutilized [3]. Among citrus fruits, sweet oranges are the most commonly grown tropical fruits worldwide [4]. USDA [5] forecasted 45.8 million tons of sweet orange production for 2015/16. Industrial orange processing, for example, orange juice production generates about 50–60% residue of the original mass of the orange [6]. This vast quantity of waste is high in organic matter content (approx. 95% of total solids) and water (approx. 80–90%) and has a low pH (3-4) and inappropriate handling could cause severe damage to the environment [6]. Orange waste also contains pectin, soluble sugars, hemicelluloses, cellulose, starch, protein, lignin, ash, fat, and flavonoids [7, 8], which have been shown to be beneficial to many yet imperfect disposal and recovery applications [6]. These compounds on the other hand could be interesting for bioplastics applications.

Orange waste has already been applied as a reinforcement in petrochemical or biobased matrices [9–11]. In all of these cases, the authors reported increased mechanical properties of the products compared to the neat polymer. The increased mechanical properties are most probably the direct effect of the present cellulosic fibres. However, pectin, the major component of orange peel, seems to have no significant effect in the above-mentioned composites. Nevertheless, pectin-based composites have been prepared with different reinforcing substances [12] and cellulosic plant fibres have certainly been of great interest because of their favourable mechanical properties as a potential substitute for glass fibres [13] in biocomposites. Cellulose reinforced pectin composites have been developed, for example, for tissue engineering applications [14] and for food packaging applications [15] from commercial sources. However, there is no study on directly using cellulosic fibres and pectin obtained from citrus waste to prepare a biofilm.

This study investigated transforming orange waste into a biobased film as well as evaluating its properties. As a result of our study, a new function was given to orange waste and biofilms were developed without prior chemical modification of the raw material. Films were prepared with a film-casting method. The orange waste film (OWF) is plant-based and biodegradable and represents competitive mechanical properties with some of the commodity plastics. Therefore, with further improvements, OWF could be potentially used for nonstructural applications, for example, as a sustainable packaging material for the food industry.

2. Materials and Methods

2.1. Materials. Orange waste (OW) was obtained from Brämhults Juice AB (Borås, Sweden). Until further processing, the OW was stored at −20°C. Citric acid (monohydrate, >99.5%, Duchefa Biochemie, Netherlands), glycerol (>99%, ARCOS Organics, Belgium), organic antifoam 204 (Sigma-Aldrich, USA), and pectins (P9311, P9436, and P9561, Sigma-Aldrich, USA) were other materials used in this study.

2.2. Pretreatment of Orange Waste for Film Preparation. OW was washed with water to extract soluble sugars. Firstly, the material was soaked in tap water overnight; then, two further washing steps followed. The water (L) to OW (kg) ratio for all steps was 1.5 : 1. Each washing included shaking the OW in a water bath at 115 rpm for 20 min at 35°C. After each washing, the OW was collected from the flask using a metal sieve and rinsed under tap water. Size of the washed OW particles was reduced with a knife and they were dried for 16 h at 40°C. Dried OW was milled to a fine powder using a ball mill (Retsch MM 400, Germany) at a frequency of 30 Hz, for a total of 40 min, allowing the equipment to cool down after each 10 min milling (preliminary tests were also performed for 10, 20, and 30 min).

2.3. Formation of Orange Waste Films. A mixture of 2% (w/v) of OW powder was prepared in 1% (w/v) citric acid solution under constant magnetic stirring while heating up to 70°C. The acid solution also contained 7% (w/w) glycerol and 1 drop of organic antifoam/100 mL solution. The suspension was sieved through a metal sieve to eliminate nascent air bubbles before it was poured onto PTFE plates and dried at 40°C. Each plate contained 30 g of suspension and the diameter of the plates was 100 mm. The drying process was performed in either a laboratory drying oven (Termaks TS9026, Norway) or an incubator shaker (New Brunswick™ Scientific Excella® E24, USA) rotating at 50 rpm. In the following, films dried in the oven are referred to as oven-dried (OD) and in the incubator as incubator-dried (ID). Properties of OD and ID films were compared in the study.

2.4. Characterization of Orange Waste and Biofilms. Soluble sugars were removed from the orange waste before any compositional analyses.

2.4.1. Dry and Moisture Contents. Dry content of wet OW and moisture content of dry OW powder were determined by drying the samples at 105°C until constant weight. Tests were performed in duplicate.

2.4.2. Carbohydrate Content. Structural carbohydrate content of the OW powder was determined according to NREL/TP-510-42618 [17] via hydrolysis with hot sulphuric acid. The different carbohydrate fractions were identified and quantified by high-performance liquid chromatography (HPLC) (Waters 2695, Waters, Milford, USA), using a lead(II)-based column (HPX-87P, BioRad) with two Micro-Guard Deashing (Bio-Rad) precolumns operated at 85°C with 0.6 mL/min ultrapure water as eluent.

2.4.3. Pectin Content. Pectin content of the OW was measured by extraction. Pectin was extracted using a laboratory scale high-performance microwave digestion system (Milestone Ethos UP MA182, Italy). Size-reduced OW (20 g) was mixed with 45 mL of acidified water (25 mL of distilled water and 25 mL of 0.1 M HNO_3) to gain pH 2 [4]. The mixture was transferred into a PTFE vial with a total volume of 100 mL and was treated at 120°C for 10 min at a maximum power of 500 W. The treatment followed a 10 min ramping while the mixture was continuously stirred with a magnet stirrer. After the extraction, the mixture was cooled to room temperature and the solids were separated using vacuum filtration through a grade 3 filter paper. Pectin precipitation was performed according to Srivastava and Malviya [18] with slight modifications. The filtrate was adjusted to pH 4.5 [4] with 2 M of NaOH and cooled to 4°C before it was precipitated with ethanol under continuous stirring for 15 min. Ethanol to filtrate ratio was 2 : 1. The mixture was kept at 4°C for another 2 hours to allow pectin floatation, before being filtered through a cheesecloth to collect the extracted pectin. The pectin was then washed with absolute ethanol and dried to constant weight at 40°C.

2.4.4. Degree of Esterification of Pectin. A Fourier transform infrared (FTIR) spectrometer (Thermo Scientific, Nicolet iS10, USA) was used to measure the degree of esterification (DE) of the isolated pectin. Nicolet OMNIC 4.1 software was used to generate spectral data, which were analysed by Essential FTIR® (eFTIR, USA) software. The analysis was performed according to Oliveira et al. [19] with some modifications. Pectins with known values of DE (27 ± 7%, 62.5 ± 7.5%, and ≥85%) were used as standard. Determination of DE was based on the band areas at 1700–1750 cm^{-1}, corresponding to methyl esterified galacturonic acids (EGA), and at 1600–1630 cm^{-1}, corresponding to the free galacturonic acids

(FGA), and calculated according to the equation obtained from the calibration curve after plotting the standard values:

$$\mathrm{DE} = 75.308R_A + 29.923,$$
$$R^2 = 0.9588, \tag{1}$$
$$R_A = \frac{A_{\mathrm{EGA}}}{A_{\mathrm{EGA}} + A_{\mathrm{FGA}}}.$$

2.4.5. Morphology. The morphology of film surface and transversal sections was investigated by FE-SEM imaging (Zeiss, Sigma, Germany). For surface visualisation, films were attached to a carbon tape and covered with gold. For transverse visualisation, images of cross section of the films were also taken. Each film was immersed in liquid nitrogen for one minute; then it was broken and immediately attached to a carbon tape on a stub. Following that, the samples were coated with gold and images were taken. Photomicrographs were taken at 1000 magnifications, using an accelerating voltage of 10 and 25 kV.

2.4.6. Mechanical Testing. A tensile test was performed according to ISO 527, using Tinius Olsen H10KT universal tester and QMat software package. A moving cross-head was used to pull the dumbbell shaped specimens apart with a load cell of 250 N and a test speed of 10 mm/min. Specimens with a gauge length of 22 mm and a width of 4 mm were tested in triplicate and the averages are reported as tensile strength (MPa), elongation at break (%), and elongation at maximum tensile strength (%).

2.4.7. Thermal Analyses. Thermogravimetric analysis (TGA) (Q500 TA instruments, Waters LLC, USA) was performed to determine chemical film properties, such as decomposition. Approximately 6 to 8 mg of each sample was heated from room temperature to 700°C at a rate of 10°C/min. The analysis was performed under nitrogen atmosphere. Tests were done in triplicate, and an average is reported.

A differential scanning calorimetry (DSC) analysis (Q2000 TA Instruments, Waters LLC, USA) was performed to determine thermal properties. Approximately 6 to 8 mg of the biofilm sample was heated in an aluminium pan from −20 to 200°C at a rate of 10°C/min in the first scan and in the rescanning. The analysis was done under nitrogen atmosphere. Tests were done in triplicate and an average is reported.

A dynamic mechanical thermal analysis (DMTA) (Q800, TA Instruments, Waters LLC, USA) was performed to determine dynamic mechanical properties, operated in a multifrequency strain mode, using a tension film clamp. The samples were cut in a typical width of 5.3 mm by a length of 8 to 9 mm and examined at a heat rating of 3°C/min from 0 to 140°C, at a strain frequency of 1 Hz, and at amplitude of 15 μm. Tests were done in triplicate.

2.4.8. Antimicrobial Test. The antimicrobial activity of the films was investigated by cultivating *Escherichia coli* (ATTC 25922, received from Södra Älvsborgs Sjukhus Borås, Sweden), a food-related and easily growing bacteria, and *Aspergillus oryzae* (CBS 819.72, Centraalbureau voor Schimmelcultures, Netherlands), a type of mould, in 250 mL Erlenmeyer-flasks. The cultivation flasks contained 50 mL LB (10 g/L tryptone, 5 g/L yeast extract, 10 g/L NaCl) and YPD (10 g/L yeast extract, 20 g/L peptone, 20 g/L glucose) broth, as well as 5 mL phosphate buffer (pH 7) to maintain the optimum pH for the mentioned microorganisms, respectively. A sample of 0.50 g of the film was added to each flask, except for the reference flask that contained only the microorganism. The test was carried out in duplicate for 24 and 36 hours, respectively, in an incubator shaker at 125 rpm and 30°C. LB broth was inoculated with one colony, using an inoculation loop obtained from previously grown agar plates. YPD broth was inoculated with 3 mL spore solution of 5.41×10^5 spores/mL obtained from previously grown agar plates. Bacterial growth was determined by measuring the optical density value of the broth with a spectrophotometer (Biochrom, WPA Spectrawave S1200 Diode Array Spectrophotometer, UK) at a wavelength of 600 nm. Fungal growth was determined by measuring the sugar consumption in the broth with HPLC.

2.4.9. Biodegradability Test. Anaerobic digestion was performed in order to test the biodegradation of the films. Digestion was carried out in batch reactors according to a previous study [21]. Bacterial inoculum was obtained from a large-scale thermophilic biogas plant (Borås Energi & Miljö AB, Borås, Sweden). Total solids (TS) content and volatile solids (VS) content of the films were determined using a gravimetric method. Each 120 mL glass bottle of a working volume of 50 mL contained 35.5 mL inoculum, and the rest was water. Each bottle contained 0.15 g sample VS. The reactors were then flushed with a gas mixture consisting of 80% N_2 and 20% CO_2 for 2 min. The reactors were incubated for 30 days at 55°C. Reactors containing only water and inoculum were used as blanks, and pure cellulose (Avicel®, Sigma-Aldrich) was also used as a reference. The experiment was performed in triplicate. 0.2 mL gas samples were taken with a 0.25 mL syringe (VICI, USA). The gas composition was analysed using a gas chromatograph (Clarus 500, Perkin-Elmer, USA) equipped with a packed column (Perkin-Elmer, $6' \times 1.8''$ OD, 80/100, Mesh, USA) and a thermal conductivity detector (Perkin-Elmer, USA) with an injection temperature of 150°C. The carrier gas was nitrogen, and the equipment was operated at a 20 mL/min flow rate at 60°C. At the end of the digestion, the pH of the digestates was also measured.

2.4.10. Statistical Analysis. Statistical analysis of the data obtained from mechanical and thermal tests was performed using MINITAB 17 Statistical Software. Results were compared by obtaining a p value via t-test, with a 95% confidence interval.

3. Results and Discussion

The results of the compositional analyses of orange waste are summarized in Table 1. Although values depend on the source and on the conditions of treatment and isolation methods, the major component of OW is pectin, followed by

TABLE 1: Chemical composition of OW according to the literature [20] and the results obtained from material characterization (% (w/dry w)).

Compound/content	[20] (%)	Characterization (%)
Pectin	42.5	29.84 ± 0.29
Cellulose	9.2	18.66 ± 0.48
Hemicelluloses	10.5	20.89 ± 0.89
Soluble sugars	16.9	
Protein	6.5	
Lignin	0.8	
Fat	2.0	
Ash	3.5	
Starch	3.8	
Others	4.4	about 30
DE of pectin		42.72 ± 1.10
Dry matter of OW		9.84 ± 1.21
Moisture content of OW powder		9.58 ± 0.03

hemicellulose and cellulose. In this study, pectin, cellulose, and hemicellulose fibres gained main focus for biofilm preparation. Biofilms from OW were produced by the direct conversion of orange waste (Figure 1) with no prior chemical treatment of the waste. The physical and thermal properties of the films are influenced by both the cellulose and hemicelluloses fractions and the surrounding pectin phase. This new approach could be a solution to the problems associated with orange waste handling.

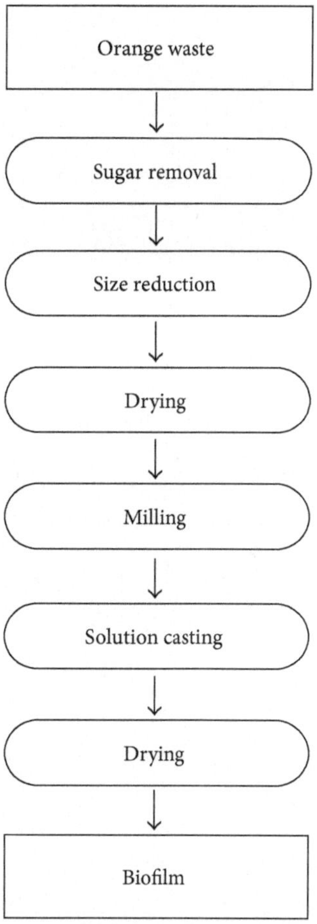

FIGURE 1: Flow chart of the developed method for production of orange waste films.

3.1. Details of the Method for Production of Biofilms from Orange Waste. Soluble sugars were removed from the OW during the pretreatment steps because preliminary experiments showed that when the sugars were not removed, the film surface was heterogeneous and cracked. To obtain a homogeneous OW powder to result in the most compatible film structure, different milling times (10, 20, 30, and 40 min) were tested. 40 min milling showed satisfactory results based on visual observation. Particle sizes were mostly between 125 and 75 μm. Since particle size could not be controlled with ball milling, a variety of the sizes were present in the OW powder.

Films were prepared by a casting method in which cellulose and hemicelluloses fibres were suspended in the pectin solution and further dried to a film. The orientation of the cellulosic fibres in the films could not be aligned and therefore, the particles were randomly distributed. Pectin film formation as well as previous studies on OWF formation showed that pectin forms foam on the surface of the mixture (data not shown); therefore, organic antifoam was added to the solvent liquid. Initial investigations on drying at 30, 40, and 50°C as well as at room temperature were performed. Drying temperature for solvent evaporation seemed not to be sufficient when films were dried at room temperature or at 30°C. When 50°C was applied, cracks were formed in the films. The higher temperature resulted in faster evaporation of the solvent at the edges of the film, thus creating a presumably more intensive replacement of liquid from the centre towards the edges, as explained by Routh [22]. In order to reduce the liquid flow in which the dispersed particles could be carried towards the edges (coffee-ring effect) [22], the temperature was reduced to obtain a slower evaporation of the solvent. Oven drying at 40°C resulted in no cracks, only a few small holes in the films were visible, and about 7 hours were enough for complete drying. Once the temperature was optimized, another drying method was also evaluated to eliminate the holes. An incubator shaker was then used to provide continuously rotating movements. The continuous movement of liquid and particles avoided the so-called coffee-ring effect as explained above. As a result, ID films without holes were produced, though the thickness of those films was uneven. The thickness of OD films was 0.09–0.10 mm, while ID films were more irregular. For mechanical characterization, the most uniform part of the ID films was chosen (0.08–0.10 mm), and the average thickness was used for analysis. By appearances, the biofilms are opaque and have a yellowish colour. The method for production of biofilms from orange waste is schematically presented in Figure 1.

3.2. Morphology of Orange Waste Films. The FE-SEM images shown on Figure 2 reveal that a "second layer" (highlighted

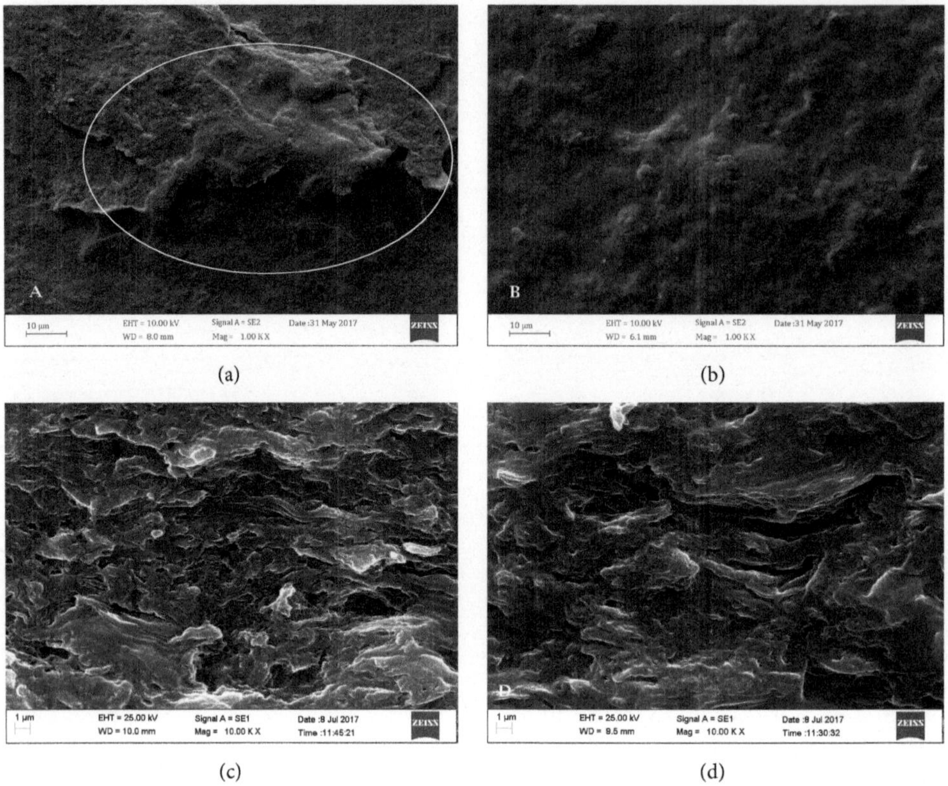

FIGURE 2: FE-SEM images taken from surface of OD (a) and ID (b) films and from transversal sections of OD (c) and ID (d) films. On image (a) a circle is highlighting the unevenness of the surface compared to the more uniform surface shown on image (b).

on the image) was formed on the surface of OD films. In drying films, where there are more than one type of particles, a common phenomenon is self-stratification, which means that one particle ends on top of another [22]. As OWFs are made of a complex mixture of dissolved and undissolved polymers this phenomenon under the "normal" drying conditions seems to occur. When the "normal" drying conditions were disturbed with continuous rotation, this phenomenon was not visible; instead it resulted in a smooth and uniform surface of the ID films. The transversal sections confirm that continuous rotation resulted in a more homogenous material: pores are visible in the structure of OD films, while there are no clear pores in ID films. The few voids that can be seen could be the result of pulling the cellulosic fibres out during fractioning. Since there was no alignment of the fibres, the magnitude of these voids is low.

3.3. Mechanical Analysis of Orange Waste Films. Tensile strengths of the OWFs produced in this study were measured between 28 and 36 MPa (Figure 3). Yu et al. [15] varied the ratio between LM pectin and CMC to prepare nanocomposites by adding up to 8% w/w of montmorillonite reinforcement to the solutions. The tensile strengths of the unreinforced polymers were measured from approximately 16 to 32 MPa, and with reinforcement the tensile strength increased. The tensile strengths of the OWFs are also in the same range for some of the most commonly used polymers, such as low density polyethylene (LDPE), high density polyethylene (HDPE), polytetrafluorethylene (PTFE), polypropylene (PP), and polystyrene (PS) [16]. A comparison is shown in Figure 3. Elongation however of the same commodity plastics (LDPE 100–650%; HDPE 10–1200%; PTFE 200–400%; PP 100–600%) is much higher than of OWF, except for PS (1.2–2.5%) [16].

3.4. Thermal Characteristics of Orange Waste Films. The results of the thermogravimetric analysis are represented as Table 2 and Figure 4. In both drying cases, the trend was similar, and thermal degradation consisted of three stages. The first thermal event was a weight loss up to 3.79–3.94% at an average temperature of 80.38–80.61°C due to the evaporation of water in the samples. This could be explained by biopolymers and glycerol being hygroscopic substances, and moisture content is dependent on the ambient conditions. The second thermal event occurred at a maximum temperature of 251.89–253.20°C, and the decomposition ratio was 28.54–29.28%, which was related to the depolymerisation of the pectin present in the films [23]. The third thermal event was observed at a maximum temperature of 351.89–354.34°C; the decomposition of the films showed a further 24.10–25.15% that could be referred to the decomposition of the cellulose [24]. The remaining 21.97–23.20% of the films after thermal treatment is ash.

TABLE 2: TGA thermogram showing decomposition rate of the OWFs: moisture loss, decomposition ratio 1, decomposition ratio 2 and residue and the maximum temperatures related to the thermal events.

Sample	Moist. loss (%)	MLT (°C)	D1 ratio (%)	DT 1 (°C)	D2 ratio (%)	DT 2 (°C)	Residue (%)
OD 1	4,07	81,44	27,65	252,58	25,63	351,5	24,66
OD 2	3,94	78,90	26,77	252,6	26,7	350,27	19,03
OD 3	3,82	81,49	33,43	250,5	23,12	353,91	22,22
Average	3,94	80,61	29,28	251,89	25,15	351,89	21,97
St. dev.	0,13	1,48	3,62	1,21	1,84	1,85	2,82
ID 1	3,50	84,83	29,07	252,37	22,68	352,7	24,58
ID 2	3,87	79,93	29,31	254,09	24,96	356	20,97
ID 3	4,02	76,38	27,25	253,14	24,66	354,31	24,05
Average	3,79	80,38	28,54	253,20	24,10	354,34	23,20
St. dev.	0,26	4,24	1,13	0,86	1,24	1,65	1,95

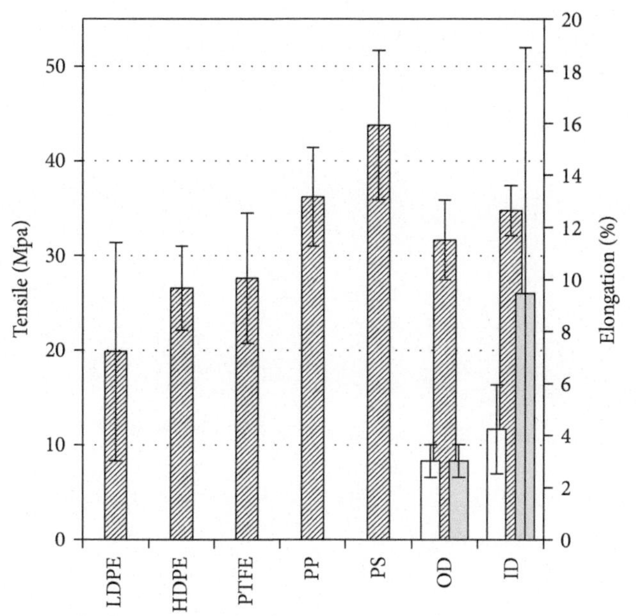

FIGURE 3: Left axis: tensile strength of different commodity plastics [16] and tensile strength at maximum load of OD and ID films are shown; right axis: elongation at maximum tensile strength (white bars) and elongation at break (grey bars) belonging to OD and ID films are shown (elongation at break of commodity plastics is not shown in the image due to the great difference of values).

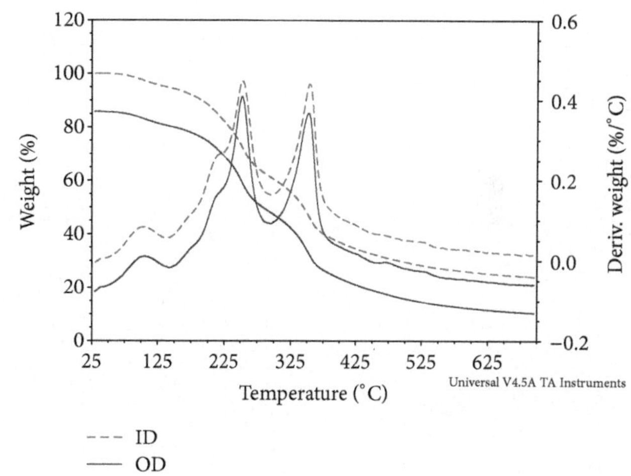

FIGURE 4: TGA thermogram showing weight loss curves and derivative weight curves of OD and ID films.

The films were also analysed using differential scanning calorimetry (Figure 5) to determine glass transition temperature (Tg). An endothermic transition was detected around 83.29 and 88.12°C, with an onset temperature of 61.21 and 66.22°C during the first scanning of the ID and OD samples, respectively. After the rescanning of the samples, the endothermic transition disappeared. Aguilera et al. [25] also reported that endothermic curves of biopolymers had disappeared after rescanning the samples and thus there was no Tg observed. As there was no Tg observed in the thermograms of OWFs, there was no melting point either. It is then assumed that the OWFs have less crystals present and they are mostly amorphous.

DMTA characterization was carried out in order to compensate DSC analysis and to monitor the temperature dependence of the storage modulus, tan δ, and the loss modulus. The results are shown in Figures 6 and 7. The storage modulus (Figure 6) is a measurement of the stiffness of the material, and both ID and OD films showed similar behaviours. The sample OD had a storage modulus of 1.4 GPa at 10°C. As the temperature rose, there was a clear decrease of the storage modulus. At 70°C, the storage modulus decreased to 0.6 GPa. In other words, the sample showed a more elastic behaviour at higher temperatures and a more plastic behaviour (storing energy rather than returning it) at low temperatures, which is a typical characteristic of polymers. The other sample, ID, had a storage modulus of 1.0 GPa at 10°C and 0.4 GPa at 70°C. While OD generally had a somewhat higher storage modulus, the difference between the two samples was not significant. The temperature dependence of tan δ can be seen in the same figure. Under the chosen test conditions, tan δ continuously increased with the increasing temperature. For the sample OD, a weak transition could be seen at 98°C. For ID, no clear transition was observed.

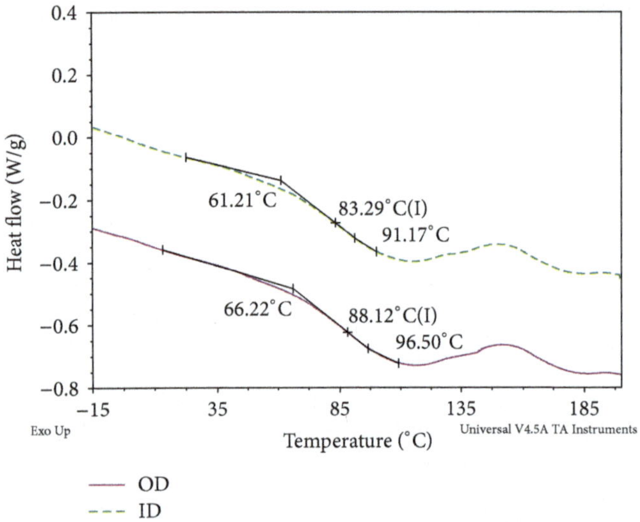

FIGURE 5: DSC analyses showing an endothermic transition of OD and ID films.

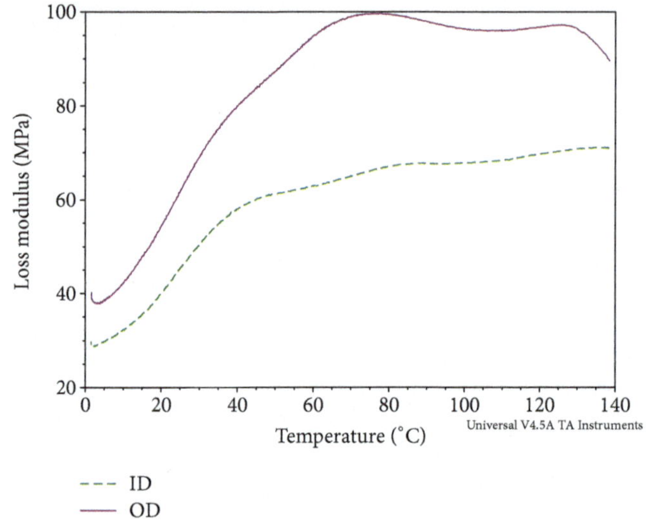

FIGURE 7: DMTA thermogram is showing the loss modulus of OD and ID films.

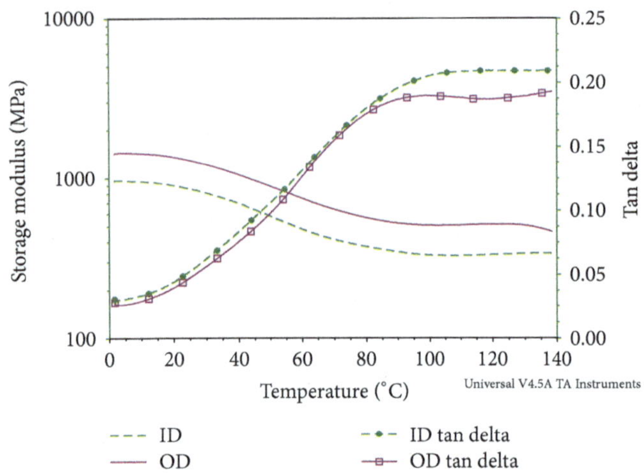

FIGURE 6: Storage modulus and tan δ are shown of OD and ID films.

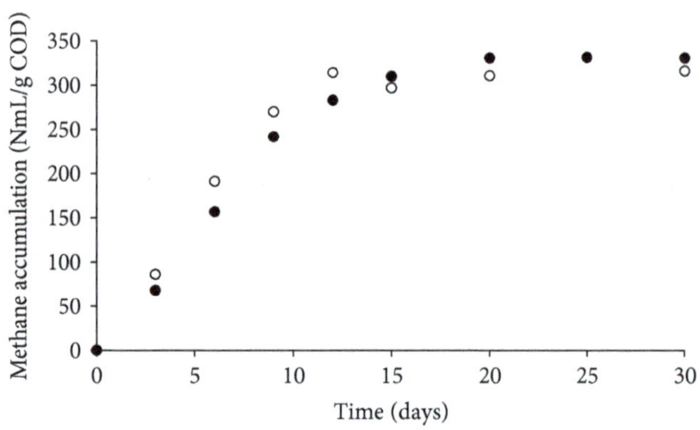

FIGURE 8: Anaerobic degradation of OD (full circles) and ID (empty circles) films, expressed in the accumulation of methane gas/g COD over a period of 30 days at 55°C.

The loss modulus describes the viscous properties of the material (Figure 7). As mentioned in the DSC analysis, there was no obvious glass transition in the DSC curve. The reason the loss modulus curve does not show a maximum could be the same why no Tg was observed in the DSC curve. For the sample OD, loss modulus increased continuously and reached a maximum at 76°C. Again, for ID, the transition could not be seen so clearly.

3.5. Antimicrobial Activity of Orange Waste Films. D-limonene, the major component of orange oil, is in turn well-known for its protecting behaviour against pathogens [26]; thus it would indicate an antimicrobial property of the films. However, the results obtained from the optical density measurements and glucose consumption, showed that films did not have antimicrobial property. In both cases, the length of the lag phase and the exponential phase of the microorganisms were the same for blanks and for the samples containing the films. One reason could be that orange oils were removed during the sugar removal pretreatment step, which was similarly performed as the cold pressing method, a preferred method for extracting essential oils from citrus fruits [27], and is usually done by soaking the citrus fruit in warm water and pressing the rind with a sponge to collect the oil.

3.6. Biodegradability of Orange Waste Films. Biodegradability is a key feature of next generation plastics and the European Committee for Standardization specifies a 90% degradation within six months [28]. In the anaerobic degradation study, films reached 90% degradation in about 15 days (Table 3). Figure 8 shows the methane accumulation based on the theoretical maximum (350 NmL/g COD) of a substance by measuring the COD.

TABLE 3: Degradation (%) of ID and OD films (results based on average) by days.

Deg. time (d)	3	6	9	12	15	20	25	30
ID	25	55	77	90	85	89	95	90
OD	20	69	81	89	95	95	95	94

4. Conclusions

The biofilms made from orange waste show similar physical properties as do some of the commodity plastics. It was observed that the drying method did not make a difference for mechanical and thermal properties of the films but resulted in a more uniform surface. The films were also biodegradable under anaerobic conditions. The main constituents of the biofilms, pectin and cellulosic fibres, suggest an application where biodegradability is as important as strength, such as a short-lived packaging material. Generally, the properties are promising, although further characterization and improvements are necessary in order to achieve the desired features, less hygroscopic characteristics, for example. Finally, production of biofilms from orange waste not only opens up opportunities for the production of environmentally friendly biomaterials but also could be a solution for the challenges arisen upon disposal of orange waste.

Acknowledgments

The authors would like to express their gratitude to ÅForsk Foundation for financial support and to Brämhults Juice AB for the raw material. The authors are also grateful for the technical help received from Lukitawesa, Ilona Sárvári Horváth, Ramkumar B. Nair, and Jorge A. Ferreira and from Christine Räisänen for feedback on the structure and language.

References

[1] Statista, *Global plastic production from 1950 to 2015 (in million metric tons)*, Statista, 2016.

[2] E. Bioplastics, "Biobased plastics - Fostering a resource efficient circular economy; Benefits, feedstock types, sustainable sourcing, land use," in *Proceedings of the in 10th European Bioplastics Conference*, Berlin, Germany: European Bioplastics, 2015.

[3] A. M. Balu, V. Budarin, P. S. Shuttleworth et al., "Valorisation of orange peel residues: Waste to biochemicals and nanoporous materials," *ChemSusChem*, vol. 5, no. 9, pp. 1694–1697, 2012.

[4] M. Boukroufa, C. Boutekedjiret, L. Petigny, N. Rakotomanomana, and F. Chemat, "Bio-refinery of orange peels waste: A new concept based on integrated green and solvent free extraction processes using ultrasound and microwave techniques to obtain essential oil, polyphenols and pectin," *Ultrasonics Sonochemistry*, vol. 24, pp. 72–79, 2015.

[5] USDA, *Citrus: World Markets and Trades*, United States Department of Agriculture Foreign Agricultural Service, USA, 2016.

[6] J. Ángel Siles López, Q. Li, and I. P. Thompson, "Biorefinery of waste orange peel," *Critical Reviews in Biotechnology*, vol. 30, no. 1, pp. 63–69, 2010.

[7] F. R. Marín, C. Soler-Rivas, O. Benavente-García, J. Castillo, and J. A. Pérez-Alvarez, "By-products from different citrus processes as a source of customized functional fibres," *Food Chemistry*, vol. 100, no. 2, pp. 736–741, 2007.

[8] K. Rezzadori, S. Benedetti, and E. R. Amante, "Proposals for the residues recovery: Orange waste as raw material for new products," *Food and Bioproducts Processing*, vol. 90, no. 4, pp. 606–614, 2012.

[9] R. U. Abass, "Mechanical behaviour of natural material (orange peel) reinforced polyester composite," *International Journal of Engineering Sciences Research Technology*, vol. 4, no. 3, pp. 166–172, 2015.

[10] V. S. Aigbodion, C. U. Atuanya, E. A. Igogori, and P. Ihom, "Development of High-Density Polyethylene /Orange Peels Particulate Bio-Composite," *Gazi University Journal of Science*, vol. 26, no. 1, pp. 107–117, 2013.

[11] N. Hall, "Bioplastics made from coffee and orange?" *3D Printing Industry*, 2016.

[12] V. Bátori, D. Åkesson, A. Zamani, and M. J. Taherzadeh, "Pectin based composites," in *Handbook of composites from renewable materials*, M. K. Thakur, M. R. Kessler, and V. K. Thakur, Eds., pp. 487–517, John Wiley & Sons Inc, 2016.

[13] F. Ansari, M. Skrifvars, and L. Berglund, "Nanostructured biocomposites based on unsaturated polyester resin and a cellulose nanofiber network," *Composites Science and Technology*, vol. 117, pp. 298–306, 2015.

[14] N. Ninan, M. Muthiah, I.-K. Park, A. Elain, S. Thomas, and Y. Grohens, "Pectin/carboxymethyl cellulose/microfibrillated cellulose composite scaffolds for tissue engineering," *Carbohydrate Polymers*, vol. 98, no. 1, pp. 877–885, 2013.

[15] W.-X. Yu, Z.-W. Wang, C.-Y. Hu, and L. Wang, "Properties of low methoxyl pectin-carboxymethyl cellulose based on montmorillonite nanocomposite films," *International Journal of Food Science & Technology*, vol. 49, no. 12, pp. 2592–2601, 2014.

[16] J. W. D. Callister, *Materials science and engineering: an introduction*, John Wiley & Sons, Inc., USA, 2007.

[17] A. Sluiter, B. Hames, R. Ruiz et al., "Determination of Structural Carbohydrates and Ligning in Biomass," Technical Report NREL/TP-510-42618, Golden, Colorado, USA, 2008.

[18] P. Srivastava and R. Malviya, "Extraction, characterization and evaluation of orange peel waste derived pectin as a pharmaceutical excipient," *Natural Products Journal*, vol. 1, no. 1, pp. 65–70, 2011.

[19] T. Í. S. Oliveira, M. F. Rosa, F. L. Cavalcante et al., "Optimization of pectin extraction from banana peels with citric acid by using response surface methodology," *Food Chemistry*, vol. 198, pp. 113–118, 2016.

[20] B. Rivas, A. Torrado, P. Torre, A. Converti, and J. M. Domínguez, "Submerged citric acid fermentation on orange peel autohydrolysate," *Journal of Agricultural and Food Chemistry*, vol. 56, no. 7, pp. 2380–2387, 2008.

[21] T. L. Hansen, J. E. Schmidt, I. Angelidaki et al., "Method for determination of methane potentials of solid organic waste," *Waste Management*, vol. 24, no. 4, pp. 393–400, 2004.

[22] A. F. Routh, "Drying of thin colloidal films," *Reports on Progress in Physics*, vol. 76, no. 4, pp. 46603–46632, 2013.

[23] A. Ghaffari, K. Navaee, M. Oskoui, K. Bayati, and M. Rafiee-Tehrani, "Preparation and characterization of free mixed-film of pectin/chitosan/Eudragit® RS intended for sigmoidal drug delivery," *European Journal of Pharmaceutics and Biopharmaceutics*, vol. 67, no. 1, pp. 175–186, 2007.

[24] M. Semsarilar, J. Tom, V. Ladmiral, and S. Perrier, "Supramolecular hybrids of cellulose and synthetic polymers," *Polymer Chemistry*, vol. 3, no. 12, pp. 3266–3275, 2012.

[25] J. M. Aguilera, T. R. Cuadros, and J. M. Del Valle, "Differential scanning calorimetry of low-moisture apple products," *Carbohydrate Polymers*, vol. 37, no. 1, pp. 79–86, 1998.

[26] G. Forgács, M. Pourbafrani, C. Niklasson, M. J. Taherzadeh, and I. S. Horváth, "Methane production from citrus wastes: process development and cost estimation," *Journal of Chemical Technology & Biotechnology*, vol. 87, no. 2, pp. 250–255, 2012.

[27] M. A. Ferhat, B. Y. Meklati, and F. Chemat, "Comparison of different isolation methods of essential oil from Citrus fruits: cold pressing, hydrodistillation and microwave 'dry' distillation," *Flavour and Fragrance Journal*, vol. 22, no. 6, pp. 494–504, 2007.

[28] A. Soroudi and I. Jakubowicz, "Recycling of bioplastics, their blends and biocomposites: A review," *European Polymer Journal*, vol. 49, no. 10, pp. 2839–2858, 2013.

21

Electrical, Thermal, and Morphological Properties of Poly(ethylene terephthalate)-Graphite Nanoplatelets Nanocomposites

Basheer A. Alshammari,[1] Arthur N. Wilkinson,[2] and Ghzzai Almutairi[3]

[1] *Material Research Institute, King Abdulaziz City for Science and Technology, P.O. Box 6086, Riyadh 11442, Saudi Arabia*
[2] *Materials Science Centre and North West Composites Centre, School of Materials, The University of Manchester, Grosvenor Street, Manchester M13 9PL, UK*
[3] *Water and Energy Research Institute, King Abdulaziz City for Science and Technology, P.O. Box 6086, Riyadh 11442, Saudi Arabia*

Correspondence should be addressed to Basheer A. Alshammari; bshammari@kacst.edu.sa

Academic Editor: Bernabé L. Rivas

Graphite nanoplatelets (GNP) were incorporated with poly(ethylene terephthalate) (PET) matrix by melt-compounding technique using minilab compounder to produce PET-GNP nanocomposites, and then the extruded nanocomposites were compressed using compression molding to obtain films of 1 mm thickness. Percolation threshold value was determined using percolation theory. The electrical conductivity, morphology, and thermal behaviors of these nanocomposites were investigated at different contents of GNP, that is, below, around, and above its percolation threshold value. The results demonstrated that the addition of GNP at loading >5 wt.% made electrically conductive nanocomposites. An excellent electrical conductivity of ~1 S/m was obtained at 15 wt.% of GNP loading. The nanocomposites showed a typical insulator-conductor transition with a percolation threshold value of 5.7 wt.% of GNP. In addition, increasing screw speed enhanced the conductivity of the nanocomposites above its threshold value by ~2.5 orders of magnitude; this behavior is attributed to improved dispersion of these nanoparticles into the PET matrix. Microscopies results exhibited no indication of aggregations at 2 wt.% of GNP; however, some rolling up at 6 wt.% of GNP contents was observed, indicating that a conductive network has been formed, whereas more agglomeration and rolling up could be seen as the GNP content is increased in the PET matrix. These agglomerations reduced their aspect ratio and then reduced their reinforcement efficiency. NP loading (>2 wt.%) increased degree of crystallinity and improved thermal stability of matrix slightly, suggesting that 2 wt.% of GNP is more than enough to nucleate the matrix.

1. Introduction

Graphite nanoplatelets (GNP) are also called expanded graphite (EG) or exfoliated graphite. In graphite form, the graphene sheets are held together by van der Waals interaction which allows some molecules, atoms, or ions to be intercalated between these sheets. Usually, GNP is prepared by modifying graphite using graphite intercalation method (chemical exfoliation method). In this method, graphite is immersed in a concentrated mixture of sulfuric acid with other oxidizers agents. The resulting products are called graphite intercalation compounds (GIC) or expandable graphite. Typically, the GIC materials are heated to a high temperature of ~1000°C for few seconds that generates huge amounts of gases such as sulfur dioxide and water resulting in extra expansion of graphene sheets, which is known as GNP, exfoliated graphite, or expanded graphite (EG). This procedure is achieved under nitrogen or in air atmospheres. Further exfoliation of these materials can produce a single atomic layer of graphite known as graphene. Typically, high ultrasonication process will be used for breaking down the GNP to smaller sheets. Both GIC and GNP are commonly modified graphite forms that have been used for preparing polymer-carbon composites [1–4].

During the previous years, polymer-GNP nanocomposites have been widely investigated to develop materials to be used where electrical, thermal, and mechanical properties are required. According to a review by Li and Zhong [2] and an investigation study by Alasvand Zarasvand and Golestanian [5], many researchers prepared various conducting polymer nanocomposites containing various GNP weight fractions and sizes under different preparing conditions. Most results indicate significant enhancement in mechanical, thermal, and electrical properties of final nanocomposites. One of the main problems for polymer nanocomposite materials is the appropriate dispersion and distribution of nanofillers into the matrix that needed to be addressed before taking the full advantages of particulate nanoscale fillers.

Furthermore, PET is a thermoplastic polymer. It is used in textile fibers, films, and packaging technologies. This is because PET has good properties such as mechanical and chemical resistance, ability to be spun, and thermal and dimensional stabilities [6, 7].

Despite the above-stated properties, improvement of its electrical conductivity is necessary for electronic applications [8], for example, replacing of indium tin oxide (ITO) electrodes as these materials have poor mechanical and lower conductivity when compared to the conductive carbon nanofillers such as graphene nanoplatelets or carbon nanotubes [2]. The electrical conductivity values higher than 10^{-8} S/m are required for numerous electronic applications, for instance, semiconducting, antistatic, and electromagnetic shielding materials for electronic devices [9].

To achieve electrically conductive pathways within the polymer composites with high conductivity value and acceptable mechanical properties and processability, a reduction of the amount of conductive fillers is required. Therefore, percolation theory is frequently used to determine the percolation threshold value which is the minimum content of conductive fillers associated with the conversion in the electrical behavior of the composite from insulate to conductive phase. This theory also describes the insulator-conductor transitions in polymer composites made of conductive fillers and an insulating polymer matrix. Previous published studies have shown that the percolation threshold value strongly depends on the aspect ratio (length-to-diameter) of the filler particles [10–12].

There are three common adopted methods for the production of composites: solution, in situ polymerization, and melt compounding. The latter method is an environmentally friendly technique. Therefore, it was used in this study. Relatively, there are few investigations reported on PET-GNP nanocomposites. For instance, a study by Zhang et al. [4] prepared PET-GNP nanocomposites using melt blending. A homogenous dispersion of GNP in the PET matrix was noted in this study after some shearing has been used to improve interaction between PET matrix and GNP. Similar work done by Li and Jeong, [8] who prepared PET nanocomposites with diverse concentrations of GNP using melt mixing method, reported good dispersed GNP within the PET matrix, leading to improvements in the mechanical, electrical, and thermal properties of nanocomposites compared with the unfilled PET. The electrical percolation threshold was found to be ~5 wt.% Al-Jabareen et al. [13] added GNP into PET by melt mixing to enhance its oxygen barrier properties. The authors observed a uniform distribution and dispersion of GNP into PET. The results indicated an increase in crystallinity, thermal stability, and Young's modulus of the resultant composite compared to pure PET matrix. However, tensile strength and elongation at break showed reductions of 56% and 40%, respectively. A mother work carried out by Paszkiewicz et al. [14] on the electrical conductivity of PET-GNP nanocomposites prepared by suit polymerization approaches proved good dispersion of GNP in the PET matrix. The transition from insulator to conductor system was started at amount of 0.05 wt.% GNP. Such a low percolation threshold is related to the large surface area, high aspect ratio, and uniform dispersion of the GNP as well as manufacturing method used where monomers polymerized in the existence of the GNP, therefore allowing stronger interactions.

Although the above-mentioned studies have examined properties of PET-GNP nanocomposites, only a few studies have characterized their properties after determining the minimum amount of GNP required for forming conductive pathways within the PET-GNP nanocomposites system. Therefore, in this study, a nanocomposite was prepared using PET as polymer matrix and GNP as nanoscale filler. The effect of GNP on electrical conductivity, morphology, and thermal behavior of PET matrix is investigated as a function of GNP loadings. At first stage of this study, the percolation threshold (Φ_c) value was determined using percolation theory. This value is important for improving fabrication of the polymer composite (PCs) and their properties. The development of PCs based on conductive filler has concentrated on reducing Φ_c in order to reduce cost, improve the fabrication or processability, and enhance mechanical properties of the final composite products [9]. Therefore, finding such value is very important to indicate the performance of the final composite products.

2. Experimental Work

2.1. Materials. PET (matrix) was purchased from Equipolymers Company (Schkopau, Germany) in the form of resin (Grade: LIGHTER C93). It has a glass transition temperature of 75°C, melting temperature of 247°C, and bulk density of 0.88 g/cm^3. XG Sciences, Lansing, MI, USA (xGnPs-15), supplied the GNP (nanoscale filler) in the form of powder. It has average thickness; diameters are ~6–8 nm and 15 μm, respectively.

2.2. Preparation of Nanocomposites. Both as-received materials of PET and GNP dried at about 120°C in vacuum oven overnight before melt compounding was carried out. The PET-GNP nanocomposites were prepared through the melt-compounding technique by mixing or/and compounding PET and GNP in a molten state using a Thermo-Haake Minilab compounder. The melt-compounding technique has been described in detail somewhere else [1]. In the current study, the time of mixing was 5 minutes at processing temperature of 280°C and speed of screws was 45 rpm. Then extruded samples were compressed at the same processing temperature, that is, 280°C for 10 min using compression molding machine and followed by quenching in ice bath to reduce the crystallinity and then chopped, dried at 40°C for 24 h, and stored for the further characterization and testing.

2.3. Characterization of Nanocomposites

2.3.1. Electrical Conductivity and Percolation Threshold Measurements. An impedance spectroscopy, model NumetriQ PSM1735, was used for determining both electrical conductivity values (σ) and percolation threshold values (Φ) for PET and its nanocomposites. Samples plates of dimension (~10 × 10 × 1 mm^3) were cut from compressed films and then coated using sliver paint. The coating is to reduce the contact resistance between the samples and copper wires attached to the specimen using silver epoxy adhesion. Resistance of all tested samples was measured in a range of 1 Hertz up to 10×10^{-6} Hertz at 1 volte as an amplitude voltage. For checking the accuracy, each sample was run five times at room temperature.

2.3.2. Morphology Observations. The morphology state of GNP and the PET nanocomposites was characterized using a SEM (Model: Philips SEM XL30, at an accelerating voltage of 10–20 kV). The nanocomposites were cryogenically frozen in a liquid nitrogen for 5 min and then fractured. The samples were mounted on pin stubs using adhesive tape and then coated with a layer of gold using an Edwards S150B sputter coater, to avoid any charge during scan observation.

TEM (Philips CM200) was applied also at an accelerating voltage of 200 kV. Nanocomposites samples were embedded into epoxy resin for slices purpose using an ultramicrotome and then diamond blade was used to cut slices of ~50 nm thick from the mid of the nanocomposites. In addition, GNP were mixed with ethanol and sonicated using an ultrasonic bath at room temperature for 20 min, and little drops were dropped onto a copper grid using a micropipette. The grid was left to dry in a fume cupboard to evaporate the solvent.

2.3.3. Thermal Behaviors. Crystallization behaviors of PET and its nanocomposites were examined using DSC (TA Instrument DSC Q-100). Samples were scanned from room temperature 270°C under nitrogen condition using a heat-cool-heat run at a heating and cooling rate of 10°C/min. To confirm results, different three specimens of each sample were measured. Thermal stability was investigated using a thermogravimetric analyzer (TGA, TA Q-500). Samples were scanned from room temperature to 700°C under nitrogen and air gas condition.

3. Results and Discussions

3.1. Electrical Conductivity and Percolation Threshold Measurements. From impedance spectroscopy measurements, the σ values are obtained as a function of frequencies. In this study, frequency at 10 Hz has been chosen to obtain the values of σ for comparative purpose for all samples. This is because, at low frequencies, insulator-conductor transitions can be only noticed [11].

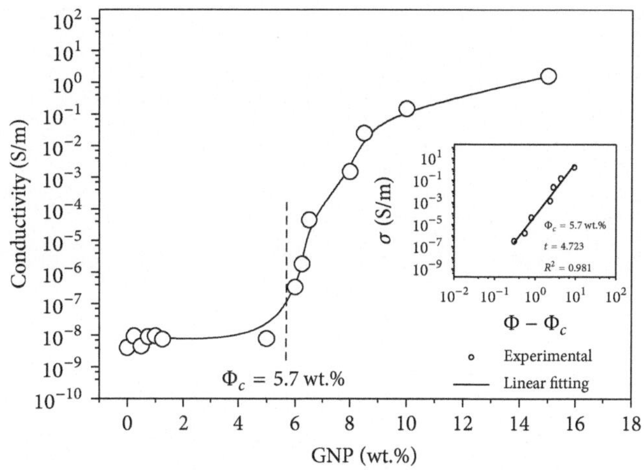

FIGURE 1: Electrical conductivity of PET-GNP nanocomposites. The inset plot is log σ versus log $(\Phi - \Phi_c)$ for determination of percolation threshold value.

Usually, conductivity values were determined using the following equation [11, 15]:

$$\sigma = \frac{L}{RA}, \quad (1)$$

where σ is the electrical conductivity (S/m), L is the distance between the electrodes (m), A is the cross sectional area (m^2), and R is the measured electrical resistance (Ω).

Figure 1 displays the values of σ of PET matrix and its nanocomposites as a function of GNP loadings. It is clear that σ of the nanocomposites having <5 wt.% of GNP is nearly constant and close to σ of pure PET. Then, a great increase in σ occurs between 5 and 9 wt.% of GNP; this is followed by a small improvement in σ with increasing the amount of GNP. Thus, Φ_c of these nanocomposites is expected between 5 and 9 wt.% loadings of the GNP. Typically, the variation in the conductivity of PCs as a function of the conductive filler content exhibits an S-shaped curve for transition from insolating to conductive system [16]. Therefore, the PET-GNP nanocomposites exhibited a typical transition curve (Figure 1) in this study.

To illustrate transitions of insulator to conductor phase or behavior of PCs reinforced with conductive fillers, percolation theory is applied. This theory will assist the determination of dimensions of conductive network in polymer matrices and percolation threshold values of PCs system. According to this theory, σ of the PCs is calculated by the following equation [11–19]:

$$\sigma = \sigma_o (\Phi - \Phi_c)^t, \quad \text{for } \Phi > \Phi_c, \quad (2)$$

where σ_o is coefficient constant, Φ_c is percolation threshold value (vol.%), Φ is the filler loading (vol.%), and t is dimension of the conductive network path. In theory, t values of 1.33 and 2.2 represent 2D and 3D systems, respectively. However, experimental values are also stated outside this range somewhere else [17, 18]. The density of the nanofiller (such as GNP) is frequently an approximate value. Therefore, weight fractions have been used for determining the values of Φ_c and

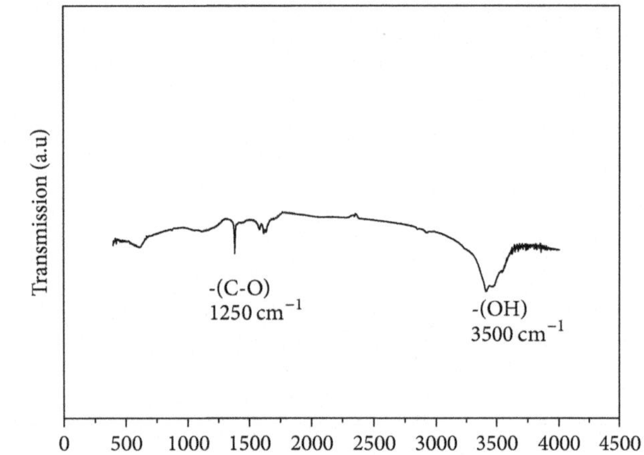

FIGURE 2: FTIR spectra of GNP.

t in numerous published studies [12, 15, 16]. In this study, all determinations will be in expression of weight fractions.

According to percolation theory, the best-fit data to the percolation threshold model (2) was achieved at the values of 5.7 wt.% and 4.7 for Φ_c and t, respectively, as shown in the insert graph in Figure 1. These results are in good agreement with the results published in the literature. For example, Zhang et al. [4] obtained a Φ_c at 3.4 vol.% of expanded graphite (~6 wt.%) for PET-expanded graphite nanocomposites using melt-compounding technique. Moreover, Li and Jeong [8] used the same preparation method and reported that the electrical conductive path of exfoliated graphite in the PET matrix was developed at about 5 wt.% of exfoliation graphite filler. These studies reported that there are some functional groups on the surface of the GNP fillers.

Generally, in PCs system, high aspect ratio of such nanofiller is found to reduce Φ_c values. However, based on the data sheets provided by the supplier, the calculated aspect ratio for GNP is ~ 1875. Thus, value of Φ_c is expected to be lower than the value obtained in these studies. The reasons for this could be due to the fact that the GNP has unreduced graphite oxide in the commercial products and then some functional groups are still on its surface after the preparation process. Therefore, FTIR technique was used in the current study to identify the functional groups attached to the surface of GNP during exfoliation techniques (Figure 2). This result is in agreement with those obtained by Li and Jeong [8] who reported some functional groups on the surface of the GNP used. However, a good attraction can occur between the ester groups (-C=O) of PET (see Figure 3) and functional groups such as hydroxyls acids (-OH) on the surface of GNP which leads to improvements in the desperation state of these nanoparticles.

In addition, there is a possibility of an agglomeration and rolling up of GNP filler during melt-compounding process; this could reduce aspect ratio of GNP sheets. Additionally, t ~ 4.72 was much higher than the theoretical values mentioned earlier. Similar value (t ~ 4.25) has been obtained by Zhang et al. [4] for PET-EG nanocomposites. The authors proposed the reasons for this deviation, that is, tunneling transmission, the microstructure, and anisotropic behavior of the composites.

TABLE 1: Electrical conductivity of PET with GNP loading of 5 wt.% and 6 wt.% measured at different processing condition.

Processing conditions	σ at 5 wt.%	σ at 6 wt.%
45 rpm/5 min	7.7×10^{-9}	3.4×10^{-7}
45 rpm/10 min	7.4×10^{-9}	3.2×10^{-7}
90 rpm/5 min	6.8×10^{-9}	3.1×10^{-5}
90 rpm/10 min	7.4×10^{-9}	4.7×10^{-5}

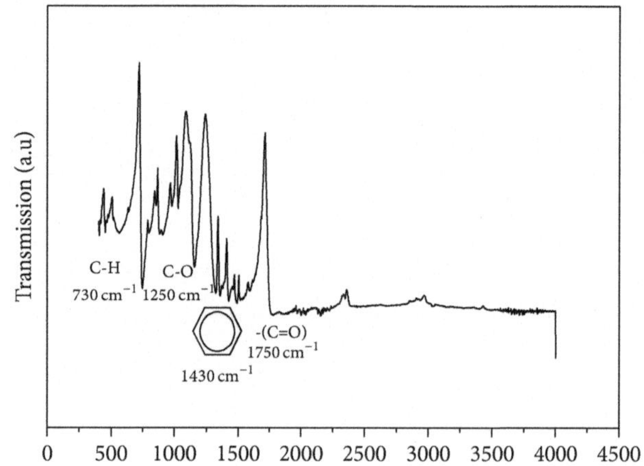

FIGURE 3: FTIR spectra of PET.

Additionally, the screw speed and residence time during melt process are expected to impact the values of σ of PCs [20, 21]. Therefore, in the present study, the impact of these two processing parameters on σ was investigated. Two samples were selected just before and after ϕ_c value of the nanocomposites to note the effect of processing parameter on its conductivity.

Table 1 shows σ data for the nanocomposites processed at two different screw speeds and residence times. It is clear that the screw speed has an important effect on σ for the samples above that of ϕ_c only, where σ increased by ~2.5 orders of magnitude for the nanocomposites processed at 90 rpm instead of 45 rpm. In contrast, there is no obvious effect of screw speed for samples having loading below ϕ_c. This could be due to the amount of GNP being insufficient to form a conductive path whatever its dispersion state in the composite. In contrast, increasing the residence time of the nanocomposites mixing from 5 to 10 minutes had no important effect on their electrical conductivity.

3.2. Morphological Characterization. The morphology of GNP was examined before its incorporation into the PET matrix in order to understand their size and shapes. Figure 4 shows SEM image of as-received GNP. As can be seen, it shows the separated platelets resulting from break down of expandable graphite particles during the preparation process of GNP [22].

Similar morphologies for GNP were observed in previous studies [8, 22, 23]. On the other hand, Figures 5, 6, and 7 show the PET-GNP nanocomposites at 2, 6, and 8 wt.% of GNP loading. The nanocomposites at 2 wt.% of GNP show uniform

TABLE 2: DSC data for PET-GNP nanocomposites.

GNP (wt.%)	T_{cc} (°C)	T_{m1} (°C)*	T_{m2} (°C)**	T_{mc} (°C)	X_c (%)***
0	133.6 ± 0.2	253.4 ± 0.4	245.9 ± 0.9	209.2 ± 0.1	11.8 ± 1.8
2	120.1 ± 0.3	253.8 ± 0.5	247.1 ± 1.0	220.0 ± 0.2	19.3 ± 0.3
4	118.8 ± 0.7	251.5 ± 0.4	248.3 ± 1.7	219.9 ± 0.2	19.6 ± 2.4
6	116.6 ± 0.2	251.2 ± 1.0	248.8 ± 0.5	220.7 ± 0.4	21.1 ± 1.3
8	114.7 ± 0.3	251.9 ± 1.4	249.2 ± 0.3	220.9 ± 0.7	22.0 ± 1.8
10	114.7 ± 1.2	251.5 ± 0.4	249.3 ± 0.5	220.7 ± 0.6	22.4 ± 0.6

*Melting temperatures values obtained from the first heating runs. **Melting temperatures values obtained from the second heating runs. ***Degree of crystallinity after quenching process ($X_c = X_m - X_{cc}$), where X_m is crystallinity related to melting processes and X_{cc} is crystallinity linked with cold crystallization process.

FIGURE 4: SEM image of GNP.

distribution and no sign of aggregations/agglomeration (Figure 5). Nevertheless, at higher GNP loadings, the aggregation and rolling up of GNP are observed.

For instance, the nanocomposites containing 6 wt.% of GNP, which are just above Φ_c, showed some rolling up and agglomeration (Figure 6), indicating that a conductive network has been formed which allows the electron to move.

More agglomeration could be observed in Figure 7 as the GNP content is increased. Similar morphology observations have been reported for polypropylene nanocomposites by Kalaitzidou et al. [24]. In addition, Zhang et al. [4] considered the relationship between the morphology and σ of PET-EG nanocomposites. In the SEM results, it has been observed that when adding EG above Φ_c, particles contacted with each other, indicating that conductive path has been developed. However, below Φ_c, the conductive particles were observed far away from each other; thus, there is no path for electron to move in the nanocomposite system.

Figure 8 shows the TEM images of both pure GNP and PET-GNP nanocomposites. The folding and rolling up of nanoplatelets in the composite are clearly observed. Similar observations have been reported by Kalaitzidou et al. [24] for same grade of GNP in a PP matrix.

3.3. *Thermal Analyses of PET-GNP Nanocomposites*

3.3.1. Crystallization Behavior. Pure PET and the nanocomposites with varying contents of GNP (2, 4, 6, 8, and 10 wt.%) were exposed to heat/cool/heat cycles at heating and cooling of 10°C/min. The DSC curves in Figures 9(a)–9(d) show cold crystallization peaks, melting peaks from the first run, melt crystallization peaks, and melting peaks from the second cycle, respectively. The data obtained from these peaks are presented in Table 2.

From Figure 9(a) and Table 2, it is clear that the incorporation of 2 wt.% GNP into PET matrix decreased the cold crystallization temperature (T_{cc}) by 14°C, followed by slight decrease in T_{cc} with increasing GNP loadings. This reduction in the value of T_{cc} indicates that the cold crystallization of the nanocomposites was initiated earlier than that of pure PET matrix. During the cooling run (Figure 9(c)), the PET-GNP nanocomposites displayed roughly 10°C higher melt crystallization temperature (T_{mc}) than pure PET (Table 2). In addition, degree of supercoiling is ($T_m - T_{mc}$), where T_m is melting temperature. This difference in temperatures reflects the rate of crystallization. From Table 2, it is clear that ($T_m - T_{mc}$) decreased with increasing amount of GNP. It should be noted that addition of >2 wt.% of GNP results only in slight changes in the values of T_{cc}, T_{mc}, and ($T_m - T_{mc}$), indicating that 2 wt.% loading of GNP is more than enough to nucleate the PET matrix. Such an effect in the crystallizations of PET is in agreement with previous work done by [25, 26].

X_c can be calculated using the enthalpies of both crystallization and melting, according to the following equation [27, 28]. X_c of PET and its nanocomposite was calculated from the first-heat run as these data reflect the thermal history of the samples.

$$X_c = \frac{\Delta H_m - \Delta H_{cc}}{\left(1 - w_f\right) \Delta H_o} * 100, \qquad (3)$$

where ΔH_m is the melting enthalpy (J/g), ΔH_{cc} is the cold crystallization enthalpy (J/g), ΔH_o is the theoretical enthalpy of 100% crystalline PET ($\Delta H_o = 140$ J/g) [26], and w_f is the weight fraction of GNP.

The addition of GNP was found to affect X_c as presented in Table 2. There is an improvement of ~7.5% in the value of X_c upon adding of 2 wt.% of GNP. However, additional adding of GNP does not show any considerable alteration in the values of X_c. Similar behavior was observed by Al-Jabareen et al. [13], who studied PET-GNP nanocomposites and reported an enhancement of 4% in value of X_c at 1.5 wt.% GNP concentration. In contrast, it was reported that GNP did not change the values of X_c of a PP matrix [25].

Furthermore, the addition of GNP appears to have no noticeable effect on values of T_m within the content of GNP used in this study (Table 2). Comparative peaks of melting for two the heating runs are shown in Figures 9(b) and 9(d). From these curves and Table 2, it can be observed that the

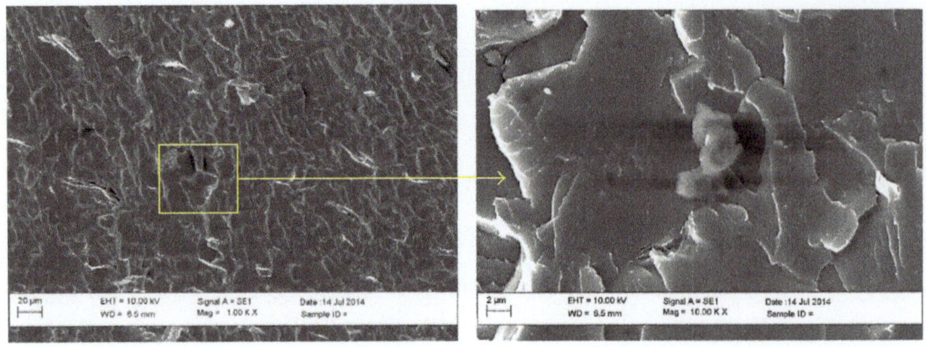

FIGURE 5: SEM images at low and high magnifications of fracture surfaces of PET-GNP nanocomposites containing 2 wt.% of GNP.

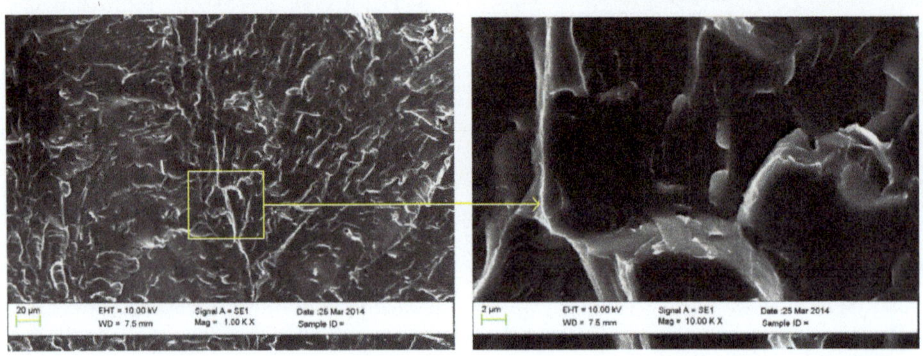

FIGURE 6: SEM images at low and high magnifications of fracture surfaces of PET-GNP nanocomposites containing 6 wt.% of GNP.

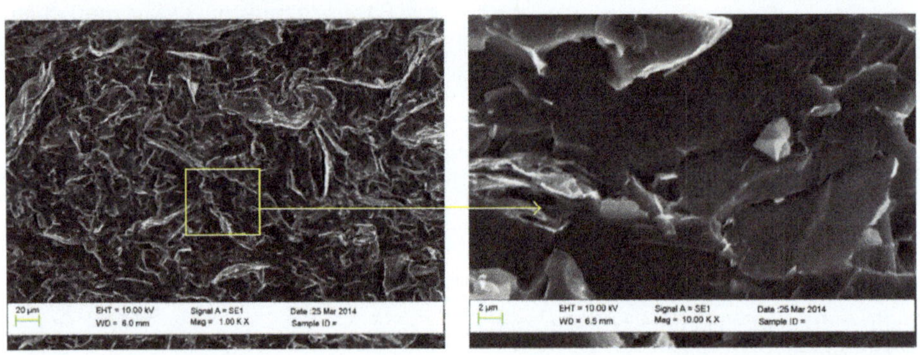

FIGURE 7: SEM images at low and high magnifications of fracture surfaces of PET-GNP nanocomposites containing 8 wt.% of GNP.

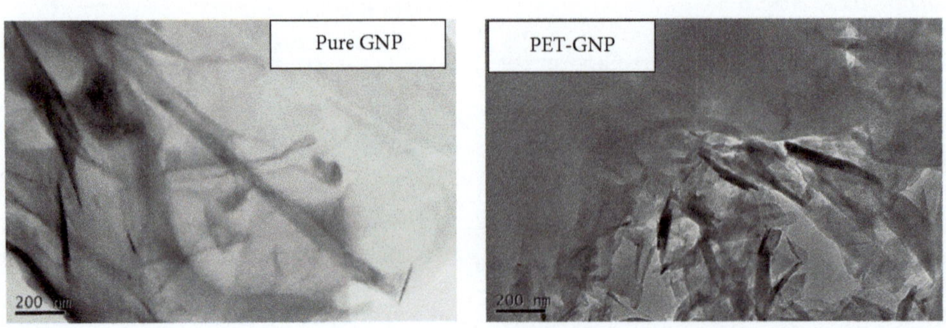

(a) (b)

FIGURE 8: TEM images of as-received GNP (a) and a PET-GNP nanocomposite containing 6 wt.% of GNP (b).

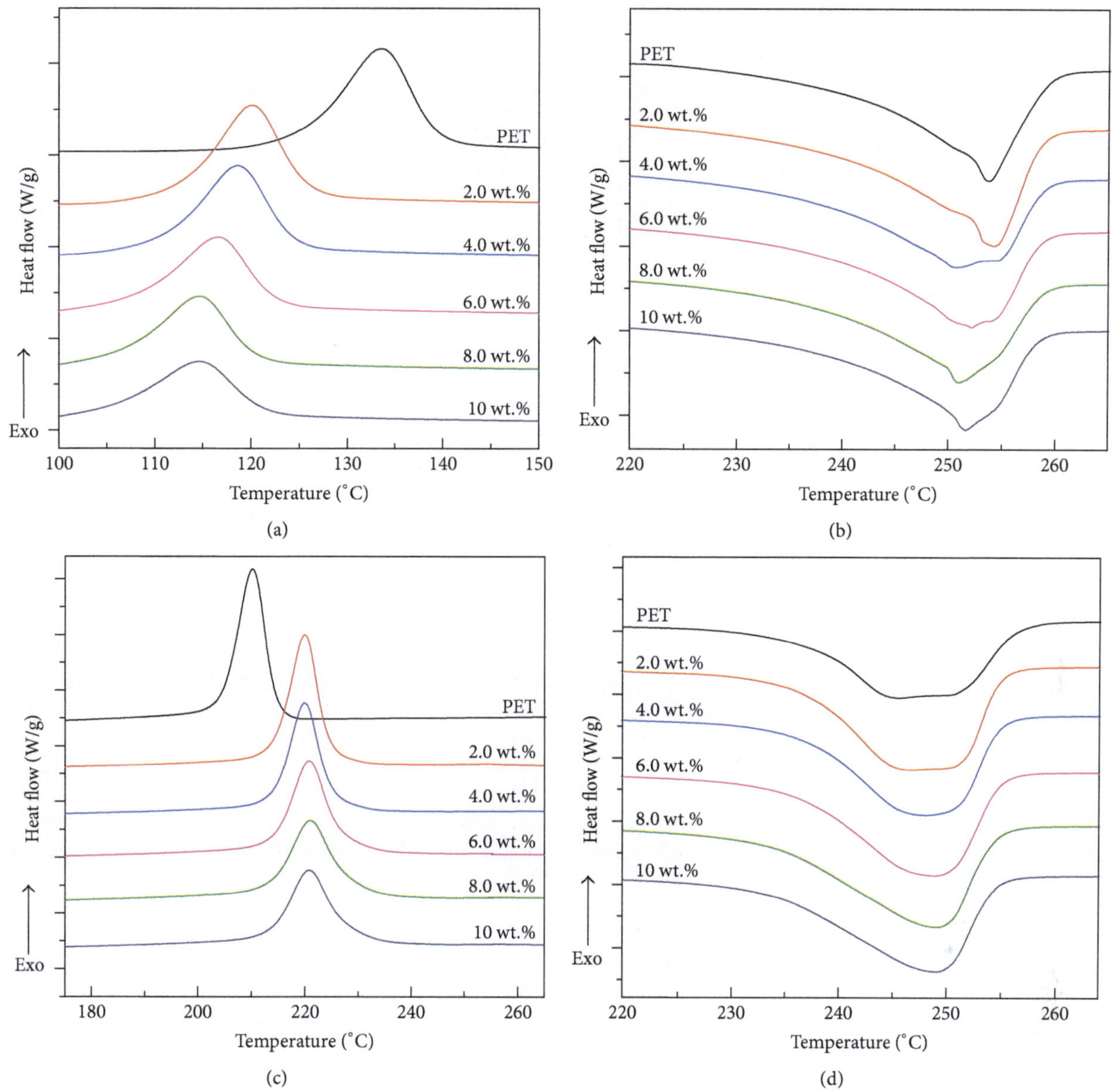

FIGURE 9: DSC curves of PET-GNP nanocomposites showing cold crystallization peaks (a), melting peaks from the first scan (b), subsequent cooling curves (c), and melting peaks from second heating run (d).

values of T_{m2} are lower than that of T_{m1}. Similar observations were reported by Kalaitzidou et al. [25] who studied the nucleation effects and crystallization behavior of polypropylene- (PP-) GNP nanocomposites. They observed that GNP changed the melting behavior of PP matrix. This behavior was attributed to both the increased homogeneity and thinning of crystal size.

3.3.2. Thermal Stability. Modified graphite forms such as GNP have been used to improve thermal stability of polymer matrices. For example, Al-Jabareen et al. [13] reported that the adding of GNP into PET matrix delays the onset temperature of degradation, indicating the improvement of thermal stability. The authors attributed this improvement to 2D structure of GNP that acts as barrier for oxygen diffusion through the nanocomposites. A similar result was given by Li and Jeong [8]. In other studies, thermal stability enhancements have been reported for polycarbonate-GNP nanocomposites [29] and poly(methyl methacrylate) as a polymer matrix [30] using the same grade of GNP as used in the current study.

The results of the present study agree with these studies; as in the current work, the incorporation of GNP into PET matrix enhanced its thermal stability of PET under nitrogen and air as shown in Figures 10 and 11, respectively. $T_{5\%}$ values taken from these curves are presented in Table 3. It is clear that the values of $T_{5\%}$ for the nanocomposites are higher than that of pure PET in both atmospheres, indicating that GNP enhanced the thermal stability. $T_{5\%}$ values for neat PET are ≈382°C and ≈370°C in nitrogen and air, respectively. However, $T_{5\%}$ values for the nanocomposites having 2 wt.%

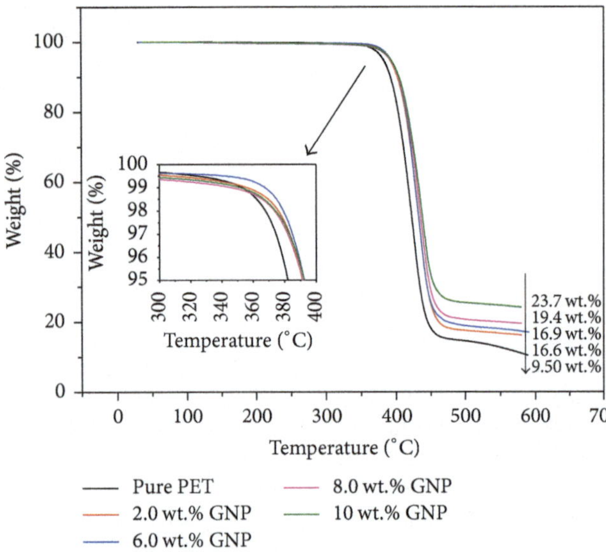

FIGURE 10: TGA thermograms of nanocomposites examined under nitrogen atmosphere.

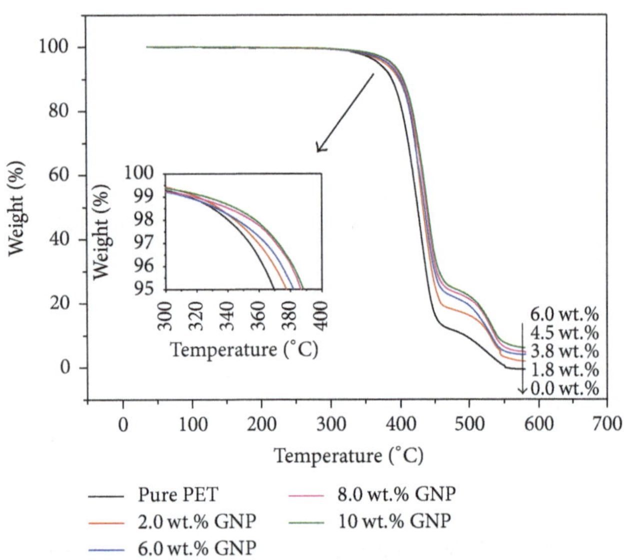

FIGURE 11: TGA thermograms of nanocomposites examined under air atmosphere.

GNP are ~392°C and ~378°C in nitrogen and air, respectively. Addition of >2 wt.% GNP improved the thermal stability slightly. This could be due to the agglomeration and rolling up of GNP during melt mixing at higher loadings.

Furthermore, pure PET and the nanocomposites have lower $T_{5\%}$ values when heated in air (Figure 11) compared with when heated in nitrogen. However, the improvement of $T_{5\%}$ values in air is higher than that in nitrogen when compared to pure PET. For example, with the addition of 10 wt.% of GNP, $T_{5\%}$ values are higher than unfilled PET matrix by ~11°C and ~18°C in nitrogen and air atmospheres, respectively (Table 3).

The enhanced thermal stabilities in the previous studies explained by a barrier effect associated with GNP filler that obstructs the diffusion of oxygen into the nanocomposites and hinders diffusion of volatiles materials.

TABLE 3: Thermal degradation temperatures (T_5%) for PET-GNP nanocomposites under nitrogen and air atmospheres.

GNP (wt.%)	N_2 $T_{5\%}$ (°C)	Air $T_{5\%}$ (°C)
0	382.0 ± 1.4	370.0 ± 1.0
2	391.9 ± 1.3	378.2 ± 0.5
6	393.2 ± 1.3	384.9 ± 3.0
8	392.3 ± 1.1	386.9 ± 0.7
10	393.1 ± 0.9	388.5 ± 0.1

4. Conclusion

In the present study, a large number of nanocomposites were prepared using PET matrix and GNP nanofiller by melt-compounding method in order to determine the percolation threshold value for PET-GNP nanocomposites system. The electrical conductivity, morphology, thermal stability, crystallization behaviors, and degree of crystallinity of these nanocomposites were characterized. In particular, at ~6 wt.% of GNP, an enhancement in conductivity was observed. Such loading marks the insulator-conductor transition having a percolation threshold value of 5.7 wt.%. The conductivity that is required for antistatic applications was observed just above the percolation threshold value. Screw speed during mixing exhibited strong influence on the conductivity of the nanocomposites above such value. Addition of ~2 wt.% GNP into PET matrix caused an increase in the degree of crystallinity, accelerating both cold and melt crystallizations along with good improvements in the thermal stability. However, further addition (>2 wt.%) improves thermal properties slightly. This attributed to higher agglomeration, pulling out, and rolling up of GNP sheets during mixing process, which reduced their aspect ratio and then reduced reinforcement efficiency.

Acknowledgments

The authors would like to thank King Abdulaziz City for Science and Technology (KACST) for their funding throughout this project.

References

[1] R. Sengupta, M. Bhattacharya, S. Bandyopadhyay, and A. K. Bhowmick, "A review on the mechanical and electrical properties of graphite and modified graphite reinforced polymer composites," *Progress in Polymer Science (Oxford)*, vol. 36, no. 5, pp. 638–670, 2011.

[2] B. Li and W.-H. Zhong, "Review on polymer/graphite nanoplatelet nanocomposites," *Journal of Materials Science*, vol. 46, no. 17, pp. 5595–5614, 2011.

[3] M. Li and Y. G. Jeong, "Influences of exfoliated graphite on structures, thermal stability, mechanical modulus, and electrical resistivity of poly(butylene terephthalate)," *Journal of Applied Polymer Science*, vol. 125, no. 1, pp. E532–E540, 2012.

[4] M. Zhang, D.-J. Li, D.-F. Wu, C.-H. Yan, P. Lu, and G.-M. Qiu, "Polyethylene terephthalate/expanded graphite conductive composites: Structure, properties, and transport behavior," *Journal of Applied Polymer Science*, vol. 108, no. 3, pp. 1482–1489, 2008.

[5] K. Alasvand Zarasvand and H. Golestanian, "Investigating the effects of number and distribution of GNP layers on graphene reinforced polymer properties: Physical, numerical and micromechanical methods," *Composites Science and Technology*, vol. 139, pp. 117–126, 2017.

[6] J. Brydson, *Plastic Materials*, Butterworth-Heineman Linacre House, Oxford, UK, 1999.

[7] C. A. Harper and E. M. Petrie, *Plastics Materials and Processes: a Concise Encyclopedia*, John Wiley & Sons, 2003.

[8] M. Li and Y. G. Jeong, "Poly(ethylene terephthalate)/exfoliated graphite nanocomposites with improved thermal stability, mechanical and electrical properties," *Composites Part A: Applied Science and Manufacturing*, vol. 42, no. 5, pp. 560–566, 2011.

[9] P.-C. Ma, N. A. Siddiqui, G. Marom, and J.-K. Kim, "Dispersion and functionalization of carbon nanotubes for polymer-based nanocomposites: a review," *Composites A: Applied Science and Manufacturing*, vol. 41, no. 10, pp. 1345–1367, 2010.

[10] F. Xin, L. Li, S. H. Chan, and J. Zhao, "Influences of carbon fillers on electrical conductivity and crystallinity of polyethylene terephthalate," *Journal of Composite Materials*, vol. 46, no. 9, pp. 1091–1099, 2012.

[11] Z. Římská and V. Křesálek, "AC conductivity of carbon fiber - polymer matrix," *Polymer Composites*, vol. 23, 2002.

[12] J. K. W. Sandler, J. E. Kirk, I. A. Kinloch, M. S. P. Shaffer, and A. H. Windle, "Ultra-low electrical percolation threshold in carbon-nanotube-epoxy composites," *Polymer*, vol. 44, no. 19, pp. 5893–5899, 2003.

[13] A. Al-Jabareen, H. Al-Bustami, H. Harel, and G. Marom, "Improving the oxygen barrier properties of polyethylene terephthalate by graphite nanoplatelets," *Journal of Applied Polymer Science*, vol. 128, no. 3, pp. 1534–1539, 2013.

[14] S. Paszkiewicz, A. Szymczyk, M. Soccio, J. Mosnáček, T. A. Ezquerra, and Z. Rosłaniec, "Electrical conductivity of poly(ethylene terephthalate)/expanded graphite nanocomposites prepared by *in situ* polymerization," *Journal of Polymer Science, Part B: Polymer Physics*, vol. 50, no. 23, pp. 1645–1652, 2012.

[15] J. Sandler, M. S. P. Shaffer, T. Prasse, W. Bauhofer, K. Schulte, and A. H. Windle, "Development of a dispersion process for carbon nanotubes in an epoxy matrix and the resulting electrical properties," *Polymer*, vol. 40, no. 21, pp. 5967–5971, 1999.

[16] M. L. Clingerman, E. H. Weber, J. A. King, and K. H. Schulz, "Development of an additive equation for predicting the electrical conductivity of carbon-filled composites," *Journal of Applied Polymer Science*, vol. 88, no. 9, pp. 2280–2299, 2003.

[17] J. O. Aguilar, J. R. Bautista-Quijano, and F. Avilés, "Influence of carbon nanotube clustering on the electrical conductivity of polymer composite films," *Express Polymer Letters*, vol. 4, no. 5, pp. 292–299, 2010.

[18] R. J. Young and P. A. Lovell, *Introduction to Polymers*, CRC Press, Boca Raton, Fla, USA, 3rd edition, 2011.

[19] W. Bauhofer and J. Z. Kovacs, "A review and analysis of electrical percolation in carbon nanotube polymer composites," *Composites Science and Technology*, vol. 69, no. 10, pp. 1486–1498, 2009.

[20] M. Guehenec, V. Tishkova, S. Dagreou et al., "The effect of twin screw extrusion on structural, electrical, and rheological properties in carbon nanotube poly-ether-ether-ketone nanocomposites," *Journal of Applied Polymer Science*, vol. 129, no. 5, pp. 2527–2535, 2013.

[21] T. Villmow, P. Pötschke, S. Pegel, L. Häussler, and B. Kretzschmar, "Influence of twin-screw extrusion conditions on the dispersion of multi-walled carbon nanotubes in a poly(lactic acid) matrix," *Polymer*, vol. 49, no. 16, pp. 3500–3509, 2008.

[22] S. Kim, J. Seo, and L. T. Drzal, "Improvement of electric conductivity of LLDPE based nanocomposite by paraffin coating on exfoliated graphite nanoplatelets," *Composites Part A: Applied Science and Manufacturing*, vol. 41, no. 5, pp. 581–587, 2010.

[23] M. Li and Y. G. Jeong, "Preparation and characterization of high-performance poly(trimethylene terephthalate) nanocomposites reinforced with exfoliated graphite," *Macromolecular Materials and Engineering*, vol. 296, no. 2, pp. 159–167, 2011.

[24] K. Kalaitzidou, H. Fukushima, and L. T. Drzal, "A route for polymer nanocomposites with engineered electrical conductivity and percolation threshold," *Materials*, vol. 3, no. 2, pp. 1089–1103, 2010.

[25] K. Kalaitzidou, H. Fukushima, P. Askeland, and L. T. Drzal, "The nucleating effect of exfoliated graphite nanoplatelets and their influence on the crystal structure and electrical conductivity of polypropylene nanocomposites," *Journal of Materials Science*, vol. 43, no. 8, pp. 2895–2907, 2008.

[26] J.-Z. Xu, G.-J. Zhong, B. S. Hsiao, Q. Fu, and Z.-M. Li, "Low-dimensional carbonaceous nanofiller induced polymer crystallization," *Progress in Polymer Science*, vol. 39, no. 3, pp. 555–593, 2014.

[27] S. Yesil and G. Bayram, "Poly(ethylene terephthalate)/carbon nanotube composites prepared with chemically treated carbon nanotubes," *Polymer Engineering and Science*, vol. 51, no. 7, pp. 1286–1300, 2011.

[28] G. Antoniadis, K. M. Paraskevopoulos, D. Bikiaris, and K. Chrissafis, "Melt-crystallization mechanism of poly(ethylene terephthalate)/multi-walled carbon nanotubes prepared by in situ polymerization," *Journal of Polymer Science, Part B: Polymer Physics*, vol. 47, no. 15, pp. 1452–1466, 2009.

[29] G. Gedler, M. Antunes, V. Realinho, and J. I. Velasco, "Thermal stability of polycarbonate-graphene nanocomposite foams," *Polymer Degradation and Stability*, vol. 97, no. 8, pp. 1297–1304, 2012.

[30] T. Ramanathan, S. Stankovich, D. A. Dikin et al., "Graphitic nanofillers in PMMA nanocomposites—an investigation of particle size and dispersion and their influence on nanocomposite properties," *Journal of Polymer Science Part B: Polymer Physics*, vol. 45, no. 15, pp. 2097–2112, 2007.

22

Effect of Heat Drawing Process on Mechanical Properties of Dry-Jet Wet Spun Fiber of Linear Low Density Polyethylene/Carbon Nanotube Composites

Jong Won Kim and Joon Seok Lee

Department of Textile Engineering and Technology, Yeungnam University, Gyeongsan 38541, Republic of Korea

Correspondence should be addressed to Joon Seok Lee; leejs@ynu.ac.kr

Academic Editor: Peng He

Polyethylene is one of the most commonly used polymer materials. Even though linear low density polyethylene (LLDPE) has better mechanical properties than other kinds of polyethylene, it is not used as a textile material because of its plastic behavior that is easy to break at the die during melt spinning. In this study, LLDPE fibers were successfully produced with a new approach using a dry-jet wet spinning and a heat drawing process. The fibers were filled with carbon nanotubes (CNTs) to improve the strength and reduce plastic deformation. The crystallinity, degree of orientation, mechanical properties (strength to yield, strength to break, elongation at break, and initial modulus), electrical conductivity, and thermal properties of LLDPE fibers were studied. The results show that the addition of CNTs improved the tensile strength and the degree of crystallinity. The heat drawing process resulted in a significant increase in the tensile strength and the orientation of the CNTs and polymer chains. In addition, this study demonstrates that the heat drawing process effectively decreases the plastic deformation of LLDPE.

1. Introduction

Polyethylene (PE) is one of the most widely used thermoplastic materials because of its high toughness, near-zero moisture absorption, excellent chemical inertness, low coefficient of friction, ease of processing, and low electrical conductivity. The mechanical properties of PE are impacted by the extent and type of branching, the crystal structure, and the molecular weight [1]. PE polymers are classified into several different categories, but mostly these do not depend on the density or branching. The main forms of PE are high density polyethylene (HDPE), high molecular weight HDPE (HMWHDPE), ultrahigh molecular weight polyethylene (UHMWPE), linear low density polyethylene (LLDPE), and very low density polyethylene (VLDPE). These categories are then subdivided into groups related to the density and branching. In general, the most used PE grades are HDPE, LDPE, and medium-density polyethylene (MDPE) [2]. LDPE is considered as one of the commercially important thermoplastics. In particular, it is used for a wide range of applications that require low density, good processability, and easy molding.

LLDPE has better tensile strength, heat tolerance, and low-temperature properties than LDPE. It also has better environmental stress cracking resistance and impact strength and less intensity difference in various directions than HDPE. Although it has suitable properties for use as a textile material, to date, LLDPE has only been used for laminate and multilayer films and for injection molding (in HDPE/LDPE blends) [3, 4]. Some companies developed spinnable LLDPE for textile application, but it is a blend of PE and other additive polymers. Therefore, LLDPE was made as staple fiber form which was used in nonwoven fiber [5, 6] and bonding fiber [7, 8]. However, it has never been manufactured as thin fiber before [9]. In order to manufacture thin LLDPE fibers, high-speed winding is necessary during the spinning process, but the fibers are easily broken during the spinning process due to the plastic behavior of LLDPE. In some studies, to prevent the breakage of LLDPE fibers, a heat drawing process was applied after winding without any tension in the melt spinning process. Although the LLDPE fibers have been made, the draw ratio of the fibers was very low so only thick fibers with the diameter of few hundred microns were produced, which could not be used in most textile applications [10, 11]. In addition, for use as a precursor for the production of carbon fibers, the diameter of the LLDPE fibers should be several tens of microns or less [12].

New composites with PE as the matrix are used in many applications and have better mechanical and physical properties than the pure polymer alone. PE composites can be used in packaging, electrical, thermal energy storage, automotive, biomedical, and space applications [13–17]. Although PE is a relatively soft material, it has broad applications due to its resistance to moisture, corrosion, wear, and chemicals and an extremely low cost. PE composites with clay and multiwalled carbon nanotubes have been fabricated to improve the toughness, tensile strength, and other chemical, thermal, and barrier properties [18, 19]. Because of high Young's modulus of 1–5 TPa and anisotropic properties, CNTs have been proposed as a material that can increase the elastic modulus and tensile strength of composite materials [20, 21]. In addition, PE has been used as a matrix for preparing composites with CNTs using UHMWPE [22–25], HDPE [26, 27], and LDPE [19, 28] and as a thin film [29–31].

In this study, to apply LLDPE on the textile application, thin LLDPE fiber was manufactured by dry-jet wet spinning. After manufacturing, heat drawing process was applied. Then manufactured LLDPE fiber analyzed the degree of crystallinity, the degree of orientation, the mechanical properties (strength to yield, strength to break, elongation at break, and initial modulus), electrical conductivity, and thermal properties.

2. Materials and Methods

2.1. Materials. LLDPE (LLDPE-3126, melt flow index = 1 g/10 min, Mw/Mn = 127,028/35,091, Hanwha Chemical Co., Seoul, Korea) was used as the polymer matrix and decahydronaphthalene (Decalin, $C_{10}H_{18}$, Mw = 138.25, Kanto Chemical Co., Seoul, Korea) was used as solvent. Multiwalled carbon nanotubes (MWCNT, aspect ratio around 500, Carbon Nanotech Co., Pohang, Korea) were used as the filler.

2.2. Preparation of LLDPE Fiber. The mixture of 80 wt.% decalin and 20 wt.% LLDPE was spun into fibers in distilled water (air gap: 10 mm) at room temperature through a spinneret die. A dry-jet wet spinning machine (KNFI-1000, Koen Co., Ansan, Korea) was maintained at 170°C. The die consisted of spinneret hole with a diameter of 250 μm, at a spinneret pressure of 2 kgf/cm^2 and take-up speed of 15.07 m/min. After 24 h dipping in n-hexane (C_6H_{14}, Duksan Co., Ansan, Korea) to elute the decalin, in the fiber, the LLDPE fiber was rinsed three times with distilled water. The LLDPE fiber was drawn at 80°C with a draw ratio of 4 using a batch type hot draw machine (UL-FD300, UL Sic Co., Daegu, Korea) and then cooled down gradually. The diameter of LLDPE fiber was calculated on the basis of density (ρ = 0.921 g/cm^3) and constant length system (1 Denier = 1 g/9000 m). The diameter of undrawn fibers and drawn fibers was approximately 63 μm and 32 μm for each. The composite LLDPE fibers were prepared by decalin with MWCNT dispersed by ultrasonication. The spinning method is the same as that of neat-LLDPE. The content of MWCNT was controlled by the amount of neat-decalin and decalin dispersed with MWCNT. Finally, the content of MWCNT in the composite LLDPE fibers ranged from 0 wt.% to 2 wt.%.

2.3. Characterization. To investigate the crystallinity of the LLDPE fiber as a function of MWCNT content and draw ratio, X-ray diffraction (XRD) was undertaken using a MPD diffractometer (PANalytical, Eindhoven, Netherlands). Differential scanning calorimetry (DSC, Q-200, TA Instruments, DW, USA) patterns of the LLDPE fibers were monitored between 50 and 200°C at a heating rate of 10°C/min in a nitrogen atmosphere. The degree of crystallinity (X_c) was determined from

$$X_c = \frac{\Delta H_m}{\Delta H^{\circ}_m} \times \frac{100}{w}, \quad (1)$$

where ΔH_m is the melting enthalpy calculated from the experimental data and ΔH°_m is the enthalpy of fusion of 100% crystalline PE (ΔH°_m = 290 J/g) and w is the weight fraction of LLDPE in the composite fiber [31, 32].

The degree of orientation was measured from the parallel and perpendicular axes using a polarizing microscope (AXSZ-400B, Axis Co., Minbo, China). The birefringence (Δ_n) was calculated from the measured refractive index using

$$\Delta_n = n_{\parallel} - n_{\perp}, \quad (2)$$

where $n_{\parallel}$ is the refractive index in the parallel direction and $n_{\perp}$ is the refractive index in the perpendicular direction.

The tensile properties were measured on five different specimens with a length of 100 mm, holding the end of specimen with silicone grip at a cross head speed of 300 mm/min, using a tensile tester (OTT-05, Oriental Co., Seoul, Korea) with a 3 N load cell. Both ends of the fiber, 10 cm in length, were coated with a silver paste and the electrical resistivity was measured using an electrical resistivity meter (MI-2077, Metrel Co., Horjul, Slovenia) to calculate the electrical conductivity (σ) according to

$$\sigma\,(\text{S/cm}) = \frac{d}{(A \times R)}, \quad (3)$$

where d is the distance between the electrodes (cm), A is the cross-sectional area of the fiber (cm^2), and R is the electrical resistivity (Ω).

The cross-sections of the LLDPE fibers were observed by scanning electron microscopy (SEM, S-4100, Hitachi, Tokyo, Japan). Before SEM observation, specimens were mounted with aluminum plate and then sputter coated with platinum. Specimens were observed at an accelerating voltage of 10 kV.

3. Results and Discussion

3.1. XRD and DSC. Figure 1 shows XRD patterns of LLDPE as a function of CNT content. Figure 1(a) shows undrawn

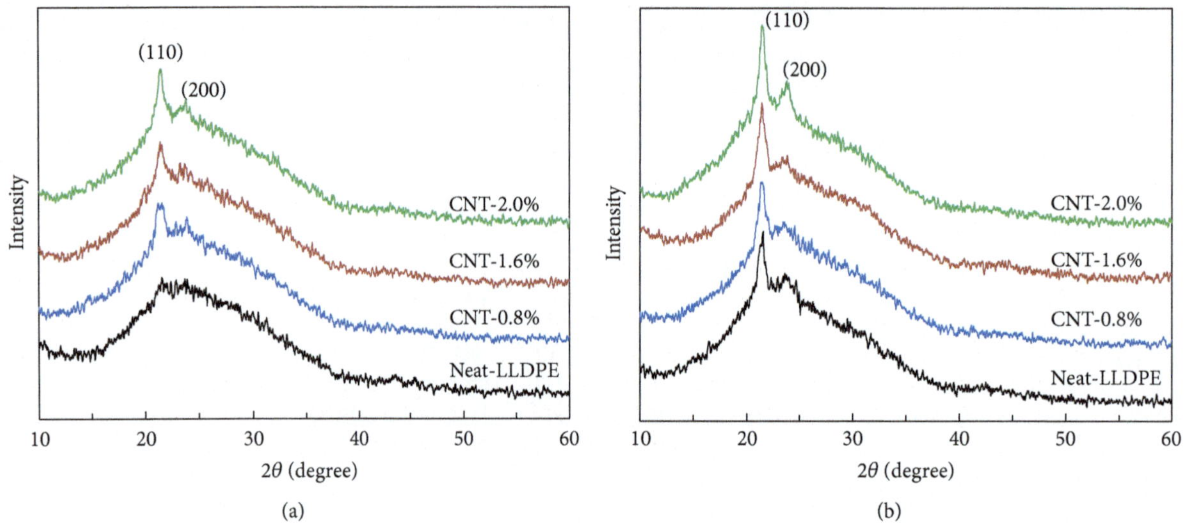

FIGURE 1: X-ray diffraction patterns of LLDPE fiber as a function of CNTs content: (a) undrawn LLDPE fiber; (b) heat drawn LLDPE fiber (DR 4).

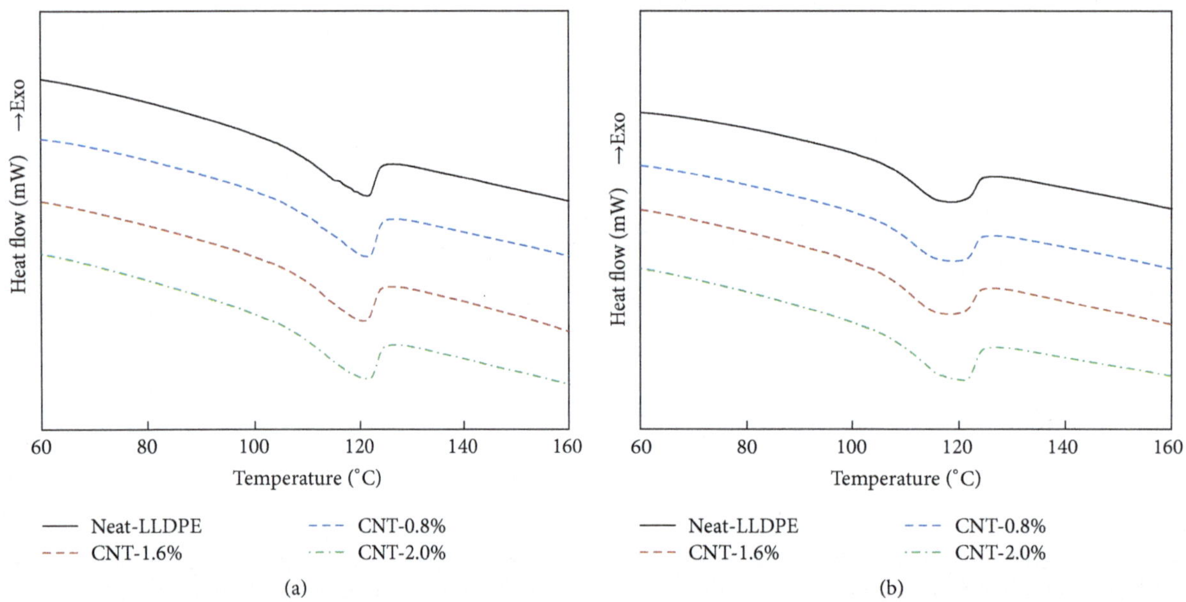

FIGURE 2: DSC patterns as a function of CNTs content: (a) undrawn and (b) heat drawn LLDPE fibers with a draw rate of 4.

LLDPE fiber. The XRD peaks at 21.6° and 24.0° were assigned to the 110 and 200 reflections of PE [31]. The intensity of diffraction peaks at 21.6° and 24.0° was increased with increasing CNTs content, which suggests the crystallinity of the LLDPE fiber was increased by filled CNTs. The addition of CNTs to the polymer provided a large interface area between the CNTs and polymer, which in turn created very suitable conditions for the nucleation and crystallization process during heating [2, 33]. Figure 1(b) shows the XRD pattern of LLDPE treated by the heat drawing process as a function of the CNTs content. Compared to Figure 1(a), the intensities of the diffraction peaks increased greatly with heat drawing process. This is because the crystallinity was increased by orientation of polymer chain of LLDPE fiber due to the heat drawing process. Also, it shows the crystallinity increased with increasing CNTs content [31, 34]. Figure 2 shows differential scanning calorimetry (DSC) patterns of LLDPE fibers as a function of CNT content. Other studies have demonstrated that the melting temperature of the polymer changed with the addition of CNTs [35]. However, in this study, it can be seen that the melting temperature of both the undrawn and drawn LLDPE fiber was similar. However, comparing Figures 2(a) and 2(b), the melting feature of the drawn LLDPE is broader than that of the undrawn LLDPE. This is because the crystal size became broader with increasing crystallinity of LLDPE fiber by heat drawing process [2, 30, 36].

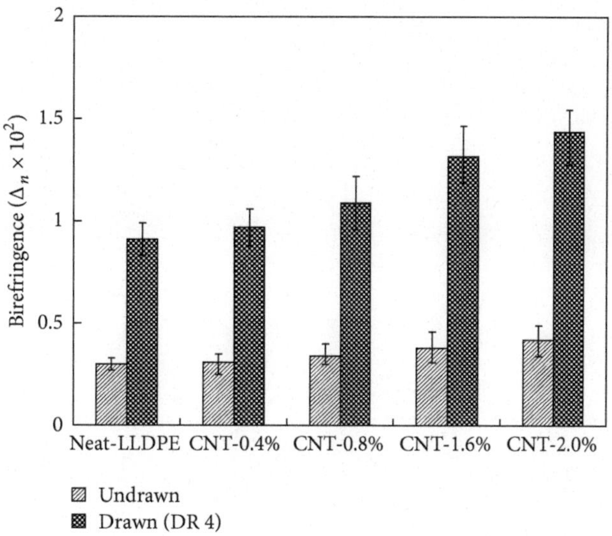

FIGURE 3: Birefringence (Δ_n) of undrawn and drawn LLDPE fibers as a function of CNTs content.

TABLE 1: Degree of crystallinity of undrawn and drawn LLDPE fibers as a function of CNTs content.

	Neat-LLDPE	CNT-0.8 wt.%	CNT-1.6 wt.%	CNT-2.0 wt.%
Undrawn LLDPE (%)	22.58	24.76	26.32	27.91
Drawn LLDPE (DR 4) (%)	29.40	33.52	37.00	39.72

Table 1 shows values for the degree of crystallinity calculated using (1) from the melting enthalpy of the DSC patterns of the drawn and undrawn LLDPE fibers. In the case of undrawn LLDPE fibers, when the CNTs content increased to 2.0 wt.%, the crystallinity was increased slightly from 22.58% to 27.91%. On the other hand, in the case of the drawn LLDPE fibers, when the CNTs content increased to 2.0 wt.%, the crystallinity increased greatly from 29.40% to 39.72%. As shown in XRD results, heat drawing process is more effective than filled CNTs for increasing the crystallinity [13, 31, 37].

3.2. Degree of Orientation. Figure 3 shows the birefringence (Δ_n) of undrawn and drawn LLDPE fibers as a function of CNTs content. The Δ_n values of the undrawn LLDPE fiber increased slightly with CNTs content. This shows that although the LLDPE fiber contained anisotropic CNTs, it did not have a great effect on the orientation undrawn LLDPE fiber. However, in the case of the neat-LLDPE, the birefringence increased by about 203.3%, from 0.0030 to 0.0091, for the undrawn and drawn fibers. It follows that the orientation of polymer chains increased greatly according to the draw ratio (DR) of 4. In addition, in the case of the LLDPE-CNT 2.0 wt.% fiber, the increase in the birefringence between the undrawn and drawn fibers was about 242.1%, from 0.0042 to 0.0143. This was a much larger increase than the neat-LLDPE fiber, which indicates that the anisotropic CNTs are oriented along the polymer chain by heat drawing [38, 39].

3.3. Tensile Properties. Figure 4 shows tensile strength for undrawn and drawn LLDPE fibers as a function of CNTs content. As shown in Figures 4(a) and 4(b), the strength at yield and strength at break were increased with increasing the content of CNTs. However, the strength at break of LLDPE fiber decreased at CNTs contents of above 1.6 wt.%. The reason may be due to the agglomeration of CNTs. As shown in Figure 4(a), the strength at yield of the neat-LLDPE fiber increased by 419% from 8.11 MPa to 42.17 MPa by heat drawing process. On the other hand, the strength at yield of 1.6 wt.% LLDPE-CNT fiber increased by 267% from 13.94 MPa to 51.18 MPa. The increase of LLDPE-CNTs 1.6 wt.% fiber is less than that of the neat-LLDPE fiber, suggesting that the strength at yield depends more on the increased orientation of the polymer chains than the CNTs content. However, in the case of the tensile strength at break (Figure 4(b)), the neat-LLDPE fiber increased by 180% from 16.54 MPa to 46.33 MPa, whereas that of the LLDPE-CNTs 1.6 wt.% fiber increased by 191% from 22.78 MPa to 66.41 MPa after heat drawing process. Compared to the strength at yield, the tensile strength at break depends on both the CNTs contents and the orientation of the polymer chains. As shown in Figure 4(c), the elongation at break continuously decreased with increasing the CNTs contents. In the case of undrawn LLDPE fiber, the plastic deformation of neat-LLDPE fiber is about twice that of the LLDPE-CNTs 2.0 wt.% fiber, 410% and 214%, respectively; it indicates that the elasticity of the

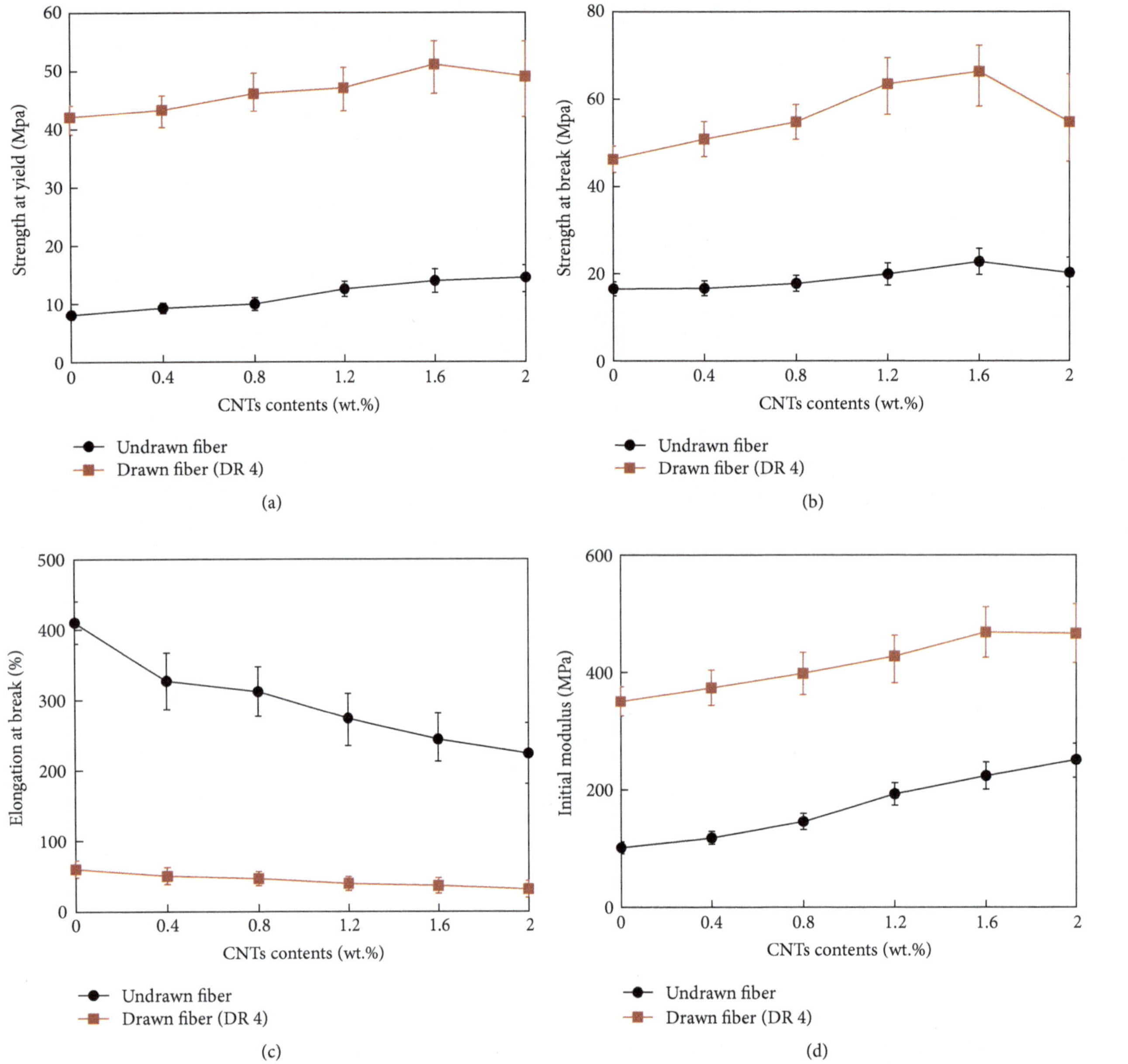

FIGURE 4: Tensile property of undrawn and drawn LLDPE fiber as a function of CNTs contents: (a) strength at yield; (b) strength at break; (c) elongation at break; and (d) initial modulus.

CNT results in decreased plastic deformation of the LLDPE fiber. Also, after heat drawing process, the plastic deformation of the LLDPE fiber decreased rapidly. Comparing the values of the undrawn and drawn fibers, the neat-LLDPE fiber decreased from 410% to 61%, and the LLDPE-CNTs 2.0 wt.% fiber decreased significantly from 214% to 32%. Hence, these results show that the heat drawing process is very effective in decreasing the plastic deformation of LLDPE fibers. As shown in Figure 4(d), the initial modulus increased continuously with increasing CNTs contents. This means that the increase in the crystallinity of the LLDPE fiber made the LLDPE fiber stiffer. Therefore, the heat drawing process increases the motility of the macromolecular chains and leads them to better rearrangement in the fiber axis direction. In other words, the heat drawing process results in straightening and sliding of the strained interfibrillar tie molecules, which allow easy alignment of the polymer segments to the crystals [40, 41]. This can be confirmed by an increase in crystallinity.

3.4. Electrical Conductivity. Figure 5 shows the electrical conductivity (σ) of undrawn and drawn LLDPE fiber as a function of CNT content compared to tensile strength. The electrical conductivity of the drawn neat-LLDPE fiber was higher than that of the undrawn neat-LLDPE fiber. This is because the measuring range of the electrical resistivity meter

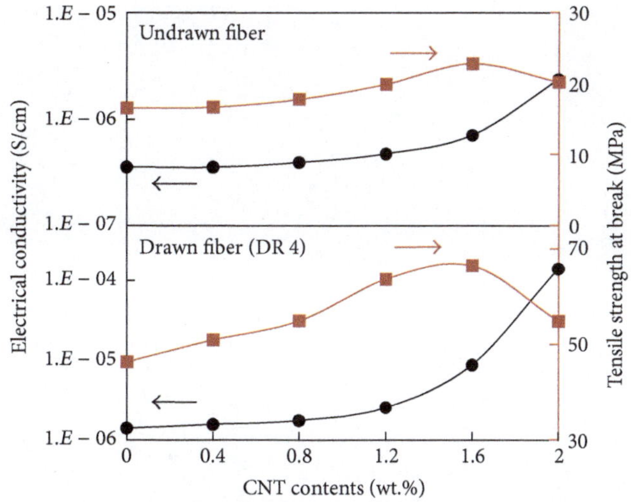

FIGURE 5: Electrical conductivity (σ) and tensile strength of undrawn and drawn LLDPE fiber as a function of CNTs content for a draw ratio of 4.

is only up to $10^{12}\ \Omega$. Although the electric resistances of the drawn and undrawn fibers were the same as $10^{12}\ \Omega$, the cross-section area of the drawn LLDPE fiber is smaller than that of the undrawn LLDPE fiber. The electrical conductivity increased greatly with increasing CNTs content. The increase of electrical conductivity started to increase from the decrease of tensile strength of undrawn and drawn LLDPE fiber. This is due to the agglomeration of CNTs and formation of a network with increasing content of CNTs. Also, in the same CNT content, the electrical conductivity of drawn LLDPE fiber was higher than that of undrawn LLDPE fiber. This is considered that the CNTs were networked by heat drawing process due to anisotropic CNTs aggregation and orientation in the axial direction and decrease of the diameter of LLDPE fiber [42–44].

3.5. SEM and TEM. Figure 6 shows cross-sectional SEM and TEM images of undrawn and drawn LLDPE fibers with CNTs 1.6 wt.% and 2.0 wt.%. As shown in energy dispersive spectroscopy (EDS) (Figures 6(a_3), 6(b_3), and 6(c_3)), the spot is CNT. It can be seen that the distance between the CNTs inside LLDPE fiber is shorter after heat drawing with a draw ratio of four as shown in Figures 6(a_2) and 6(b_2). The materials which were formed of fiber were shrunk in the center axis direction by heat drawing process and drawn in the axial direction. Therefore, the filled CNTs were aggregated and oriented with each other. The aggregation of CNTs can be seen in the LLDPE fiber in the Figure 6(c_4).

4. Conclusions

In this study, thin LLDPE fiber was manufactured with dry-jet wet spinning. To decrease plastic deformation and increase tensile strength, CNT was filled and heat drawing process was applied. The following conclusions were drawn.

(i) As the CNTs content increased to 2.0 wt.%, the crystallinity of undrawn LLDPE fiber was increased, but that of drawn LLDPE fiber was increased greatly. The value of the birefringence for the heat drawn LLDPE fiber was significantly higher than that of the undrawn LLDPE fiber.

(ii) The strength at yield and strength at break of LLDPE fibers increased with CNTs concentration up to 1.6 wt.% and then decreased. The initial modulus increased with increasing CNT content, but the elongation at break decreased continuously with increasing CNTs contents.

(iii) The electrical conductivity of both the undrawn and drawn LLDPE fibers increased with increasing CNTs content. In the same CNTs content, electrical conductivity of drawn LLDPE fibers is higher than that of undrawn LLDPE fiber.

(iv) The heat drawing process increased the crystallinity and the orientation of the polymer chain, and the filling of CNT further increased the crystallinity. As a result, thin LLDPE fibers with high tensile strength and low plastic deformation were manufactured, which can be applied to the textile field based on the characteristics of good softness compared to other high strength olefin fibers.

Disclosure

Current affiliation of Jong Won Kim is as follows: School of Chemical & Biomolecular Engineering, Georgia Institute of Technology, Atlanta, GA 30332, USA.

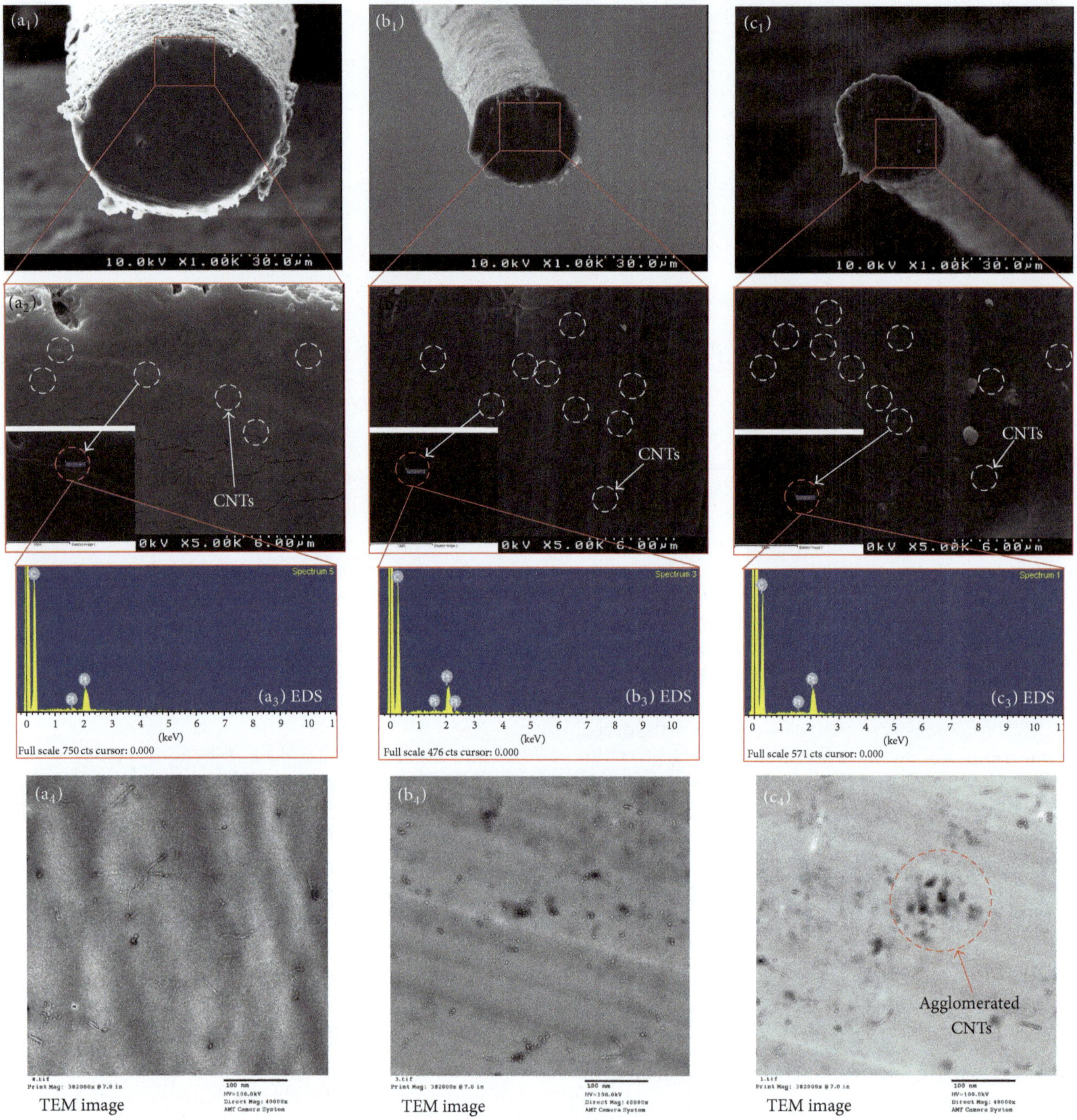

(a) Undrawn fiber (CNT 1.6%) (b) Drawn fiber (CNT 1.6%) (c) Drawn fiber (CNT 2%)

FIGURE 6: Cross-sectional SEM and TEM images of undrawn and drawn LLDPE fibers with CNTs 1.6 wt.% and 2.0 wt.%: (a) undrawn LLDPE-CNT 1.6 wt.%; (b) drawn LLDPE-CNT 1.6 wt.%; (c) drawn LLDPE-CNT 2.0 wt.%.

References

[1] X. Huang, Q. Ke, C. Kim et al., "Nonisothermal crystallization behavior and nucleation of LDPE/Al nano- and microcomposites," *Polymer Engineering and Science*, vol. 47, no. 7, pp. 1052–1061, 2007.

[2] P. N. Khanam and M. A. A. Almaadeed, "Processing and characterization of polyethylene-based composites," *Advanced Manufacturing: Polymer & Composites Science*, vol. 1, no. 2, pp. 63–79, 2015.

[3] J.-Z. Liang, "Tensile, flow, and thermal properties of CaCO3-filled LDPE/LLDPE composites," *Journal of Applied Polymer Science*, vol. 104, no. 3, pp. 1692–1696, 2007.

[4] L. Qiu, W. Chen, and B. Qu, "Morphology and thermal stabilization mechanism of LLDPE/MMT and LLDPE/LDH nanocomposites," *Polymer*, vol. 47, no. 3, pp. 922–930, 2006.

[5] S. L. Gessner, J. D. Gillespie, J. A. Austin, D. D. Newkirk, and W. Fowells, "Nonwoven fabrics and fabric laminates from multi-constituent polyolefin fiber," U. S. Patent, no. 6,465,378, Oct. 2002.

[6] K. Masuda, Y. Kawakami, and H. Morimoto, "Fibers, nonwoven fabric and uses thereof," U. S. Patent, no. 9,074,303, Jul. 2015.

[7] J. A. Lu, R. H. Turner, L. M. Reynolds et al., "Laminates with bonded webs," U. S. Patent, no. 8,728,051, May, 2014.

[8] T. E. Quantrille, H. E. Thomas, B. D. Meece et al., "Extensible composite nonwoven fabrics," U. S. Patent, no. 6,506,698, Jan. 2003.

[9] R. Pérez, E. Rojo, M. Fernández, V. Leal, P. Lafuente, and A. Santamaría, "Basic and applied rheology of m-LLDPE/LDPE blends: Miscibility and processing features," *Polymer*, vol. 46, no. 19, pp. 8045–8053, 2005.

[10] K. Mezghani, M. Farooqui, S. Furquan, and M. Atieh, "Influence of carbon nanotube (CNT) on the mechanical properties of LLDPE/CNT nanocomposite fibers," *Materials Letters*, vol. 65, no. 23-24, pp. 3633–3635, 2011.

[11] H. Mahfuz, A. Adnan, V. K. Rangari, S. Jeelani, and B. Z. Jang, "Carbon nanoparticles/whiskers reinforced composites and their tensile response," *Composites Part A: Applied Science and Manufacturing*, vol. 35, no. 5, pp. 519–527, 2004.

[12] J. W. Kim and J. S. Lee, "Preparation of carbon fibers from linear low density polyethylene," *Carbon*, vol. 94, pp. 524–530, 2015.

[13] A. Sarı, "Form-stable paraffin/high density polyethylene composites as solid–liquid phase change material for thermal energy storage: preparation and thermal properties," *Energy Conversion and Management*, vol. 45, no. 13-14, pp. 2033–2042, 2004.

[14] T. K. Dey and M. Tripathi, "Thermal properties of silicon powder filled high-density polyethylene composites," *Thermochimica Acta*, vol. 502, no. 1-2, pp. 35–42, 2010.

[15] L. Fang, Y. Leng, and P. Gao, "Processing of hydroxyapatite reinforced ultrahigh molecular weight polyethylene for biomedical applications," *Biomaterials*, vol. 26, no. 17, pp. 3471–3478, 2005.

[16] Q. Zhang, S. Rastogi, D. Chen, D. Lippits, and P. J. Lemstra, "Low percolation threshold in single-walled carbon nanotube/high density polyethylene composites prepared by melt processing technique," *Carbon*, vol. 44, no. 4, pp. 778–785, 2006.

[17] R. K. Kaul, A. F. Barghouty, and H. M. Dahche, "Space radiation transport properties of polyethylene-based composites," *Annals of the New York Academy of Sciences*, vol. 1027, pp. 138–149, 2004.

[18] S. L. Ruan, P. Gao, X. G. Yang, and T. X. Yu, "Toughening high performance ultrahigh molecular weight polyethylene using multiwalled carbon nanotubes," *Polymer*, vol. 44, no. 19, pp. 5643–5654, 2003.

[19] S. H. Abbasi, A. A. Al-Juhani, A. Ul-Hamid, and I. A. Hussein, "Effect of aspect ratio, surface modification and compatibilizer on the mechanical and thermal properties of LDPE-MWCNT nanocomposites," *Polymers*, vol. 67, Article ID 067, pp. 1–17, 2011.

[20] M.-F. Yu, B. S. Files, S. Arepalli, and R. S. Ruoff, "Tensile loading of ropes of single wall carbon nanotubes and their mechanical properties," *Physical Review Letters*, vol. 84, no. 24, pp. 5552–5555, 2000.

[21] E. W. Wong, P. E. Sheehan, and C. M. Lieber, "Nanobeam mechanics: elasticity, strength, and toughness of nanorods and nanotubes," *Science*, vol. 277, no. 5334, pp. 1971–1975, 1997.

[22] Y. Bin, A. Yamanaka, Q. Chen, Y. Xi, X. Jiang, and M. Matsuo, "Morphological, electrical and mechanical properties of ultrahigh molecular weight polyethylene and multi-wall carbon nanotube composites prepared in decalin and paraffin," *Polymer Journal*, vol. 39, no. 6, pp. 598–609, 2007.

[23] A. Mierczynska, M. Mayne-L'Hermite, G. Boiteux, and J. K. Jeszka, "Electrical and mechanical properties of carbon nanotube/ultrahigh-molecular-weight polyethylene composites prepared by a filler prelocalization method," *Journal of Applied Polymer Science*, vol. 105, no. 1, pp. 158–168, 2007.

[24] Q. Zhang, D. R. Lippits, and S. Rastogi, "Dispersion and rheological aspects of SWNTs in ultrahigh molecular weight polyethylene," *Macromolecules*, vol. 39, no. 2, pp. 658–666, 2006.

[25] J.-T. Yeh, Y.-C. Lai, H. Liu et al., "Ultradrawing properties of ultrahigh- molecular-weight polyethylene/carbon nanotube fibers prepared at various formation temperatures," *Polymer International*, vol. 60, no. 1, pp. 59–68, 2011.

[26] Y. Zou, Y. Feng, L. Wang, and X. Liu, "Processing and properties of MWNT/HDPE composites," *Carbon*, vol. 42, no. 2, pp. 271–277, 2004.

[27] R. Haggenmueller, J. E. Fischer, and K. I. Winey, "Single wall carbon nanotube/polyethylene nanocomposites: Nucleating and templating polyethylene crystallites," *Macromolecules*, vol. 39, no. 8, pp. 2964–2971, 2006.

[28] K. Q. Xiao, L. C. Zhang, and I. Zarudi, "Mechanical and rheological properties of carbon nanotube-reinforced polyethylene composites," *Composites Science and Technology*, vol. 67, no. 2, pp. 177–182, 2007.

[29] W. Tang, M. H. Santare, and S. G. Advani, "Melt processing and mechanical property characterization of multi-walled carbon nanotube/high density polyethylene (MWNT/HDPE) composite films," *Carbon*, vol. 41, no. 14, pp. 2779–2785, 2003.

[30] J. Aalaie, A. Rahmatpour, and S. Maghami, "Preparation and characterization of linear low density polyethylene/carbon nanotube nanocomposites," *Journal of Macromolecular Science Part B: Physics*, vol. 46, no. 5, pp. 877–889, 2007.

[31] H. Mahfuz, A. Adnan, V. K. Rangari, and S. Jeelani, "Manufacturing and characterization of carbon nanotube/polyethylene composites," *International Journal of Nanoscience*, vol. 4, no. 1, pp. 55–72, 2005.

[32] B. Wunderlich, *Macromolecular Physics*, Academic Press, New York, NY, USA, 1976.

[33] T. McNally, P. Pötschke, P. Halley et al., "Polyethylene multiwalled carbon nanotube composites," *Polymer*, vol. 46, no. 19, pp. 8222–8232, 2005.

[34] K. P. Ryan, S. M. Lipson, A. Drury et al., "Carbon-nanotube nucleated crystallinity in a conjugated polymer based composite," *Chemical Physics Letters*, vol. 391, no. 4-6, pp. 329–333, 2004.

[35] S. Kanagaraj, F. R. Varanda, T. V. Zhil'tsova, M. S. A. Oliveira, and J. A. O. Simões, "Mechanical properties of high density polyethylene/carbon nanotube composites," *Composites Science and Technology*, vol. 67, no. 15-16, pp. 3071–3077, 2007.

[36] L. Valentini, J. Biagiotti, J. M. Kenny, and S. Santucci, "Morphological characterization of single-walled carbon nanotubes-PP composites," *Composites Science and Technology*, vol. 63, no. 8, pp. 1149–1153, 2003.

[37] D. Milicevic, M. Micic, G. Stamboliev, A. Leskovac, M. Mitric, and E. Suljovrujic, "Microstructure and crystallinity of polyolefins oriented via solid-state stretching at an elevated temperature," *Fibers and Polymers*, vol. 13, no. 4, pp. 466–470, 2012.

[38] G. Terife and K. A. Narh, "Properties of carbon nanotube reinforced linear low density polyethylene nanocomposites fabricated by cryogenic ball-milling," *Polymer Composites*, vol. 32, no. 12, pp. 2101–2109, 2011.

[39] P. Pötschke, H. Brünig, A. Janke, D. Fischer, and D. Jehnichen, "Orientation of multiwalled carbon nanotubes in composites

with polycarbonate by melt spinning," *Polymer*, vol. 46, no. 23, pp. 10355–10363, 2005.

[40] J. Dong, C. Yin, J. Lin, D. Zhang, and Q. Zhang, "Evolution of the microstructure and morphology of polyimide fibers during heat-drawing process," *RSC Advances*, vol. 4, no. 84, pp. 44666–44673, 2014.

[41] A. Rudolf, J. Geršak, and M. S. Smole, "The effect of heat treatment conditions using the drawing process on the properties of PET filament sewing thread," *Textile Research Journal*, vol. 82, no. 2, pp. 161–171, 2012.

[42] J. Shen, M. F. Champagne, R. Gendron, and S. Guo, "The development of conductive carbon nanotube network in polypropylene-based composites during simultaneous biaxial stretching," *European Polymer Journal*, vol. 48, no. 5, pp. 930–939, 2012.

[43] M. Park, H. Kim, and J. P. Youngblood, "Strain-dependent electrical resistance of multi-walled carbon nanotube/polymer composite films," *Nanotechnology*, vol. 19, no. 5, Article ID 055705, 2008.

[44] H. Deng, R. Zhang, C. T. Reynolds, E. Bilotti, and T. Peijs, "A novel concept for highly oriented carbon nanotube composite tapes or fibres with high strength and electrical conductivity," *Macromolecular Materials and Engineering*, vol. 294, no. 11, pp. 749–755, 2009.

Genome Structure of *Bacillus cereus* tsu1 and Genes Involved in Cellulose Degradation and Poly-3-Hydroxybutyrate Synthesis

Hui Li,[1] Suping Zhou,[1] Terrance Johnson,[1] Koen Vercruysse,[1] Ouyang Lizhi,[1] Parthasarathy Ranganathan,[1] Nsoki Phambu,[1] Alexander J. Ropelewski,[2] and Theodore W. Thannhauser[3]

[1] *College of Agricultural, Human and Natural Sciences, Tennessee State University, Nashville, TN, USA*
[2] *Pittsburgh Supercomputing Center, Pittsburgh, PA, USA*
[3] *R.W. Holley Center for Agriculture and Health, Plant, Soil and Nutrition Research Unit, USDA, ARS, Ithaca, NY, USA*

Correspondence should be addressed to Suping Zhou; zsuping@tnstate.edu

Academic Editor: Raffaele Cucciniello

In previous work, we reported on the isolation and genome sequence analysis of *Bacillus cereus* strain tsu1 NCBI accession number JPYN00000000. The 36 scaffolds in the assembled tsu1 genome were all aligned with *B. cereus* B4264 genome with variations. Genes encoding for xylanase and cellulase and the cluster of genes in the poly-3-hydroxybutyrate (PHB) biosynthesis pathway were identified in tsu1 genome. The PHB accumulation in *B. cereus* tsu1 was initially identified using Sudan Black staining and then confirmed using high-performance liquid chromatography. Physical properties of these PHB extracts, when analyzed with Raman spectra and Fourier transform infrared spectroscopy, were found to be comparable to the standard compound. The five PHB genes in tsu1 *(phaA*, *phaB*, *phaR*, *phaC*, and *phaP)* were cloned and expressed with TOPO cloning, and the recombinant proteins were validated using peptide mapping of in-gel trypsin digestion followed by mass spectrometry analysis. The recombinant *E. coli* BL21 (DE3) (over)expressing *phaC* was found to accumulate PHB particles. The cellulolytic activity of tsu1 was detected using carboxymethylcellulose (CMC) plate Congo red assay and the shift towards low-molecular size forms of CMC revealed by gel permeation chromatography in CMC liquid culture and the identification of a cellulase in the secreted proteome.

1. Introduction

Since 1960s, driven by public concerns about environmental pollution by petroleum-derived plastics [1–6] and the escalating crude oil price due to the depletion of fossil oil resources, bioplastics have attracted widespread attention, as eco-friendly, biodegradable, and sustainable alternatives [4, 7]. Among all the biodegradable plastics, the polyhydroxyalkanoates (PHAs) family has unique properties like insolubility in water, biocompatibility, oxygen permeability, and ultraviolet (UV) resistance [8]. Because of these advantageous characteristics, comprehensive applications have been discovered and developed using PHAs-derived materials for packaging plastics, medical materials, chiral monomer, and others [9, 10]. Also stable engineered industrial microbial strains have been developed overexpressing genes in PHAs biosynthesis pathway with additional functions in regulating cellular metabolisms and stress resistance [11, 12]. The main member of the PHAs family is polyhydroxybutyrate (PHB). These polymers are accumulated intracellularly in PHB producing bacteria when cultured under carbon-excess and other nutrients-limited conditions [13].

A large number of microorganisms have been found to accumulate PHA as lipoidic storage materials in the cytosol [14–17]. These microorganisms are mainly divided into four classes (I, II, III, and IV) based on the type of PHA synthases, which are the key enzymes for PHA biosynthesis [18]. While a single subunit PhaC was found in class I (e.g., *Ralstonia eutropha*) and class II (e.g., *Pseudomonas aeruginosa*) synthases, two subunits, PhaE and PhaC, or PhaR and PhaC, were suggested to be used, respectively, in type III (e.g., *Allochromatium vinosum*) and type IV (e.g.,

Bacillus megaterium) synthases [19, 20]. Classes I, III, and IV synthases act on polymerase short-chain monomers (C3–C5) whereas class II synthase acts on medium-chain length (mcl) monomers (C6–C14). The most recently discovered class IV PHA synthase is only present in *Bacillus* sp. There is little information about the capacity of PHAs production and the substrate specificity of class IV PHA synthase.

The higher production cost compared to petroleum-derived plastics is the primary factor limiting practical application of these biodegradable polymer materials. The following two approaches have been taken to make the PHAs mass production economically feasible. Firstly, engineered *E. coli* strains producing higher yield of PHAs [21, 22] have been developed using recombinant DNA technology. The optimization of the intermediate substrates and fermentation conditions is the key step in utilizing this technology in large scale PHAs production. The second approach focuses on looking for cheaper raw materials. Carbon source for PHB production accounts for up to 50% of the total production costs. Agricultural byproducts like soybean cake [23], biogas methane [24], and palm oil [25] as inexpensive carbon feedstock exhibit a high potential to accelerate the commercialization of PHAs. Rapeseed (canola seed) oil is one of the preferred oil stocks for biodiesel production, partly because rapeseed produces more oil per unit of land area compared to other oil sources, such as soybeans [26]. Rapeseed cake is generated as a byproduct during the oil extraction process from rapeseed (canola seed), which will be tested in this study for its potential use as a substrate for PHB production. The physical properties of PHB products produced by the bacteria cultured in aqueous extracts of rapeseed cake were confirmed with Raman spectrum and Fourier transform infrared spectroscopy (FTIR).

Previously, we reported the genome sequence of *B. cereus* tsu1 [27]. This paper reports on the assembly of the genomic structure, and characterization of cellulolytic and PHB producing activities of this strain. PHB biosynthesis pathway genes were cloned and (over)expressed in *E. coli* BL21 (DE3) using TOPO cloning system. The recombinant bacterial clones were confirmed to accumulate PHB granules.

2. Materials and Methods

2.1. Genomic Structure Analysis. In previous research, draft genome of *Bacillus cereus* tsu1 was generated using next generation sequencing analysis [27]. To generate an alignment map of the assembled tsu1 scaffolds (GenBank: KN321896-KN321931) using MUMmer (version 3.0) on the galaxy working station (biou.psc.edu/galaxy) [28], the genome sequence of *B. cereus* B4264 (NCBI GenBank: CP001176.1) with the highest identity similarity clustered by phylogenetic COG (PCOGR) was downloaded from NCBI database and used as the reference (Figure S1 in Supplementary Material, available online at https://doi.org/10.1155/2017/6192924) [29]. Based on the alignment, 20 scaffolds of tsu1 were selected to construct a circular genomic map using DNAPlotter (version 10.2) [30, 31]. Localizations of annotated genes on these scaffolds and on the reconstructed circular map were described in Supplementary Table S1.

2.2. Cloning and Expression of PHB Pathway Genes. Putative genes encoding for enzymes in PHB biosynthesis pathways were identified in the annotated genome. Five genes were amplified from tsu1 genomic DNA using the polymerase chain reaction (PCR) with primers designed against the assembled gene sequences (see Supplementary Table S2). The PCR program was conducted as follows: after a hot start cycle of 94°C for 2 mins, there were 35 cycles of denaturation at 94°C for 30 s, annealing at a melting temperature (T_m) for each individual primer, and extension at 72°C for 1 min, followed by a final cycle of 72°C for 10 mins. PCR products were separated on a 0.7% agarose gel. After staining with ethidium bromide, DNA fragments were isolated from the gel, purified using Qiagen Gel Extraction Kit (Cat. number 28704), and then cloned into TOPO pET101 vector (Invitrogen, CA). Plasmid with gene inserts was sent to GenHunter (624 Grassmere Park Drive, St 17, Nashville, TN 37211) for Sanger sequencing with primers (T7 forward and reverse) that flank the insertion site. Recombinant plasmids carrying full-length gene sequences were transformed into *E. coli* BL21 (DE3) cells and expression of recombinant proteins was induced by the addition of isopropyl β-D-1-thiogalactopyranoside (IPTG) in LB broth to a final concentration of 0.5–0.8 mM. Cells were harvested after 4-hour induction by centrifugation. Cell pellets were frozen in liquid N_2 and stored at −20°C before conducting protein analyses.

For two-dimensional (2D) protein gel electrophoresis of the recombinant proteins, cell pellets were homogenized in 500 μL dissolution buffer consisting of 7 M urea, 2 M thiourea, and 4% (3-((3-cholamidopropyl) dimethylammonio)-1-propanesulfonate) (CHAPS). After centrifugation at 16,000 ×g, 4°C for 20 min, supernatants were collected. Protein concentration was assayed following the Bradford method using BSA as the standard protein (Bio-Rad). Supernatants containing 200 μg protein samples were mixed with 5 μM dithiothreitol (DTT), and 1.25 μL IPG buffer (pH 3–10 NL, GE Healthcare). Upon bringing to a final volume of 250 μl using DeStreak Rehydration Solution (GE), proteins were loaded onto 13 cm pH 3–10 NL Immobiline DryStrips (GE). After an overnight passive rehydration at room temperature, proteins were focused on an Ettan IPGphor II (Amersham Biosciences) until reaching 24,000 total voltage hours (VhT). Prior to second dimensional electrophoresis, IPG strips were reduced with 1% DTT and then neutralized in 2.5% iodoacetic acid (IAA). Both of these steps were performed in a buffer containing 50 mM Tris-HCl, pH 8.8, 6 M urea, 30% glycerol, and 2% SDS. The second dimensional separation was performed on 12.5% sodium dodecyl sulfate polyacrylamide gel electrophoresis (SDS-PAGE) gels (1 mm in thickness) using a SE 600 Ruby Standard Vertical Unit (GE). Gels were run at a constant current (20 mA/strip after an initial run of 10 mA/strip for 30 min) until the bromophenol front reached the bottom of the gel and then stained with Colloidal blue staining kit (Invitrogen, LC6025). Gel images were captured by scanning on a Typhoon 9400 variable mode imager (GE). Recombinant proteins were localized to the spots on 2D gels according to their hypothetical isoelectric point and molecular weight. These protein spots and spots at the same position on gels loaded with proteins without IPTG induction were picked followed by in-gel tryptic digestion [32].

Tryptic peptide samples were reconstituted in 15 μL of 3% acetonitrile with 0.1% trifluoroacetic acid. Nano-liquid chromatography (LC) separation of tryptic peptides was performed using a nanoAcquity UPLC (Waters, Manchester), equipped with a Symmetry C_{18} 5 μm, 20 mm × 180 μm trapping column and a bonded ethyl hybrid C_{18} 1.7 μm, 15 cm × 75 μm analytical column (Waters). Mobile phase A consisted of water with 0.1% formic acid (FA) in water and mobile phase B acetonitrile with 0.1% FA. Samples, at 5 μL injection volume, were transferred to the trapping column at a flow rate of 7 μL/min 100% mobile phase A for 5 min. Following desalting and concentrating, the trapping column was eluted to the analytical column equilibrated with 2% mobile phase B at 300 nL/min. The eluent from the analytical column was delivered to a Xevo G2 Q-TOF mass spectrometer (MS) via a nanolockspray ion source (Waters). Data dependent acquisition (DDA) mode was used to obtain one 0.25 s MS survey scan. MS survey scans were acquired from m/z 300–1500, while product ion scans were acquired from m/z 50–2000. All data were acquired using MassLynx 4.1 SCN 862 (Waters). ProteinLynx Global Server v.2.5 was used to convert raw spectral data files for each injection into a peak list (.pkl format). To identify the matching sequences, the peak list from each protein spot was compared to theoretical tryptic digestion fragments of recombinant proteins.

2.3. Cellulolytic Activity Assay. Bacteria were cultured on the double-layered carboxymethyl cellulose sodium salt- (CMC-Na-) containing plates of which the bottom M9 minimal salt (11%; wt/vol) medium was overlaid with soft-agar containing 1% (wt/vol) CMC-Na (Sigma, St. Louis, MO). A cellulolytic bacterial strain *Paenibacillus polymyxa* 25A2^T [33–35] was obtained from the Bacillus Genetic Stock Center (Columbus, OH). In these assays, *P. polymyxa* 25A2^T was used as the positive control for testing cellulase activity and *E. coli* was used as the negative control. After incubation at 37 ± 1°C for 2 days, the CMC agar plates were stained with 0.1% Congo red solution following the method described previously [36, 37]. Plates were recorded for the formation of a clear zone around colonies, which indicates extracellular cellulolytic activity of the bacteria.

To further characterize the bacterial strain, a single colony of *B. cereus* tsu1 was inoculated into LB broth and incubated at 37°C under constant agitation at 200 rpm. Aliquots of an overnight culture of *B. cereus* tsu1 (1 mL) were inoculated into 10 mL M9 minimal salt medium supplemented with 1% (wt/vol) CMC-Na, and the control culture used the broth solution only. Cultures were continued under the same conditions. Two culture period treatments of 2 days and 6 days each with three replicates were conducted. At the end of each treatment period, bacterial cultures were centrifuged at 16,000 ×g for 1 min to collect supernatants. For gel permeation chromatography (GPC) assays of cellulose degradation, 2 mL of the supernatant from each sample was filtered through a 0.22 μm sterile filter (EMD Millipore, Massachusetts) and eluted into a clean 2 mL Eppendorf centrifuge tube. Gel permeation chromatography (GPC) analyses were performed on a Varian Prostar chromatography system (Walnut Creek, CA) equipped with a Waters Ultrahydrogel 2000 column (Milford, WA). Each 800 μL sample was diluted in 200 μL GPC solution (25 mM Na-acetate : methanol; 9 : 1). For each sample, 20 μL was injected each time. Analyses were performed at room temperature with a mobile phase consisting of 25 mM Na-acetate : methanol (9 : 1) at a flow rate of 0.75 mL/min and the UV/Vis absorbance was monitored at 210 nm using a photodiode array (PDA) detector. The collected data was used to compare the molecular weight changes between samples.

Supernatant collected for GPC analysis was mixed with acetone (1 : 3; v/v) followed by incubation overnight at −20°C for secreted protein precipitation. After centrifugation at 16,000 ×g for 1 min at 4°C, supernatant was removed. Protein pellets were air-dried, solubilized in a 1x Laemmli protein sample buffer (Biorad), and denatured by boiling for 5 min. Proteins were separated on a SDS-PAGE protein gel [27]. After staining with Coomassie Blue, protein bands were isolated from the gel and digested with trypsin [32], followed by liquid chromatography (LC)/mass spectrometry (MS) as described above. The generated peptides were searched against annotated protein database of *B. cereus* tsu1 (download from https://www.ncbi.nlm.nih.gov/protein/?term=bacillus±cereus±tsu1).

2.4. Poly-3-Hydroxybutyrate (PHB) Producing Activity Assay. The intracellular accumulation of PHB in *B. cereus* tsu1 was determined using the Sudan Black B staining method [38]. Bacterial smears were prepared using a four-day culture in LB broth and stained in a 0.3% Sudan Black stain solution (w/v) in 60% ethanol for 10 min. After rinsing with water, bacterial cells were counter-stained with 0.25% safranin for 1 min. Stained bacterial cells were observed and photographed under a Nikon Eclipse E600 Pol microscope (Japan).

Rapeseed cake samples (25 g) were soaked in 1 L water overnight under constant stirring. The supernatant was filtered through a 0.22 μm filtration system (EMD Millipore), and the rapeseed cake substrate (RCS) was used for bacterial cultures without any supplements. The total protein concentration in RCS was quantified using Bradford protein assay method. Bacterial cultures were incubated overnight in RCS and harvested by centrifugation at 3,220 ×g, 24°C, for 10 min. Cell pellets were oven-dried at 70°C to a constant weight and lysed by dispersing in 6% sodium hyperchlorite. After incubation at 37°C for 1 h, cell pellets were washed in 5 mL alcohol and then in 5 mL acetone. PHB was extracted by bathing the pellet in chloroform at 60°C for 1 h [39–41]. Then, chloroform was evaporated to obtain PHB crystals. PHB extracts were digested in 1 mL concentrated sulfuric acid at 100°C for 30 min, chilled to room temperature, and then diluted in 0.001 N H_2SO_4 to a final concentration of 0.8 mg/mL adipic acid, which are the 250x stock solutions.

The digested PHB-containing mixture was fractionated using high-performance liquid chromatography (HPLC) equipped with an Aminex HPX-87H ion-exclusion resin for organic acid analysis column (300 by 7.8 mm). The poly-(R)-3-hydroxybutyric acid (Sigma, MO) was used as the standard.

The crotonic acid formed from PHB acid digestion was detected by the absorbance peak at 210 nm [42] (Supplementary Table S4). Aliquots of PHB extracts were also analyzed on a spectrophotometer (SpectraMax M5, Molecular Devices, CA). The absorbance of the diluted sample (1 : 250) from acid digestion was measured at 235 nm. A standard curve was constructed using commercial PHB (Sigma) [43]. After confirmation of both methods giving nearly the same results, PHB content from bacterial batch culture was assayed using the spectrophotometric method (Supplementary Table S4).

2.5. Physical Structural Property Analysis of tsul-PHB. An Xplora Raman spectrometer (LabRAM; HORIBA Jobin Yvon, NJ) was used with a NIR diode laser ($\lambda = 785$ nm, power = 2.5 mW) as an excitation source. The instrument settings were 100 μm confocal hole, 100 μm wide entrance slit, 600 gr/mm grating, and Olympus SLM Plan N 10x objective lens. Samples were mounted on a computer-controlled, high-precision x-y stage. An exposure time of 40 s and 5 accumulations were used to collect the spectra. Baseline fitting was performed using a LabSPEC 5 (HORIBA Jobin Yvon).

Samples were placed on the diamond crystal top plate of an attenuated total reflectance (ATR) accessory (Thermo Scientific Nicolet IS10, Thermo Scientific, Waltham, MA). Thumbscrew pressure was used to ensure that samples were in contact with the crystal. Data from sixteen scans were averaged over the spectral range of 4000 to 650 cm^{-1}, with a resolution of 4 cm^{-1}. Ambient air was used as the reference for the background spectrum before each sample. Between samples, the ATR crystal was cleaned using distilled water and dried. All spectra were recorded at room temperature. No data processing was performed on the raw spectra. The PHB standard (Sigma) was used as the reference for these analyses.

3. Results

3.1. Characterization of Microbial Genome of B. cereus tsul. The 36 assembled scaffolds of *B. cereus* tsul were each aligned to a distinct region of the genome from *B. cereus* B4262 [27]. Regions of scaffolds 4, 6, and 15 did not match the reference genome (Figure 1). The circular genome map of *B. cereus* tsul (Figure 2) was constructed using the annotation of the assembled scaffolds in reference to *B. cereus* B4264 genome (see Table Sl in Supplementary Material).

3.2. Characterization of the PHB Biosynthesis Pathways. Six PHA synthesis related genes were located on scaffold 9. *phaR* (PHB synthase subunit), *phaB* (acetoacetyl-coA reductase), and *phaC* (PHB synthase subunit) are divergently transcribed as a tricistronic operon; *padR/phaQ* (transcription regulator), *phaP* (Phasin protein), and *phaJ* are transcribed in one direction. The *padR* gene is a PHB-responsive repressor controlling expression of *phaP* and *phaR*. Phasins are proteins that accumulate during PHA synthesis; they bind to PHA granules and promote further PHA synthesis. Gene *phaA* is located on a separate scaffold. The *phaJ* gene encodes for (R)-specific enoyl-CoA hydratase which is involved in fatty acid metabolism (Figure 3(a)). PHB granules were seen inside bacterial cells (Figure 3(b)). The five genes in PHB synthesis pathway *(phaA, phaB, phaC, phaP,* and *phaR)* were cloned. The size and sequences of these genes are identical to the predicted gene sequences in the annotated genome sequences (Figure 3(c)). The recombinant proteins matched the predicted proteins in both molecular size and isoelectric points (pI value). Furthermore, mass spectrometry (MS) analysis of the tryptic digests of recombinant proteins showed that the peptide coverage ratio (identified/predicted) is 70%, 95%, 82%, 78%, and 100% for PhaA, PhaB, PhaC, PhaR, and PhaP, respectively (Figure 3(d); see Table S3 in the Supplementary Material). When the *phaC*-(over)expressing recombinant *E. coli* cells were cultured overnight under the IPTG induction condition, PHB granules were observed in Sudan Black stained cells (Figure 3(e)). These results confirmed that the PHB genes from *B. cereus* tsul can drive biosynthesis of the polymer in recombinant *E. coli* clones.

3.3. Characterization of Cellulolytic Pathways. The conversion of cellulose into glucose consists of two steps. During the first step, beta-1,4 glucanase breaks the glucosidic linkage to cellobiose. Subsequently, this beta-1,4 glucosidic linkage of cellobiose is broken down by beta-glucosidase to produce glucose. Both endo-beta-glucanase and beta-glucosidase were found in the tsul genome. One xylanase gene for the degradation of hemicellulose was identified in the genome (Figure 4(a)).

The Congo red test showed that the *B. cereus* tsul colonies formed a clear distinct yellow halo, which is an indication of extracellular cellulase enzymes produced by the bacterium. The same reaction was observed in *Paenibacillus polymyxa*, but not in *E. coli* colonies (Figure 4(b)). Gel permeation chromatography (GPC), also known as size-exclusion chromatography, is often used to characterize the molar mass distribution of natural and synthetic polymers. Various incubation-period products from CMC-containing substrates after culturing *B. cereus* tsul were compared. With longer treatment time (in 6-day sample), a lower intensity of high molecular weight component (between 11.5 and 12.5 min) was observed whereas a higher peak intensity of small molecules (between 13 and 14 min) emerged. The clear shift to longer retention times of the bacterial-treated products relative to the untreated products clearly demonstrates CMC degradation into smaller molecules in the bacterial-treated samples (Figure 4(c)).

The bacterial secreted proteins were separated into five major bands on SDS-PAGE gels (Supplementary Figure S2). Among all the proteins identified by searching the tryptic digestion peptides against the *B. cereus* tsul protein dataset annotated by prokaryotic genome annotation pipeline, one endo-glucanase matching (GenBank ID: KGT43479.1) was identified (Table S5 in the Supplementary Material) [27].

Conclusively, the Congo red test, GPC analysis, and secreted proteome analysis all supported the extracellular cellulase activity of *B. cereus* tsul which concurs with its genome structure.

3.4. PHB Production Efficiency of Bacillus cereus tsul on Rapeseed Cake Substrate (RCS). In this study, bacterial cell cultures were grown in RCS without any additional materials.

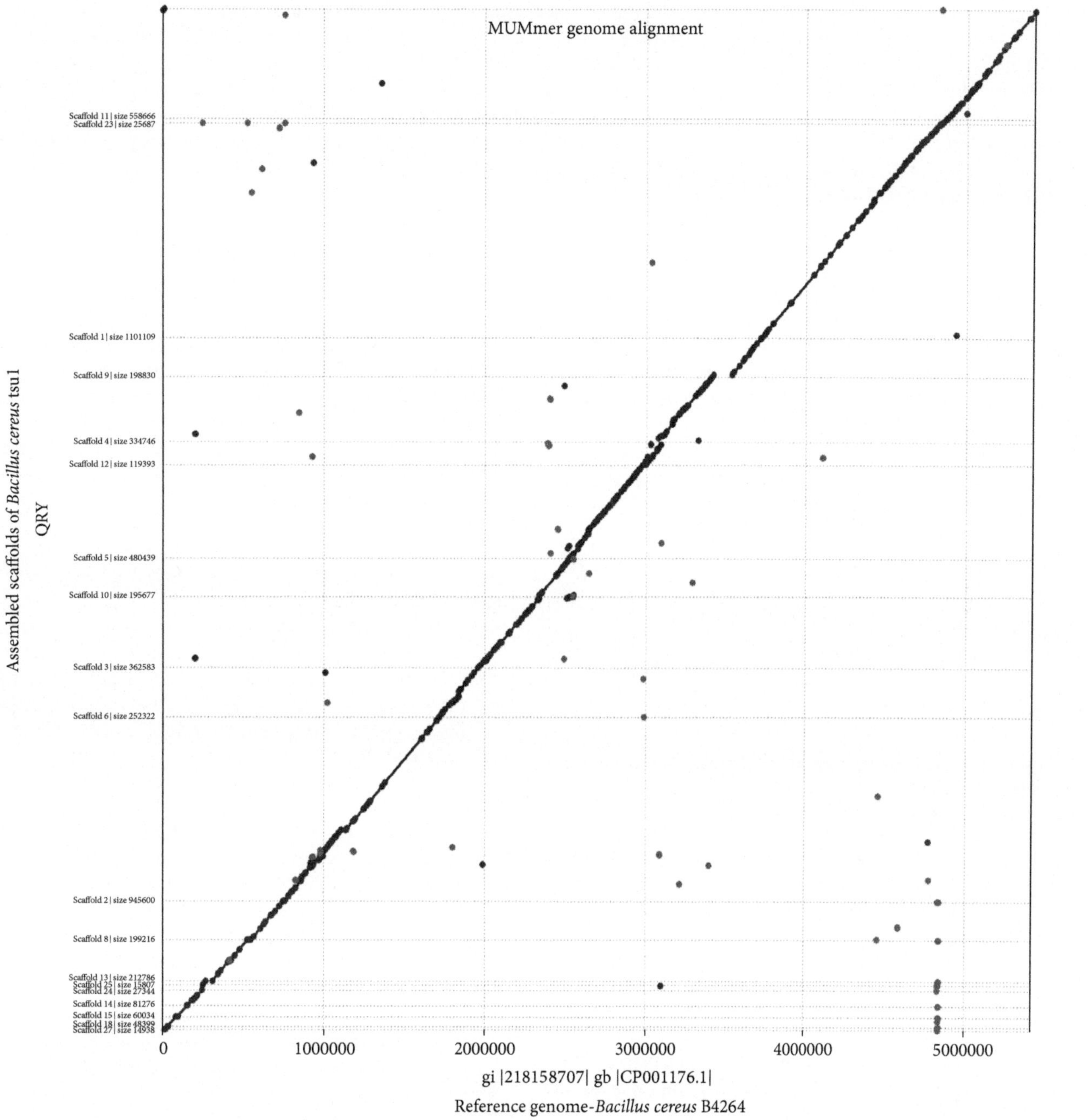

FIGURE 1: Alignment of assembled scaffolds from *Bacillus cereus* tsul with the reference genome of *B. cereus* B4262 using MUMmer on biou.psc.edu/galaxy.

After overnight batch culture, the cell dry weight reached 540–575 mg/L, which produced a 13-14% PHB content per cell biomass (Table 1).

3.5. Physical Structural Properties of PHB. PHB extracts from tsul cultures in RCS were compared to poly-(R)-3-hydroxybutyric acid (Sigma, MO) as the standard. The spectroscopic data of Raman and FTIR were used to determine the physical properties of PHB polymers [8]. The Raman spectra from regions 300–2200 cm^{-1} and from regions 2200–3000 cm^{-1} showed that the tsul PHB and the standard were both crystalline. The presence of sharp and narrow peaks in the Raman spectra at the major positions 434 and 841 cm^{-1} also indicated the crystallinity of the tsul PHB. However, peaks corresponding to C=O stretching and to CH_3 stretching occur at 1731.2 and 2935.4 cm^{-1}, respectively, showing that there were some amorphous regions in the tsul PHB extracts (see Table S6 in the Supplementary Material). Both Raman and FTIR spectra of tsul PHB extracts matched those from the standard (Figure 5). The mean square deviation is 3.7

Table 1: *Bacillus cereus* tsu1 cell propagation and PHB production on RCS substrate[†].

Items	Batch 1	Batch 2	Average
Cell biomass (mg/L culture)	575.17 ± 56.02	540.89 ± 18.43	558.03 ± 28.58
PHB mixture extract (mg/L)[a]	139.44 ± 14.18	112.78 ± 12.54	126.11 ± 9.88
Pure PHB content (mg/L)[b]	74.66 ± 8.54	80.61 ± 10.32	77.64 ± 6.45
PHB content in cell biomass (%)[c]	13%	14.9%	13.95%
PHB conversion efficiency of rape seed cake (g/kg)	2.99	3.22	3.10

[†]*Bacillus cereus* tsu1 was batch cultured for 24 hours in one-liter bottle containing (25 g/L) aqueous extracts of RCS. Two independent experiments each containing six replicates were conducted; [a]PHB was extracted from dried bacterial cells. Six replicates were conducted; [b]PHB content in crude extracts was determined by spectrophotometry analysis; [c]PHB content measured using the spectrometric method is similar to the HPLC method (37).

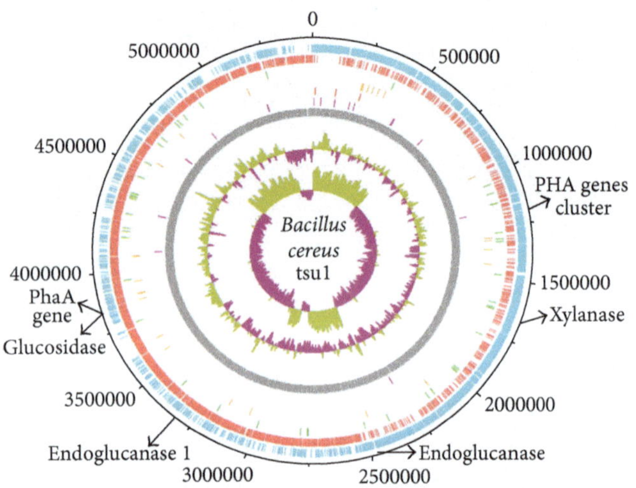

Figure 2: Circular genome map of *Bacillus cereus* tsu1. The map was generated using DNAPlotter (version 10.2) based on annotation data of the assembled scaffolds of *B. cereus* tsu1 (NCBI accession KN321896-KN321931). The draft genome sequence is 5.81 Mb, and 5.4 Mb was aligned to create the circular genome map. Genes in PHB biosynthesis pathway phaA, phaR, phaB, phaC, phaQ, phaP, and phaJ and cellulase genes for endoglucanase, glucosidase, and hemicellulase gene xylanase were indicated on the genome map. Locations of genes on the assembled scaffolds and their linkage to the circular map were provided in Supplementary Table S1.

to 7.1 cm^{-1}, which is within the range of discrepancy between Raman and FTIR spectra of PHB reported in previous studies [44–46].

4. Discussions

In this paper, we are reporting the characterization of *B. cereus* tsu1, a bacterium that has the ability to produce PHB polymer and degrade cellulose. In order to understand the molecular basis for these biochemical activities and evaluate the biotechnological potential of this bacterial strain, we proceeded to analyze its genome structure and test its extracellular cellulase and PHB producing ability. Analysis of cellulose degradation activity using GPC and Congo red staining methods suggests that the bacteria can degrade cellulose into smaller molecular products, but the end-products of degradation were not identified [47–49]. Cellulose has been used as a low-cost substrate in PHAs fermentation studies. Gao et al. reported that the successful transformation of cellulase and PHB synthesis pathway genes into *E. coli* resulted in the production of PHB directly from cellulose; however, the conversion efficiency and PHB yield were very low [50, 51]. In our study, the annotated *B. cereus* tsu1 genome contained cellulase (GenBank ID: KGT42715.1; KGT43479.1) and xylanase (GenBank ID: KGT44235.1) genes. A cellulase was also identified in secreted proteomes in CMC liquid culture. In addition, genes in each step of PHB biosynthesis pathway were identified in *B. cereus* tsu1 genome including several *phaA* genes and a gene cluster with six PHA genes: *phaR* (PHB synthase subunit), *phaB* (acetoacetyl-coA reductase), *phaC* (PHB synthase), and a *phaJ,* the downstream *phaP* (Phasin protein), and the *padR* (PhaQ transcription regulator) (see Table S7 in the Supplementary Material).

Based on the PHB gene cluster, *B. cereus* tsu1 should express class IV PHA synthase, which is composed of subunits PhaC and PhaR. PhaC is the key enzyme involved in the polymerization process; it determines the types of monomers (R-hydroxyacyl-CoAs) incorporated into the PHA polymer chain based on the enzyme's substrate specificity, as well as controlling PHA chain length and polydispersity. Previous researches consistently indicate that class IV synthases favor short-chain-length monomers such as 3-hydroxybutyrate (C4) and 3-hydroxyvalerate (C5) for polymerization, but it can also polymerize some unusual monomers as minor components [52–54]. There is an increasing interest in class IV PHA synthase, due to the possible alcoholysis activity as an inherent feature among these enzymes [55]. This alcoholysis reaction is useful not only for the regulation of PHA molecular weight but also for the modification of the PHA carboxy terminus, which can be manipulated to produce more promising PHA materials with more beneficial properties [56, 57]. The discovery of these genes has increased our understanding of the PHB synthesis pathway because they have a specific role that would affect the efficiency and the types of PHB polymers being synthesized. These genes found in *B. cereus* tsu1 genome related to PHB synthesis and cellulose degradation will contribute in building a library of information for constructing more efficient PHB fermenter hosts in future work.

The high cost associated with the production of PHAs remains a major barrier for large scale use of bioplastics [58]. In order to produce PHAs economically, a reliable and economical supply of raw material is essential; meanwhile, new bacterial strains and genes need to be discovered in order

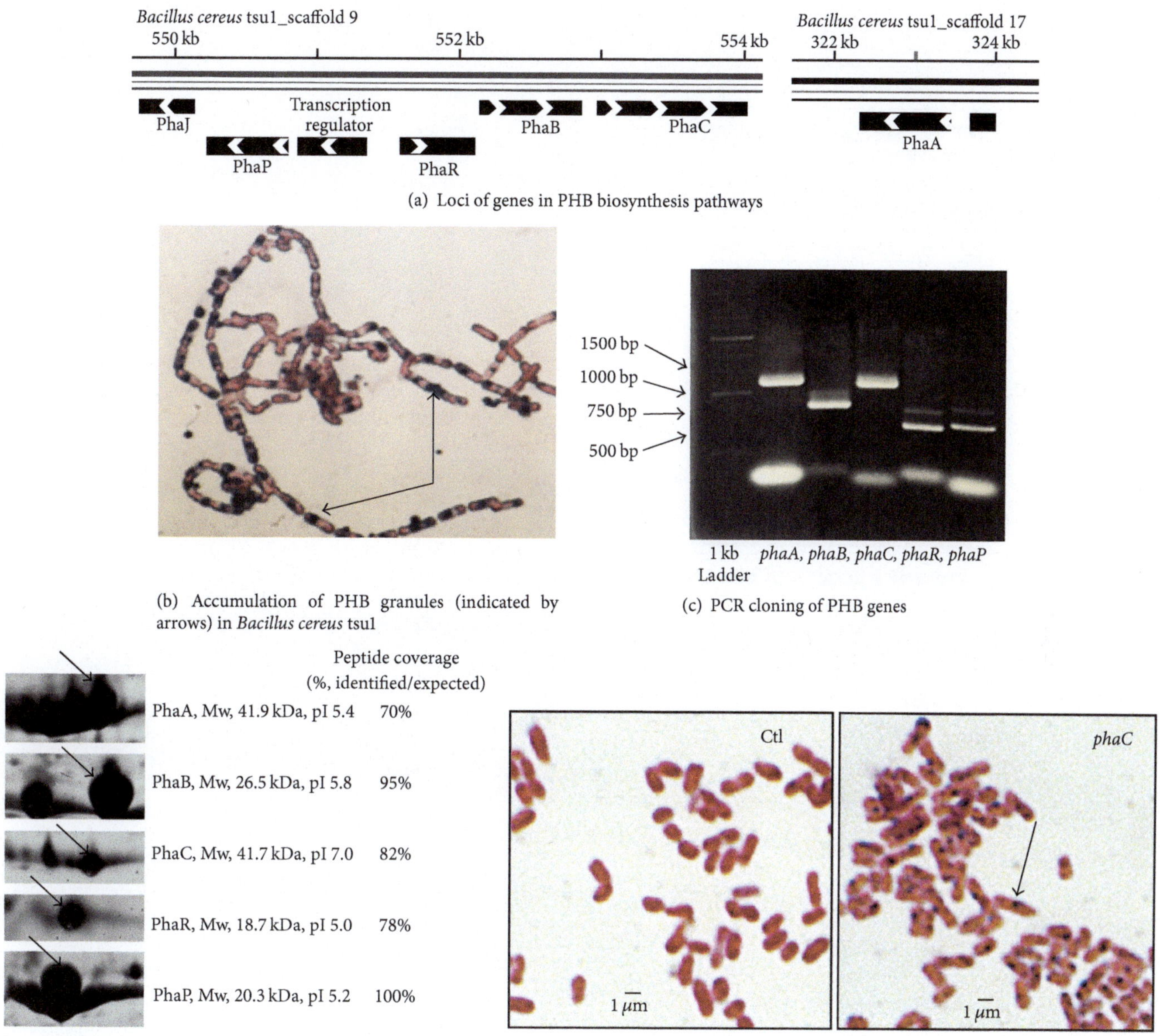

(a) Loci of genes in PHB biosynthesis pathways

(b) Accumulation of PHB granules (indicated by arrows) in *Bacillus cereus* tsu1

(c) PCR cloning of PHB genes

(d) Recombinant PHB proteins (indicated by arrows) on 2D gels and confirmation using peptide mapping

(e) *E. coli* with *phaC* genes displayed PHB producing activity

FIGURE 3: Characterization of genes in the poly-3-hydroxybutyrate (PHB) biosynthesis pathways in *Bacillus cereus* tsu1. (a) Genes in the PHB biosynthesis pathways: *phaR*, *phaB*, and *phaC* on the same operon, *phaJ*, *phaP*, and *phaQ* on reverse direction, and *phaA* located on a separate locus. (b) Sudan Black staining of *B. cereus* tsu1 showing accumulation of PHB granules in the bacterial cells. (c) The PHB gene fragments amplified from the genomic DNA of *Bacillus cereus* tsu1 using polymerase chain reaction (PCR) and separated on a 0.7% agarose gel. (d) Two-dimensional gel electrophoresis of recombinant proteins of PhaA, PhaB, PhaC, PhaR, and PhaP. The PCR amplified gene fragments showing identical sequence matches with the PHB synthesis genes were cloned into TOPO pET101 vector (Invitrogen, CA); recombinant proteins were expressed in DE3 cells. Recombinant proteins were separated on 2D gels. Protein spot with matching molecular size and isoelectric point (pI) of each predicted protein was picked from the gel. Protein identity was confirmed using the peptide fingerprinting (70–100% coverage with identical predicated peptides) using mass spectrometry analysis of tryptic digests of these proteins. (e) Sudan Black staining of the recombinant *E. coli* (over)expressing phaC gene showing intracellular PHB granules; *E. coli* transformed with the empty vector had no PHB accumulation. The arrow refers to the black spots inside of the bacterial cell, which is the PHB accumulation stained with Sudan Black.

to use a wider variety of substrates [59, 60]. The aqueous extract of rapeseed cake (RCS) can be a promising raw material for PHB production. In this study, PHB content was 13-14% per cell dry weight in overnight batch culture in RCS. Taking into consideration that no other nutrients were added in RCS, the bacterial growth performance and PHB content are relatively lower than other well-developed nutrient substrates. Meanwhile, in this batch culture for PHB production, no pH buffer solution was added, which may limit the biomass production, and further limit the PHB production

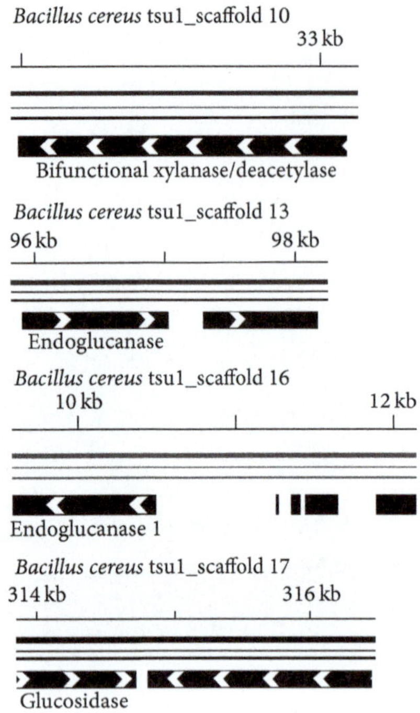

(a) Gene sequences

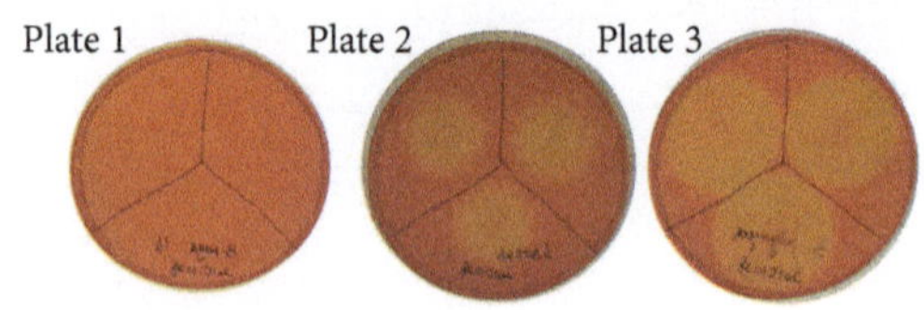

(b) Congo Red staining assay on CMC plates. Plate 1, *Escherichia coli (negative control)*; Plate 2, *Bacillus cereus* tsul; Plate 3, *Paenibacillus polymyxa (positive control)*

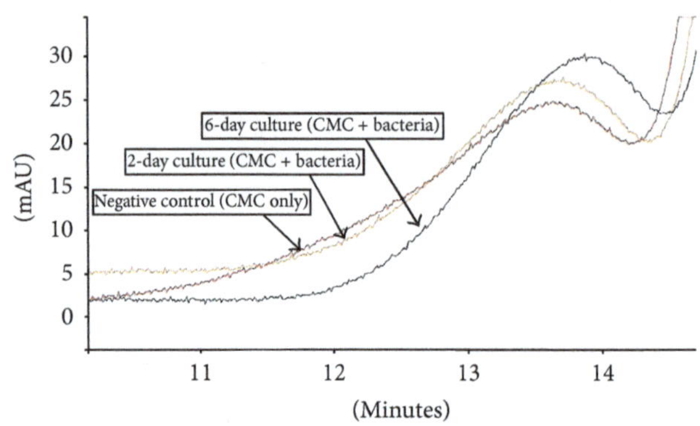

(c) Gel permeation chromatography assay of products derived from CMC cultured with *Bacillus cereus* tsul

FIGURE 4: Characterization of cellulase gene and enzymatic activity of *Bacillus cereus* tsul. (a) Cellulase genes annotated in the *Bacillus cereus* tsul genome. (b) Congo red plate assay of cellulase activity. Bacteria were cultured on the double-layered carboxymethylcellulose sodium salt- (CMC-Na-) containing plates of which the bottom M9 minimal salt (11%; wt/vol) agar medium was overlaid with soft-agar containing 1% (wt/vol) CMC-Na (Sigma, St. Louis, MO). Plates from two-day incubation at 37°C were stained with 0.1% Congo red. The yellowish halo around the bacterial colony indicates degradation of CMC. Plate 1, *E. coli* showing no CMC degradation activity (negative control); Plate 2, *B. cereus* tsul; Plate 3, *Paenibacillus polymyxa* 25A2^{T} with CMC degradation activity (a positive control of cellulolytic bacterial strain from the Bacillus Genetic Stock Center, Columbus, OH). (c) Gel permeation assay of CMC derived products after incubation with *B. cereus* tsul for 2 days and 6 days. The right shifts of the peaks indicate that the CMC derived molecules after the digestion with *B. cereus* tsul were smaller in size and therefore they were eluted at a delayed time-frame than the original CMC. These results confirmed the extracellular cellulase activity of the bacterial strain.

[61]. RCS was estimated to have 8.98 g/L crude proteins, and it also contained all the essential amino acids for bacterial growth, and some α- and γ-amino-butyric acid (AIB, GABA) (see Figure S3 in Supplementary Material). Some of these amino acids can potentially be converted into PHB through various alternative pathways (see Figure S4 in Supplementary Material). In a previous study [62], GABA was reported as an alternative route of catabolism in *Saccharomyces cerevisiae*. It is involved in the conversion of GABA into succinate-semialdehyde (SSA) by 4-aminobutyrate aminotransferase; meanwhile, glutamate can be converted into GABA by glutamate decarboxylase. In 2007, Valappil et al. reported that *B. cereus* 14579 genome contained all the alternative pathway genes involved for the conversion of succinyl-CoA from

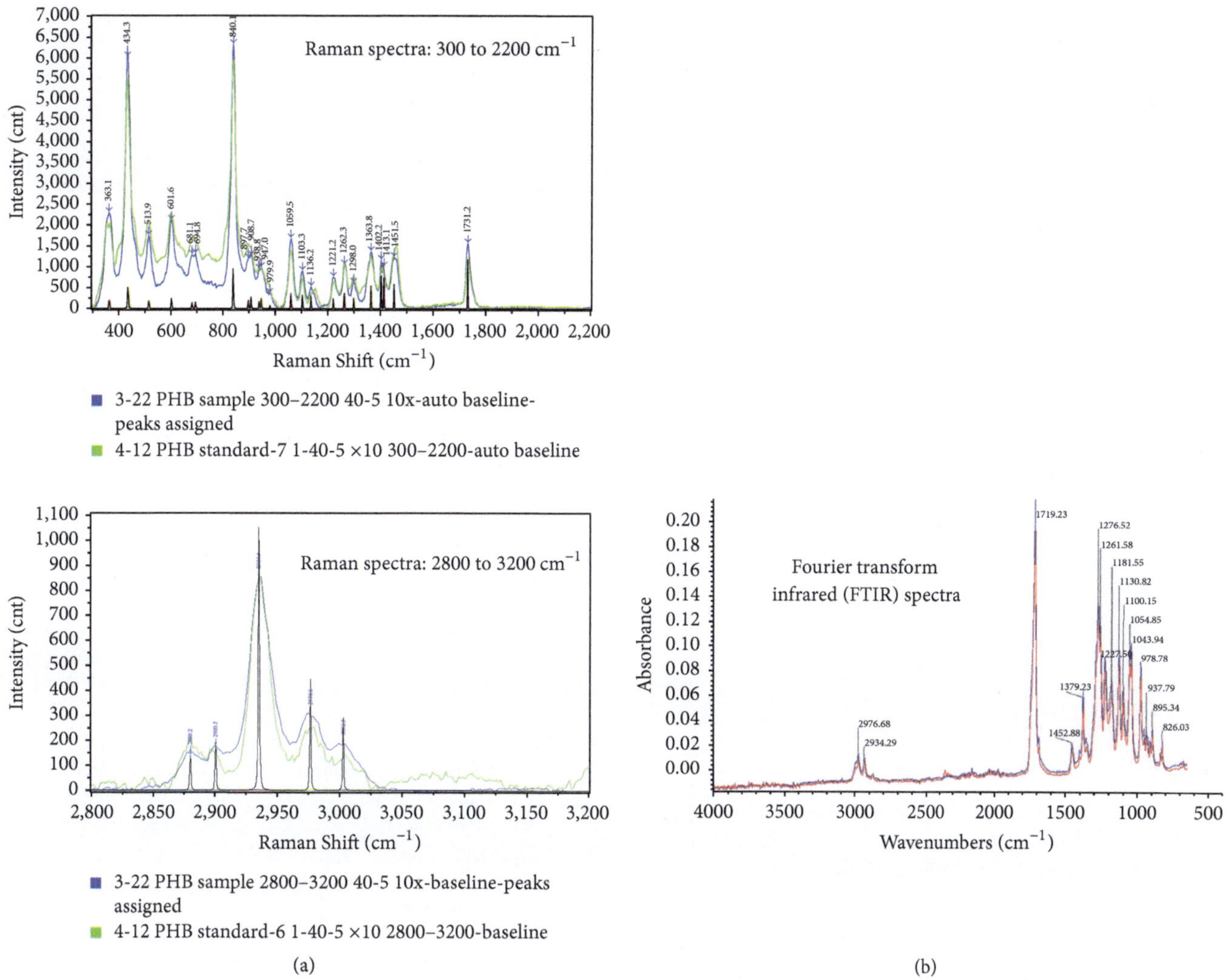

FIGURE 5: Physical properties of polyhydroxybutyrate PHB produced by *Bacillus cereus* tsul on rapeseed cake substrate (RCS). The PHB standard (Sigma) was used as the reference for these analyses. Raman spectra of PHB produced by *B. cereus* tsul and standard (a). FTIR spectra of PHB produced by *B. cereus* tsul and standard (b).

TCA cycle into P4HB/PH3B/P(3HB-co-4HB) [63]. P(3HB-co-4HB)-like copolymer was found in *B. cereus* SPV. In this alternative pathway, succinyl-CoA is first converted to SSA by SSA dehydrogenase followed by reduction of SSA into GABA by 4-hydroxybutyrate dehydrogenase. GABA is activated to 4-hydroxybutyryl-CoA by a Co-A transferase enzyme. The R-4-hydroxybutyryl-CoA and/or R-3-hydroxybutyryl-CoA are then to be polymerized to form P4HB/PH3B/P(3HB-co-4HB) by PHB synthase.

The (R)-specific enoyl-CoA hydratase/MaoC-like protein (PhaJ) is a monomer supplying enzyme from fatty acid β-oxidation. Tajima et al. demonstrated that PhaJ may make a channeling route from β-oxidation to PHA biosynthesis and PhaC synthases from *B. cereus* and its relatives have the ability to incorporate both scl and mcl PHAs [64]. This alternative metabolic pathway was confirmed [65] in a recombinant *E. coli* strain which utilizes two substrate-specific enoyl-CoA hydratases, R-hydratase (PhaJ) and S-hydratase (FadB). In these bacterial strains, the PhaJ works in coordination with S-specific hydratases to provide (R)-3HB-CoA for PHA synthesis from crotonyl-CoA. In 2012, Cai reported that *B. cereus* may be another example harboring *phaJ* in the pha locus [66]. In *B. cereus* tsul, the *phaJ* gene encoding R-specific enoyl-CoA hydratase (GenBank ID: KGT44860.1) may function coordinately with an acyl-CoA dehydrogenase to form (R)-3HB-CoA via crotonyl-CoA intermediate which is derived from fatty acid β-oxidation (see Figure S4 in Supplementary Material). Genes in this pathway were found in the genome of *B. cereus* tsul. The production of PHB from RCS may support this alternative catabolism, which may shed new light on the PHB biosynthesis pathway and open up new opportunities for its industrial applications.

Disclosure

Mention of trade names or commercial products in this publication is solely for the purpose of providing specific information and does not imply recommendation or endorsement by the U.S. Department of Agriculture. The funders of the project had no role in study design, data collection and interpretation, or the decision to submit the work for publication.

Acknowledgments

The authors wish to thank Dr. Charles Lee at USDA/ARS for his advice on gene cloning strategies, Dr. Peter A. Schweitzer at Cornell University Institute of Biotechnology for sequencing the bacterial genome, Dr. Roger Sauve for reviewing this manuscript, Dr. Korsi Dumenyo for suggestions of bacterial cellulase activity assay, Dr. Fur-Chi Chen, Kevin Howe, and Tara Fish for protein analysis, and Dr. Jason P. de Koff for kindly providing canola seed cake materials and Mrs. Sarabjit Bhatti for assisting in the experiments. Amino acid analysis of rapeseed cake was conducted by Dr. John Schulze, Molecular Structure Facility at University of California at Davis. PHB analysis using HPLC was conducted by Dr. Allison Pieja at Mango Materials Co., California, USA.

References

[1] J. Hammer, M. H. S. Kraak, and J. R. Parsons, "Plastics in the marine environment: the dark side of a modern gift," *Reviews of Environmental Contamination and Toxicology*, vol. 220, pp. 1–44, 2012.

[2] M. Claessens, L. Van Cauwenberghe, M. B. Vandegehuchte, and C. R. Janssen, "New techniques for the detection of microplastics in sediments and field collected organisms," *Marine Pollution Bulletin*, vol. 70, no. 1-2, pp. 227–233, 2013.

[3] L. Van Cauwenberghe, A. Vanreusel, J. Mees, and C. R. Janssen, "Microplastic pollution in deep-sea sediments," *Environmental Pollution*, vol. 182, pp. 495–499, 2013.

[4] A. L. Lusher, A. Burke, I. O'Connor, and R. Officer, "Microplastic pollution in the Northeast Atlantic Ocean: validated and opportunistic sampling," *Marine Pollution Bulletin*, vol. 88, no. 1-2, pp. 325–333, 2014.

[5] L. C. Woodall, A. Sanchez-Vidal, M. Canals et al., "The deep sea is a major sink for microplastic debris," *Royal Society Open Science*, vol. 1, no. 4, Article ID 140317, 2014.

[6] D. Eerkes-Medrano, R. C. Thompson, and D. C. Aldridge, "Microplastics in freshwater systems: a review of the emerging threats, identification of knowledge gaps and prioritisation of research needs," *Water Research*, vol. 75, pp. 63–82, 2015.

[7] R. Jain and A. Tiwari, "Biosynthesis of planet friendly bioplastics using renewable carbon source," *Journal of Environmental Health Science and Engineering*, vol. 13, no. 1, article 11, 2015.

[8] S. Muhammadi, M. Afzal, and S. Hameed, "Bacterial polyhydroxyalkanoates-eco-friendly next generation plastic: production, biocompatibility, biodegradation, physical properties and applications," *Green Chemistry Letters and Reviews*, vol. 8, no. 3-4, pp. 56–77, 2015.

[9] S. J. Park, S. Y. Lee, and Y. Lee, "Biosynthesis of R-3-hydroxyalkanoic acids by metabolically engineered *Escherichia coli*," *Applied Biochemistry and Biotechnology*, vol. 113-116, pp. 373–379, 2004.

[10] G.-Q. Chen, "A microbial polyhydroxyalkanoates (PHA) based bio- and materials industry," *Chemical Society Reviews*, vol. 38, no. 8, pp. 2434–2446, 2009.

[11] J. Zhang, N. Hao, and G.-Q. Chen, "Effect of expressing polyhydroxybutyrate synthesis genes (*phbCAB*) in *Streptococcus zooepidemicus* on production of lactic acid and hyaluronic acid," *Applied Microbiology and Biotechnology*, vol. 71, no. 2, pp. 222–227, 2006.

[12] P. Gu, J. Kang, F. Yang, Q. Wang, Q. Liang, and Q. Qi, "The improved L-tryptophan production in recombinant *Escherichia coli* by expressing the polyhydroxybutyrate synthesis pathway," *Applied Microbiology and Biotechnology*, vol. 97, no. 9, pp. 4121–4127, 2013.

[13] B. S. Saharan, A. Grewal, and P. Kumar, "Biotechnological production of polyhydroxyalkanoates: a review on trends and latest developments," *Chinese Journal of Biology*, vol. 2014, Article ID 802984, 18 pages, 2014.

[14] C.-S. Ha and W.-J. Cho, "Miscibility, properties, and biodegradability of microbial polyester containing blends," *Progress in Polymer Science*, vol. 27, no. 4, pp. 759–809, 2002.

[15] Y. K. Leong, P. L. Show, C. W. Ooi, T. C. Ling, and J. C.-W. Lan, "Current trends in polyhydroxyalkanoates (PHAs) biosynthesis: insights from the recombinant *Escherichia coli*," *Journal of Biotechnology*, vol. 180, pp. 52–65, 2014.

[16] V. Urtuvia, P. Villegas, M. González, and M. Seeger, "Bacterial production of the biodegradable plastics polyhydroxyalkanoates," *International Journal of Biological Macromolecules*, vol. 70, pp. 208–213, 2014.

[17] P. Kumar, S. Ray, S. K. S. Patel, J.-K. Lee, and V. C. Kalia, "Bioconversion of crude glycerol to polyhydroxyalkanoate by *Bacillus thuringiensis* under non-limiting nitrogen conditions," *International Journal of Biological Macromolecules*, vol. 78, pp. 9–16, 2015.

[18] B. H. A. Rehm, "Polyester synthases: natural catalysts for plastics," *Biochemical Journal*, vol. 376, no. 1, pp. 15–33, 2003.

[19] C. Yang, W. Zhang, R. Liu et al., "Analysis of polyhydroxyalkanoate (PHA) synthase gene and PHA-producing bacteria in activated sludge that produces PHA containing 3-hydroxydodecanoate," *FEMS Microbiology Letters*, vol. 346, no. 1, pp. 56–64, 2013.

[20] K. Ushimaru, Y. Motoda, K. Numata, and T. Tsuge, "Phasin proteins activate aeromonas caviae polyhydroxyalkanoate (PHA) synthase but not *Ralstonia eutropha* PHA synthase," *Applied and Environmental Microbiology*, vol. 80, no. 9, pp. 2867–2873, 2014.

[21] P. I. Nikel, A. De Almeida, E. C. Melillo, M. A. Galvagno, and M. J. Pettinari, "New recombinant *Escherichia coli* strain tailored for the production of poly(3-hydroxybutyrate) from agroindustrial by-products," *Applied and Environmental Microbiology*, vol. 72, no. 6, pp. 3949–3954, 2006.

[22] Z. Lin, Y. Zhang, Q. Yuan et al., "Metabolic engineering of *Escherichia coli* for poly(3-hydroxybutyrate) production via threonine bypass," *Microbial Cell Factories*, vol. 14, no. 185, 2015.

[23] F. C. Oliveira, M. L. Dias, L. R. Castilho, and D. M. G. Freire, "Characterization of poly(3-hydroxybutyrate) produced by *Cupriavidus necator* in solid-state fermentation," *Bioresource Technology*, vol. 98, no. 3, pp. 633–638, 2007.

[24] C. S. Criddle, S. L. Billington, and C. W. Frank, "Renewable Bioplastics and Biocomposites from Biogas Methane and Waste-Derived Feedstock: Development of Enabling Technology, Life

Cycle Assessment, and Analysis of Costs," Tech. Rep., California Department of Resources Recycling and Recovery, Sacramento, Claif, USA, 2014.

[25] S. L. Riedel, J. Bader, C. J. Brigham et al., "Production of poly(3-hydroxybutyrate-co-3-hydroxyhexanoate) by *Ralstonia eutropha* in high cell density palm oil fermentations," *Biotechnology and Bioengineering*, vol. 109, no. 1, pp. 74–83, 2012.

[26] M. Frier and G. Roth, *Canola or Rapeseed Production in Pennsylvania*, Department of Crop and Soil Sciences, Pennsylvania, Pa, USA, 2006.

[27] H. Li, S. Zhou, T. Johnson, K. Vercruysse, A. J. Ropelewski, and T. W. Thannhauser, "Draft genome sequence of new *Bacillus cereus* strain tsu1," *Genome Announcements*, vol. 2, no. 6, article 2014, 2014.

[28] E. Afgan, D. Baker, M. van den Beek et al., "The Galaxy platform for accessible, reproducible and collaborative biomedical analyses: 2016 update," *Nucleic Acids Research*, vol. 44, no. W1, pp. W3–W10, 2016.

[29] R. L. Tatusov, D. A. Natale, I. V. Garkavtsev et al., "The COG database: new developments in phylogenetic classification of proteins from complete genomes," *Nucleic Acids Research*, vol. 29, no. 1, pp. 22–28, 2001.

[30] T. Carver, N. Thomson, A. Bleasby, M. Berriman, and J. Parkhill, "DNAPlotter: circular and linear interactive genome visualization," *Bioinformatics*, vol. 25, no. 1, pp. 119-120, 2009.

[31] D.-H. Lee, H. R. Kim, H. Y. Chung et al., "Complete genome sequence of *Bacillus cereus* FORC_005, a food-borne pathogen from the soy sauce braised fish-cake with quail-egg," *Standards in Genomic Sciences*, vol. 10, no. 1, article 97, 2015.

[32] S. Zhou, R. Sauvé, and T. W. Thannhauser, "Proteome changes induced by aluminium stress in tomato roots," *Journal of Experimental Botany*, vol. 60, no. 6, pp. 1849–1857, 2009.

[33] M. I. Alshelmani, T. C. Loh, H. L. Foo, W. H. Lau, and A. Q. Sazili, "Characterization of cellulolytic bacterial cultures grown in different substrates," *The Scientific World Journal*, vol. 2013, Article ID 689235, 6 pages, 2013.

[34] E. B. Górska, U. Jankiewicz, J. Dobrzyński, S. Russel, S. Pietkiewicz et al., "Degradation and colonization of cellulose by Diazotrophic Strains of *Paenibacillus polymyxa* isolated from soil," *Journal of Bioremediation & Biodegradation*, vol. 6, no. 2, 2015.

[35] R. López-Mondéjar, D. Zühlke, T. Větrovský, D. Becher, K. Riedel, and P. Baldrian, "Decoding the complete arsenal for cellulose and hemicellulose deconstruction in the highly efficient cellulose decomposer *Paenibacillus* O199," *Biotechnology for Biofuels*, vol. 9, no. 1, article 104, 2016.

[36] R. M. Teather and P. J. Wood, "Use of Congo red-polysaccharide interactions in enumeration and characterization of cellulolytic bacteria from the bovine rumen," *Applied and Environmental Microbiology*, vol. 43, no. 4, pp. 777–780, 1982.

[37] A. Sazci, K. Erenler, and A. Radford, "Detection of cellulolytic fungi by using Congo red as an indicator: a comparative study with the dinitrosalicyclic acid reagent method," *Journal of Applied Bacteriology*, vol. 61, no. 6, pp. 559–562, 1986.

[38] Y.-H. Wei, W.-C. Chen, C.-K. Huang et al., "Screening and evaluation of polyhydroxybutyrate-producing strains from indigenous isolate *Cupriavidus taiwanensis* strains," *International Journal of Molecular Sciences*, vol. 12, no. 1, pp. 252–265, 2011.

[39] S. K. Hahn, Y. K. Chang, B. S. Kim, and H. N. Chang, "Optimization of microbial poly(3-hydroxybutyrate) recover using dispersions of sodium hypochlorite solution and chloroform," *Biotechnology and Bioengineering*, vol. 44, no. 2, pp. 256–261, 1994.

[40] B. Aslim, F. Çalişkan, Y. Beyatli, and U. Gündüz, "Poly-β-hydroxybutyrate production by lactic acid bacteria," *FEMS Microbiology Letters*, vol. 159, no. 2, pp. 293–297, 1998.

[41] P. Singh and N. Parma, "Isolation and characterization of two novel polyhydroxybutyrate (PHB)-producing bacteria," *African Journal of Biotechnology*, vol. 10, no. 24, pp. 4907–4919, 2011.

[42] D. B. Karr, J. K. Waters, and D. W. Emerich, "Analysis of poly-β-hydroxybutyrate in *Rhizobium japonicum* bacteroids by ion-exclusion high-pressure liquid chromatography," *Applied and Environmental Microbiology*, vol. 46, no. 6, pp. 1339–1344, 1983.

[43] J. H. Law and R. A. Slepecky, "Assay of poly-beta-hydroxybutyric acid," *Journal of bacteriology*, vol. 82, pp. 33–36, 1961.

[44] T. Furukawa, H. Sato, R. Murakami et al., "Raman microspectroscopy study of structure, dispersibility, and crystallinity of poly(hydroxybutyrate)/poly(L-lactic acid) blends," *Polymer*, vol. 47, no. 9, pp. 3132–3140, 2006.

[45] J. De Gelder, D. Willemse-Erix, M. J. Scholtes et al., "Monitoring poly(3-hydroxybutyrate) production in *Cupriavidus necator* DSM 428 (H16) with Raman spectroscopy," *Analytical Chemistry*, vol. 80, no. 6, pp. 2155–2160, 2008.

[46] M. Ramezani, M. A. Amoozegar, and A. Ventosa, "Screening and comparative assay of poly-hydroxyalkanoates produced by bacteria isolated from the Gavkhooni Wetland in Iran and evaluation of poly-β-hydroxybutyrate production by halotolerant bacterium *Oceanimonas* sp. GK1," *Annals of Microbiology*, vol. 65, no. 1, pp. 517–526, 2015.

[47] B. Saake, S. Horner, T. Kruse, J. Puls, T. Liebert, and T. Heinze, "Detailed investigation on the molecular structure of carboxymethyl cellulose with unusual substitution pattern by means of an enzyme-supported analysis," *Macromolecular Chemistry and Physics*, vol. 201, no. 15, pp. 1996–2002, 2000.

[48] M. Melander and T. Vuorinen, "Determination of the degree of polymerisation of carboxymethyl cellulose by size exclusion chromatography," *Carbohydrate Polymers*, vol. 46, no. 3, pp. 227–233, 2001.

[49] A. Cohen, H. Schagerlöf, C. Nilsson, C. Melander, F. Tjerneld, and L. Gorton, "Liquid chromatography-mass spectrometry analysis of enzyme-hydrolysed carboxymethylcellulose for investigation of enzyme selectivity and substituent pattern," *Journal of Chromatography A*, vol. 1029, no. 1-2, pp. 87–95, 2004.

[50] C. Du, J. Sabirova, W. Soetaert, and S. K. Carol Lin, "Polyhydroxyalkanoates production from low-cost sustainable raw materials," *Current Chemical Biology*, vol. 6, no. 1, pp. 14–25, 2012.

[51] D. Gao, Y. Luan, Q. Wang, Q. Liang, and Q. Qi, "Construction of cellulose-utilizing *Escherichia coli* based on a secretable cellulase," *Microbial Cell Factories*, vol. 14, no. 1, article 159, 2015.

[52] G. J. McCool and M. C. Cannon, "PhaC and PhaR are required for polyhydroxyalkanoic acid synthase activity in *Bacillus megaterium*," *Journal of Bacteriology*, vol. 183, no. 14, pp. 4235–4243, 2001.

[53] S. Tomizawa, M. Hyakutake, Y. Saito et al., "Molecular weight change of polyhydroxyalkanoate (PHA) caused by the PhaC subunit of PHA synthase from *Bacillus cereus* YB-4 in recombinant *Escherichia coli*," *Biomacromolecules*, vol. 12, no. 7, pp. 2660–2666, 2011.

[54] T. Tsuge, M. Hyakutake, and K. Mizuno, "Class IV polyhydroxyalkanoate (PHA) synthases and PHA-producing *Bacillus*," *Applied Microbiology and Biotechnology*, vol. 99, no. 15, pp. 6231–6240, 2015.

[55] M. Hyakutake, Y. Saito, S. Tomizawa, K. Mizuno, and T. Tsuge, "Polyhydroxyalkanoate (PHA) synthesis by class IV PHA synthases employing *Ralstonia eutropha* PHB^{-}4 as host strain," *Bioscience, Biotechnology and Biochemistry*, vol. 75, no. 8, pp. 1615–1617, 2011.

[56] M. Hyakutake, S. Tomizawa, K. Mizuno, H. Abe, and T. Tsuge, "Alcoholytic cleavage of polyhydroxyalkanoate chains by Class IV synthases induced by Endogenous and exogenous ethanol," *Applied and Environmental Microbiology*, vol. 80, no. 4, pp. 1421–1429, 2014.

[57] M. Hyakutake, S. Tomizawa, K. Mizuno, T. Hisano, H. Abe, and T. Tsuge, "A common active site of polyhydroxyalkanoate synthase from *Bacillus cereus* YB-4 is involved in polymerization and alcoholysis reactions," *Applied Microbiology and Biotechnology*, vol. 99, no. 11, pp. 4701–4711, 2014.

[58] H. Yue, C. Ling, T. Yang et al., "A seawater-based open and continuous process for polyhydroxyalkanoates production by recombinant *Halomonas campaniensis* LS21 grown in mixed substrates," *Biotechnology for Biofuels*, vol. 7, no. 1, article 108, 2014.

[59] H. Chua and P. H. F. Yu, "Production of biodegradable plastics from chemical wastewater—a novel method to reduce excess activated sludge generated from industrial wastewater treatment," *Water Science and Technology*, vol. 39, no. 10-11, pp. 273–280, 1999.

[60] C. F. Budde, S. L. Riedel, L. B. Willis, C. Rha, and A. J. Sinskey, "Production of poly(3-hydroxybutyrate-co-3-hydroxyhexanoate) from plant oil by engineered *Ralstonia eutropha* strains," *Applied and Environmental Microbiology*, vol. 77, no. 9, pp. 2847–2854, 2011.

[61] X. Zhang, T. Bao, Z. Rao et al., "Two-stage pH control strategy based on the pH preference of acetoin reductase regulates acetoin and 2,3-butanediol distribution in *Bacillus subtilis*," *PLoS ONE*, vol. 9, no. 3, Article ID e91187, 2014.

[62] B. Bach, E. Meudec, J.-P. Lepoutre et al., "New insights into γ-aminobutyric acid catabolism: evidence for γ-hydroxybutyric acid and polyhydroxybutyrate synthesis in *Saccharomyces cerevisiae*," *Applied and Environmental Microbiology*, vol. 75, no. 13, pp. 4231–4239, 2009.

[63] S. P. Valappil, A. R. Boccaccini, C. Bucke, and I. Roy, "Polyhydroxyalkanoates in Gram-positive bacteria: insights from the genera *Bacillus* and *Streptomyces*," *International Journal of General and Molecular Microbiology*, vol. 91, no. 1, pp. 1–17, 2007.

[64] K. Tajima, T. Igari, D. Nishimura, M. Nakamura, Y. Satoh, and M. Munekata, "Isolation and characterization of *Bacillus* sp. INT005 accumulating polyhydroxyalkanoate (PHA) from gas field soil," *Journal of Bioscience and Bioengineering*, vol. 95, no. 1, pp. 77–81, 2003.

[65] S. Sato, C. T. Nomura, H. Abe, Y. Doi, and T. Tsuge, "Poly[(R)-3-hydroxybutyrate] formation in *Escherichia coli* from glucose through an enoyl-CoA hydratase-mediated pathway," *Journal of Bioscience and Bioengineering*, vol. 103, no. 1, pp. 38–44, 2007.

[66] S. Cai, L. Cai, H. Liu et al., "Identification of the haloarchaeal phasin (PhaP) that functions in polyhydroxyalkanoate accumulation and granule formation in *Haloferax mediterranei*," *Applied and Environmental Microbiology*, vol. 78, no. 6, pp. 1946–1952, 2012.

Permissions

All chapters in this book were first published in IJPS, by Hindawi Publishing Corporation; hereby published with permission under the Creative Commons Attribution License or equivalent. Every chapter published in this book has been scrutinized by our experts. Their significance has been extensively debated. The topics covered herein carry significant findings which will fuel the growth of the discipline. They may even be implemented as practical applications or may be referred to as a beginning point for another development.

The contributors of this book come from diverse backgrounds, making this book a truly international effort. This book will bring forth new frontiers with its revolutionizing research information and detailed analysis of the nascent developments around the world.

We would like to thank all the contributing authors for lending their expertise to make the book truly unique. They have played a crucial role in the development of this book. Without their invaluable contributions this book wouldn't have been possible. They have made vital efforts to compile up to date information on the varied aspects of this subject to make this book a valuable addition to the collection of many professionals and students.

This book was conceptualized with the vision of imparting up-to-date information and advanced data in this field. To ensure the same, a matchless editorial board was set up. Every individual on the board went through rigorous rounds of assessment to prove their worth. After which they invested a large part of their time researching and compiling the most relevant data for our readers.

The editorial board has been involved in producing this book since its inception. They have spent rigorous hours researching and exploring the diverse topics which have resulted in the successful publishing of this book. They have passed on their knowledge of decades through this book. To expedite this challenging task, the publisher supported the team at every step. A small team of assistant editors was also appointed to further simplify the editing procedure and attain best results for the readers.

Apart from the editorial board, the designing team has also invested a significant amount of their time in understanding the subject and creating the most relevant covers. They scrutinized every image to scout for the most suitable representation of the subject and create an appropriate cover for the book.

The publishing team has been an ardent support to the editorial, designing and production team. Their endless efforts to recruit the best for this project, has resulted in the accomplishment of this book. They are a veteran in the field of academics and their pool of knowledge is as vast as their experience in printing. Their expertise and guidance has proved useful at every step. Their uncompromising quality standards have made this book an exceptional effort. Their encouragement from time to time has been an inspiration for everyone.

The publisher and the editorial board hope that this book will prove to be a valuable piece of knowledge for researchers, students, practitioners and scholars across the globe.

List of Contributors

Nina Graupner and Jörg Müssig
The Biological Materials Group, Department of Biomimetics, HSB-City University of Applied Sciences, Neustadtswall 30, 28199 Bremen, Germany

Michael Wolf, Tobias Kleffel, Christoph Leisen and Dietmar Drummer
Friedrich-Alexander-University (FAU), Institute of Polymer Technology, AmWeichselgarten 9, 91058 Erlangen, Germany

Federica Pirani
Dipartimento di Scienza Applicata e Tecnologia, Politecnico di Torino, Corso Duca degli Abruzzi 24, 10129 Torino, Italy
Centro per le Tecnologie Future Sostenibili, Istituto Italiano di Tecnologia, Corso Trento 21, 10129 Torino, Italy

Angelo Angelini, Francesca Frascella and Emiliano Descrovi
Dipartimento di Scienza Applicata e Tecnologia, Politecnico di Torino, Corso Duca degli Abruzzi 24, 10129 Torino, Italy

Maria P. Sokolova
Saint Petersburg State University, Universitetskaya Nab. 7-9, Saint Petersburg 199034, Russia
Lappeenranta University of Technology, Skinnarilankatu 34, 53850 Lappeenranta, Finland

Erkki Lahderanta
Lappeenranta University of Technology, Skinnarilankatu 34, 53850 Lappeenranta, Finland

Igor A. Kasatkin and Alexander M. Toikka
Saint Petersburg State University, Universitetskaya Nab. 7-9, Saint Petersburg 199034, Russia

Michael A. Smirnov
Institute of Macromolecular Compounds, Russian Academy of Sciences, Bolshoy Pr. 31, Saint Petersburg 199004, Russia
ITMO University, Saint Petersburg 197101, Russia

Ivan Yu. Dmitriev, Natalia N. Saprykina and Galina K. Elyashevich
Institute of Macromolecular Compounds, Russian Academy of Sciences, Bolshoy Pr. 31, Saint Petersburg 199004, Russia

Araceli Flores-Sánchez, Ulises Figueroa López, José Mauricio Martín-Bufájer and Berenice Vergara-Porras
Escuela de Ingeniería y Ciencias, Tecnológico de Monterrey, Campus Estado de México, Carretera Lago de Guadalupe Km 3.5, Margarita Maza de Juárez, Atizapán de Zaragoza, MEX, Mexico

Ma. del Rocío López-Cuellar
Cuerpo Academico de Biotecnología Agroalimentaria, Instituto de Ciencias Agropecuarias, Universidad Autónoma del Estado de Hidalgo, Pachuca, HGO, Mexico

Fermín Pérez-Guevara
Departamento de Biotecnología y Bioingeniería, Centro de Investigación y Estudios Avanzados (CINVESTAV), Avenida IPN 2508, Zacatenco, Gustavo A. Madero, Ciudad de México, Mexico

Quan Xu, Qingqiao Li, Xu Wu, Bofan Li, Erdong Yao, Yuan Li, Fujian Zhou and Wei Cai
State Key Laboratory of Heavy Oil Processing, China University of Petroleum (Beijing), Beijing 102249, China

Yan Luo
Department of Chemical Engineering,West Virginia University, Morgantown, WV 26506, USA

Salah Al-Enezi
Petroleum Research Centre, Kuwait Institute for Scientific Research, Safat, Kuwait

Jin Liu, Qiao Feng, Faming Zhang, Changqing Qi, and Jihong Wei and Yong Wang
School of Earth Sciences and Engineering, Hohai University, Nanjing 210098, China

Yi Lu
Key Laboratory of Earth Fissures Geological Disaster, Ministry of Land and Resource, Geological Survey of Jiangsu Province, Nanjing 210049, China

Debi Prasanna Kanungo
CSIR-Central Building Research Institute (CBRI), Roorkee 247667, India

Toratane Munegumi
Department of Science Education, Naruto University of Education, Naruto, Tokushima 772-8502, Japan

Takafumi Yamada
Department of Chemistry, University of Tsukuba, Tsukuba, Ibaraki 305-857, Japan

T. V. Travinskaya, A. N. Brykova, Yu. V. Savelyev, N. V. Babkina and V. I. Shtompel
Institute of Macromolecular Chemistry, NAS of Ukraine, Kharkovskoe Shosse 48, Kiev 02160, Ukraine

Sang-Pyuk Woo and Soon-Jong Yoon
Department of Civil Engineering, Hongik University, Seoul 04066, Republic of Korea

Sun-Hee Kim and Wonchang Choi
Department of Architectural Engineering, Gachon University, 1342 Seongnam-daero, Sujeong-gu, Seongnam-si, Gyeonggi-do 13120, Republic of Korea

Kinga Pielichowska, Ewelina Kilian, Beata Macherzyńska and Stanisław Błahewicz
Department of Biomaterials, Faculty of Materials Science and Ceramics, AGH University of Science and Technology, Al. Mickiewicza 30, 30-059 Kraków, Poland

Dariusz Bieliński
Polymer Institute, Technical University of Lodz, ul. Zwirki 36, 90-924 Łódź, Poland

Michał Dworak
Department of Biomaterials, Faculty of Materials Science and Ceramics, AGH University of Science and Technology, Al. Mickiewicza 30, 30-059 Kraków, Poland
Institute of Materials Science, Faculty of Computer Science and Materials Science, University of Silesia, ul. 75 Pułku Piechoty 1A, 41-500 Chorzów, Poland

PaweB SwdBak, Tomasz Stawicki and Piotr Kostencki
Department of Agrotechnical Systems Engineering, West Pomeranian University of Technology, Papieza Pawla VI Street 1, 71-459 Szczecin, Poland

Beata BiaBobrzeska
Department of Materials Science, Welding and Strength of Materials, Wrocław University of Technology, Smoluchowskiego Street 25, 50-370Wrocław, Poland

David Cisneros-Rosado and Jorge Alonso Uribe-Calderon
Centro de Investigación Cient´ıfica de Yucatán, A.C., Unidad de Materiales, Calle 43 No. 130 x 32 y 34, Colonia Chuburn´a de Hidalgo, 97205 M´erida, YUC, Mexico

Nguyen Thi Hiep and Vo Van Toi
Tissue Engineering and Regenerative Medicine Laboratory,Department of Biomedical Engineering, International University, Vietnam National University, Ho Chi Minh City 700000, Vietnam

Nguyen Dai Hai
Institute of Applied Materials Science, Vietnam Academy of Science and Technology, 01 Mac Dinh Chi, District 1, Ho Chi Minh City, Vietnam
Graduate University of Science and Technology, Vietnam Academy of Science and Technology, Hanoi, Vietnam

Odinei Fogolari, Claudia Sayer and Pedro Henrique Hermes De Araújo
Department of Chemical Engineering and Food Engineering, Federal University of Santa Catarina, 88010 970 Florianópolis, SC, Brazil

Arlindo Cristiano Felippe
Federal University of South Frontier (UFFS), Chapecó, SC, Brazil

Fernanda Vitória Leimann and Odinei Hess Gonçalves
Post-Graduation Program of Food Technology (PPGTA), Federal University of Technology–Paraná (UTFPR), Campo Mourão, PR, Brazil

Guohui Huang and Xiaoxuan Liu
School of Materials and Energy, Guangdong University of Technology, Guangzhou, Guangdong 510006, China

Yiling Bei, and Huiqing Ma
School of Chemistry and Chemical Engineering, Shandong University, Jinan, Shandong 250100, China

Chan Kyu Lim and Su Yong Nam
Department of Graphic Arts Information Engineering, Pukyong National University, Busan 48547, Republic of Korea

Yo Seb Lee and Sung Hoon Choa
Graduate School of NID Fusion Technology, Seoul National University of Science and Technology, Seoul 01811, Republic of Korea

Deuk Young Lee
ELK Corporation, Daejeon 34013, Republic of Korea

Lee Soon Park
School of Material Science and Engineering, Ulsan National Institute of Science and Technology (UNIST), Ulsan 44919, Republic of Korea

Laksamon Raksaksri, Saowaroj Chuayjuljit and Anyaporn Boonmahitthisud
Department of Materials Science, Faculty of Science, Chulalongkorn University, Bangkok 10330, Thailand

Phasawat Chaiwutthinan
MTEC, National Science and Technology Development Agency (NSTDA),Thailand Science Park, Khlong Luang, PathumThani 12120,Thailand

Veronika Bátori, Mostafa Jabbari, Dan Åkesson, Patrik R. Lennartsson, Mohammad J. Taherzadeh and Akram Zamani
Swedish Centre for Resource Recovery, University of Borås, 50190 Borås, Sweden

Basheer A. Alshammari
Material Research Institute, King Abdulaziz City for Science and Technology, P.O. Box 6086, Riyadh 11442, Saudi Arabia

Arthur N. Wilkinson
Materials Science Centre and North West Composites Centre, School of Materials, The University of Manchester, Grosvenor Street, Manchester M13 9PL, UK

Ghzzai Almutairi
Water and Energy Research Institute, King Abdulaziz City for Science and Technology, P.O. Box 6086, Riyadh 11442, Saudi Arabia

Jong Won Kim and Joon Seok Lee
Department of Textile Engineering and Technology, Yeungnam University, Gyeongsan 38541, Republic of Korea

Hui Li, Suping Zhou, Terrance Johnson, Koen Vercruysse, Ouyang Lizhi, Parthasarathy Ranganathan and Nsoki Phambu
College of Agricultural, Human and Natural Sciences, Tennessee State University, Nashville, TN, USA

Alexander J. Ropelewski
Pittsburgh Supercomputing Center, Pittsburgh, PA, USA

Theodore W. Thannhauser
R.W. Holley Center for Agriculture and Health, Plant, Soil and Nutrition Research Unit, USDA, ARS, Ithaca, NY, USA

Index